Fundamentals of Physical Geography

D0351405

GT
KD1
19-99

To Ann and Gwen

Fundamentals of Physical Geography

David Briggs and Peter Smithson

ROUTLEDGE

First published by Hutchinson Education in 1985
Fifth impression 1989 published by the Academic Division of Unwin Hyman

Reprinted 1992, 1993
by Routledge
11 New Fetter Lane, London EC4P 4EE

© 1985 David Briggs and Peter Smithson

Set in Linotron Palatino by
Wyvern Typesetting Limited, Bristol
Printed and bound in Great Britain by
Butler & Tanner Ltd, Frome and London

All rights reserved. No part of this book may be reprinted or
reproduced or utilized in any form or by any electronic, mechanical, or
other means, now known or hereafter invented, including photocopying and
recording, or in any information storage or retrieval system, without
permission in writing from the publishers.

British Library Cataloguing in Publication Data is available

Library of Congress Cataloging in Publication Data also available

ISBN 0–415–08394–X

Contents

Preface

The basic aim of physical geography has always been to understand how the world works. In the past this aim was pursued through careful description and classification of features, and the search for pattern in their distribution across the world. Explanation was then achieved by relating these patterns to what were believed to be the causative phenomena—often in a non-quantitative, largely subjective fashion. Thus, theories of landscape development were proposed which stressed the role of time (or stage) and saw the convergence of the landscape towards relatively flat surfaces (peneplains). Similarly, patterns of soil and vegetation were explained in terms of the overriding influence of climate, and much attention was given to definition and and classification of climatic regions, weathering regions and morphogenetic regions.

Without doubt these approaches had their value, and there is still scope for some of these ideas. But in the last twenty years or so physical geographers have begun to adopt different approaches and to ask different questions. Increasingly, emphasis has been placed on monitoring and understanding processes and predicting change in the physical world. At the same time, we have started to consider the world in terms of physical systems – dynamic, and closely inter-related sets of objects which can only be understood when viewed as a whole. To a great extent, these changes have occurred in association with technical developments, such as the production and adoption of powerful computers, the evolution of sophisticated statistical methods, advances in satellite imagery, and improvements in the design of laboratory and field equipment. The consequences, though, are far-reaching and fundamental. Physical geography today is more quantitative, more analytical and more rigorously scientific than ever before.

There are other consequences, one of them being that the subject has become more specialized. We now have more information to deal with, and this information is much more technical and complex. We cannot hope to master it all, so we have to concentrate our energies on selected parts. As a result, the discipline tends to have split more and more into all its different '-ologies': climatology, meteorology, hydrology, various types of geomorphology (fluvial, glacial etc.), pedology, ecology and so on. A glance at the range of specialist books on the average student reading list, or on the shelves of the typical academic bookshop, shows this clearly. In many ways this is no bad thing. It encourages us to become experts rather than generalists. But it also has a number of adverse implications. In the first place, specialism may narrow our breadth of knowledge in physical geography. Breadth of understanding, however, has a value; it gives us one of our unique skills, the ability to step back from the details and see the wider context and significance. Second, it runs counter to that very search for integration and wholeness which a systems approach requires. Third, it can make physical geography a more exclusive subject, barred to all but the experts.

It was, therefore, these considerations which motivated us to write this book. We wished to begin to bring the '-ologies' back together; to build a foundation from which those who are interested could step more easily into their selected specialisms, yet which would also serve for those who required only the fundamentals of the subject as a whole. We wished also to emphasize the links within the environment, and thus we have adopted a systems approach. At the same time we wanted to show how important humankind is in the physical world and how physical processes in turn affect our daily life; consequently we have tried to integrate the human and applied issues. Additionally, it is our belief that physical geography lies outside our own front door, and we have therefore stressed the everyday examples and occurrences of the features and processes we discuss. Finally, we needed to build the whole thing in a logical and rational manner, piece by piece, without losing sight of the broader 'wholeness' of the physical world. For this reason, we begin by presenting an overview – a framework – of the earth and its subsystems. Then we examine each of these subsystems in turn: the atmosphere, the hydrosphere, the lithosphere and the biosphere. In the process we consider each systematically, showing how the various components are related and recalling the links with other parts of the world; showing, too, how humankind fits into the scheme. By the end, we hope that the reader will not only have a grasp of the fundamentals of physical geography, but will also be aware of the excitement and importance of the subject, and thus be stimulated to take it further.

Many years ago, we were given this stimulation by our own teachers of geography – Gordon Roberts at Lewes Grammar School and Allan Griffiths at West Leeds Boys' High School. To them we owe a great debt. If we can pass on their sense of excitement and interest to some of our readers, we will consider this book a success.
D.B. and P.S.

1

Physical geography: a starting point

In the Peak District of Derbyshire, a few miles west of Sheffield, there is a valley. It is not a spectacular valley, like that of the Colorado or Rhine. It is like many thousands of others. Grindsbrook Clough, which rises a little way to the north, on the flanks of Kinder Scout, tumbles amid the gritstone boulders of the Pennines before flowing through the village of Edale and on to join the River Noe. From a few hundred metres north of the village, it is possible to stand beside the stream and look up the valley towards the craggy edge of Hartshorn (Plate 1.1). What we can see is a small segment of a most unexceptional Peak District valley.

And yet it is a scene which can tell us a great deal; which contains the key to understanding the physical world around us. If we could start to understand this valley – if we could begin to appreciate how the individual features which we see are linked together, how they function – then we would have opened a door on our physical environment. For this small valley, like almost any other we might have chosen, demonstrates many of the fundamental processes and factors which make up the environment. If we can in some way recognize and comprehend these things, we would have a basis for answering a vital question: how does our natural world work?

Let us look a little more closely at the view. The river, we can see, splashes over the rocks and rapids which line its channel. It is bounded in front of us by a fairly flat area – the floodplain of the river. On the right, the valley side rises to the bracken-covered flanks of the hills; on the left, there is a small scar in the hillside and, above that, a patch of woodland. Here and there, boulders and outcrops of bedrock protrude through the vegetation and we can see the pale line of the Pennine Way footpath where it follows the valley towards Kinder Scout. It is a scene typical of countless parts of upland Britain. How can we start to learn from it?

Processes in the Grindsbrook valley

What becomes clear if we inspect this scene is that a great deal is happening. Some of the events are obvious. The air is constantly moving, bending the grass, loosening the leaves from the trees, sweeping clouds across the sky. At times rain or snow falls and storms unleash their fury on the land. All the time, too, water is flowing through the stream channel, carrying with it logs and twigs which have been washed from the banks and woods, pebbles and boulders which have slid or rolled down the valley side. The boulders can be seen in periods of flood bouncing and shuffling along the channel floor; they come to rest in shoals and bars at the margins of, or within, the channel, and are moved on again as the river rises. Sheep are moving around on the floodplain, grazing the vegetation. Occasionally a hiker walks along the path.

Other events are less apparent. Beneath the soil, the rocks are being weathered and broken down.

Plate 1.1 Grindsbrook Clough, Edale, Derbyshire: an example of an environmental system (photo: Dave Briggs)

On the faces of the cliffs, ice freezes in the crevices and loosens small fragments of rock. Plant roots extend into larger fissures and open up the cracks. The soil is also invisibly active. Within the soil the rock fragments are being broken down further. Leaves and twigs from the vegetation are being chewed up and decomposed by the soil organisms. The soil itself is moving, pushed and prodded downslope by the development of ice crystals during the winter, carried by the wash of water over and through the soil, splashed by the impact of raindrops. Water and dissolved substances are being drawn from the soil by the plants; the water is being released to the atmosphere from the leaves. Slowly, the vegetation is changing, advancing in some areas, retreating in others. Unseen, energy is being passed through the atmosphere to the land and vegetation, and returned to the air and space.

A model of the Grindsbrook valley

These processes are the lifeblood of the Grindsbrook valley. They are the means by which the valley is being shaped and is developing. They show the pathways by which material and energy is being moved through the valley. They thus provide us with a way of understanding the scene. We can start to build up a picture of the way the valley is functioning; we can start to account for the features we see in the valley at the present time and even begin to predict some of the changes which may occur in the future.

We might go even further, for we can try to represent this picture in diagrammatic form. We might use the technique employed in motor car manuals, for example, to show the workings of the valley as we have interpreted it. Just as the car designer explains the position and operation of all the components of the engine by drawing exploded diagrams of the vehicle, so we can construct a similar picture of the Grindsbrook valley. We can show the various features of the valley, use arrows and lines to indicate the main processes and movements affecting them, annotate the diagram to show what is happening (Figure 1.1). What we end up with is a picture – a model – of the valley.

This model, however, is a little limited at present.

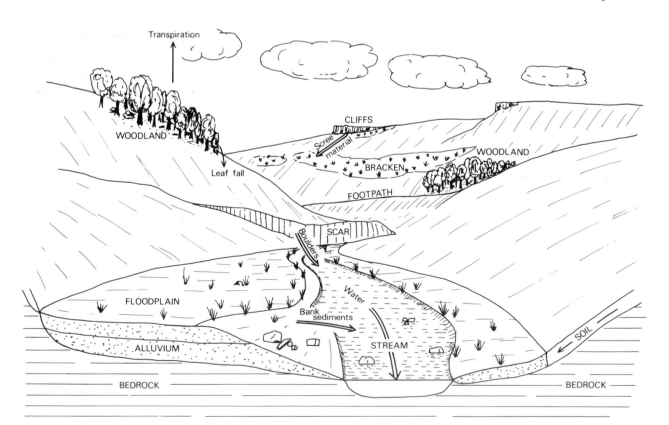

Figure 1.1   The Grindsbrook valley

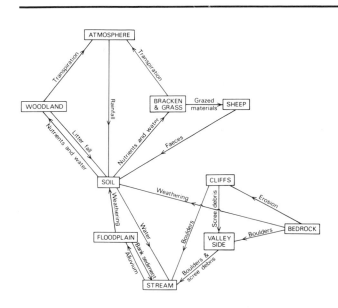

Figure 1.2 A preliminary model of the Grindsbrook valley, showing general relationships between the main features of the environment

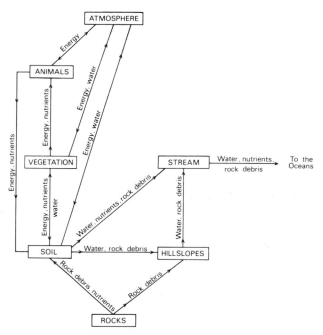

Figure 1.3 A generalized model of the Grindsbrook valley showing the major flows of energy, water, rock debris and nutrients

It helps us to see what is happening within the Grindsbrook valley, but it does not tell us much about the rest of the world. The reason is that the model is at present too specific; it retains too much of the detail of this particular valley. What we need is a more general, a more abstract model. How can we produce it?

From Grindsbrook Clough to the world!

The first thing we must do is to omit some of the unnecessary detail. After all, it matters little when we are trying to understand the broad aspects of our environment that the river in this little valley twists in this way here, or that the slope has a little undulation in it there. We can dispose of much of this local detail and concentrate instead on the most important features of the area. We can show the clouds, the stream, the floodplain, the hillsides, the woods and moorland – in fact all the features that we feel characterize the valley. We can represent all these features as boxes and we can show the processes operating within and between them by arrows (Figure 1.2). As a consequence, we obtain a more abstract, yet simpler and clearer, picture of the Grindsbrook valley.

We now have a model which is less specific to the particular view before us. Nevertheless, we can still not use it to understand the world as a whole. For not every valley or area we might be interested in contains the features we have recognized here. To progress further, to get to the basic structure and character of the scene, we need to take a rather different approach. We need to ask two even more fundamental questions. What does this valley (or indeed any other part of the natural world) consist of? And what is it that links together these components?

The atmosphere. The water. The rocks and landforms. The soil, animals and vegetation. These are the basic building blocks of our environment. And the mortar between them is provided by the flow of energy and matter: the movement of heat and other forms of energy from the atmosphere to the ground, through the soil, vegetation and landscape and ultimately back to space; the movement of water from the atmosphere to the oceans and back again; the movement of rock debris from the rocks through the landscape; the flow of nutrients from the soil to the vegetation and hence back to the soil. If we describe the Grindsbrook valley in this way, as we have done in Figure 1.3, we indeed have a model which is relevant not just to this area but to anywhere in the world.

9

Figure 1.4 The dangers of modelling!

We have described a system

We have made some real progress, therefore. We have discovered what we set out to find: a key to understanding our environment. And we have done more than that, for we have recognized and described a system.

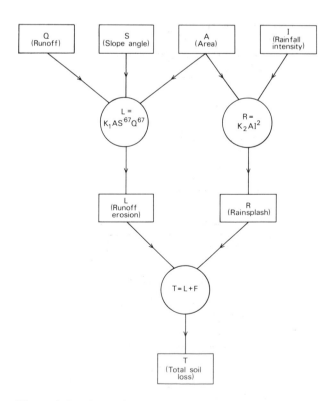

Figure 1.5 A mathematical model: soil erosion by water (based on Meyer and Wischmeier 1969)

Let us define what we mean. A system is simply a set of objects and the relationships between them. Almost anything, at any scale, can be seen as a system, from a small droplet of water on a table top to the whole universe. But the concept of a system gives us a way of looking at those objects. It is an approach which helps us to focus on the way the objects interact. This is important because the natural world is highly interactive, and if we are to understand – or, even more critically, to manage – the world effectively, we need to be aware of these interactions.

The other important step we have made is to represent this system in terms of a model. What we have done is to simplify the reality into a more manageable form. The need for this is all too apparent. The world is an extremely complex place, and we cannot hope to comprehend all the details of its complexity. We need, instead, to produce models which reflect the most important components and linkages of the system, but allow us to ignore (at least for a while) the less important details. Subsequently we might wish to extend our models, and bring them closer to reality, though there is always a danger in doing so (Figure 1.4).

There are, in fact, many different sorts of models which we can construct: **hardware models** in which we actually build a physical likeness of the system we are concerned with; or **mathematical models** in which we represent all the components of the system by mathematical symbols and the relationships between them by equations (Figure 1.5); and more general, **conceptual models** like the one we have built of the Grindsbrook valley.

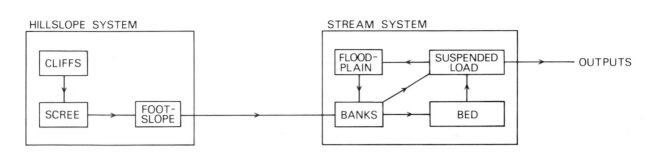

Figure 1.6 The relationship between two connected systems. Sediment moves through the hillslope system and into the stream system; there it is transferred through the banks, bed, suspended load and floodplain deposits before being lost as outputs. Many other examples which fit this general structure can be found

A systems approach

The structure of systems

Our example of a system in Figure 1.1 leaves something to be desired. It is not, in truth, a very clearly defined system; its boundaries are determined simply by the limits of our view. In most cases we are dealing with rather more distinctive systems, bounded by more easily definable features. We may, for example, take a whole river valley, the edges of which are marked by the line of highest ground which encircles the valley, and treat this as a system. Or we may take an area covered by a particular vegetation type, a field or farm, a glacier or cloud.

None the less, although the boundaries of these systems may appear obvious, they are rarely impermeable. The systems are not self-contained. Instead, matter and energy flow into and out of the system, across its boundaries as **inputs** and **outputs** (Figure 1.6). We can see this in our example: inputs of water, sediment and other materials enter the system from upstream and leave it downstream; energy enters from the sun and is lost by reradiation back to the atmosphere and space. This flow of inputs and outputs is typical of all natural systems (except, perhaps, at the scale of the whole earth or universe). Such systems are called **open systems**.

Sometimes that is all we know about a system – its boundaries and the inputs and outputs. We can detect a relationship between the inputs and outputs, but we do not understand what goes on inside. In this case we are looking at the system as a **black box** (Figure 1.7). As our knowledge progresses, however, we may be able to discover what lies inside the system. What we find is a series of smaller systems, or **subsystems**, each linked by a series of flows of energy and matter. We then have a view of the system as a **grey box**. But delve a little deeper and we may be able to see the whole internal working of the system: its individual **components**, the pathways by which the energy and matter flow between them, the **storages** where the energy and matter may be held for certain periods of time. We now have a view of the system as a **white box** (Figure 1.7).

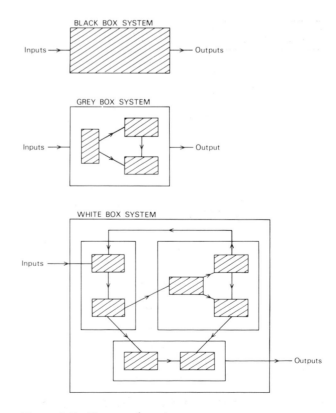

Figure 1.7 Types of system

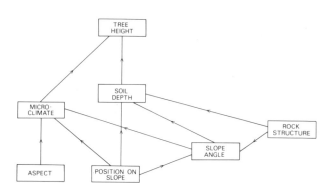

Figure 1.8 Factors affecting rate of tree growth in Grindsbrook valley: an example of a morphological system

Types of system

A system, therefore, is a set of components linked by flows of energy and matter. But what does that mean? If we take our example in Plate 1.1 we could, for example, measure all the main features and attributes of the valley and look at the relationships between them. We might see how the size of the trees relates to the soil depth, how soil depth relates to slope angle, how slope angle relates to rock structure, and so on (Figure 1.8). Is that a system?

The answer is yes. It is what is referred to as a **morphological system**. In looking at the system in this way we are concerned not with the dynamics of

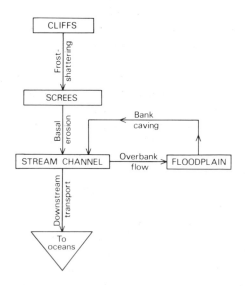

Figure 1.9 Sediment movement through the Grindsbrook valley: an example of a cascading system

the interactions and flows, but merely their morphological expression. Another way of looking at the system, however, could be to focus attention on the flows of matter and energy through the valley. We might, for example, represent the movement of sediment through the system as in Figure 1.9. In this case we have described the valley as a **cascading system**: one in which there is a cascade of energy and matter through the environment from one component to another. It is a particularly useful way of dealing with systems and we will use it to look at the flow of solar energy through the atmosphere (Chapter 4) and the cascade of sediment through the landscape (Chapters 17–20).

There are other ways of looking at environmental systems. We can combine morphological and cascading systems to define what are called **process–response systems**. Figure 1.10 depicts part of the Grindsbrook valley in these terms. As this shows, the morphology of the system is related to the flows of energy and matter. In other words, the form of the system is a function of the processes operating within it. This is a vital concept and one we will use repeatedly throughout this book for it helps us to see the ways in which the environment develops and is maintained.

Finally, we can also define what are referred to as **ecosystems**. These are concerned with the biological relationships within the environment – the interactions between the plants, animals and their physical surroundings. It is an approach we will adopt in the last section of this book when we look at soil–vegetation–animal relationships.

System change and system stability

The environment is dynamic. It is constantly changing. What causes this change? The immediate answer is energy. As we will see in the next chapter the earth obtains most of its energy from the sun (as **radiant energy**). The energy within the environment tends to be unevenly distributed, however, and so it tends to flow through the environment, in an attempt to produce a more equitable distribution. In the process it is itself changed but it also carries out work. It alters the environment.

If we were to monitor an environmental system such as that we have defined in the Grindsbrook valley we would be able to detect some of these changes. We might see, for example, that the stream changed its course, that the woodland extended as seedlings grew amid the surrounding grassland, that the weather changed from day to

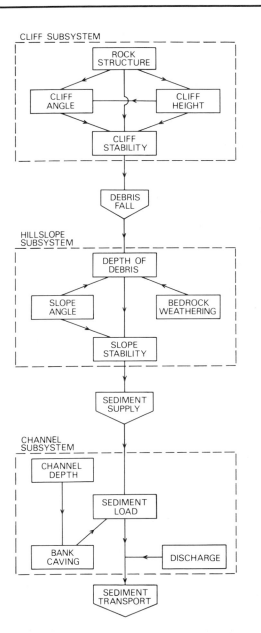

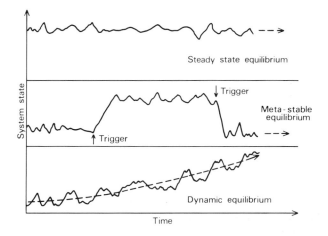

Figure 1.11   Types of equilibrium

Figure 1.10   The cliff–hillslope–channel system of Grindsbrook valley: an example of a process–response system

day and season to season. Much of this change would often appear haphazard and random. Changes in one direction at one time might be reversed at the next. The reason for this is that although the system is active, it is also in a state of **equilibrium**; over time it tends to maintain its general structure and character in sympathy with the processes acting upon it. Thus, although the channel position varies, on average it follows the

same route down the valley. Although the extent of the woodland fluctuates, its mean area and position are constant. Though the weather varies, the long-term average climate is more or less consistent. Systems such as this are said to be in a condition of **steady state equilibrium** (Figure 1.11).

Not all systems behave like this, however. Some are subject to much more marked and often irreversible changes. These changes may be triggered off by certain events which knock the system out of equilibrium. Hillslopes provide an excellent example. Within certain limits they remain stable and their form does not change. But if those limits are exceeded – if the ground becomes too wet or the pressures exerted on the slope are too great – slope failure may occur and a new slope form is created. The same is true of many atmospheric systems. The air near the ground may be stable within limits, but if the system is disturbed (for example if the air is over-heated) so that it passes these limits, it may become unstable: the air may start to rise until it reaches a new, equilibrium position. Systems of this type are said to be in a condition of **metastable equilibrium**.

In other cases, environmental systems may change more gradually and progressively over time. Short-term, random fluctuations may still be detectable due to the effects of minor variations in environmental conditions. But in the longer term a distinctive and consistent trend may be visible. This is characteristic of climatic change. As we will see in Chapter 10, the climate is gradually changing (though with minor fluctuations) and over the last 6000 years average summer temperatures in NW

Europe have fallen by about 2°C. Under these circumstances, the system is said to be in a state of **dynamic equilibrium**.

All these types of equilibrium are illustrated in Figure 1.11. Although they vary in detail they all reflect a fundamental principle: that the form of any system adjusts towards a state of equilibrium with the processes and conditions to which it is subject. However, these examples also illustrate a further property of environmental systems – the operation within them of feedback processes. **Feedback** refers to the ability of a system to modify earlier links in the chain of inter-relationships so that an initial change in the system is either amplified (**positive feedback**) or damped down (**negative feedback**).

We can see how these feedback processes operate with an example. Suppose the Peak District were affected by a prolonged drought (or even a more permanent change in climate) which resulted in the area receiving much less rainfall than it presently does. What would be the effect on Grindsbrook Clough?

In the short term the consequences are fairly obvious. Less rainfall would mean that there was less water available for runoff. Less water would therefore enter the stream channel, and stream discharge would decline. At the same time, less sediment would be brought into the stream by water flowing down the hillside, so the sediment load of the stream would fall. In addition, the stream itself would be unable to transport its load so effectively, and would deposit much of its material on the channel bed. The reduction in the amount of water in the channel and in the sediment load of the stream would, in turn, diminish the erosive capacity of the stream which would then be unable to erode its banks. This, too, would reduce the amount of sediment entering, or passing down, the channel. Thus, the initial decline in the amount of rainfall would have triggered off a cycle of positive feedback which results in progressively less sediment being transported by the stream (Figure 1.12).

But the story does not end there, for in the longer run other changes would be taking place. The increased aridity might lead to changes in the vegetation cover, for the trees and heather would start to experience moisture stress. In time, much of the vegetation might wilt and die (we saw this in England in the well-known drought of 1976). This would reduce the amount of evapotranspiration and would leave areas of bare soil. The rainfall which did fall, therefore, would not be trapped or taken up by the vegetation, but would be free to run off across the unprotected surface. This would ultimately help to increase the quantity of sediment being washed into the stream. Moreover, because there would be less vegetation to intercept or slow down the water after rainfall, much of the water would reach the stream channel relatively swiftly, so that the stream would respond more quickly to storm events. At the peak of its flow, it would be able to carry large amounts of material and would have the capacity once more to erode its bed and banks. Thus, a longer-term cycle is operating to restabilize the system; to halt the progressive reduction in the rate of sediment movement. This is a process of negative feedback (Figure 1.12).

This example is simple but it illustrates a number of important principles which are worth emphasizing. First, environmental systems are dynamic and intimately inter-related. Changes in one part of the system work through to affect other parts. Second, change tends to be exacerbated by processes of positive feedback, inhibited by negative feedback. Third, over time systems may undergo alternating periods of positive and negative feedback, as they change in response to some external impulse then restabilize (this is characteristic of systems which are in a state of metastable equilibrium).

In addition, we may draw two further implications about change and stability in environmental systems. The first is that because conditions change, and because the systems adjust to that change, many of the features we see in the environment at any moment might, at least in part, be inherited from previous times when conditions and processes were different. We can certainly see this in the case of the Grindsbrook valley. Several of the features there are due not to any present-day process but to the effects of glacial conditions several thousands of years ago. The glaciers have gone, but the features do not simply disappear. They remain and are slowly modified by new processes. Their present form is thus a product of several superimposed periods of development during which different processes have been at work.

The final point is that, because of their complex inter-relationships and their potential for change, environmental systems are difficult to manage. All too easily man may trigger off changes in these systems, more far-reaching and destructive perhaps than he intends simply because he does not appreciate all the ramifications of his actions. We will see many examples of this in the rest of this

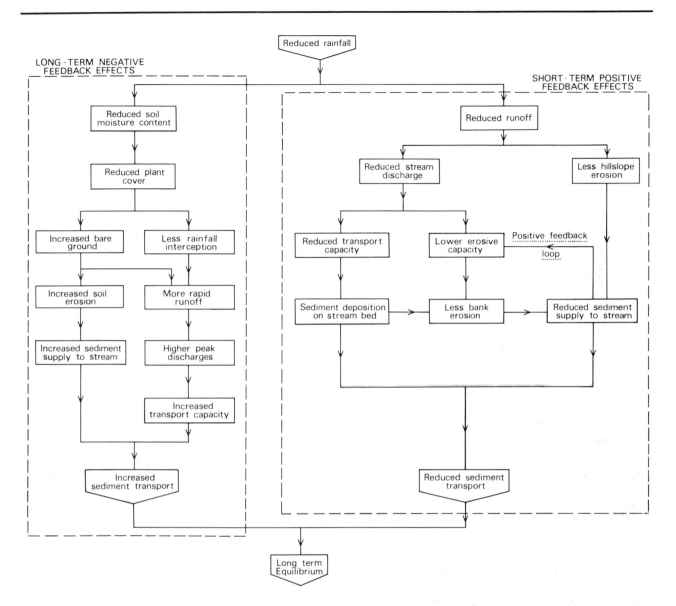

Figure 1.12　Potential effects of reduced rainfall on Grindsbrook valley showing the operation of feedback relationships

book. In almost every case the lesson will be the same. We need to understand how environmental systems work, how they change, how they respond to external conditions, if we are to manage them effectively. For this reason if no other a systems approach to physical geography is useful.

2

The global energy system

The nature of inputs and outputs of energy

Imagine the earth from 300,000 km into space. An isolated sphere; blue, patched with brown and green and wreathed in white. A world of water, dotted with land, partly clothed in swirling cloud (see cover illustration). It is a view of the global system. Into this system pour the inputs of solar energy; from it come reflected and reradiated energy which are its outputs. From our privileged vantage point, we could measure the inputs to the earth and, with suitable equipment, monitor the global outputs. We could therefore draw up a simple model of the globe as an energy system (Figure 2.1), showing the inputs and outputs, but this gives no idea of what happens inside. It is a picture of the globe as a **black box system**. It is the simplest view of the system that we can obtain but it tells us nothing about the internal components or subsystems, nor the relationships between them, only what enters and leaves the globe.

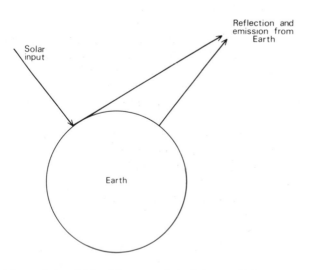

Figure 2.1   A black box model of the earth's energy system

Let us start by looking at those energy flows which we can examine. Without doubt, the main input of energy to the global system comes from the sun. Compared with the solar contribution, all other inputs are almost negligible. The gravitational effect of the moon and of the sun, and reflected and radiant energy from the moon provide some energy with even smaller inputs from the impact of objects such as meteorites and comets. Many of these extraterrestrial objects burn up upon entering the atmosphere and do not reach the surface, although they do act as a minute input of energy to the atmosphere.

Just as the input of energy is dominated by radiant energy from the sun, so the output of energy from the earth is almost entirely radiant energy, although this time with somewhat different properties. Much of the energy has been emitted by the earth and its atmosphere, being modified in the process, but some of the output is represented by solar radiation which has been reflected from clouds or from the earth's surface without any major modification. As the overall energy level of the earth is not changing we can assume that there must be a balance between the energy input to and energy output from the globe as a whole.

Concepts of energy

Before discussing the quantities of energy received by the earth any further, it is necessary to consider, briefly, the nature of energy and the ways in which we measure it.

Energy exists in a variety of forms. We are familiar with electrical energy in the home and increasingly with nuclear (or atomic) energy. Neither of these has any great significance with regards to environmental processes. More important as far as environmental processes are concerned are **radiant**, **thermal**, **kinetic**, **chemical** and **potential energy**.

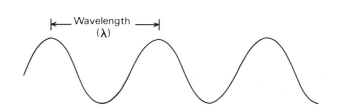

Figure 2.2 Electromagnetic radiation: the distance from one crest to one crest or from one trough to one trough is known as the wavelength (λ). It is an important indicator of the properties of the electromagnetic radiation

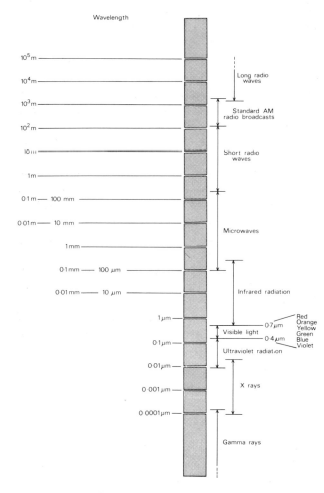

Figure 2.3 The electromagnetic spectrum (after Neiburger *et al.*, 1982)

Radiant energy is the most relevant to our discussion here, for it is in this form that the sun's energy is transmitted to the earth. The heat from the sun excites or disturbs electric and magnetic fields, setting up a wave-like activity in space. The length

of these waves – that is, their distance apart (Figure 2.2) – varies considerably, so that solar radiation comprises a wide range of electromagnetic wavelengths (Figure 2.3). Only a very small proportion of these are visible to the human eye reaching us as light, but all transmit energy from the sun to the earth. Assuming a mean distance from sun to earth of about 15×10^7 km, it takes this energy about $8\frac{1}{3}$ minutes to reach us.

On passing through the atmosphere which surrounds the earth, some of this radiant energy is reflected or absorbed. Because of this interception not all the radiant energy finds its way to the earth's surface. That which does, and that which is absorbed by the atmosphere, is converted from radiant to other forms of energy. Much is altered to **thermal (heat)** energy. It warms the earth's surface and the atmosphere by exciting the molecules of which they are composed. In simple terms, the radiant energy (which involves disturbance of magnetic and electric fields) is transmitted into the molecules making up the earth and atmosphere.

Thermal energy can therefore be considered as energy involved in the motion of extremely small components of matter. The energy of motion is referred to as **kinetic energy** (and thus thermal energy is sometimes described as the kinetic energy of molecules). Any moving object possesses kinetic energy, and it is through the utilization of this energy that, for example, a stone thrown into a lake can disturb the water to the extent of producing waves. It is also through the exploitation of kinetic energy that turbines and engines are able to produce heat, light and so on.

Chemical energy represents a form of electrical energy bound up within the chemical structure of any substance. It is released in the form of thermal or kinetic energy when the substance breaks down. Coal, when it is burnt, releases heat. Food, when it is digested, is used to provide the body with heat and movement.

Potential energy is related to gravity. Because of the apparent pull that the earth exerts upon objects within its gravitational field, material is drawn towards the earth's centre. Thus, objects lying at greater distances from the earth's centre (for example, rocks on a hillside, water at the top of a waterfall or the air near a mountain summit) possess more potential energy. This energy is converted to kinetic energy when the rock, the water or the air descends to lower levels; some energy is converted to heat through friction.

Thermal, kinetic, chemical and potential energy are important to the earth's system but operate internally and so cannot be observed directly from space. To understand what is happening as a result of all these different flows of energy, we must look in more detail at them, concentrating on the forms of energy which have significance to the physical geography of the earth.

General patterns and principles of radiation

Solar energy is transmitted to the earth in the form of radiant energy. How does this energy reach us? Why do we receive the amount we do? Why does the energy have the particular properties it has? To answer these questions we need to examine some of the principles of radiation.

Principles of radiation

Radiant energy consists of electromagnetic waves of varying lengths. Any object whose temperature is above absolute zero (0K or −273.15°C) emits radiant energy. The intensity and the character of this radiation depend upon the temperature of the emitting object. As the temperature rises, the radiant energy increases in intensity but its wavelength decreases; as the temperature falls the intensity decreases and the wavelength expands (Figure 2.4). In addition, the amount of radiation reaching any object is inversely proportional to the square of the distance from the source (Figure 2.5). This distance decay factor accounts for the difference in solar inputs to the various planets in our solar system.

To a certain extent radiation is able to penetrate matter, as for example X-rays which can pass through the human body, but most radiant energy is either absorbed or reflected by objects in its path. Absorption occurs when the electromagnetic waves penetrate but do not pass through the object; reflection involves the diversion or deflection of the waves from the surface of objects. The ability of an object to absorb or reflect radiant energy depends upon a number of factors, including the detailed physical structure of the material, its colour and surface roughness, the angle of the incident radiation and the wavelength of the radiant energy.

An object which is able to absorb all the incoming

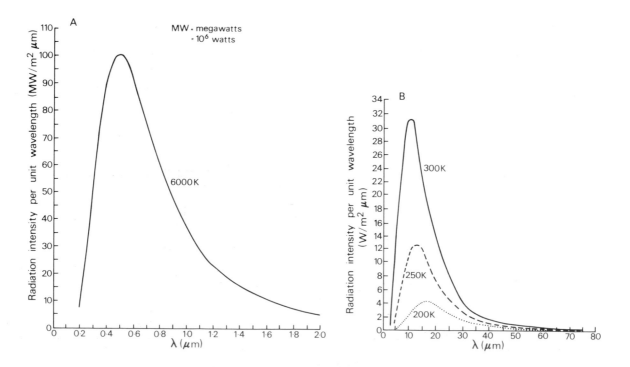

Figure 2.4 Variation of intensity of black-body radiation with wavelength: (A) T=6000 K (approximately the emission temperature of the sun); (B) T=200, 250, and 300 K (range of earth emission temperatures). Note the differences in scales (after Neiburger *et al.*, 1982)

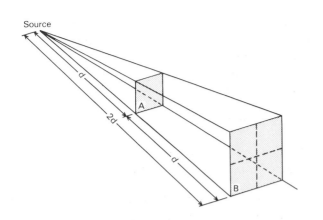

Figure 2.5 The inverse square law. Intercept area at distance d from source is just one-fourth that at distance 2d; energy passing through area A is spread over an area four times as large at B

radiation is referred to as a **black body**. Although this has conceptual value, a perfect black body does not exist in reality. All objects absorb a proportion of the incoming energy and reflect the remainder. Differences also occur according to the wavelength of the energy. Thus snow and sand both absorb long-wave radiation (5–50 μm) quite efficiently, but

they reflect relatively large proportions of short-wave radiation (0.4–0.8 μm). Indeed, under constant conditions, it is possible to define the wavelengths which specific materials selectively absorb, and this knowledge can be used to characterize or identify materials. It is frequently used in astronomy to determine the gases in stellar atmospheres. Whereas solid substances usually absorb most wavelengths of radiation, gases tend to be very selective in their absorption and emission wavelengths (Figure 2.6). This property is very important to the earth as it means that the atmosphere only absorbs and emits in certain wavelengths. At other wavelengths, radiation is able to pass right through the atmosphere without modification.

When radiant energy is reflected by an object, very little change in the nature of the radiation occurs, although the effect may be to **scatter** the radiation. Scattering changes the direction of the incoming radiation without directly affecting its wavelength. Shorter waves, however, tend to be scattered more easily so that blue light, at the shorter end of the spectrum, tends to be scattered more than red light to give us blue sky – when it is not cloudy. Absorption of radiant energy has more

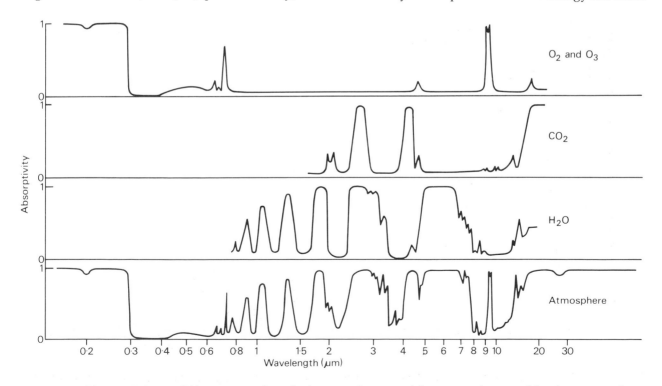

Figure 2.6 Absorptivity at different wavelengths by constituents of the atmosphere and by the atmosphere as a whole (after Fleagle and Businger, 1963)

far-reaching consequences than reflection or scattering. As an object absorbs energy its temperature rises because the radiant energy is converted to heat. Reradiation of this energy tends to occur at a temperature different to that of the initial, radiating object, and thus the radiation emitted is at a different wavelength. The earth, for example, is considerably cooler than the sun; thus the energy it emits is characteristic of longer wavelengths than the initial solar inputs.

We can summarize the radiation laws as follows:

1 All substances emit radiation as long as their temperature is above absolute zero (−273.15°C or 0K).

2 Some substances emit and absorb radiation at certain wavelengths only. This is mainly true of gases.

3 If the substance is an ideal emitter (black body), the amount of radiation given off is proportional to the fourth power of its absolute temperature. This is known as the Stefan–Boltzmann law and can be represented as $E = \sigma T^4$ where σ is a constant (the Stefan–Boltzmann constant) which has a value of 5.67×10^{-8} W m^{-2} K^{-4}, and T is the absolute temperature.

4 As substances get hotter, the wavelength at which radiation is emitted will become shorter (Figure 2.4). This is called Wien's displacement law which can be represented as $\lambda_m = a/T$ where λ_m is the wavelength at which the peak occurs in the spectrum, T is the absolute temperature of the body and a is a constant with a value of 2898 if λ_m is expressed in micrometres.

5 The amount of radiation passing through a particular unit area is inversely proportional to the square of the distance of that area from the source - $(1/d^2)$ (Figure 2.5).

Solar radiation input

Now that the principles of radiation have been outlined, we can look at the details of the solar radiation input to the earth in a more meaningful manner.

Because we know the mean distance of the earth from the sun we can work out, from law 5 above, how much radiation the earth should receive. This amount is called the **solar constant** and has a value of about 1370 W m^{-2} at the top of the atmosphere. Recent work from satellites has suggested that the solar constant may vary slightly depending upon solar activity. As the change measured is only

slightly larger than instrumental uncertainty, it is still unclear whether the variation is real or not. By measuring how much radiation reaches the top of the atmosphere and knowing the size of the sun as well as the earth's mean distance, the emission temperature of the sun can be determined from law 3. For the outer-radiating surface of the sun, this value works out to be approximately 6000K. This figure then enables us to determine at what wavelength most radiation will be emitted from the sun from law 4, i.e.

$$\lambda_m = 2898/6000$$
$$= 0.48 \ \mu m$$

From Figure 2.3, we can see that this value is in the middle of the visible part of the spectrum.

From the radiation laws it has been possible to determine how much radiation the earth ought to receive as well as the amount and properties of solar radiation. Similar calculations can be made for the earth when we are considering outputs.

The input of energy to the earth at its mean distance from the sun is only an average value, for changes are taking place all the time. For example, the earth is rotating on its axis once in 24 hours, the earth is orbiting the sun once in 365 days and, as its axis of rotation is at an angle of $23\frac{1}{2}°$ to the vertical, the distribution of radiation at the top of the atmosphere is constantly changing. Over even longer periods of time, the nature of the earth's orbit and

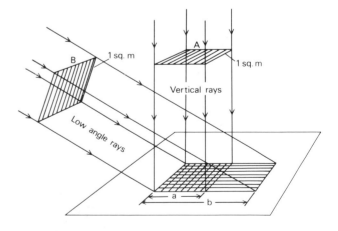

Figure 2.7 Energy distribution on an intercepting surface depends upon the angle of the incoming energy rays. Energy distribution is more concentrated on a perpendicular surface (a) than on a surface at a lower angle (b)

the angle of tilt of the earth also change, thus affecting the amount and distribution of radiation over the earth. These, however, are only important on a time scale of thousands of years and so will be discussed more fully in Chapter 10. Let us look in more detail here at the diurnal and seasonal effects.

Diurnal variation

As the earth rotates on its axis, a different portion of the top of the atmosphere will be exposed to the **incoming solar radiation** (often abbreviated to **insolation**). At dawn, the sun will be low in the sky and the amount of radiation passing through a unit area normal to the line from the sun will be spread over a large area (Figure 2.7). As the sun rises in the sky the surface area decreases and so intensity increases. If our surface is eventually at right angles to the solar beam it will receive the maximum intensity of radiation – the surface area is at its smallest. As well as the angle between the sun's rays and the top of the atmosphere, the length of daylight will also affect the amount of radiation received. At the equator daylength remains at 12 hours throughout the year. At the poles it varies between 0 and 24 hours depending upon the time of year.

Seasonal variation

Seasonal variations in insolation arise from the changing axial tilt of the earth throughout the year (Figure 2.8) and the eccentricity of the earth's orbit. The orbit is an ellipse not a circle so that the earth is slightly nearer the sun (147,300,000 km) on 4 January and at its furthest distance (152,100,000 km) on 4 July.

Being nearer the sun means that the radiation input will be slightly higher. As the earth orbits around the sun with its axis of rotation pointing in a constant direction, the area that is illuminated by the sun and the angle between the sun's rays and the top of the atmosphere will change. At the June solstice, the sun is above the horizon throughout the 24 hours for all latitudes north of the Arctic Circle, while south of the Antarctic Circle the sun would not be visible. Between the autumn equinox (23 September) and the winter solstice (22 December), the latitude at which the midday sun is overhead gradually moves southwards from the

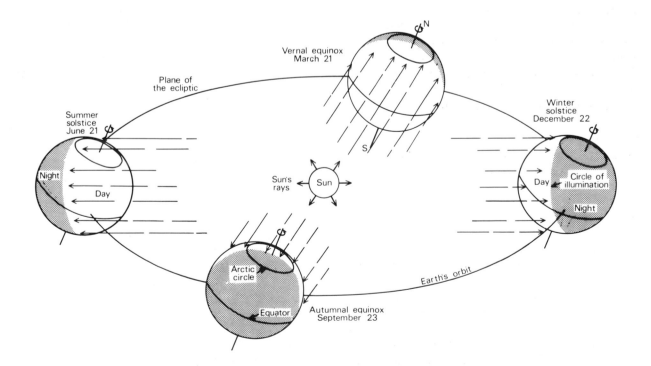

Figure 2.8 The revolution of the earth around the sun. The seasons result because the tilted earth's axis keeps in a constant orientation in space as the earth revolves around the sun

equator to the Tropic of Capricorn (23½°S). By 22 December insolation will be at a maximum at that latitude and be zero north of the Arctic Circle. Between 22 December and 21 March, the sequence is reversed and then, in the period leading up to the summer solstice, the latitude of the overhead sun moves northwards from the equator to the Tropic of Cancer, and insolation increases in the northern hemisphere, and the southern pole is thrown progressively into shadow (Figure 2.9).

The presence of the earth's atmosphere has a dramatic effect on the input of radiant energy but we can illustrate the essentially astronomic controls of the input either at the top of the atmosphere or by assuming there was no atmosphere; the result is the same (Figure 2.10). It is clear from the diagram that, taking an annual figure, the tropics would receive most radiation as the input never falls to low values, unlike the situation at the poles where 24 hours of daylight in summer becomes 24 hours of darkness in winter.

Endogenetic energy

Energy supplied by the sun is known as exogenetic, in that it is derived from outside the earth (Greek *exo* = outside + *genos* = creation). It can be argued that, indirectly, almost all the earth's energy is exogenetic and all but a small amount is derived from the sun; the remainder comes from other cosmic bodies, including the attractive force of the moon which causes tidal activity (a form of kinetic energy). In addition, a minute proportion of the total energy comes from within the earth and is thus endogenetic (Greek *endon* = within + *genos*).

The most obvious source of endogenetic energy is from the hot interior of the earth. The outer core of the globe consists of molten materials at immense pressures, and at temperatures up to 2600°C. There is an almost immeasurably small and continuous conduction of this heat to the earth's surface which adds to the energy inputs acting upon the landscape. This can be detected in deep mines and caves. Locally the decay of radioactive minerals can provide energy to the surface. More dramatic leakages of this endogenetic energy are seen in the form of volcanoes, hot springs and various other tectonic activities. Taken together, all sources of endogenetic energy contribute no more than 0.0001 per cent of the total energy supply averaged over the total earth's surface.

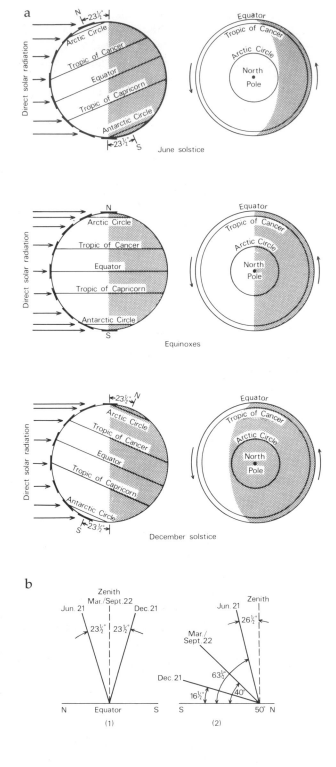

Figure 2.9 (a) Exposure of the earth to the sun's radiation at the solstices and the equinoxes; (b) position of the midday sun at the equator (1) and at 50°N (2) at the solstices and the equinoxes

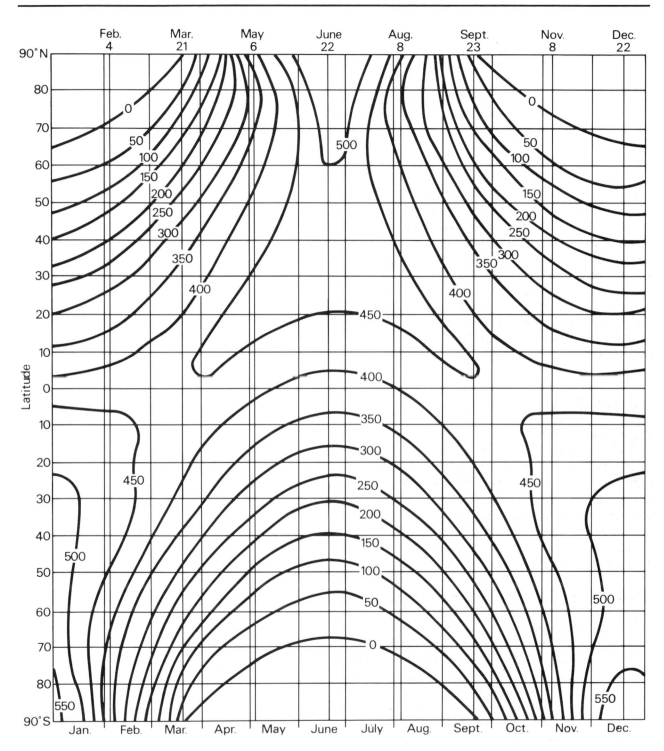

Figure 2.10 Solar radiation (W m⁻²) falling on a horizontal surface at the outside of the atmosphere (after Neiburger *et al.*, 1982)

Endogenetic energy, of course, is present within the global system and, strictly speaking, is not an input.

Energy outputs of the globe

The output of energy from the earth is in radiant

form but it is not identical to the input of radiant energy from the sun. The earth has modified the input by a variety of processes. Some of the original solar energy input is reflected by clouds or the ground surface and returned to space with little change in its radiative properties; it is still short-wave radiation. As insolation passes through the atmosphere it interacts with the molecules of the atmosphere and is scattered, much of it towards the earth, but a small proportion back to space as an output.

Of much greater importance is the emission of radiant energy from the earth itself. As a result of the absorption of solar energy in the atmosphere and at the surface, the earth will have gained energy which will be converted into heat. In turn, the earth and its atmosphere emit radiation, but at longer wavelengths than that of solar radiation because of their lower emission temperatures following Wien's Law. The average temperature of the earth is about 290 K, while that of the atmosphere is a chilling 250 K. Consequently the energy emission will reach a maximum at a wavelength of 2898/290 or 2898/250 which is 9.99 μm or 11.59 μm; and overall emission is entirely within the infra-red range.

In this form the energy is susceptible to absorption by the atmosphere, so very little escapes directly to space; most is repeatedly reabsorbed and re-emitted before it is able to leave the system. The ability of the globe to trap energy in this way helps to keep the temperature of the earth and atmosphere higher than it would otherwise be. In other words, it promotes energy storage within the system.

At a global scale these processes lead to energy outputs of which about 36 per cent are in the short wavelengths derived from reflected insolation and about 64 per cent in the long wavelengths largely from emission by the atmosphere.

Spatial and temporal variations in outputs

Radiation outputs from the globe vary considerably over time and across the global surface. Spatial fluctuations depend upon a number of factors, including the character of the atmosphere (e.g. its temperature and the degree of cloudiness) and the nature of the earth's surface (e.g. vegetation cover and topography). We find that differences in output from the earth are less pronounced than the input of insolation. From the polar regions an out-

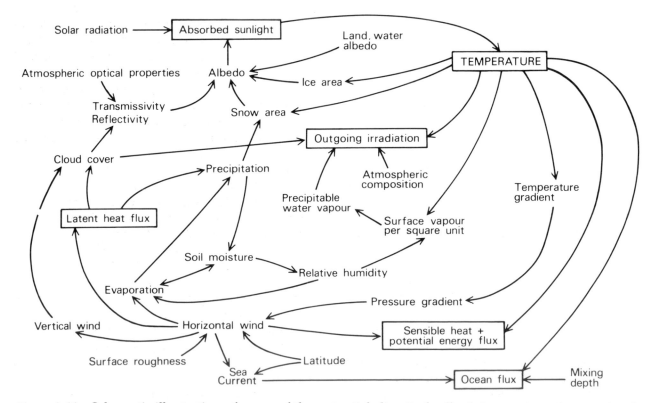

Figure 2.11 Schematic illustration of many of the potential climatic feedback interactions that need to be considered in a climatic model (after Kellogg and Schneider, 1974)

put of about 140 W m⁻² compares with 250 W m⁻² from equatorial areas – a ratio of about 2:1 – whereas the ratio for the short wavelength input is about 6:1. These aspects will be dealt with more fully in Chapter 4.

Over the long term the fluctuations in global energy outputs possibly relate to outside influences; a change in input may lead to an adjustment in the output. The ways in which these adjustments take place are complex, and involve interactions called **feedback relationships** (Figure 2.11). Vegetation cover, atmospheric conditions (including moisture content and cloud cover), the extent of polar and mountain snow cover, the area of the sea surface and even soil cover and roughness may change in response to alterations in energy inputs. Through such changes the earth is able to adjust its energy outputs in the event of any long-term variation in inputs by altering the balance between absorption, retention, emission and reflection of energy.

The question, however, is whether long-term variations of this kind occur.

Certainly over geological time quite marked fluctuations in climate have taken place, as attested by the evidence of ice ages and tropical conditions contained in, for example, the rocks of eastern USA, Britain and many other areas. Some of these changes are due to movement of the continental plates but some may be related to alterations in energy inputs and, if so, it is clear that outputs, too, must have changed. As the snow cover was extended during the ice ages, reflection must have increased while absorption (and hence reradiation) must have been reduced. We will examine some of the possible consequences of this process in Chapter 10. Ultimately, however, a new equilibrium seems to be established as energy outputs decline to match the new, lower level of inputs.

It is an intriguing question, also, to ask whether changes in global conditions could arise due to adjustments in the outputs independently of change in energy inputs. Any event which significantly alters the reflectivity of the earth's surface might trigger off such changes. An increase in the extent of the oceans relative to land due, perhaps, to major earth movements; increased snow cover as a result of mountain-building; changes in vegetation cover due to these events (or even due to human activity); or changes in the atmosphere brought about by massive volcanic eruptions – all could lead to significant changes in the global climate and hence in energy outputs. The implications for the world's climate are very important.

What is certain is that marked variations in global energy outputs do occur in the long term. Many of these variations are probably cyclical, related to changes in solar inputs such as those resulting from differences in the tilt and orbit of the earth. It is also apparent that such variations in outputs are critical if the earth is to adjust to alterations in the energy inputs which are known to occur, and thereby maintain a steady state condition. An unanswered question is, to what extent man can change these outputs and upset the equilibrium?

Transfers of energy

The inputs and outputs of energy so far considered do not have a uniform distribution over the globe. Both the earth and the air experience major inequalities in energy receipts and emissions. As a result of these differences, spatial transfers of energy take place, for energy is redistributed to minimize the inequalities, to maintain (or to achieve) an equilibrium.

To understand how energy is transferred we need to consider a little further the principles of energy transformation and modification. We have seen already that energy can exist in a number of forms, and as a general principle energy will be transferred from areas of high energy status to areas of low energy status in an attempt to eradicate the differences. Thus, energy differences expressed by the level of temperature in two bodies, such as the air and the soil, tend to be reduced over time as heat is transferred from the hotter to the cooler body. In this way the soil is heated during the day when the air is warm and it loses heat energy back to the air at night when the atmosphere is cool (Figure 2.12).

In the case of thermal energy, three main methods of transfer can be identified: radiation, convection and conduction. **Radiation** is the process by which energy is transmitted through space, mainly in the form of electromagnetic waves. **Convection** involves the physical transfer of substances containing heat, such as water or air, and is not possible in a solid. **Conduction** is the transfer of heat through a medium, from molecule to molecule.

These three processes of transfer are often closely related. Thus, energy may be conducted through the soil to the surface and then radiated or

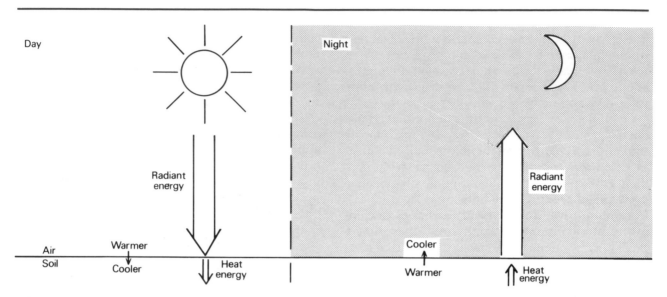

Figure 2.12   Energy exchange at the soil surface

convected into the atmosphere. Similarly, in the air, convection currents may raise warm air masses to higher levels, from whence conduction to surrounding cooler atmosphere (see Chapter 5) may occur, while condensation of water vapour releases latent heat. Convection is very important as an energy transfer mechanism because it transfers energy in two forms. The first is the **sensible heat** content of the air, which is transferred directly by the rising and mixing of warmed air. It can also be transferred by conduction. The other form of energy transfer by convection is less obvious as there is no temperature change involved, hence its name of **latent heat**. The evaporation of water into vapour or the melting of ice into water involves a supply of heat to allow the change to take place. When the reverse process operates, from vapour to liquid, or liquid to ice, this heat is released. We shall return to these mechanisms in more detail in Chapters 4 and 6.

Transfers also occur between other forms of energy. If two objects with different kinetic energies are brought together, a transfer takes place between the two which tends to equalize the energy levels. A rapidly flowing stream, for example (high kinetic energy), which comes into contact with a static boulder (no kinetic energy) tends to push the boulder into motion. In so doing the stream loses energy by friction but imparts some of this energy to the boulder in the form of motion (kinetic energy). Similar principles apply to chemical energy.

One way of looking at these transfers of energy is

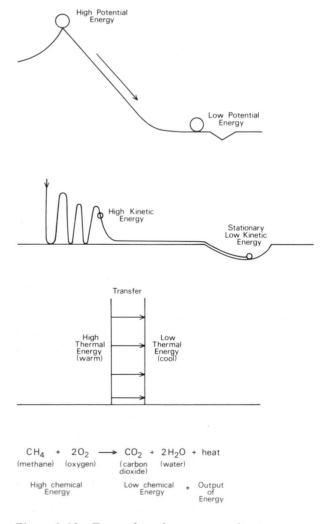

Figure 2.13   Examples of energy gradients

to regard them as movements down an energy gradient. It is easier to see this principle in the example of heat energy, for we can all appreciate that heat moves from hotter to colder areas. Heat the end of a metal bar in a fire and the heat will move along the metal until it burns our fingers! Heat energy in this case moves down the energy gradient in the bar. The same general processes operate with other forms of energy (Figure 2.13).

Energy transformations

During these transfers of energy it is clear that the nature of the energy often changes, although the total quantity of energy involved remains constant. Radiant energy heats the objects which it meets; it is converted from radiant energy to heat energy. Kinetic energy may similarly be converted to heat energy; the friction of a moving body against another liberates heat as we can demonstrate by filing or sawing a piece of metal – or even by rubbing a finger over a slightly rough surface.

Under natural conditions the range of probable transformations is fairly limited. That is to say, the various forms of energy are normally able to be converted to all other forms, but follow relatively well-defined pathways (Figure 2.14) towards the lowest level of energy – that of heat.

Energy and work

The transfer and conversion of energy are associated with the performance of work. The sun performs work in heating the earth through its provision of radiant energy. A river uses kinetic energy to perform the work of moving boulders. The weathering of rocks or the decomposition of plant debris involves work carried out largely by chemical energy. Indeed, it is the work which is done in these ways that characterizes the myriad of processes operating in the environment.

When this work is carried out, therefore, energy is transferred from one body to another and, in some cases, it is also converted from one form to another. In the process, the total energy content remains the same, it is only changed in form. When a river or glacier cuts a valley, the energy they use is not destroyed but transferred or converted to other forms – some to heat energy, some to potential energy, some remaining as kinetic energy. When a plant grows, it takes in energy from the sun, from the air and from the soil and stores it in the plant; the energy is not lost, merely transferred and transformed.

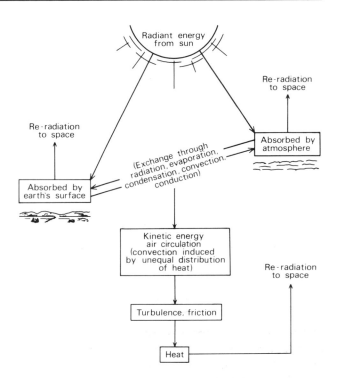

Figure 2.14 Energy transformations in the atmosphere (after Miller, 1966)

Résumé

Let us recap for a moment.

Energy represents the driving force for all the processes operating in the global system. It performs the work which is involved in processes such as moving rocks, cutting valleys, lifting mountains, making water flow, the wind blow and plants grow.

This work is performed through the transfer and transformation of energy. These transformations tend to follow well-defined routes (Figure 2.14).

The work is carried out because of differences in the energy status of different objects or conditions. Inequalities in the distribution of available energy (i.e. that which is capable of performing work) lead to energy transfers; in the process of these transfers work is done.

The energy involved in these transfers is not destroyed; it merely changes form.

Thermodynamics

The basic principles of energy are embodied in the laws of thermodynamics. These were initially developed in 1843 by Prescott Joule to explain processes seen in steam engines. Since then it has been appreciated that they have far wider significance, and they now represent one of the basic precepts of science.

The first two laws of thermodynamics state that

1 Energy can be transformed but not destroyed.
2 Heat can never pass spontaneously from a colder to a hotter body; a temperature change can never occur spontaneously in a body at uniform temperature.

The first law therefore defines the conservation of energy. The second law leads to the principle that energy transfers are a result of inequalities in the energy distribution and that energy is always transferred from areas of high energy status to areas of low energy status; i.e., down the energy gradient.

The third law of thermodynamics is a little less easy to understand. In very general terms, it says that systems tend towards equilibrium; i.e., a random distribution of energy over time.

Nature of global energy transfers

Every feature and every part of the globe is at some stage or another involved in energy transfers and transformations, and, as conditions change, so the nature of the transfers and conversions operating at any one place also change. We cannot attempt, therefore, to describe the processes operating throughout the global system in any detail. We can, however, try to identify the dominant transfers operating at a global scale and indicate, within this general pattern, the roles played by the various subsystems.

We have already noted that the balance between incoming and outgoing radiation is such that marked disparities occur between the energy status of different parts of the globe. The most obvious effect of this is the range in temperature which we find when travelling from pole to equator, a range amounting to 30–60°C depending upon the time of year and the hemisphere. In very simple terms it is these differences which drive the global energy circulation. In order to achieve equilibrium, energy is transferred from the warmer parts of the globe to the cooler. If someone turned off the sun these transfers would result eventually in a more or less uniform distribution of energy across the globe; the fact that the sun continues to supply this unequal distribution of energy, however, maintains the imbalance and makes the attempt to gain uniformity a losing battle. On the other hand, if this battle was not fought, the fact that the equatorial areas are constantly gaining more energy than they lose, while the polar areas are losing more than they gain, would result in a massive accumulation of heat in lower latitudes and indescribable cold in higher latitudes.

Thus, there exists a net poleward transfer of energy, and this transfer maintains the existing pattern of energy distribution; it feeds the higher latitudes and drains the lower latitudes (Figure 2.15). This transfer is performed by a variety of processes. Undoubtedly the main transfers occur in the atmosphere. Winds carry warm air and water vapour away from the tropics. The warm air thus transfers sensible heat to the cooler latitudes. The water vapour carries energy in the form of latent heat. When the water vapour condenses, this energy is released as heat and warms the surrounding atmosphere. The oceans, too, transfer significant amounts of energy polewards. Heating of the sea in equatorial areas creates a temperature gradient between the higher and lower latitudes. Ocean currents are therefore set up which carry the warmer waters down this gradient by a process of lateral convection. Equally important, surface winds move this water polewards, a reflection of the close interaction between atmosphere and ocean.

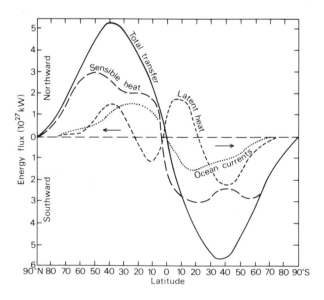

Figure 2.15 The average annual latitudinal distribution of the components of the poleward energy transfer in the earth–atmosphere system (after Sellers, 1965)

In both air and sea, however, the transfers are not one way. If this were the case we would be faced with a build up of air and water in the higher latitudes and a slow emptying of the tropical areas. Clearly this does not happen; the warm air and ocean currents which flow towards the poles are replaced by a counter flow of cooler air and water moving from the poles. In the case of the sea, the flow tends to occur at depth, for the cool water sinks. In the case of the air, the pattern is a little more complex. The transfer of latent heat (i.e. energy tied up within water vapour) occurs mainly in the lowest 2–3 km of the atmosphere. It is closely related to the surface wind network which we will discuss in Chapter 5. The sensible heat transfer (i.e. of warm air masses) occurs both close to the earth surface and at high altitudes (c. 10 km). The reasons for this we will see in later sections (Chapter 4). Both flows, however, are reflected by counter flows of cooler air from higher latitudes.

The three main processes of energy transfers at a global scale are therefore:

1   The horizontal transfer of sensible heat by warm air masses and currents.
2   The transfer of latent heat in the form of atmospheric moisture.
3   The horizontal convection of sensible heat by ocean currents.

Of these three processes, the first is the most important, accounting for about 50 per cent of the total annual energy flow. The other two processes account for about 20 per cent and 30 per cent respectively.

We will consider the detailed processes involved in these transfers in Chapter 5, and will see there the factors that lead to the spatial distribution of these transfer mechanisms, but it is worth noting here that marked latitudinal variations in the three processes occur. Sensible heat transfers by the atmosphere, for example, are at a maximum between 50° and 60° north and south of the equator, and again at 10° to 30° north and south. This pattern reflects the two types of transfer referred to earlier; the higher level transfers are dominant in the subtropical zone while surface transfers are most active in middle latitudes.

The transfer of latent heat also shows a complex pattern, related to the distribution of water vapour in the atmosphere and the dominant, lower air wind patterns. Thus its main effects are seen between 20° and 50° north and south of the equator, where winds blowing outwards from the subtropics carry moist air polewards. Nearer the equator the pattern is reversed, and winds carry this air into lower latitudes. As we will see in Chapter 5, therefore, this pattern is closely related to the global wind system.

Oceanic transfers of energy are most important either side of the equator, reflecting the outward movement of warm water from the tropical region. We will discuss the processes involved more fully in Chapter 15.

In total, these processes of energy transfer maintain a condition of steady state within the global system; they replenish energy losses in areas where outputs exceed inputs (the higher latitudes) and they remove energy from areas where inputs are in excess (the lower latitudes).

Local and regional energy transfers

While these atmospheric and oceanic processes account for the spatial redistribution of energy at a global scale, they are not the only means of energy transfer in the global system. At a more local level, numerous other transfers are taking place.

Atmospheric transfers
Within the atmosphere, local and regional winds, convection currents and air masses carry energy as sensible heat and as latent heat. The uplift of the air

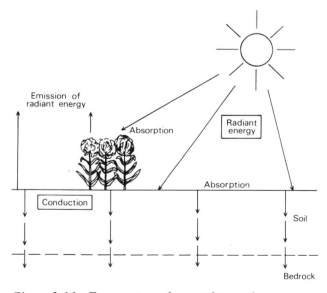

Figure 2.16   Energy transfers at the surface

and the water contained within it transforms some of this energy into potential energy, which is released when the air sinks or the water falls as rain. Small, local transfers of energy to the earth's surface occur due to friction, while the kinetic energy of the wind is transmitted to soil and rock particles as these are picked up and blown along. Heat energy from the atmosphere is also transferred to soils and plants through conduction and radiation (Figure 2.16).

Hydrological transfers

Water similarly takes part in a variety of transfer processes (Figure 2.17). Water condensing in the atmosphere releases latent heat; this warms up the surrounding atmosphere. The potential energy, which is derived from the initial uplift of the water vapour into the atmosphere, is transformed into kinetic energy as the raindrops fall, and some of this kinetic energy is transmitted to the earth's surface as rock and soil particles are splashed into motion. Further potential energy is expended and converted to kinetic energy as the water percolates through the soil, runs into streams and flows down to the sea. The flowing water again imparts some of its kinetic energy to material which it catches up and carries along.

The water also takes part in chemical processes of weathering and thus chemical energy is transferred to heat energy, given off during the chemical reactions. In the sea, the currents transfer energy laterally, while the upwelling and sinking of water masses leads to vertical transfers. Finally, the evaporation of water from the sea, from rivers and lakes and from the soil, involves the conversion of heat or radiant energy to kinetic and potential energy as the water is again raised from its original position and carried to higher levels in the atmosphere.

Landscape transfers

Many of these transfers impinge upon the landscape processes, for the movement of water through the landscape is one of the main means by which the earth's surface is altered and moulded. The potential energy possessed, for example, by boulders on a slope is a product of the erosion of the valley by water and ice. Potential energy is also derived from earth movements, for mountain building lifts the rock to leave it higher than the surrounding earth surface. Since these mountain-building processes are powered by heat energy within the earth, they represent the transformation of heat energy to kinetic and, ultimately, potential energy. The potential energy is subsequently converted to kinetic energy as the rock particles tumble, sludge or wash downslope. Friction with the surface and between the particles releases further energy in the form of heat.

Ecological transfers

On land, the formation of soil, the growth in the soil of plants and the support given by this vegetation to animals all reflect further energy transfers and conversions.

In the case of terrestrial ecosystems (Figure 2.18), the development of a soil cover involves weathering, which in turn reflects the transfer of chemical energy from rocks to soil. Plants take up substances from the soil and store the chemical energy in their plant tissue. They also use radiant energy from the sun, and chemical and heat energy from the atmosphere, all three forms being converted to chemical energy by the plant. As the vegetation dies, or animals devour the plant materials, this energy is cycled through the environment. Animals convert the chemical energy to heat for bodily warmth and to kinetic energy for motion. They return some to the soil and the atmosphere as chemical energy.

Similar processes operate in aquatic ecosystems, although in this case much of the initial input of chemical energy is derived from organic matter washed into the waters from the land.

On a global scale it is almost impossible to quantify precisely the effects of all these processes.

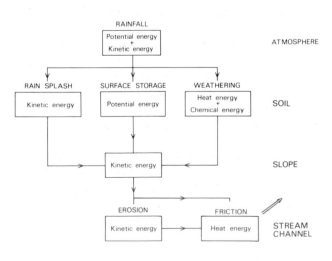

Figure 2.17 Transfer processes involving water

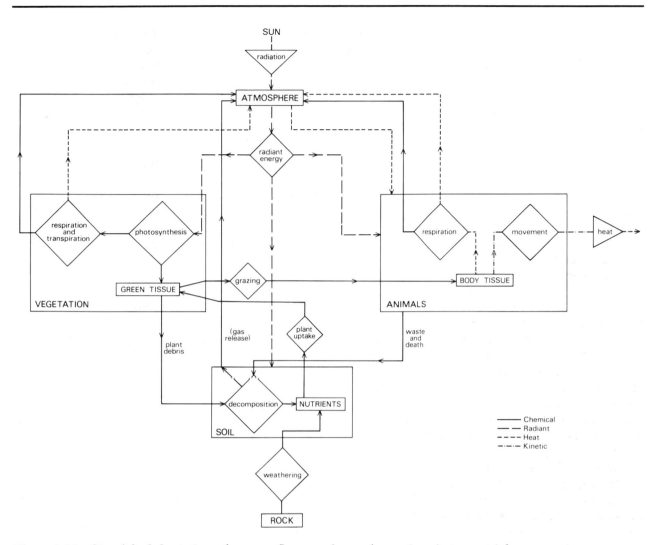

Figure 2.18 Simplified depiction of energy flows and transformations in terrestrial ecosystems

What is clear is that energy transfers create a fabric of relationships which bind the global system together, that provide the motive power for the processes operating within our world, and which are the very foundation of our existence.

Summary

The energy transfers which operate in the global system derive from inequalities in inputs and outputs across the world. At a global scale, they involve the movement of energy as sensible heat and latent heat by the atmosphere and as sensible heat by the oceans.

At a more detailed level, these transfers permeate every part of the global system. They involve transfers from rocks to soil, soil to plants, plants to atmosphere; in fact all the components of the world are interconnected by these transfers. They also involve transformations of energy from one state to another.

Together these transfers and transformations provide the power for all the processes operating in our environment. They bind the global system into a unified whole. They are the life-blood of our planet.

The atmospheric system

In the opening chapters we have seen the ways in which energy reaches the earth from the sun and how a balance is achieved between the inputs and outputs of energy. In the case of the earth's energy system, we can take measurements of what goes on within the atmosphere and at the surface. In this chapter, we shall examine the role of these energy flows in generating atmospheric movement and in deriving our climates.

The solar energy cascade

What happens to the energy that pours into the atmosphere from the sun? We can start at the top of the atmosphere with the input of solar radiation (Figure 3.1). Following our input of energy, the first important effect is **absorption** by the atmosphere. Energy which is not absorbed proceeds to the next step, which is **reflection**. If reflection takes place either from clouds or from **scattering** by gas molecules, then the short-wave radiation returns to space as part of the earth's output. The remaining radiation will reach the ground surface where some of it is again reflected by the earth. The proportion of radiation which is not reflected is absorbed by the ground in the earth subsystem.

Over a long time period, we can assume that the net gain to the ground is zero. The amount of energy which is received here must eventually be lost. It is lost partly as **long-wave radiation** and partly during **evaporation** and **convection**. The energy from evaporation and **sensible heat**, together with long-wave radiation, is absorbed by the atmosphere, which in turn radiates some of its energy back to space to complete the system.

The main components of this system are the atmosphere itself and the earth. Both these serve two functions, however; they trap and store energy and they divert or reflect it. They act, therefore, as both stores and regulators of the energy which flows between them.

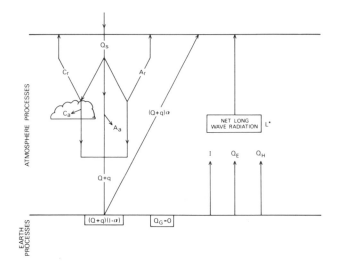

Figure 3.1 The solar energy cascade in diagrammatic form. Q_S is incoming short-wave radiation; C_r is short-wave radiation reflected from clouds; A_r is short-wave radiation reflected and scattered by the atmosphere; C_a is the short-wave radiation absorbed by the clouds; A_a is the short-wave radiation absorbed by the atmosphere; α is the surface albedo; $Q+q$ is the direct and diffuse short-wave radiation reaching the ground surface and $(Q+q)\alpha$ is the amount reflected; Q_G is the net heat flow into the earth; Q_E is the heat transfer from surface to atmosphere by latent heat of vaporization; Q_H is the sensible heat transfer from surface to atmosphere and I is the long-wave emission from the surface; L^* is the net long-wave radiation from the atmosphere

This model of heat energy transfers in the atmosphere (Figure 3.1) is highly simplified; it is no more than a summary of what we call the energy cascade. What it has not yet allowed for is the fact that the energy in our atmosphere occurs in other forms. Some is due to the movement of the air – **kinetic energy**; some depends upon its position above the earth's surface – **potential energy**. There are even small amounts of **chemical energy** that could be included in a larger but more realistic model of the earth–atmosphere system.

The subsystems

The interface between the atmosphere and the earth plays a vital role in energy exchanges. It is also a very diverse zone with surfaces ranging from simple forms like smooth water or ice surfaces, to complex ones such as forests or urban areas. Each surface affects the way in which energy is apportioned (Figure 3.2). Some surfaces, such as water, use the majority of available energy for evaporation; at the other extreme we have urban surfaces and deserts where most of the energy is used for heating (Plate 3.1). The influence exerted by the surface will depend upon its areal extent. A small pool of water surrounded by an extensive grassland surface will have little effect on the local climate – it is too small. A lake the size of Ontario or Superior is sufficient to produce major modifications of energy use and therefore in climate.

Water, ice and desert surfaces may be thought of as simple interfaces between air and earth, but for the majority of land surfaces we are dealing with a zone, a three-dimensional layer from canopy top to the roots, where a myriad of complicated subsystems may exist. In Figure 3.3 we can see possible flows of energy within a forest, where the canopy top acts as a major interface between air and ground. As well as the canopy top we also have to consider the trunk space where further energy transfers take place: the two-way transfer of infrared radiation to the soil and the canopy, the effects of latent and sensible heat losses and the influence of transpiration. The man-made surface of the city presents an even more complicated situation (Figure 3.4). Buildings vary in height and surface material, while heat is released in vast quantities but with an irregular distribution, so that radiation exchanges are difficult to determine or predict. Yet it is this type of surface on which an increasing proportion of the world's population are dependent.

Spatial and temporal variations

Clearly our system is beginning to get more complex. However, the diagrams used so far show the system at a single point assuming average conditions. All the individual inputs and outputs of the system, and the character of each of the sub-

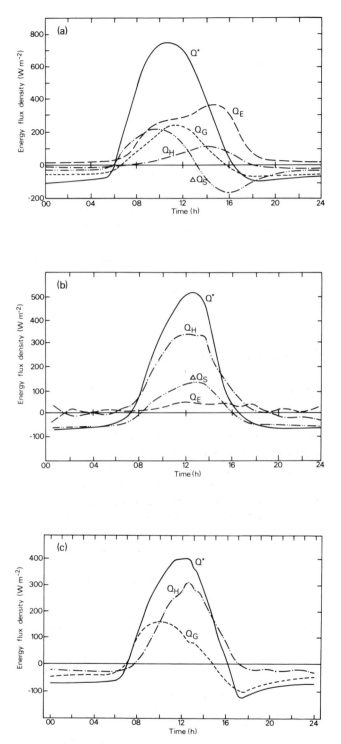

Figure 3.2 Diurnal variation of the energy balance components for (a) a shallow water layer in September; (b) the top of an urban surface and (c) a desert. Q^* is available energy; ΔQ_S is net energy storage ((a) and (b) after Oke, 1978, (c) after Stearns, 1969)

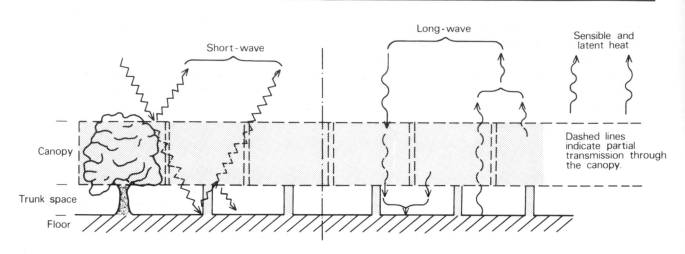

Figure 3.3 Schematic model of radiation exchanges above and within a forest

systems, vary according to a variety of conditions – time of day, time of year, regional climate, state of the ground and so on. To understand the role and function of the atmosphere and its reactions with the surface more fully, we need to move away from the individual sites to consider the global interactions.

In Figure 3.2 we can see that each surface possesses features in common. On a daily basis, **available energy** changes from being slightly negative at

Plate 3.1 Meteosat image for 4 September 1982, infra-red waveband. Light areas are relatively cool and dark areas are relatively warm. The strong heating over the Saharan Desert is evident by its dark appearance in this waveband. The visible image taken at the same time is shown in Plate 6.2 where the Sahara looks light because of its high albedo

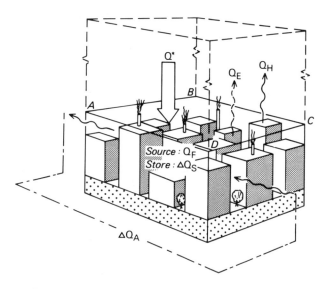

Figure 3.4 Schematic depiction of the fluxes involved in the energy balance of an urban building-air volume. Q^* is available energy (net radiation); Q_H is sensible heat; Q_E is latent heat; Q_F is anthropogenic heat; $\triangle Q_S$ is the net energy storage by the buildings; $\triangle Q_A$ is the net energy (sensible and latent) advected into and out of the city (after Oke, 1978)

night to strongly positive values by day before returning to negative amounts in the evening. The intensity of the daytime surplus will depend upon such factors as latitude and amount of cloud. The available energy can then be converted into latent heat, sensible heat or heat exchange with the soil. Over water or ocean surfaces the situation may differ (Figure 3.2a). The vast energy store of deep water means that latent and sensible heat flows can be maintained throughout the day and night, even when more energy is being lost than is being gained.

Extending our time period to one year, the detailed appearance of the energy balance components will depend upon the nature of the climate.

In equatorial areas (Figure 3.5), available energy is high throughout the year. The two maxima are associated with the passage of the overhead sun at the equinoxes. The soil is permanently moistened by abundant precipitation so that most of the surplus energy is used in evapotranspiration. Values for sensible heat are low apart from a small increase towards the end of the drier season. Few parts of the world are as well watered as São Gabriel. Yuma, Arizona is one of the driest parts of the United States. The clear desert air gives high values of energy input but there is little moisture to evaporate so the curve for sensible heat (Q_H) closely follows that for available energy (Q^*) (Figure 3.6). Moving polewards, the period of surplus radiant energy

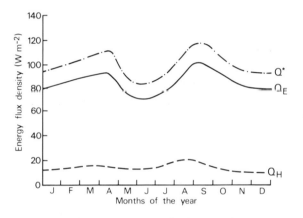

Figure 3.5 Components of the surface energy budget for São Gabriel, Brazil, in an equatorial continental climate

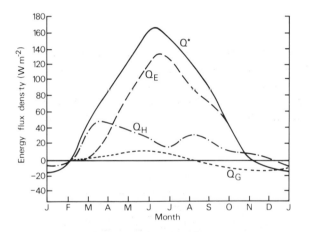

Figure 3.7 Components of the surface energy budget for Madison, Wisconsin in a humid mid-latitude climate (after Sellers, 1965)

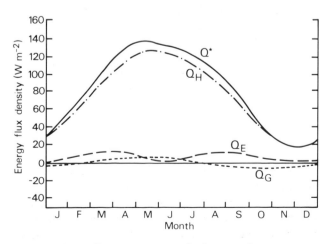

Figure 3.6 Components of the surface energy budget for Yuma, Arizona, USA in a desert climate (after Sellers, 1965)

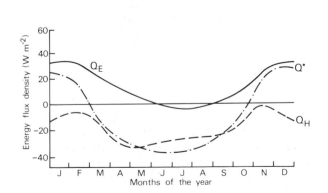

Figure 3.8 Components of the surface energy budget for Mirny, Antarctica (66° 33'S) in a polar climate

becomes more concentrated into the summer period (Figure 3.7), and more strongly negative in winter. The subdivision into latent and sensible heat fluxes will depend upon the moistness of the climate. On the polar ice caps we find yet another change (Figure 3.8). First, the energy balance components are small in absolute terms throughout the entire year. Second, the annual cycle of the latent heat flow is similar to that of available energy, but the energy use for evaporation is greater than that from radiation. Third, throughout the year, the flow of sensible heat is from the atmosphere to the surface because of the semi-permanent surface temperature inversion (Chapter 8).

The variety of surface energy balances is important because it affects the overlying atmosphere to a marked degree. Across the globe, the atmosphere experiences a variation of energy input from the ground surface to produce an almost infinite mosaic of microclimates. These are very important for any particular location but their effects would be insignificant at the global scale. However, a consistent pattern can be observed. More energy is received in tropical areas than in temperate and polar regions, an energy difference which generates horizontal temperature gradients between tropics and poles. In turn, the temperature gradients drive the major airflows and the weather systems which develop within them. Together these flows of air help to overcome the imbalance of energy and so represent a basic element of our global climate.

Of course, the atmosphere does not act in isolation. It is influenced by – and in turn influences – the other main components of the global system: the **hydrosphere** (water), **cryosphere** (ice), **lithosphere** (rocks) and **biosphere** (life). Changes in any one of these components may lead to changes in atmospheric – and therefore climatic – conditions.

Some effects may be persistent due to the fact that the oceans respond much more slowly than does the atmosphere; once a temperature anomaly has been produced it takes a long time for it to be eradicated. Moreover, relatively small changes in the hydrosphere may have drastic implications for atmospheric conditions.

Similarly, rapid changes in the biosphere affect the atmosphere. The nature and extent of the vegetation control the amount of energy reflected by the earth. In this context, man's impact upon the vegetation, through forest clearance and agriculture, may be particularly important.

Feedbacks

Changes in energy absorption at the earth's surface may produce a response in the atmosphere. In turn, further responses by the atmosphere may be initiated to accentuate or suppress the initial change. **Positive feedback** has a snowballing effect which could produce a potentially damaging situation whereas **negative feedback** tends to restore conditions to their original state. A simple example of negative feedback is shown by the effect of an increase in the mean temperature of the lower atmosphere. If this occurred, it would be accompanied by an increase in emission of long-wave radiation to space which would tend to cool the air. A positive feedback is shown in Figure 3.9. These feedback processes will be discussed further in Chapter 10 where their importance in long-term changes of climate is outlined.

Some feedback mechanisms in the atmosphere, like the ones described above, are quite common and simple to understand. Problems arise when we try to anticipate their consequences. A feedback is likely to affect other parts of the atmospheric system. We have to appreciate this point when attempting to understand and model our atmospheric circulation. While the essential controls of the global atmosphere are all identified in general terms, some of the feedback interactions are so difficult to determine that we do not even know whether they are positive or negative (Table 3.1). For example, an increase in the amount of cloud will decrease the input of short-wave radiation to the surface and reduce the loss of long-wave radiation to space; the former should produce a decrease in surface temperature and the latter an increase. Hence the net effect will depend upon the precise

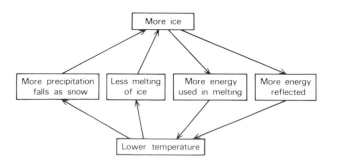

Figure 3.9 Positive feedback system in operation initiated by a decrease in temperature

Table 3.1 *Climate feedback mechanisms*

Mechanisms	Linkages	Feedback
Temperature – radiation	Increased $T_{a,s}$→Increased IR loss→ Cooling	–
Water vapour – greenhouse	Increased T_a →Increased IR trapping→ Warming (at constant RH)	+
Snow/ice albedo – temperature	Decreased T_s →Increased albedo→ Cooling	+
Cloudiness – temperature[1]	Change of cloud amount→Radiative fluxes→ Temperature change	?0 or –
	Decreased T_s→Decreased cloud altitude→Less radiation gets through→Cooling	+
Lapse rate – surface temperature	Local convective adjustment→Large-scale baroclinic effects	?
Sea-temperature anomalies – atmospheric circulation[2]	Wind field forcing of surface water drift and divergence Heat fluxes	?

Notes
  T = equilibrium temperature of atmosphere (a) or surface (s)
  IR = infra red radiation
  RH = relative humidity.
[1] Cloud effects are complex due to several factors involved: cloud amount, height of cloud tops, thickness and optical properties. The responses will vary with season and latitude.
[2] Air–sea interactions probably differ according to time scale. On the short term, sea temperature anomalies appear to be forced by the atmospheric circulation.

After R. G. Barry, *Progress in Physical Geography*, **3** no. 1 (1979), pp. 119–31.

changes in short-wave input and long-wave output, which in turn will depend upon the type, height, thickness and global distribution of cloud. We cannot easily tell which way the change will be.

An illustration of a major feedback problem in the atmosphere is a feature known as **'the Southern Oscillation'**. As well as the Hadley cell around the intertropical convergence zone, there is also a circulation in the Pacific between the Indonesian equatorial low and the south-east Pacific subtropical high. In general terms, when pressure is high in the Pacific Ocean, it tends to be low in the Indian Ocean from Africa to Australia. Where pressure is low, rainfall is high and vice versa. The pressure gradient tends to maintain strong equatorial easterlies which in turn maintain an upwelling of cold water, so the process is self-sustaining or even a positive feedback. Periodically the situation reverses to give high rainfall in the eastern Pacific and low rainfall in Indonesia (Table 3.2). The effect was pronounced in 1983. Coastal Ecuador received 300 per cent of average precipi-tation between November 1982 and May 1983, amounting to 3258 mm, whereas Indonesia had received only 30 per cent of its average.

The cause of the persistence is reasonably well established and is a good example of a feedback mechanism. The problem is to understand what causes a breakdown of the system to give the reversal, with its characteristic period of between two and five years (Figure 3.10).

Long-term changes

Finally we must consider how our system changes over longer periods of time than the few years already mentioned. It is clear that the local surface conditions vary considerably from year to year, season to season, and even from day to day. In addition, we have stated that the major driving force of the atmospheric circulation – the input of solar radiation – changes when a sufficiently long time period is considered. The balance between the

Table 3.2 *Circulation characteristics for two successive years*

	Dec. 1962–Feb. 1963	Dec. 1963–Feb. 1964
Sea surface temperatures at Canton Island (Pacific)	cold	warm
Surface pressure anomaly		
Canton Island	+1.2 mb	−0.8 mb
Djakarta (Indonesia)	−0.6 mb	+0.3 mb
Rainfall anomaly		
Canton Island	−166 mm	+469 mm
Djakarta	+569 mm	−409 mm
Vertical motion (500 mb level)		
Eastern Pacific	weak	strong
Indian Ocean	strong	weak
Hadley circulation	weak	strong

After R. E. Newell *et al.*, *General Circulation of the Tropical Atmosphere*, vol. 2 (1974), Cambridge, Mass.: MIT Press.

processes operating in the atmosphere, therefore, is not constant; the details of the system shown in Figure 3.11 vary over time. If we had been around 10,000 years ago, at the end of the last Ice Age, we would have seen just how different these conditions could be. A few million years earlier, in the period geologists call the Tertiary, the system would have been different again. Indeed, one of the vital aims of the climatologist must be to understand what causes changes in the atmospheric system, and what are the likely effects of these changes.

In a previous section in this chapter we showed how short-term changes in the hydrosphere and biosphere could affect the atmospheric circulation. At a longer time scale, the four major components of the global system have potential climatic consequences. Changes in any one of these components – the hydrosphere, the lithosphere, the cryosphere and the biosphere – may lead to changes in atmospheric, and therefore climatic, conditions. Let us have a look at a few examples of the changes which are likely to have occurred. We know that during the last Ice Age, sea-levels fell by up to 100 m, relative to the present. In the areas of wide continental shelves, this would have exposed large areas of dry land which would have had a very different albedo, and especially a changed heat storage capacity from the water it had replaced. The lowered sea-level could also change the positions of the ocean currents, tongues of warm water being deflected from their previous patterns, though currents would also be affected by the atmospheric circulation.

Expansion or contraction of the ice caps, even by quite small amounts, may lead to prolonged changes in climate over a wide area. Concern has been expressed recently about the apparent increase of winter snow-cover in the northern hemisphere because of these potential changes, but luckily the trend appears to have been short-lived. The effects of changes in the lithosphere may be similarly far-reaching. As we shall discover in Chapter 18, the earth's crust is highly mobile; it is constantly being raised into mountain chains, and moved across the surface of the globe. In the process the character and distribution of the continents change. As a result, the patterns of both atmo-

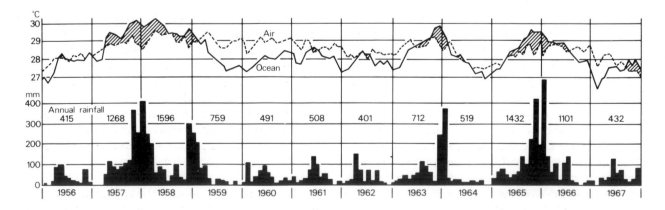

Figure 3.10 Monthly air and sea temperatures with rainfall data for Canton Island (2°48'S, 171°43'W) from 1956 to 1967 (after Boucher, 1975)

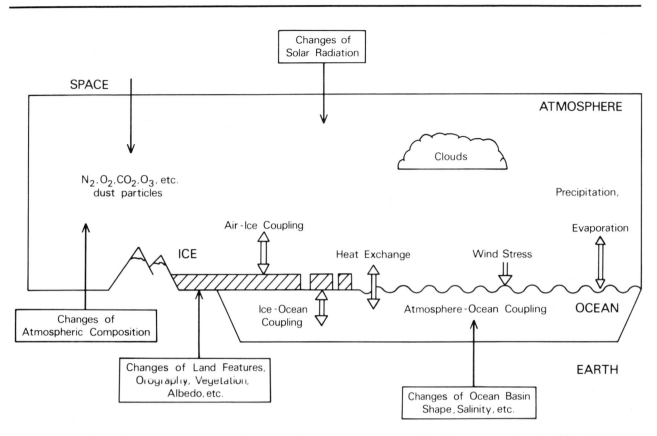

Figure 3.11 The physical processes and properties that govern the global climate and its changes (after US National Academy Report 'Understanding Climatic Change', 1975)

spheric and oceanic circulation are disturbed and the climate, too, is modified. It is interesting to speculate what would be the effect on the climate of the northern hemisphere if the Western Cordillera of the United States did not exist. On a hemispheric scale, conditions would be more like those of the southern hemisphere today with much less north–south movement of air; the south-west to north-east flow of air across the Atlantic would be greatly reduced, together with the speed of the North Atlantic Drift. More locally, rainfall would penetrate more easily into the Great Plains, the Chinook would disappear and the wide swings of temperature which can affect the Mississippi Valley would be reduced due to fewer depressions moving south-eastwards or north-eastwards. However, because of all the complex interactions involved much of this is speculation. We cannot be sure of the precise consequences, which is why proposals for widespread weather modification must be treated with great caution.

Although long-term changes in the biosphere could have an effect, many of the changes, apart from those brought about by man, are themselves the result of changes in environmental conditions of which the atmosphere is probably the most important.

Conclusion

The reason we need to know about these changes in surface conditions depends upon the nature of our atmospheric system. As we have seen, the primary driving force is the radiant energy from the sun. Some radiation is absorbed in the atmosphere, some is absorbed at the surface but in both cases energy is reradiated back to space. The difference in absorption between surface and atmosphere produces vertical temperature gradients which result in vertical flows of energy.

Similarly, the surplus of energy which builds up in the tropical areas and the deficit in the polar regions produces horizontal temperature gradients across the earth's surface. These drive the major horizontal air flows. The temperature gradients

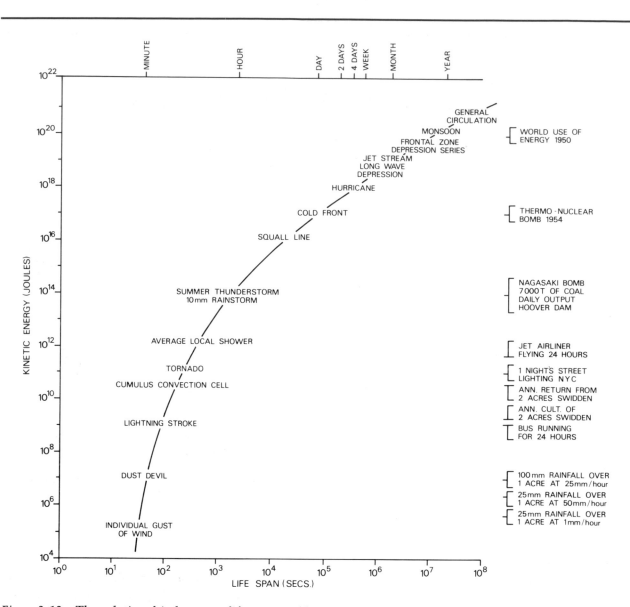

Figure 3.12 The relationship between lifespan and kinetic energy of meteorological systems, together with the kinetic energy associated with a number of human activities. (1 acre equals 0.40 hectares; swidden is a technical term for shifting agriculture) (after Bennett and Chorley, 1978)

give rise to pressure gradients which in turn result in the wind systems and the kinetic energy of atmospheric motion. Kinetic energy can operate over a wide time range (Figure 3.12) but eventually is dissipated through turbulence and friction, ending up as heat. To complete the energy cycle, this heat is returned to space as infra-red radiation emitted from the atmosphere.

This sequence represents the basic controls of the atmosphere. What is clear though is that in detail the atmospheric system is highly dynamic. Changes outside the globe (e.g. solar radiation amounts), within the atmosphere and within the rest of the global system may all cause complex reactions. It is for this reason that the atmosphere presents us with such an exciting challenge. And it is because atmospheric and climatic conditions impinge so forcefully upon our existence that it is essential we come to understand the system more thoroughly. In Chapters 4 and 5 we shall examine the components of the atmospheric system in more detail to determine how they work and interact to produce the wide range of climates which we experience.

4

Energy balance of the atmosphere

The atmosphere

A thin shell of gases surrounds the earth. This is the atmosphere. It plays a vital role in allowing life to survive. It shields us from harmful radiation. It helps to stabilize the differences in temperature between tropical and polar regions. It reduces losses of heat from the earth's surface. It provides water to sustain life. We only have to contrast the earth with the moon to see what it would be like without an atmosphere, or with Venus or Mars if we had a different type or amount of atmosphere.

The gravitational force of the earth is sufficient to retain the present atmosphere, but its density is not uniform. As we move away from the earth's surface, the density becomes progressively less until we reach the edge of space. What is the nature of this atmosphere? How does it achieve the functions outlined above? We shall be examining these points throughout this section but first the general properties of the atmosphere must be considered.

Chemical composition

The chemical composition of the atmosphere is remarkably constant in the lowest 16 km, with **nitrogen** and **oxygen** being dominant, but the composition is not static. Volcanic activity, biological and chemical reactions, radioactive decay, gravitational losses and human activity all act together to produce and maintain the atmosphere (Table 4.1). There are two climatologically important gases whose proportion is variable. The first of these is **water vapour** which gets into the atmosphere through evaporation, mainly from oceans and land surfaces. In arid areas, there is little water to be evaporated so the atmosphere is also dry, perhaps containing only 0.5 to 1 per cent of water vapour. In very humid areas this figure may rise to 3 per cent or even 4 per cent.

The second variable gas is **carbon dioxide** (CO_2). It is a by-product of combustion and photosynthesis. Increases in fuel consumption have led to

Table 4.1 *Composition of the earth's atmosphere*

Constituent	Formula	Per cent by volume
Nitrogen	N_2	78.08
Oxygen	O_2	20.95
Argon	A	0.93
Carbon dioxide	CO_2	0.03
Neon	Ne	0.0018
Helium	He	0.0005
Methane	CH_4	0.0001
Krypton	Kr	0.0001
Hydrogen	H_2	
Nitrous oxide	N_2O	
Xenon	Xe	Negligible
Carbon monoxide	CO	
Ozone	O_3	
Water vapour	H_2O	Variable (0–4)

global increases in the carbon dioxide content of the atmosphere to above 330 parts per million or 0.033 per cent (Figure 4.1). In addition, plants use carbon dioxide for growth, and plants cover a large proportion of the land hemisphere. In summer, when the plants are growing most rapidly, they consume more carbon dioxide than they release. So the atmospheric levels of carbon dioxide fall. In winter, plant growth is greatly reduced in the northern hemisphere with no corresponding increase in the southern hemisphere where so much of the surface is sea, so levels of carbon dioxide show a seasonal change.

At higher levels in the atmosphere, water vapour and carbon dioxide decrease in importance but between 20 and 55 km **ozone** becomes more common. Ozone is a form of oxygen containing three atoms instead of the more usual two. It is formed by an interaction between the ultra-violet part of the

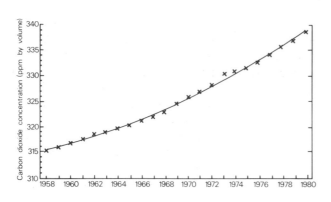

Figure 4.1 Mean annual concentration of atmospheric CO_2 at Mauna Loa Observatory, Hawaii between 1958 and 1980

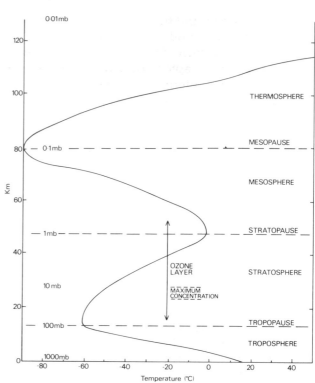

Figure 4.2 The layers of the atmosphere in relation to the mean temperature profile

sun's rays and the oxygen atoms. If it were not for this effective shielding, the harmful ultra-violet rays would reach the surface and seriously affect animal tissues and bacteria. This is why there has been concern over the use of this layer for supersonic aircraft flights and why the use of Freon in aerosol sprays has been banned in the USA. Both tend to reduce the amount of available ozone, although the magnitude of the effect has been disputed.

Temperature structure

Every day, hundreds of meteorological balloons are released into the atmosphere to take observations of temperature, wind velocity and humidity. Each ascent shows some differences in the temperature structure of the air above the release point, mainly related to local weather. In general they all show a decrease of temperature with height up to about 9 to 15 km above the surface. Then temperatures stop falling and even begin to rise (Figure 4.2) from the lowest values of about −60°C. By 30 km we are approaching the limit of most standard radio-sonde ascents and additional information about the atmospheric temperature structure has to be obtained from rockets or satellites. These show that there is a progressive increase in temperature up to about 50 km where values reach about freezing point, followed by a further decrease to the lowest atmospheric temperatures of about −100°C at about 80 km. Beyond this level, there is a steep rise of temperature in a physical sense, but as the air is so rarified, it holds little heat.

Each of the layers of the atmosphere is identified by its temperature structure (Figure 4.2), though it is only the **troposphere** and to a lesser extent the **stratosphere** that are important to our climates. There is not a simple change of temperature with height but there are three zones of heat production separated by cooler zones. The earth's surface at the base of the troposphere is where most heat is produced by conversion of solar energy. In the stratosphere heat is generated through the ozone reaction with ultra-violet light. Finally in the **thermosphere**, we are starting to encounter the particles emitted by the sun which give high temperatures but a low heat content.

Energy in the atmosphere

The energy system

The sun's energy represents the prime source for our climatic system. Energy flows from the sun and interacts with the earth's atmosphere and surface in a complex manner to give rise eventually to the climatic zones of the earth. As mentioned in Chapter 2, heat storage and transport are also involved, so the best way to understand the processes is to examine them in turn. The laws of radiation have been outlined in Chapter 2 so here we look first in some detail at the internal mechanisms of the radiation flow, then consider the spatial variability of the flows which give rise to different climates.

Perhaps the best way to explain what is happening is to follow the path of sunlight from the top of the atmosphere and describe what affects it on its journey to the earth's surface. Long-wave exchanges can then be described.

Short-wave radiation in the atmosphere

As our beam of sunlight enters the atmosphere, it first passes through the **mesosphere** with little change. In the stratosphere, the density of atmospheric gases increases. There is more oxygen available which reacts with the shortest or ultra-violet wavelengths and effectively removes them, warming the atmosphere in the process. Already about 2 per cent of the original beam strength has been lost.

It is in the troposphere that most effects take place. In the upper troposphere, the atmosphere is relatively dense with a pressure about 20 per cent of that at the surface. The size of the gas molecules of the air is such that they interact with the insolation, causing some of it to be **scattered** in many directions. This process depends on wavelength. The shorter waves are scattered more than the longer waves and so we see these scattered waves as blue sky. If the reverse were true the sky would be permanently red, and if there were no atmosphere, as on the moon, the sky would be black. Dust and haze in the atmosphere produce further scattering, but not all of this is lost. Some of the scattered radiation is returned to space, but much is directed downwards towards the surface as down-scatter or diffuse radiation. This is also the type of radiation which we experience during cloudy conditions with no direct sunlight when the solar beam is 'diffused' by the water droplets or ice particles. Without diffuse radiation, everything we see would be either very bright, when in direct sunlight, or almost black when in shadow. Lunar photographs are a good example of this.

Another type of short-wave energy loss is **absorption**. The gases in the atmosphere absorb some wavelengths (Figure 2.6, p. 19), as do clouds. In this way, we have a warming of the atmosphere though the amounts involved are small. The most important loss of short-wave radi-

ation in its path through the atmosphere is by **reflection**. The water droplets or ice crystals in clouds are very effective in reflecting insolation. Satellite evidence shows that, for the earth, a mean figure of 19 per cent of the original insolation is reflected by clouds. The degree of reflection is usually called the **albedo**. Albedo is normally expressed as the ratio of the amount of reflected radiation divided by the incoming radiation. If multiplied by 100, this can be expressed as a percentage. The lowest and thickest clouds tend to reflect most while the thin, high-level ice clouds have an albedo of only about 30 per cent.

By now, the beam has reached the ground surface with some 50 per cent of its original energy. Even then, not all of this is absorbed as the surface itself has an albedo. The global average albedo represents some 6 per cent of the radiation *at the top of the atmosphere* so the loss is not great. However, the figure may seem large when expressed as a percentage of the radiation actually reaching the surface. For example, the albedo of freshly fallen snow may reach as high as 90 per cent (Table 4.2). The greatest variability is over water. When the sun is high in the sky, water has a very low albedo. This is why the oceans appear dark on satellite photographs (Plate 4.1). At low angles of the sun, such as at dusk, at dawn or in mid winter in temperate and sub-polar latitudes, the albedo may reach nearly 80 per cent.

Table 4.2 *Albedos for the short-wave part of the spectrum*

	Per cent
Water (angles above 40°)	2–4
Water (angles less than 40°)	6–80
Fresh snow	75–90
Old snow	40–70
Dry sand	35–45
Dark, wet soil	5–15
Dry concrete	17–27
Black road surface	5–10
Grass	10–20
Deciduous forest	10–20
Coniferous forest	5–15
Crops	15–25
Tundra	15–20

The sunlight reaching the earth's surface which is not reflected by the earth is absorbed and converted

Plate 4.1 Europe and North Africa from Meteosat on 30 July 1982. The Mediterranean Sea stands in sharp contrast to the higher albedo areas of North Africa which are drier and have less vegetation than Europe which appears darker on the image

into heat energy. The distribution of energy received at the surface is shown in Figure 4.3. Thus, incoming radiation can be absorbed (in the atmosphere and at the surface), scattered (in the atmosphere) or reflected (by clouds and at the surface). When reflected, the radiation is returned to space in the short-wave form and becomes part of the outflow of energy from the earth (see Chapter 2). Similarly, some of the scattered radiation is returned to space to give a short-wave albedo for our planet of 28 per cent. The modifications of the solar beam by the atmosphere are shown diagrammatically in Figure 4.4.

Long-wave radiation

All substances emit long-wave radiation in proportion to their absolute temperatures. The earth's surface receives most short-wave radiation and therefore normally has the highest temperatures. It follows from this that most long-wave emission will be from the ground surface. The atmosphere is much more absorbent to long-wave radiation than to short-wave radiation. Carbon dioxide and water vapour are very effective absorbers of much of the longer part of the spectrum (see Figure 2.6, p. 19) except between 8 and 12 μm. As water vapour is concentrated in the lowest layers of the atmosphere this is where most absorption will take place. Clouds are also very effective at absorbing long-wave radiation and hence their temperature will be higher than otherwise. This cloud effect is most noticeable at night. With clear skies, radiation is emitted by the surface but little is received from the atmosphere and therefore the temperature falls rapidly. If the sky is cloudy, the clouds will absorb much of the radiation from the surface and, because they are also emitters, more of the radiation will be returned to the ground as counter-radiation than if the sky had been clear. This is absorbed by the ground, compensating for the emission of long-wave radiation and so reducing the rate of cooling of the ground temperature. Figure 4.5 shows temperatures on clear and cloudy nights to demonstrate this effect.

Some of the radiation given off by the surface is lost to space but the majority gets caught up in the two-way exchange between the surface and the atmosphere. Figure 4.4 shows the emission and absorption of long-wave radiation as a proportion of incoming energy. Radiation from the atmosphere is emitted spacewards as well as downwards. As there is less water vapour at higher levels absorption by the atmosphere is less and proportionally more is lost to space.

Global radiation balance

Taking the earth as a whole we know that no part is getting warmer or cooler and so there must be an overall balance. More short-wave radiation appears to be absorbed by the earth than leaves it by a mixture of short- and long-wave radiation. The surface seems to be gaining heat. Similarly, the atmosphere seems to be losing heat. If radiation were the only process operating, the earth's surface should be getting warmer and the atmosphere cooler. They do not do so because, in addition to radiation, there are **thermal energy transfers** in the form of convective heat exchanges. Much of this takes place through evaporation and is discussed later in this chapter (p. 50).

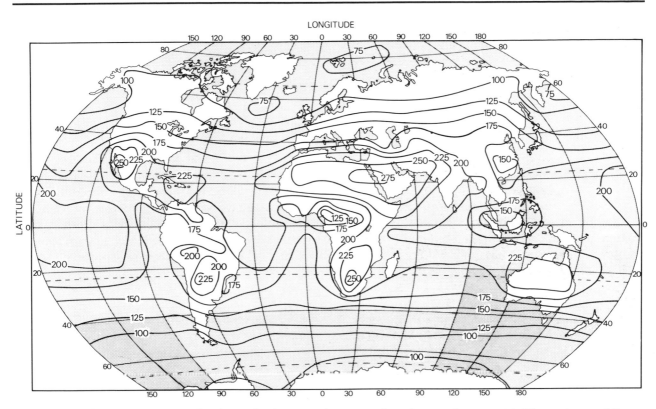

Figure 4.3 The average annual solar radiation on a horizontal surface at the ground. The units are W m⁻²
(after Sellers, 1965)

Spatial variability of radiation exchanges

The earth is a large spheroidal body which spins around on an axis tilted at $23\frac{1}{2}°$ to the vertical and has an elliptical orbit around the sun. These factors alone have a considerable influence on how radiation is distributed at the earth's surface.

In Chapter 2, we described the input of solar energy at the top of the atmosphere and how it was determined by these astronomic controls. Figure 2.10 (p. 23) showed how the radiation would be distributed at the top of the atmosphere, or at the earth's surface if we had no atmosphere. However, if we look at a map of the average annual short-wave radiation reaching the ground, it is appreciably different (Figure 4.3). The general impression of the map is of a decrease of energy input towards the poles, with local anomalies. Most of these are caused by the distribution of clouds. High values are found over the Saharan, Australian and Asian deserts. The lowest values occur in regions of high cloudiness such as Iceland, the Aleutian Islands in the north Pacific, the Congo Basin and parts of West Africa.

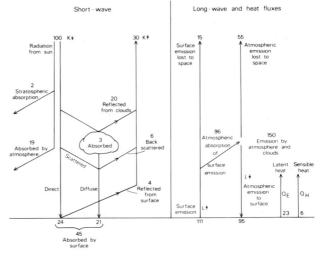

Figure 4.4 Modification of short- and long-wave radiation by the atmosphere and the surface. Figures are expressed as a percentage of incoming short-wave radiation at the top of the atmosphere, based on a global mean

This map may be a little misleading. By showing the radiation reaching the ground surface we are

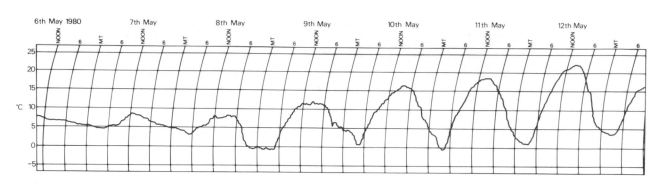

Figure 4.5 Contrasting diurnal temperature variations on cloudy and clear days. Cloudy weather prevailed for the first three days giving a small diurnal temperature range. As the skies cleared later in the week, daytime temperatures increased but night-time temperatures became lower. There is some indication of a slight, progressive warming of both day and night temperatures as a result of the storage of solar energy

omitting an important factor. The surface albedo influences how much radiation is absorbed and this is not indicated on the map. Oceanic areas absorb a similar total to that shown on the map but for ice-covered surfaces, such as Greenland, and areas with light coloured, dry soils, such as the Sahara, the total radiation absorbed may be significantly lower (Figure 4.6).

What is it that produces this spatial pattern of radiation? Obviously, the astronomic factors have a great effect giving rise to the overall decline polewards. But this decrease is far greater than one would expect from the distribution at the top of the atmosphere.

The angle of the sun's rays

Thus, we have to look for other reasons. One of the most important is the angle between the sun's rays and the earth's surface. The input is greatest whenever the surface is at right angles to the sun's rays. If the sun is overhead a horizontal surface will receive the highest intensity of radiation. When the sun is low in the sky, the steeper slopes facing the sun will receive the highest values. As the earth is a sphere at a great distance from the sun, the sun's rays appear parallel and hit the surface at different angles (Figure 2.9, p. 22).

A secondary effect which further decreases radiation intensity is the longer path through the atmosphere at higher latitudes. Scattering and absorption will be higher, though this increases diffuse radiation at the expense of direct radiation. The amount of scattering and absorption vary depending upon the degree of haziness of the atmosphere.

Where the atmosphere is very dusty, such as in semi-arid or desert areas, more radiation will be absorbed and scattered, preventing it from reaching the ground surface. As the dust particles are much larger than gas molecules, scattering is not dependent upon wavelength and the sky has a whitish hue rather than the deep blue of a clear atmosphere. This effect is also noticeable over urban areas where pollution produces the same effect.

Latitudinal radiation balance

To see how much radiant energy we have available at any location we must know how much radiation is being lost as well as how much is reaching that location. Long-wave radiation emission is proportional to the absolute temperature of the surface. It is far less variable than the input of solar radiation, ranging from about 350 W m^{-2} in the Sahara to about 100 W m^{-2} in Antarctica. The difference between incoming and outgoing radiation is known as **net radiation** or the **radiation balance**. For the earth's surface, values are shown in Figure 4.7.

If we include the effects of the atmosphere the picture changes. The atmosphere has a negative balance, even in the tropics (Figure 4.8). In fact, values differ little between equator and poles. For any particular latitude, we can sum the surface and atmospheric radiation balances to find out which areas of the earth have a radiation surplus and which areas have a deficit. This shows a surplus of energy between 38°N and 38°S and a deficit towards the poles. Naturally the magnitude of the surplus is identical to that of the deficit, but it does mean that there must be a steady transfer of energy from the tropics polewards, otherwise the tropics would get hotter and polar regions cooler. It is the winds of the world, and to a lesser extent the ocean currents, which bring about the necessary heat transfer.

Variability through time

As well as changing spatially, the components of the radiation balance at any point will be changing through time. For example, we could measure the values for a particular day. The results we obtain would depend upon the time of year and the nature of the weather conditions on that day. Atmospheric transmission, cloudiness, moisture levels, surface albedo and even the moisture content of the soil (affecting long-wave emission) would cause

changes in the components. Figure 4.9 shows two extreme examples with clear and cloudy conditions.

On an annual basis, fluctuations in weather and the changing astronomic input of solar radiation lead to further alterations of the radiation balance components (Table 4.3). In turn, these variations help to explain the seasonal changes in heat absorption at the surface and hence the climatic regimes. For longer time periods, cyclic variations in the earth's orbit and its axis of rotation produce even more changes in the input of solar radiation. This subject will be covered in more detail in Chapter 10.

The heat balance

Uses of available energy
In the previous section, we showed how the earth's surface normally receives a surplus of radiation

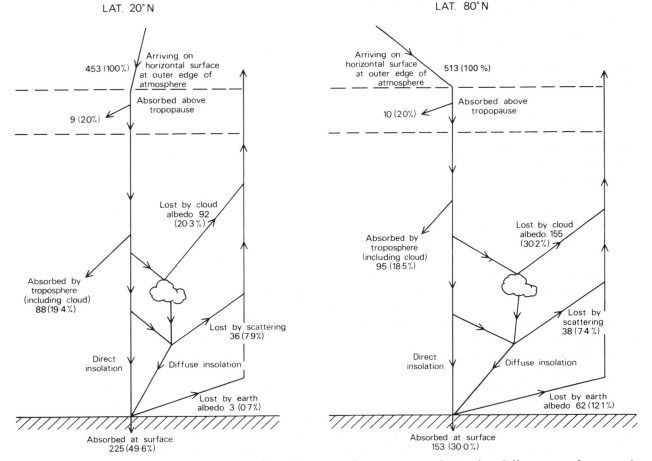

Figure 4.6 Contrasts in the disposition of insolation at the summer solstice for different surfaces and different latitudes. Units are in W m^{-2}. 'Absorption' implies the conversion of the radiation into some other form of energy. 'Lost' indicates a loss to the earth–atmosphere system (after Jackson, 1963)

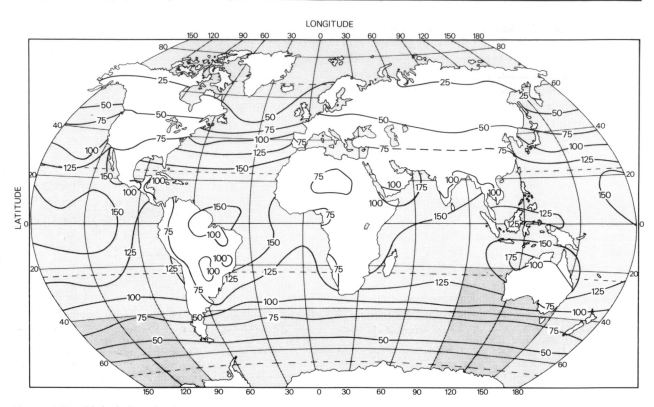

Figure 4.7 Global distribution of mean annual net radiation. Units are W m⁻² (after Budyko *et al.*, 1962)

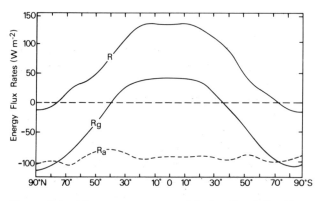

Figure 4.8 The average annual latitudinal distribution of the radiation balances of the earth's surface (R), of the atmosphere (R$_a$), and of the earth–atmosphere system (R$_g$) in W m⁻² (after Sellers, 1965)

which leads to a warming of the surface. This situation cannot last indefinitely as the temperature gradient in the air and the soil would become enormous. Energy tends to flow down a gradient and as radiation is absorbed by the surface, so heat is transmitted into the soil and into the air. This

takes place in proportion to the amount of energy originally absorbed:

$$Q^* = Q_H + Q_G$$

where Q^* is the net radiation

Q_H is the sensible heat transfer into the air

Q_G is the heat flow into or out of the soil

If the surface is damp some of the energy will be used in evaporation. Therefore:

$$Q^* = Q_H + Q_G + Q_E$$

where Q_E is the heat used for evaporation

This is a simplification, as changes in heat storage can take place and a small amount of energy is used in plant growth.

The energy transfer into the atmosphere is the final component of the radiation imbalance between surface and atmosphere. The net radiational loss in the atmosphere is counteracted by this heat transfer from the surface. So, over a long period, the atmosphere gains as much energy as it loses.

Table 4.3 *Radiative fluxes for Copenhagen*

	Range of monthly means		
	Low	*High*	*Yearly mean or total*
Downward short-wave radiation (K↓)	1.8 MJ m⁻²	28.1 MJ m⁻²	3456 MJ m⁻²
Albedo (per cent) (α)	19	50	22
Net short-wave radiation (K*)	1.4	22.3	2682 MJ m⁻²
Net long-wave radiation (L*)	−2.6	−7.0	−1649 MJ m⁻²
Net all-wave wave radiation (Q*)	−1.8	16.1	1033 MJ m⁻²

Notes

All components of radiation show a seasonal variation. In many cases it is the net effect or difference between incoming and outgoing radiation which is important. Downward short-wave radiation (K↓) is direct and diffuse radiation from the sun; some of this radiation is reflected at the surface, especially in winter, to leave us with net short-wave radiation (K*). The earth emits long-wave radiation as well as receiving it from the atmosphere, the difference between emission and absorption is the net long-wave radiation (L*). As the values are determined by the absolute temperature of the emitting body, seasonal differences are not pronounced. The value is negative because more long-wave radiation leaves the surface than reaches it. The difference between all incoming and all outgoing radiation, irrespective of wave-length, is called net all-wave radiation (Q*) and determines how much energy is available at the surface.

After D. H. Miller, *Advances in Geophysics*, **11** (1965), pp. 175–302, and D. H. Miller, pers. comm. (1985).

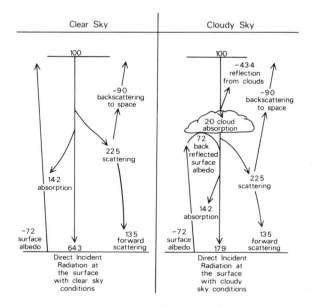

Figure 4.9 The effects of cloud on the incoming short-wave radiation balance. Figures are expressed as a percentage of insolation at the top of the atmosphere (after Bridgman, 1969)

Sensible heat

How do these processes which use the net radiational energy take place? One answer is by sensible heat transfer.

Sensible heat is the exchange of warm air down the temperature gradient. By day, this will normally be upwards, but at night there may be a weak transfer of sensible heat down to the cooler ground surface. It takes place because the air in contact with the surface becomes warmer through conduction. Being warmer, the air will be less dense than its surroundings and, like a cork in water, it will tend to rise until it has the same density (temperature) as its surroundings. Occasionally, this process can be seen operating. If the ground is being warmed intensely, the rate of sensible heat transfer is high. The rising air can then be seen as a 'shimmering' of the air layer near the ground due to the variable refractive indices of light through the air of different temperatures. Replacing the rising warm air are pockets of cooler air descending towards the ground.

The significance of sensible heat in the local heat

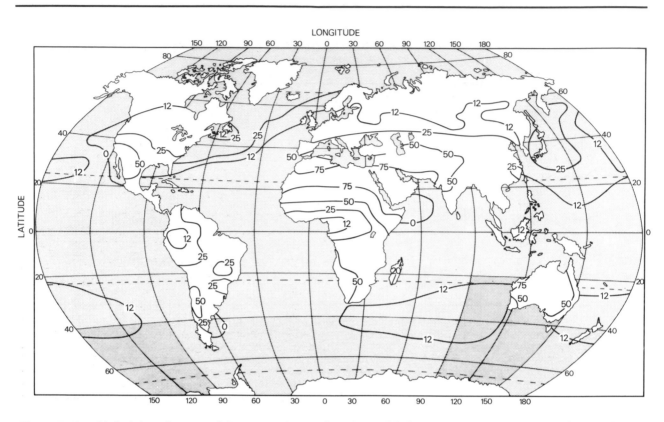

Figure 4.10   Global distribution of the vertical transfer of sensible heat. Units are in W m⁻² (after Budyko *et al.*, 1962)

budget depends upon the frequency and intensity of surface heating. Where the surface is usually hotter than the air, values may be high, but where there is little temperature difference, sensible heat transfer will be low (Figure 4.10).

Latent heat

The concept of latent heat can best be understood by conducting a small experiment. Start with a large block of ice out of a freezer and measure its temperature; perhaps it may be −10°C. Then place it in a Pyrex glass beaker and heat the beaker at a constant rate, monitoring the temperature of the ice continuously. Keep heating the beaker until all the ice has melted into water; eventually it will reach boiling point and vaporize as steam. If the temperature values are then plotted against time, we find a steady increase of temperature (representing heat input from the heater and some heat flow from the air which will be warmer than the ice) until melting starts. Despite the steady addition of heat, there is then no increase in temperature until the ice melts completely (Figure 4.11). A similar effect is found on vaporization. Where has the heat gone?

It was not being used to raise the temperature during melting or vaporization but to change the physical state of water, either from solid to liquid or liquid to vapour. As the heat appears to be hidden, it is known as latent heat.

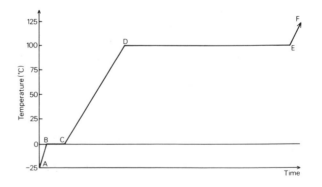

Figure 4.11   The pattern of temperature and phase changes for water. The temperature remains constant during each phase change as long as pressure remains constant. Differences in specific heat of ice and water give different gradients for the lines A–B and C–D

50

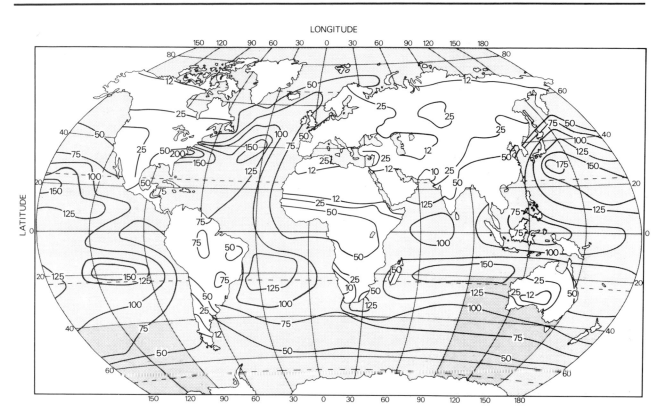

Figure 4.12   Global distribution of the vertical transfer of latent heat. Units are in W m^{-2} (after Budyko *et al.*, 1962)

Heat consumption

Change of state involves a considerable use of energy to change from solid to liquid, or from liquid to vapour. In the first case, we need 3.33×10^5 J kg^{-1}; this quantity of heat is called the **latent heat of fusion**. In the second, much more energy is needed. At 10°C the **latent heat of vaporization** is 2.48×10^6 J kg^{-1} but it falls slightly with increasing temperature. To get a better idea of this large quantity of energy needed for evaporation, the amount consumed in evaporating only 10 g of water is about the same as that needed to raise the temperature of 60 g of water from 0°C to boiling point (100°C). We tend to be most aware of evaporational cooling after swimming. The effect of evaporation leads to an extraction of heat from the skin surface; sweating works in a similar way.

Release of energy

Where the state of water changes to a lower energy level (i.e. from vapour to liquid or liquid to solid) it will release the same quantity of energy that was originally used. This is especially important in our atmospheric heat balance. Water that is evaporated from the surface will extract energy from that surface where there is usually a surplus anyway. Eventually the vapour will condense in the atmosphere, probably as a cloud droplet, releasing latent heat originally extracted from the surface and so helping to warm the atmosphere. This can also take place polewards of the original evaporation point so evaporation can transfer heat energy both into the atmosphere and polewards.

Much of the earth's surface is covered by oceans where evaporation takes place continuously. Even a large proportion of the land surface is moist much of the time. Consequently the role of latent heat in balancing the heat budget of the earth is vital. Latent heat transfer by convection carries about one-fifth of the energy of incoming solar radiation back to the atmosphere (Figure 4.12).

The heat used for evaporation over land areas depends upon the availability of moisture and of energy. In polar regions it is small, but increases equatorwards, reaching a maximum in the moist equatorial forests of South America, central Africa and Indonesia. Over the desert areas, there is little moisture available and evaporation is insignificant.

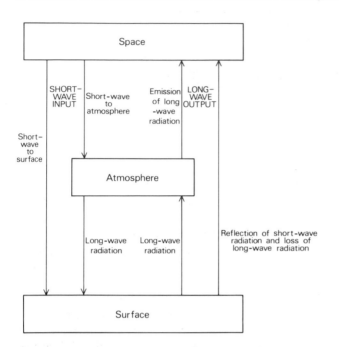

Figure 4.13 The earth–atmosphere radiant energy flow system

The atmospheric energy system

The details of the atmospheric energy system may appear complicated but the system is very important as the driving force for our present-day climates. To recap what happens; energy enters from the sun. It is reflected, absorbed and reradiated within the system but does not form a uniform distribution. Some areas receive more energy than they lose; in other areas the reverse occurs. If this situation were able to continue for long the areas with an energy surplus would get hotter and those with a deficit would get cooler. This does not happen because the temperature differences produced help to drive the wind and ocean currents of the world. They carry heat with them, either in the sensible or latent forms, and help to counteract the radiation imbalance. Winds from the tropics are therefore normally warm, carrying excess heat with them. Polar winds are blowing from areas with a deficit of heat and so are cold.

The actual flows of energy can be shown more simply if we consider them as part of a large system in which we distinguish the inputs and outputs with feedback between the different subsystems (Figure 4.13).

While the general principles of flow are known, the figures quoted are, in most cases, best estimates. Measurements have been made at a number of places, but in insufficient quantities to give a reliable global figure. It is little use giving a global average based on a few clustered observations. Satellite observations have helped (and led to appreciable changes in estimates of the earth's albedo) but there are still numerous flows which are imperfectly measured. Long-wave emission by the atmosphere, the separation into direct and diffuse radiation and sensible and latent heat transfer are the main problems as conditions vary quickly, and, until measurements are more comprehensive, some of the figures are little more than intelligent guesses. The actual value of the flow will depend, in part, on the nature of the assumptions made about them. What we can be sure about is that *what comes into the earth/atmosphere system must eventually leave.*

5

Heat and moisture in the atmosphere

On a hot day, the picture of clouds building up often signifies that a storm is imminent. We do not always appreciate what is happening, but these growing clouds represent one of the most vital processes in the atmosphere – the condensation of water as it is raised to higher levels and cooled by strong updraughts of air. The water, of course, was derived from the earth – evaporated from the oceans, rivers and soil or transpired by the vegetation. But within the atmosphere a variety of events combine to convert the water vapour, which is produced by evaporation, to water droplets. The air must rise and cool for condensation to occur. In this chapter we will be looking at the nature and consequences of these processes.

The sun and insolation

The daily march of the sun

Let us start by considering the most persistent and fundamental cause of these processes: the daily progress of the sun from east to west across the heavens. As we know, this apparent movement of the sun occurs because of the rotation of the earth as it circles the sun; a rotation that takes fractionally under 24 hours to complete. As the sun moves from the horizon to its midday position high in the sky and back to the horizon, the angle its rays make with the earth varies. We saw in the last chapter that when the sun is low in the sky the rays strike the earth obliquely and the radiation is 'spread out' over a wider area. In addition, relatively more incoming radiation is intercepted by vapour and dust and gases in the atmosphere. Consequently the average insolation per unit area is relatively low. In contrast, when the sun is high in the sky its rays strike the ground almost perpendicularly; they are concentrated in a small area and they experience relatively little interception as they pass through the atmosphere. Average insolation is at a maximum.

The effect upon temperature

The effect upon surface temperatures is obvious. If we consider a clear day in the spring in an area in, say, London, sunrise will be at about 6 a.m. Temperatures then are low, for during the night the earth has been losing heat by radiational cooling. But slowly, as the sun rises the earth warms up and, in turn, the air in contact with the surface is heated too (Figure 5.1). By about 2 p.m. the earth and air are at their warmest, the maximum temperature at the surface being earlier than that in the air because that is where the heat conversion takes place. From then on, as the sun gradually sinks, the ground surface and the overlying air will cool. The sun sets at about 6 p.m.; cooling continues throughout the night until minimum temperatures are reached just before dawn.

This daily variation in insolation and temperature is one of the most basic components of our weather. So obvious is it and so regular that we take it for granted. And yet quite marked differences in atmospheric conditions occur in response to the daily progress of the sun. As we will see later, the associated changes in temperature may lead to significant changes in humidity, and they often spark off major atmospheric processes such as

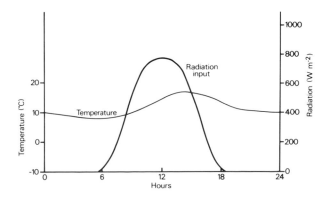

Figure 5.1 Diurnal changes in short-wave radiation input and temperatures on a clear day

vertical movements of air and even heavy storms.

It is also apparent that this daily pattern of insolation and temperature change itself varies according to atmospheric conditions. The effects are most obvious when the air is clear and still, for then heating and cooling proceed uninterrupted. If the sky is cloudy or very hazy, however, the daily pattern of temperatures is much more subdued (Figure 4.5). Similarly, the pattern varies spatially. It is less marked over the sea, for much more of the incoming energy is used to heat up and evaporate the water, and less is returned directly to heat the atmosphere. During the night, the sea cools slowly, with the result that temperatures do not fall so much as on land – one reason why coastal areas are less prone to night-time frosts (Figure 5.2). The pattern is most apparent in areas with dry air and ground surfaces, such as deserts. Here, incoming radiation is large and little energy is used for evaporation so temperatures are high, while radiational cooling at night is intense, giving rise at times to extremely low air temperatures.

The seasonal pattern

A very similar pattern of variation takes place on a seasonal scale. The cause in this case is not the earth's rotation, but its changing relationship with

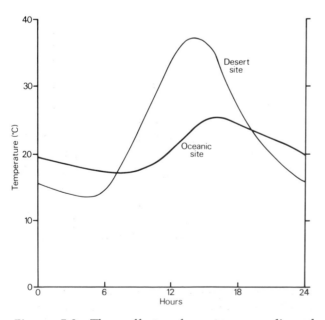

Figure 5.2 The effect of water on diurnal temperature ranges. Oceanic sites have a small diurnal temperature range whereas in a desert the range is high

the sun; the variation within its orbit that produces the apparent seasonal progress of the sun from the Tropic of Cancer to Capricorn and back.

This change in the position of the sun leads to changes both in the angle of the incoming rays and in the duration of daylight. Both factors influence the amount of insolation received by the earth and, therefore, the degree of atmospheric heating. Considering again our area in London, we would find that in the winter the maximum elevation of the sun, at midday, was about 16°, for the sun stands approximately over the Tropic of Capricorn. Thus the rays of the sun still strike the earth at a relatively low angle and the degree of midday heating is limited.

As the sun moves northwards to the equator and thence to the Tropic of Cancer its midday position rises and the rays strike the earth less obliquely. Moreover, days become longer and nights shorter. Maximum temperatures increase until, in about July, they reach their highest values, slightly after the maximum radiation in late June. From then until mid December the sun returns south, its midday position in the sky declines, the quantity of insolation received at the surface is reduced and so temperatures fall.

It is apparent that, in London, the winter months represent a period when incoming radiation is low. Outputs of energy from the earth continue, however, so the area experiences a **net radiation deficit**. During the spring, as the overhead sun moves north of the equator, radiation inputs rise to match outputs, but the degree of atmospheric warming is restricted because much of the excess energy is used to reheat the earth and oceans. By August the earth has warmed up; during autumn the sun returns to its position over the equator but now the earth still retains much of its heat gained during the summer. The air, therefore, remains relatively warm compared with spring even though the sun is at the same midday zenith angle.

This effect is most marked in coastal areas. Water heats up more slowly than the ground (it has a higher heat capacity) and, by the same token, retains its heat longer. Thus the difference between spring and autumn conditions is apparent. The Isles of Scilly, for example, are cooler on average than London in the late spring and early summer, but are warmer in autumn and winter (Figure 5.3).

The seasonal pattern of radiation and associated temperature conditions varies latitudinally. In polar areas, the sun never gets high in the sky, but

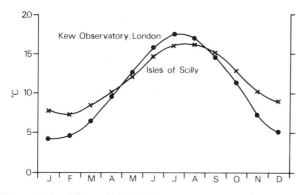

Figure 5.3 The effects of water on seasonal temperatures. At Kew (51°N), January and July are the coldest and warmest months respectively, but at the more oceanic site on the Isles of Scilly (50°N), the coldest and warmest months are February and August

day-length varies markedly, so that during summer months these areas experience perpetual daylight. Conversely, in the winter months they are in continuous darkness. The seasonal radiation balance is therefore very variable. At the north pole, for example, from April to September there is a potential continuous radiation surplus, for the sun would shine for 24 hours a day if the sky was cloud-free, so there is less night-time cooling. In contrast, for the rest of the year a radiation deficit occurs. No insolation is experienced for six months, so radiational cooling continues, interrupted only by the transfer of air from warmer latitudes (Figure 5.4).

The pattern in the tropics is very different. Here the sun never strays far from its overhead position, so seasonal variations in radiation are limited. The daily pattern of temperature and insolation is marked, but it varies little throughout the year (Figure 5.5).

The effects of heating and cooling of the atmosphere

General effects
Changes in insolation over time affect other aspects of climate as well as temperature. The atmosphere is a highly complex system, and the effects of changes in any single property tend to be transmitted to many other properties. Thus heating and cooling of the air cause adjustments in humidity and air sta..ility; they may cause condensation and evaporation, cloud formation and the development of storms.

What happens, then, when air is heated? To simplify the problem we will consider a parcel of air in contact with a warm ground. Like any other substance its temperature rises and it expands. Gases expand on heating more than either liquids or solids so this effect is quite marked. Moreover, as

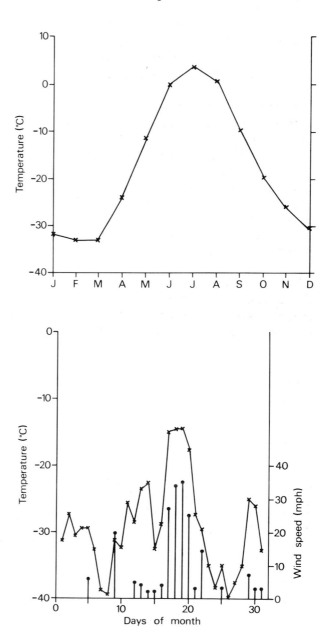

Figure 5.4 (Top) Mean daily temperature at Alert (82½°N). (Bottom) Temperatures at Alert in winter in relation to wind speed. Under calm conditions, temperatures are very low but stronger winds disturb the surface inversion and bring warmer air from more southerly latitudes

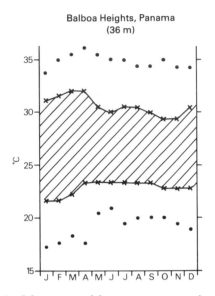

Figure 5.5 Mean monthly temperatures for Balboa Heights, Panama. Symbols from top to bottom indicate highest maximum, and mean maximum, mean minimum and lowest minimum respectively for each month

the air expands its density falls; in simple terms the same mass of air now occupies a larger volume. As its density falls so it becomes lighter than the surrounding air and it tends to rise like a bubble. Reverse the process, cool the parcel of air, and the opposite occurs. It contracts, its density increases and it sinks.

One effect of heating and cooling of the atmosphere is therefore to cause vertical movements of air. But there are other effects. As the air becomes cooler its ability to hold moisture in the form of water vapour is reduced. If it cools to the point where it can no longer hold the water vapour in vapour form, condensation occurs and water droplets form. As the air is heated, these droplets tend to evaporate and become water vapour once more. Thus, heating and cooling are intimately linked to the processes of evaporation, condensation and precipitation formation.

Let us consider these main effects in turn.

Vertical movements

The rising of warm air is a process we can see at work on a hot day by the shimmering effect of air near the surface; we can see it too if we watch the beautiful and immense towers of a convectional cloud forming. Such vertical movements usually develop if localized heating of the atmosphere takes

place, so that individual parcels of air become warmer and lighter than the air around them.

Localized heating occurs for a number of reasons. Variation in the wetness or colour of the surface may cause differences in atmospheric heating; the air above dark coloured or dry surfaces heats up more rapidly than that above light or wet surfaces. Differences in slope angle may have the same effect. But there is another factor that plays an important role in these vertical air movements. This is the vertical change in air temperature away from the ground surface. It is known as the environmental lapse rate.

Environmental lapse rate

If we measured air temperatures in the troposphere at different heights under stable, cloudless atmospheric conditions, we would find that temperature usually falls with height. The reason is quite simple. The incoming radiation heats the ground. A small proportion of this is transmitted downwards into the soil by conduction but the majority is returned as either sensible heat or long-wave radiation to heat the atmosphere. Heating is greatest close to the ground surface and declines with height.

The rate at which the temperature falls with increasing altitude is, on average, 6.4°C/1000 m, and this pattern continues as far as the tropopause. This is the **environmental lapse rate**. It is not constant, however, for it is affected by atmospheric and surface conditions. When the air is turbulent, or is being mixed by strong winds, the environmental lapse rate, at least in the lower layers of the atmosphere, is low; with strong surface heating, it is steep, meaning air temperature cools rapidly with height. Under still, calm anticyclonic conditions temperatures may even rise with height for short distances. Whatever its value, it is this environmental lapse rate that greatly influences vertical air movements.

Stability and instability

We can start to understand the importance of the environmental lapse rate by considering a simple example. Imagine localized heating of the air above an island in the sea. The island, because it converts sunlight to heat more effectively than the surrounding water, will act as a **thermal source**. Above this thermal source the air will become warmer, its density will decrease, its surface pressure will fall and the air will rise. Typically, after this bubble of air has risen a distance equal to about once or twice its own diameter it sinks back. New and larger bubbles form in its wake, however, and each rises a little higher.

What controls this movement? The answer, simply, is temperature. If the bubble of air is warmer than its surroundings it will continue to rise; if it is cooler, it will sink. We know already that the general temperature of the air declines upwards – this is the environmental lapse rate. We might imagine, therefore, that once the bubble starts to rise it will continue to do so indefinitely, for the air around it is becoming progressively cooler with height. This does not happen, however, and the reason is that as the air bubble rises it also cools. The critical factor that determines the height to which the bubble rises is the relative rate of cooling of the bubble and the surrounding air.

The next question, then, is why does the bubble get cooler? As the air bubble rises, it comes into contact with less dense surrounding air. The pressure confining the bubble is reduced and it expands. As it does so, heat is extracted from the bubble and it becomes cooler. This is in accord with the **Gas Laws** which state that:

$$PV/T = K \text{ (constant)}$$

in other words, the pressure (P), volume (V) and temperature (T) of a gas are interdependent. A change in any one of these properties tends to cause changes in the other.

The rate at which the air cools with height is a constant, at about 10°C for each 1000 m. It is known as the **dry adiabatic lapse rate** (DALR). Adiabatic means that there is no heat exchange between the bubble and its surroundings and, so long as the bubble of air rises rapidly, this condition applies. It is called dry, not because the air does not contain any moisture, but because no condensation has taken place.

We have, therefore, a framework for determining how far the bubble will rise. So long as no condensation occurs it will cool at the dry adiabatic lapse rate. The surrounding air cools at the environmental lapse rate. The bubble rises until its temperature (and therefore its density) is equal to that of the surrounding air. This is shown diagramatically in Figure 5.6.

Clearly the dry adiabatic lapse rate is a constant; the two variables in this relationship are the environmental lapse rate and the initial temperature of the air bubble. The bubble will only rise if it is warmed sufficiently to overcome the

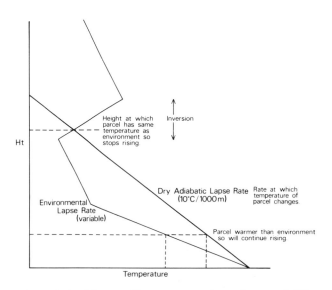

Figure 5.6 Thermal buoyancy of an air parcel. The parcel will continue to rise as long as it is warmer than the surrounding air

confining effect of the environmental lapse rate. If the bubble cannot rise it is said to be **stable**. If the temperature is raised far enough, however, or if the environmental lapse rate is great enough, the air bubble can rise a considerable distance. It is in these circumstances that part of the troposphere is said to be **unstable**.

Condensation

'So long as no condensation occurs . . . .' That was the proviso we established when considering the dry adiabatic lapse rate. But all air, even the driest desert atmosphere, contains some moisture, and if it is cooled sufficiently will experience condensation. When this happens the processes of vertical uplift are modified.

The ability of the air to hold moisture is dependent upon the temperature. As the temperature of the air increases so its moisture-holding capacity also rises; or, to put it another way, more moisture must be added to reach saturation at a higher temperature.

The amount of moisture which air can hold may be assessed in a number of ways. **Relative humidity** is the most frequently used term. It is the ratio of the amount of moisture the air contains to the amount of moisture the air could hold when saturated at that air temperature expressed as a percentage. Relative humidity may be measured indirectly from wet-bulb and dry-bulb temperature

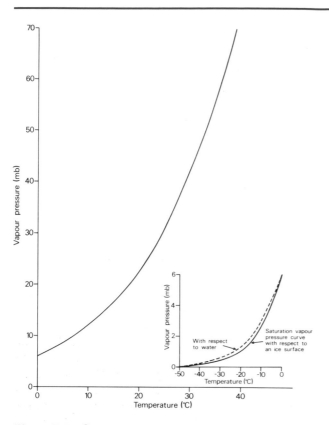

Figure 5.7 Saturation vapour pressure curve. The curves demonstrates how much moisture the air can hold at a specific temperature. Below 0°C, the curve is slightly different for an ice surface than for a super-cooled water droplet

ship between temperature and the moisture content at saturation is indicated by the **saturation vapour pressure** curve (Figure 5.7). Thus, as a rising air bubble cools, it approaches the temperature at which condensation occurs. When the air bubble reaches the temperature, it becomes saturated and condensation takes place.

If condensation was the only thing that happened on saturation then, apart from the extra weight of the droplets, the effect on the air bubble would be small. There is, however, another effect. As water changes from its vapour state to a liquid, it releases latent heat. This heat acts to warm the air and thereby counteracts the cooling resulting from expansion.

We can readily see the implications for our air bubble. Now, instead of cooling at 10°C/1000 m (its dry adiabatic lapse rate), it cools more slowly as it rises. This new, lower rate of cooling is known as the **saturated adiabatic lapse rate** (SALR). Unlike the dry rate this is not a constant, for as we can imagine it depends upon the amount of heat released by condensation, and this, in turn, depends upon the moisture content and therefore the temperature of the air. Warm air is able to hold a lot of moisture and thus, on cooling, it releases a lot of latent heat; cold air is able to hold far less

readings using humidity tables. Evaporation of moisture from the wet-bulb leads to a cooling which is inversely proportional to the relative humidity of the air. If the air is saturated, there will be no evaporation, no cooling and so no difference in temperature between the dry and wet bulbs. Although frequently used, relative humidity does have the disadvantage of being temperature dependent. For example, as air temperature rises, relative humidity will fall even though the moisture content of the air has remained constant because the air is able to hold more moisture. An absolute method of measuring moisture content is to determine the **vapour pressure**, which is that part of the total atmospheric pressure exerted by water vapour. Again it can be obtained indirectly from the wet- and dry-bulb thermometer using tables. **Humidity mixing ratio**, another absolute measure, is the ratio of the mass of water vapour to the mass of dry air with which the water vapour is associated. It is usually expressed in g kg^{-1}. The relation-

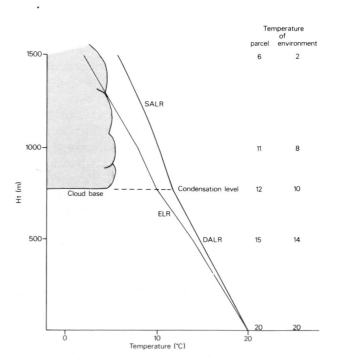

Figure 5.8 The effect of condensation on the rate of cooling of an air parcel

moisture, so the heat production during condensation is much less. This is one reason why some of the world's most severe storms, the tropical cyclones, are found in warmer climatic regions.

Let us illustrate the effect of condensation by considering a specific example. Figure 5.8 shows the path curve for the bubble. Its initial temperature is at 20°C and it will cool at 10°C/1000 m until saturation point is reached. It is at this level that we first see the visible evidence of our bubble – a small cloud will be seen forming. Above **condensation level**, the rate of cooling slows down to the saturated adiabatic lapse rate as long as the bubble's temperature is still higher than that of the environment. If it is, we get large convectional clouds building up which will probably give rain.

Whether the atmosphere is still stable or not will depend upon the relative rates of cooling of the dry bubbles, the saturated bubbles and the environment. We can summarize this in Figure 5.9. If the environmental lapse rate is cooling more rapidly than the dry adiabatic lapse rate, we have **absolute instability** as bubbles of air, even if they cool at their maximum rate (the DALR), will be cooling more slowly than the environment. If the environmental lapse rate is cooling more slowly than the saturated adiabatic lapse rate we have **absolute stability**. If the environmental lapse rate is between the DALR and the SALR we have **conditional instability**; in other words, instability depends upon the air reaching saturation point.

Stability has a considerable effect upon the degree to which convective activity will take place. If the air is unstable it will rise and may produce clouds, whereas if it is stable convection will be reduced. Sometimes, especially under anticyclonic conditions, the temperature will increase with height – a situation known as an **inversion of temperature**. If the air beneath the inversion is fairly moist, a layer of cloud may develop here. Moist air will have been brought to the inversion by convection and, as it cannot rise further, it spreads out beneath the inversion to give a dense sheet of cloud (Figure 5.10).

Absolute instability in the atmosphere is infrequent except very close to the ground – the convection it initiates helps to transfer heat upwards and so reduces the environmental lapse rate. What is much more common is for the environmental lapse rate to lie between the dry adiabatic lapse rate and the saturated adiabatic lapse rate. In this situation of conditional

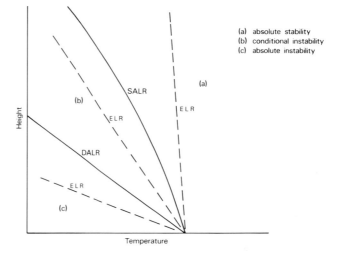

(a) absolute stability
(b) conditional instability
(c) absolute instability

Figure 5.9 The relationship between the Environmental Lapse Rate (ELR), the Dry Adiabatic Lapse Rate (DALR) and the Saturated Adiabatic Lapse Rate (SALR). If the ELR is in segment (a), the air is absolutely stable as even if the parcel is cooling at its slowest rate (SALR), the ELR indicates the atmosphere is cooling even more slowly. In segment (c), the ELR is cooling more rapidly than the quickest rate of cooling of the parcel so the parcel will always be warmer than its environment and continue rising. The air is hence absolutely unstable. In segment (b), the ELR is between the DALR and the SALR. The air *may* become unstable if it reaches the condensation level and can cool at the slower SALR. This situation is called conditional instability, i.e. it depends upon the parcel reaching the condensation level

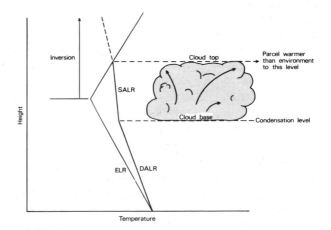

Figure 5.10 The effect of an inversion on cloud development with moist air

instability, the atmosphere is stable for air which has not reached saturation point, but is unstable for saturated air. If the air can be forced to reach the condensation level, either by ascent over hills or mountains, or by convergence associated with a depression, then the air will become unstable and assist vertical motion. The former process is one of the mechanisms which leads to higher rainfall over mountains.

The effects of condensation

Causes of condensation

Clouds are one of the most interesting aspects of the sky. They are constantly changing their shape and form to reflect the processes of formation and the environment in which they are developing. To produce clouds, we need the air to reach saturation point. It is clear that saturation can be reached either by adding water to air or by cooling the air (Figure 5.7). The first process occurs over warm water surfaces such as the Great Lakes in the autumn, or over the Arctic Ocean, where water will evaporate from the relatively warm sea surface and rapidly condense into the cold air above to give **Arctic Sea Smoke**. It is by cooling air that the majority of clouds are formed. Orographic lifting, convergent uplift near depressions or within air streams and convection will all produce vertical motion which may be sufficient to produce clouds.

Radiational cooling or contact cooling at a cold ground surface may also be sufficient to produce saturation, but as these are ground-based processes the resulting condensation is known as **fog**. It is like cloud in being composed of myriads of water droplets but the detailed mechanisms of formation are different.

Fog

Fogs are a common feature of the climate of some parts of the world. For example, they are frequent on the North Sea coast of Britain in summer and off the Grand Banks of Newfoundland. Even London used to be renowned for its fogs before air pollution controls improved the situation.

There are two weather situations which can form fog when the ground surface is cooler than the air in contact with it. First, when the ground loses heat at night by long-wave radiational cooling usually with clear skies of an anticyclone; second, when warm air flows from a warm region to cover a cold surface, particularly a melting snow surface with lots of moisture about. The first type of fog is called **radiation fog** and the second **advection fog**.

The fog consists of microscopic droplets of water between 1 and 20 μm in diameter. Visibility in a fog will depend upon the sizes and concentration of droplets in it. When they are small and numerous, visibility is poor, perhaps as little as 5 m. If pollution adds suitable nuclei condensation of water vapour is favoured. When the droplets are large or sparse, visibility is less affected. The actual formation of radiation fog represents a delicate balance between radiational cooling, air movement and condensation. It only forms when cooling occurs faster than the rate at which latent heat is added by condensation. Because vapour is converted into water droplets, the moisture content and saturation temperature fall, so further cooling is necessary to give saturation. Because of this, fog is more frequent during the long nights of autumn, winter and spring than in summer. If winds are strong, the saturated air near the ground will mix with drier air above and prevent fog forming.

By reducing visibility, fog can be a major environmental hazard. Airports may be closed for several days and road transport is hazardous and slow. Large economic losses can result from these delays, but the potential for artificial fog clearance seems low; too much energy would be needed to warm or dry out the air to prevent condensation.

Clouds

Clouds and fog are the result of similar processes which vary in intensity and duration. Clouds are composed of a mass of water droplets or ice crystals of almost microscopic size. The number of droplets per unit volume of cloud varies considerably depending upon its origins; smaller concentrations of larger droplets occur in clouds formed in the middle of large oceans, while large concentrations of smaller droplets are found in continental regions. Clearly this is a consequence of the greater availability of nuclei over the dusty continental interiors, but polluted industrial areas may have a similar effect. Studies of these condensation nuclei have shown that there are two broad classes: those with an affinity for water, called **hygroscopic particles**, like salt; and **non-hygroscopic particles** which require relative humidities above 100 per cent before they can act as centres for condensation. The role of sea salt as a source of hygroscopic nuclei has long been debated but recent results suggest that it is not as important as was once thought.

Plate 5.1 Cumulus clouds developing in an unstable westerly airstream. As the upper winds are blowing more quickly the cloud appears to be tilting forwards (photo: Peter Smithson)

We can find out much about what is happening in the atmosphere by looking at the type of clouds and especially their shape. Unfortunately there is an almost infinite variety of possible forms, and various attempts have been made to classify them. The basic division is between clouds which are dominantly layered, known as **stratiform**, and those where the vertical extent of the cloud is most important. These are known as **cumuliform** (Plate 5.1).

The stratiform types are shown in Figure 5.11 and subdivided according to their height of formation. In stratiform types of cloud, the rate of upward motion is slow, but it may take place over many hundreds or even thousands of square kilometres. At low levels, they are composed of water droplets, but at higher levels (2000 to 6000 m) we get a mixture of water droplets and ice crystals. Above about 6000 m stratiform clouds are composed mainly of ice crystals. Some of the clouds may show some signs of convection even if it is weak. These types have '*cumulus*' incorporated into their names such as stratocumulus or altocumulus.

The low- and medium-level stratiform types are

the main rain-bearing clouds of temperate latitudes. Around the centre of a depression we often see a characteristic sequence of clouds as the warm air associated with the depression approaches (Figure 5.12). The first signs of the depression are cirrus clouds which slowly expand in area to become cirrostratus. This cloud sheet thickens and becomes lower, producing altostratus, followed by nimbostratus and precipitation. By this final stage, there is a whole complex of cloud sheets which are difficult to differentiate from the ground or even from space.

The other main group of cloud, cumuliform, is the result of localized convection or instability. Bubbles of warm air, rising beyond the condensation level (if the air is unstable), are seen as cumulus clouds. The precise shape of the cloud will depend upon the degree of instability, the water vapour content of the air and the strength of the horizontal wind (Figure 5.13). There are many different types of cumulus cloud subdivided on the basis of their appearance. If an inversion of temperature exists as often happens with anticyclones, the bubbles will rise to the inversion and then start to level out or

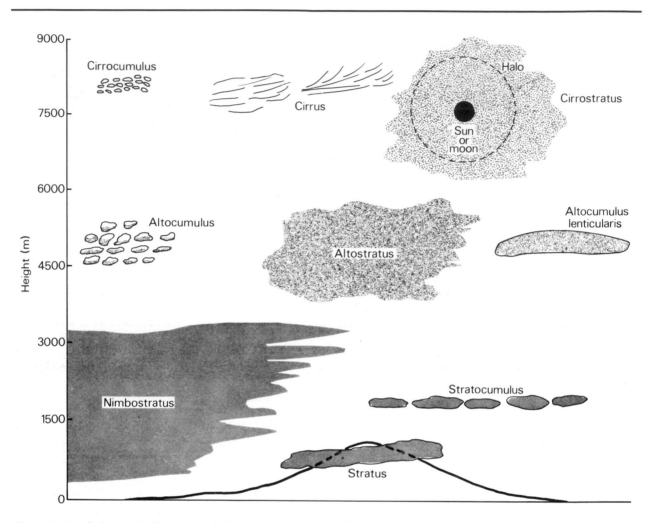

Figure 5.11   Schematic diagram of the main stratiform cloud types

descend, evaporating the cloud droplets as they do so. This type of cloud is known as **fair weather cumulus** as it never grows sufficiently to give rain (Plate 5.2).

Some cumulus clouds may grow larger and taller. The sharp and clear outlines of the cauliflower-like cumulus becomes more diffuse and ragged as the upper part of the cloud becomes a fibrous mass of ice crystals. The **cumulonimbus stage** has now been reached. At this stage of development precipitation is usually occurring, sometimes accompanied by lightning and thunder. As the mass of ice crystals develop, it is often blown downwind by the strong winds of the high atmosphere to give an anvil shape (Plate 5.3), characteristic of cumulonimbus clouds. Convection may initiate other clouds nearby or on the flanks of the parent cloud as it gradually decays and evaporates.

In this way a cloud mass may travel as a single entity for several hours.

For rain to fall, we need clouds, but many clouds survive for hours without giving rain. What special circumstances enable some clouds to produce rain whereas others give none? We will examine this problem in the next section.

Precipitation

Formation of precipitation

In parts of Snowdonia, average annual precipitation exceeds 4360 mm, and on the summit of Mt Waialeale, Kauai, Hawaii, it is 11,684 mm. In terms of the amount of water, this is equivalent to 100,000 t ha^{-1} y^{-1}. Without doubt the processes producing precipitation can be very effective when

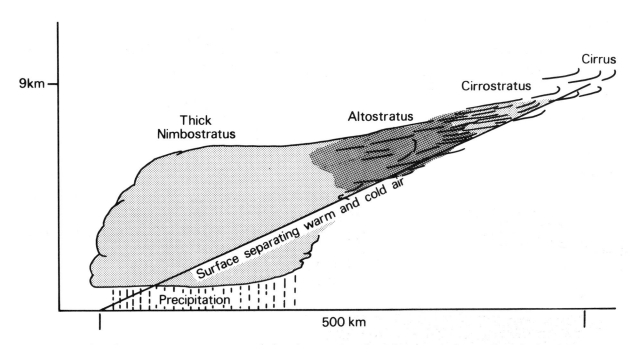

Figure 5.12 The characteristic sequence of clouds associated with a warm front ahead of a depression or mid-latitude cyclone

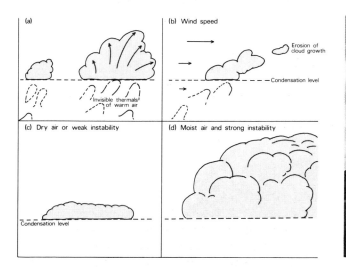

Figure 5.13 The controls affecting the form of cumulus clouds

Plate 5.2 Fair weather cumulus cloud over central Sweden. An inversion associated with an anti-cyclone is preventing the clouds from developing further (photo: Peter Smithson)

conditions are favourable. But how do these minute cloud droplets (Plate 5.4) get large enough to fall as rain within approximately twenty minutes of the cloudy air reaching saturation?

To answer this question, we must delve inside a cloud and see what is happening there. In a cloud made up entirely of water droplets there will be a variety of droplet sizes. The air will be rising within the cloud, perhaps at a rate of 10 to 20 cm per second though much more rapidly in cumulo-nimbus clouds. As it rises so the drops get larger through **collision** and **coalescence**; some will reach drizzle size. When the uplift is stronger, say 50 cm per second, the downward movement of the

Plate 5.3   Cumulonimbus cloud reaching the tropopause. Further convection is seen in the surrounding actively-growing cumulus clouds on the upward left side of the cloud mass. An unstable maritime polar air stream covers England (photo: C. J. Richards)

drops will be reduced, so there will be more time for them to grow. If the cloud is about 1 km deep small raindrops of 700 μm in diameter may be formed.

When temperatures fall below 0°C, the cloud is said to be **supercooled**. With further cooling to −10°C, ice crystals may start to develop among the water droplets. This mixture of water and ice would not be particularly important but for a peculiar property of water. The saturation vapour pressure curve for ice (Figure 5.7) is slightly different from that of water. The air can be saturated for ice when it is not saturated for water. Thus, at −10°C, air saturated with respect to liquid water is **supersaturated** relative to ice by 10 per cent and at −20°C by 21 per cent. As a result, the ice crystals in the cloud tend to grow and become heavier at the expense of the water droplets.

As the ice crystals sink into lower layers of the cloud where temperatures are only just below freezing, they have a tendency to stick together to form snowflakes. This is brought about by the

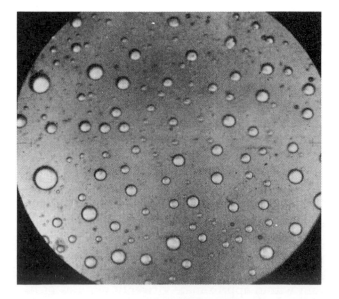

Plate 5.4   A typical sample of cloud droplets caught on an oiled slide and photographed under a microscope in the aircraft. The largest droplet has a diameter of about 30 μm (after B. J. Mason)

supercooled droplets of water in the cloud acting as an adhesive. After the snowflakes have melted the resulting drops may grow further by collision with cloud droplets before they reach the ground as rain. This method of producing rain drops is known as the **Bergeron–Findeisen process** after the developers of the theory (Figure 5.14). Beneath the base of the cloud, however, evaporation will take place in the drier air and if the drop is small it may be evaporated completely.

Precipitation formation both by collision and coalescence and by the Bergeron–Findeisen process undoubtedly occurs in the atmosphere, though clearly the Bergeron–Findeisen process can only operate when cloud temperatures are well below freezing. The rate at which vapour is converted into water droplets and precipitation depends upon three main factors: the rate of coalescence and ice-crystal growth; the cloud thickness; and the strength of the updraughts in the cloud. The total amount of rain will be determined by the lifespan of the cloud, the height of the cloud above the ground and how long these processes operate. Cloud thickness and updraught speed are largely dependent upon instability and vertical motion in the atmosphere. Precipitation has been classified in terms of

the factor which gives rise to the upward movement, so let us have a look at this in a little more detail.

Convectional precipitation

The spontaneous rising of moist air due to instability is known as convection. We have seen that upward growing clouds are associated with convection. Since the updraughts are usually strong, cooling of the air is rapid and lots of water can be condensed quickly. Collisions and coalescence are likely to be frequent so the larger droplets rapidly increase in size. Eventually, growing larger and heavier, the droplets overcome the lift provided by the updraught, and they start to fall through the cloud into the clear air beneath. As the volume of water in these big drops is large relative to their surface area, little evaporation takes place in the non-saturated air below the cloud. At the ground there will be a burst of heavy rain as the shower passes.

Unstable air which favours convectional rain is most frequently found in warm and humid areas, but even in the United Kingdom some 20 per cent of the annual rainfall is by convection. Often this is the result of very cold air moving over a warmer ground

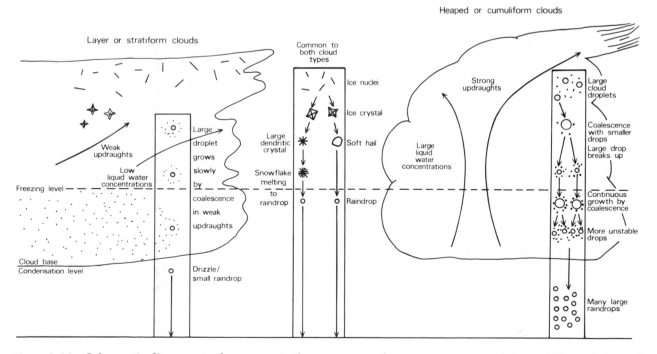

Figure 5.14 Schematic diagram to demonstrate the processes of precipitation growth in stratiform (left) and cumuliform (right) clouds

surface to give the steep lapse rates characteristic of instability and convection.

Thunderstorms

When the atmosphere is very unstable, cumulonimbus clouds develop, sometimes accompanied by lightning and thunder. At night an intense thunderstorm can provide one of the most spectacular displays of the atmosphere. Flashes of lightning shoot from cloud to earth or within the clouds (Plate 5.5) accompanied by great crashes of thunder. How can such dramatic manifestations of energy build up in a cloud? This has puzzled meteorologists for many years. In some way electrical charges in the cloud are separated (Figure 5.15). Recent observations suggest that the main way in which this occurs is by the formation, growth and electrification of pellets of **soft hail**.

As well as electricity, the thunderstorm is often accompanied by squalls of cold wind blowing *away* from the cloud. They usually originate as downdraughts of air near the main burst of rain (Figure 5.16). Even hail may fall near the centre of the storm, sometimes causing great damage.

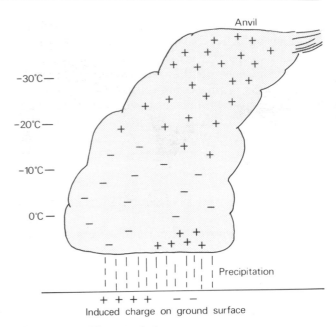

Figure 5.15 Electrical charge separation in a thunder cloud

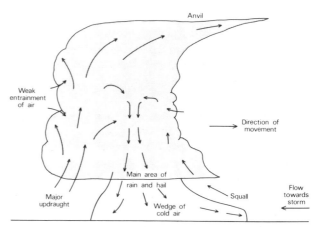

Figure 5.16 Schematic diagram of air movement in a thunderstorm

Plate 5.5 A lightning stroke from cloud to earth during a summer thunderstorm (photo: C. J. Richards)

Hail

Hail is a form of precipitation composed of spheres or irregular lumps of ice (Plate 5.6). It falls in narrow bands associated with cumulonimbus clouds and so frequently misses the observing stations. However, the destruction it can produce is dramatic. Crops can be torn to shreds, glasshouses ruined and even cars dented by the weight of half a kilogram or more of ice falling from the skies.

Splitting open a large hailstone will show that it is composed of alternating layers of clear and opaque

Plate 5.6 Giant hailstones photographed at Melksham, Wiltshire soon after falling from a summer storm (golf ball provides scale) (photo: M. E. Hardman)

Plate 5.7 A section through the centre of a large hailstone taken in reflected light showing the regions of clear ice which appear black and milky or opaque ice which appears white (photograph by courtesy of Dr K. A. Browning)

ice (Plate 5.7). It appears that the stone is involved in complex movements within the cloud, being swept up to the higher colder parts of the cloud several times. When this happens, any moisture condensing on the stone will freeze instantly, including any trapped air and producing opaque ice. At lower levels in the cloud it takes condensed water a longer time to freeze. Air bubbles can then escape to leave a layer of clear ice when it eventually freezes. The alternating layers of clear and opaque ice indicate the number of times the hailstone has been swept up by the cloud updraughts (Figure 5.17).

Cyclonic precipitation

In temperate and sub-polar latitudes, most of the precipitation comes from cyclones. The cyclone is characterized by areas of rising air. A satellite photograph of a cyclone shows the extensive areas of cloud resulting from this slow but widespread ascent of the air (Plate 5.8).

There are a number of differences from convectional precipitation. The areal extent of rising air associated with a cyclone is much larger, and the rate of upward movement and the rate of condensation in the clouds are much less. Because of this, the droplets grow more slowly and fall out of the cloud

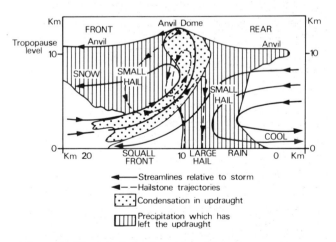

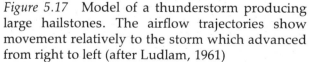

Figure 5.17 Model of a thunderstorm producing large hailstones. The airflow trajectories show movement relatively to the storm which advanced from right to left (after Ludlam, 1961)

sooner. Being small, they can be greatly affected by evaporation in the drier air beneath cloud base. For example, in an atmosphere with a relative humidity of 90 per cent, a droplet of radius 10 μm will only fall 3 cm before evaporating; drops of 100 μm and

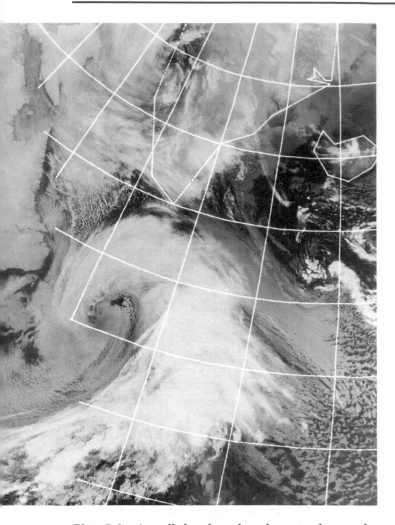

Plate 5.8 A well-developed cyclone to the northeast of Newfoundland. The extensive areas of cloud formed by widespread ascent can be clearly seen. Image taken on 10 December 1982 at 16.45 GMT in the infra-red waveband (copyright, University of Dundee)

1 mm would fall 150 m and 40 km respectively. Despite the relatively small size of the rain drops, the areas affected by rising air are vast. For a particular rainbelt, it may take several hours of steady rain before the system has passed, giving a total fall of perhaps between 5 mm and 10 mm.

In the deep and widespread clouds associated with a cyclone, it is quite common for ice-crystal clouds at higher levels to act as a source of supply to the mixed clouds of ice and water droplets at lower levels. The addition of extra ice crystals speeds up the precipitation process and leads to more intense rainfall.

Orographic precipitation

Almost all mountain areas are wetter than the surrounding lowlands. To take two examples, Hokitika on the west coast of New Zealand receives an average of 2930 mm per year. At Arthur's Pass, 740 m higher in the New Zealand Alps, the annual average has risen to 3980 mm compared with less than 670 mm for Christchurch on the more sheltered lowlands to the east. Even the Ahaggar and Tibesti mountains in the centre of the Sahara receive more rain than do the surrounding lowlands – Asekrem at 2700 m has an annual average of about 125 mm compared with only 13 mm at Silet, 720 m above sea-level. Why should this be so?

Where air meets an extensive barrier it is forced to rise. Rising, as we know, leads to cooling of the air, and cooling encourages condensation. On the mountain slopes and above the mountain summits, the clouds start to pile up, reflecting the forced ascent of air. Often they reach thicknesses sufficient to give drizzle and rain. From a distance we can see these dense clouds enveloping the mountains (Plate 5.9).

Orographic rain is also produced in another way, due to changes in the stability of the air as it rises. If the air is very moist near the ground surface but much drier above, as it rises the rates of cooling between the top and bottom of the layer will be different (Figure 5.18). The upper part will cool more quickly and so become relatively colder, leading to less stable air. The cloud development associated with instability will increase and rain may fall over the mountains. This situation is known as **convective** or **potential instability**.

Hills as well as mountains act as favourable areas for convectional showers. The slopes facing the sun will be warmed more rapidly than flatter areas, because the slopes act as thermal sources. The resulting cloud may produce rainfall which is restricted to the upland area.

The orographic effect is most pronounced when it is already raining upwind of the hills or mountains. Where air is rising, associated with a cyclone for example, the rate of uplift is increased by the extra forced ascent provided by the hills. This leads to a greater rate of condensation on the windward side, larger drops of rain being formed and so a higher rainfall at the surface. Coupled with the slowing down of the rainbelt as it passes due to increased friction, the net effect is considerably greater rainfall (Figure 5.19).

On the leeward side of the hills, subsidence or

Plate 5.9 Orographic cloud developing over an isolated, hilly island in the Faeroes. The cloud is only forming where the moist air is forced to rise, though downwind some convection is evident. Where uplift is more extensive, as over a mountain range, the cloud will be correspondingly larger (photo: J. D. Hansom)

descending air begins to dominate, so that the cloud sheet thins or even dissipates and rainfall declines. As the air descends, it gets warmer due to compression to give us the rain-shadow effect on the leeward slope of the mountains. Here, rainfall is far less than on the upslope side and sunshine amounts and daytime temperatures are normally higher.

These processes acting together ensure that rainfall normally increases with altitude, at least up to about 2500 m; above this level, the lower water content of the atmosphere becomes more significant. In Hawaii, rainfall on the windward slopes at 1000 to 1300 m is about 7500 mm yr^{-1}, but on the summit of Mauna Loa at 4000 m it is only 380 mm yr^{-1}.

As much of the precipitation in mountains is due to an intensification of pre-existing rain, it would be wrong to think of orographic precipitation as a truly separate category. It can occur as drizzle or by convective instability, but much more frequently it will depend upon cyclonic or convection processes already operating. Even these two types can occur together in cyclones, so perhaps we should identify convectional, cyclonic and orographic precipitation as interrelated mechanisms of rainfall rather than classifying them into these types.

Conclusions

In this chapter we have tried to follow and explain

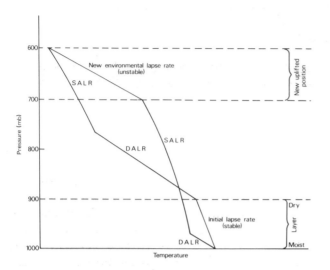

Figure 5.18 Destabilization of the lower atmosphere through uplift of a layer of air 100mb in thickness. Between 1000mb and 900mb the air is initially stable. It is moist near the base and drier aloft. As a result of uplift, the new environmental lapse rate indicates instability

the exchange and movements of moisture between the earth's surface and the atmosphere. The process of evaporation provides the supply of moisture into the lower atmosphere. The prevailing winds then circulate the moisture and mix it with drier air elsewhere. Only if we have a dry surface in areas well away from the oceans and where dry subsiding air is dominant will moisture levels be low.

Water vapour is only the first stage of the precipitation chain; we must convert the vapour into liquid form. This is usually achieved by cooling, either rapidly as in convection, or slowly as in cyclonic storms; mountains will also cause uplift but the rate will depend upon their shape and the direction of the wind. Even this is insufficient as we can tell from the large number of clouds in the sky which never give precipitation. To produce precipitation, the cloud droplets must become large enough to reach the ground without evaporating. The cloud must possess the right microphysical properties for growth of the droplets to take place. It must have ice crystals if the Bergeron–Findeisen process is to operate, or a wide spectrum of drop sizes with plenty of moisture condensing for the collision–coalescence system to work. Even these suitable conditions may be unrewarding if the cloud does not last long enough for the growth to take place. Clearly precipitation development represents a delicate balance of counteracting forces, some leading to droplet growth, others to droplet destruction. Nevertheless, where conditions are basically favourable – where air can rise high enough to produce large vertical developments of cloud – copious amounts of precipitation occur.

The effect of this upon the climate is obvious. But precipitation is also important in other ways. As we will see in Chapters 12 and 14, it is a major component of the hydrological cycle. Rainfall also takes part in many of the processes that build our landscape. And plants and animals are highly dependent upon precipitation. Therefore, in Chapter 12 we will be returning to the question of precipitation and following its progress to earth in more detail.

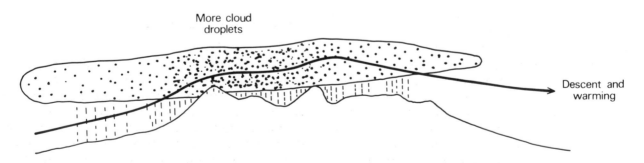

Figure 5.19 Production of greater rainfall over hills as a result of enhanced forced ascent (see also Figure 12.9)

6

The atmosphere in motion

Atmospheric circulation and winds

The earth's atmosphere is in perpetual motion; movement which is striving to eradicate the constant differences in pressure and temperature between different parts of the globe. It is this motion which produces the winds and storms with which we are all familiar. It is this circulation which plays a basic part in maintaining a steady state in the atmosphere and generating the climatic zones which characterize the earth.

So far we have considered the upward movements which transfer energy from the surface to the atmosphere. Let us now consider the more obvious horizontal movements that transfer air around the globe. With modern satellite technology we can watch and monitor these movements. We are no longer dependent solely upon balloons to provide information about the upper atmosphere. Satellites of the observing system, called Geostationary Operational Environmental Satellites (or GOES), orbit the globe with the same rate of rotation as the earth, permitting the same portion of the earth to be viewed continuously, by cameras in the day and by infra-red photography at night. The photographs show the main cloud features of the atmosphere.

Individual cloud patterns can be identified and followed and from successive photographs cloud movements can be calculated and predicted. Unfortunately, the satellite photographs only show the circulation in cloudy areas. What happens elsewhere is less clear, for we are dependent in these cloudless regions upon information supplied by surface or balloon observations. These provide us with a less detailed picture of the pattern of wind circulation over most of the continental areas of the globe (Figure 6.1).

Causes of air movement

Why do we have winds at all? Certainly the explanation is not unique to the earth, for similar patterns of atmospheric circulation have been identified on many other planets.

To answer this question it is useful to consider some of the basic principles of motion. Our understanding of these is due in large degree to Isaac Newton. Many people know the story of how Isaac Newton 'discovered' gravity when sitting beneath an apple tree. Some will know, too, that he also formulated laws of motion.

There are two main laws. The first states that:

1 A particle will remain at rest or in uniform motion unless acted upon by another force.

The second law states that:

2 The action of a single force upon a particle causes it to accelerate in the direction of the force. If there is more than one force the particle is accelerated in the direction of the **resultant** (Figure 6.2).

These forces are particularly important for movement in the atmosphere because forces are continuously acting on particles of air, causing them to accelerate or decelerate and change their direction.

Forces acting upon the air

Pressure gradient force

Let us imagine that we have a small parcel of air some distance above the ground. What forces will act upon it? The most obvious is the force of gravity which tends to attract all mass towards the earth's centre. In addition we have the pressure exerted by the air surrounding the parcel (Figure 6.3). If this pressure was the same on all sides of the parcel then its effects would cancel out. But this is not so. Pressure decreases upwards in our atmosphere, as we saw in Chapter 5. The force pushing the parcel of air upwards is greater than the downward force from the overlying atmosphere; there is a potential upward acceleration of the parcel. Luckily, this

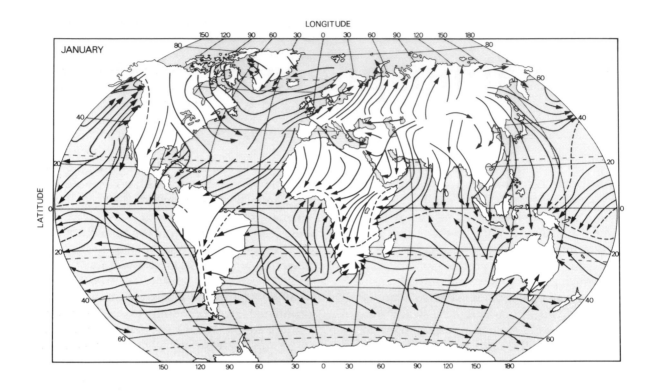

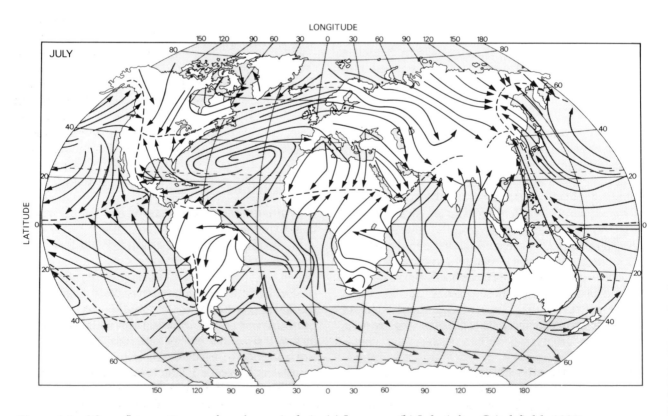

Figure 6.1 Mean flow patterns of surface winds in (a) January, (b) July (after Critchfield, 1983)

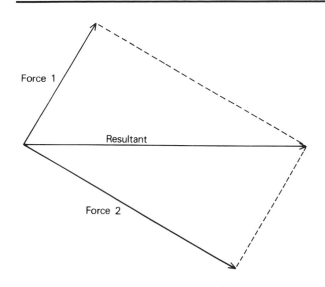

Figure 6.2 The resultant of two forces acting in different directions. The length of line is proportional to the strength of the force

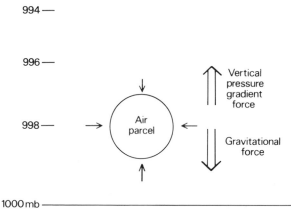

Figure 6.3 Pressure forces acting on a parcel of air

vertical force is almost exactly balanced by the force of gravity; otherwise we would have lost our atmosphere long ago. Most of the air movements that we observe are horizontal. Forces are also operating in this dimension. Where the atmosphere is dense the lateral pressure on the parcel of air is great; where the atmosphere has a low density, the lateral pressure is less. Variations in the density of the atmosphere from one part of the globe to another result in an imbalance of forces and lateral movement of the air (Figure 6.4a). The air is 'pushed' from areas of high pressure to areas of low pressure.

This, in fact, is the basic force affecting the earth's atmospheric motion. It is called the **pressure gradient force**. As we have seen, a pressure gradient exists both vertically and horizontally. Pressure decreases vertically because, as we move upwards through the atmosphere, the weight of overlying air diminishes. It varies laterally because of differences in the intensity of solar heating of the atmosphere. Where solar radiation is intense the air warms up, expands and its density declines. Air pressure falls. Where cooling occurs, the air contracts, its density increases and air pressure becomes greater.

A corollary of this principle is that the pattern of air pressure close to the surface is reversed in the upper atmosphere. Because cold air contracts, the upward decline in pressure is rapid and at any constant height above a zone of cool air the pressure is relatively low. Conversely, warm air expands and rises, so that the vertical pressure gradient is less steep. Above areas of warm air, therefore, the pressure tends to be relatively high (Figure 6.4b) The effect upon atmospheric motion is clear. At the surface the air will move from cold to warm zones; at higher altitudes the flow will be from warm to cold.

Differences in air pressure may be mapped by defining lines of equal pressure. These are known as isobars. Air movement occurs at right angles to the isobars, down the pressure gradient; that is from areas of high pressure to areas of low pressure (Figure 6.4c). The magnitude of the force causing movement (the pressure gradient force) and thus the speed of the wind, is inversely proportional to the distance between the isobars. Thus, the closer the isobars are together, and the more rapidly pressure falls with distance, the stronger is the wind.

Mathematically, this relationship can be written as follows:

$$F = -\frac{1}{\varrho} \cdot \frac{p_2 - p_1}{n}$$

where pressure values at points 2 and 1 are p_2 and p_1; n is the distance separating 2 and 1; ϱ is air density and F is the resulting acceleration. We can then use this formula to indicate how quickly the parcel ought to accelerate. The standard isobaric interval on a pressure chart is 4 mb and air density is 1.29 kg m^{-3}. Suppose the isobars are 300 km apart on a sea-level chart. What will the acceleration down the pressure gradient be? In uniform units,

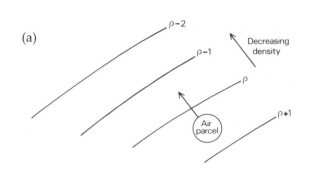

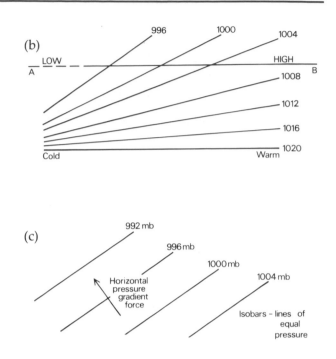

Figure 6.4 (a) Force exerted on an air parcel produced by density differences
(b) Change of pressure with height above areas of warm and cold air. At A, above the cold air, pressure is lower than at the same height, B, above the warm air
(c) Horizontal pressure gradient force acting at right angles to the isobars

the formula will become

$$F = \frac{1}{1.29} \cdot \frac{4 \times 10^2}{300 \times 10^3} = 0.00103 \text{ m s}^{-1}$$

If this rate is kept up for 1 hour (3600 seconds) we would have a value of 3.72 m s^{-1} after 1 hour. As pressure gradients of this size can last for days, we might expect very high wind speeds to develop unless other forces interfere. There are two main forces which prevent this happening. One is friction and the other is due to the earth's rotation.

If we look at the wind field on a weather map, it would be immediately apparent that air does not flow down the pressure gradient towards areas of low pressure. If it did, the low pressure areas would fill and the wind movement would stop. Instead we find that the wind is blowing parallel (or almost) to the isobars rather than across them.

This is due to the effect of the earth's rotation.

Coriolis Force

Although we are not aware of it, the earth is rotating from west to east at 15° longitude per hour. Reference back to Newton's laws shows that if we have a parcel of air moving southwards and there are no forces acting upon it, it will continue to move in the same absolute direction (i.e. a straight path as viewed from space). However, the earth is gradually turning and so, *relative to the ground surface*, this parcel will appear to have followed a curved track

towards the right (Figure 6.5). To explain this apparent deflection in Newtonian terms, we have to introduce a force to account for the movement as observed from the ground. This force is called the **Coriolis Force** after a French mathematician who formalized the concept. The value of the Coriolis Force changes with the angle of latitude and the speed of the air; mathematically for a unit mass of air it is CF = 2 ω *v* sin θ (where ω is the rate at which the earth rotates, *v* is air velocity and θ is the

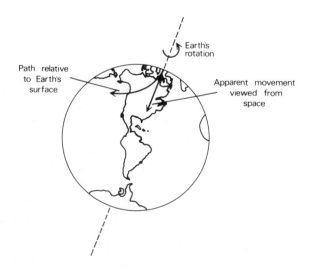

Figure 6.5 The effect of the earth's rotation on air movement

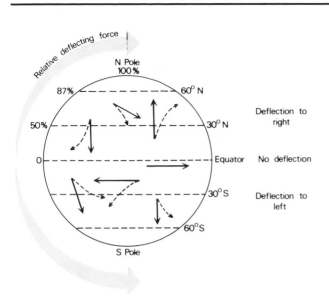

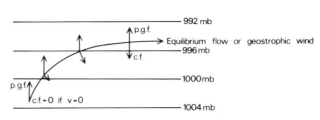

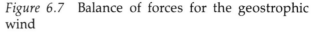

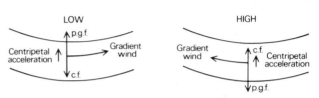

Figure 6.7 Balance of forces for the geostrophic wind

Figure 6.6 The changing magnitude of the Coriolis Force with latitude

Figure 6.8 Balance of forces for the gradient wind where curvature of the isobars is important

latitude). It is greatest at the poles where the earth's surface is at right angles to the axis of rotation, but it gets progressively less towards the equator where it reaches zero. The reason for this is shown in Figure 6.6. As one proceeds towards the equator so the earth's surface eventually becomes parallel to the axis of rotation. It can be demonstrated by pendulum experiments. Using the Foucault pendulum which portrays free motion in space as closely as can be achieved at the earth's surface, a disc will rotate under the freely swinging pendulum in one day at the poles. At 30° latitude it will take two days to rotate (sin 30° = 0.5) and at the equator it does not turn at all.

Geostrophic wind

Let us now return to our parcel of air experiencing a pressure gradient force on the rotating earth. Initially the parcel will move down the pressure gradient, but as soon as it begins to move, it will start to be affected by the Coriolis Force. It will be deflected towards the right in the northern hemisphere (Figure 6.7). As the wind accelerates, its speed will increase and, because the Coriolis Force is related to speed ($2 \omega v \sin \theta$), so the two forces pulling together eventually produce an equilibrium. This will occur when the two forces are equal and opposite, the resultant wind blowing parallel to the isobars; it is known as the **geostrophic wind**. Its velocity will be determined

primarily by the pressure gradient, though, because the value of the Coriolis Force varies with latitude, the geostrophic wind for the same pressure gradient will decrease towards the poles.

Although we have only considered two of the forces acting upon the air parcels, nevertheless the geostrophic wind is a useful concept. It approximates closely to winds observed in the atmosphere above the friction zone (i.e. in the upper atmosphere). Strictly it operates only when the isobars are straight – a rare event. Normally isobars are curved and winds are subject to another force termed centripetal acceleration which acts towards the centre of rotation. When this rotational component is included, the resultant wind is called the **gradient wind** which is closer to observed flow in the upper atmosphere (Figure 6.8).

Friction

Inspection of a surface weather map will show that, at ground level, the wind does not blow parallel to the isobars. It blows across the isobars towards the area of lower pressure. The more observant may notice that this angle between the wind flow and the isobars is greater over land areas than over seas. This may give a clue to the reasons for the change. Land surfaces are rougher than seas; they tend to slow down the wind through friction more effectively. Friction acts as a force pulling against the direction of flow. We can now rearrange our

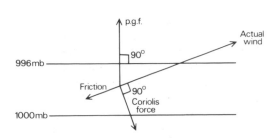

Figure 6.9 The effect of friction on the geostrophic wind. The Coriolis Force is always at right angles to the actual wind. It is smaller than the pressure gradient force because friction has reduced the speed of the wind

'balance of forces' to include friction. To achieve balance, the flow will now be across the isobars because the Coriolis pull to the right decreases as the air velocity falls (Figure 6.9).

From these forces we can now explain horizontal flows of air. They are initiated by pressure differences, then modified by friction and the effects of the earth's rotation.

The global pattern of circulation

With these principles in mind we can try to build up a picture of the global pattern of circulation in the earth's atmosphere. We can start by considering a highly simplified model of the atmospheric system: a uniform, non-rotating, smooth earth.

As we have seen, the basic force causing atmospheric motion is the pressure gradient; this gradient arises from the unequal heating of the atmosphere by solar radiation. At the equator – the 'fire-box' of the circulation as it has been called – solar radiation is converted into heat. The air

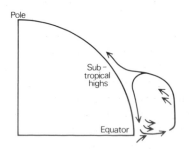

Figure 6.10 The meridional air circulation in the tropics

expands and rises and flows out towards the poles. Cool, dense air from the poles returns to replace it. We can readily demonstrate the pattern of circulation by heating a dish of water at its centre. Hot water bubbles up above the heat source and flows across the surface to the cold 'polar' areas. At depth the flow is reversed. So long as this unequal heating is continued, the cellular flow is maintained.

In reality, however, the pattern is found to be more complex for, instead of flowing directly to the poles at high altitudes, the warm air from the equator gradually cools and sinks due to radiational cooling. Most of it reaches the surface between about 20° and 30° latitude, and this subsiding air gives rise to zones of high pressure at the tropics. These used to be known as the horse latitudes because sailing ships were often becalmed and ran short of food and water. The unfortunate horses were then eaten or thrown overboard. But now the less picturesque term 'sub-tropical high pressure belts' is used.

As the descending air reaches the surface it diverges, some returning towards the equator to complete the cellular circulation of the tropics, the remainder flowing polewards (Figure 6.10).

Various other factors disrupt this pattern further, for the earth is not at rest nor uniform as we have so far assumed. It rotates. Its surface is highly variable; it has oceans and continents; it consists of a mosaic of mountains and plains. Moreover, the inputs of solar radiation vary considerably both on a seasonal and a daily basis.

The effect of the earth's rotation

The rotation of the earth causes the winds to be deflected from the simple pattern we have distinguished in Figure 6.10. The deflection is towards the right in the northern hemisphere and towards the left in the southern hemisphere. Instead of the direct meridional flow, therefore, the so-called Coriolis Force produces a surface flow similar to that shown in Figure 6.1.

This is not the only effect of the earth's rotation. Air moving towards the poles from the tropics forms a series of irregular eddies, embedded within the generally westerly flow. These can be seen on the satellite photographs as spiralling cloud patterns, similar to the patterns we can see in a turbulent river (Plate 6.1).

Again we can understand the cause of these eddies with the help of a simple experiment. A pan of water is heated at the rim and cooled at the centre. If the pan is slowly rotated it is seen that a simple thermal circulation is produced. If the rate of rotation is increased, however, the flow suddenly becomes unstable. New patterns form like those we see in the atmosphere of the temperate latitudes – eddies and waves. It seems that rapid rotation, like that of the earth, sets up forces which disturb the simple circulation of the atmosphere; particularly near to the axis of rotation (i.e. in higher latitudes) these forces destroy the simple pattern and produce more complex circulations (Plate 6.2).

Plate 6.1 A typical spiral of cloud around a mid-latitude depression. The centre of the low is situated between Iceland and Scotland and the frontal clouds are just clearing SE England. Drier air is following, accompanied by some convection over the warmer sea. NOAA image, 4 January 1980. Infra-red waveband (copyright, University of Dundee)

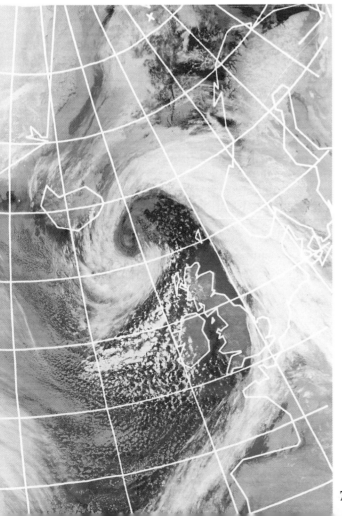

The effect of surface configuration

Even now, our picture of atmospheric circulation is far from complete. As we have noted, the earth's surface is far from uniform, and the variations in surface form cause ever more disruptions to the pattern of circulation. Friction bends the winds perpendicular to the isobars, reducing the effect of the Coriolis Force and, locally, it deflects the surface flow of air to produce highly complicated systems of movement.

It is difficult to model the effects of surface configuration, but a general indication of its influence can be obtained by comparing the northern and southern hemispheres. In the northern hemisphere, there are extensive and irregular land masses. Much of the southern hemisphere, in contrast, is ocean, except for the high ice plateau of Antarctica, where very low temperatures are experienced. As we might expect, the pattern is much simpler in the southern hemisphere. A strong westerly flow of cool polar air occurs even in the southern summer. Conversely, in the northern hemisphere, the flow is more irregular, the circulation weaker. The temperature differences between pole and equator are less marked (about 30°C compared with 60°C in the southern hemisphere) and so the driving force for the winds – the pressure gradient – is reduced.

Energy transfers and the global circulation

Forms of energy in the atmosphere

The general circulation of the atmosphere is powered by solar radiation. It operates in response to the disparities in energy inputs between different parts of the world. It acts to reduce these disparities; it is unsuccessful only because the energy differences are maintained by constant inputs of solar radiation.

The circulation, however, represents one of the main processes of energy transfer through the global system and by this process it preserves a steady state within the system, preventing the accumulation of heat in the tropics or the perpetual cooling of the poles. Energy is transferred in a variety of forms, however, and during these transfers it undergoes numerous transformations.

Four main forms of energy exist in atmospheric circulation: latent heat, sensible heat, potential energy and kinetic energy. The total energy of a unit mass of air can therefore be described as follows:

$$E_t = L_q + C_p T + gz + V^2/2$$

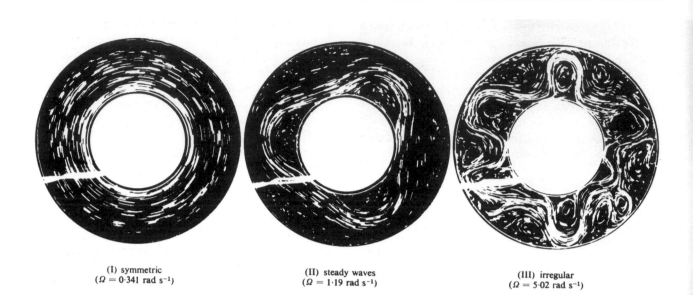

(I) symmetric
($\Omega = 0\cdot341$ rad s^{-1})

(II) steady waves
($\Omega = 1\cdot19$ rad s^{-1})

(III) irregular
($\Omega = 5\cdot02$ rad s^{-1})

Plate 6.2 Streak photographs giving an example of the top-surface flow patterns in a rotating fluid subject to heating at the outer side-wall and cooling at the inner side-wall. At low rates of rotation (left) the flow is symmetric about the axis of rotation. As the rotation rate increases (centre) the flow develops jet streams and waves. At higher rates (right) the flow is highly irregular with resemblance to the cyclonic and anticyclonic eddies found in the westerly circulation (photograph by courtesy of Dr R. Hide)

where L_q = latent heat content
C_pT = sensible heat content
gz = potential energy
and $V^2/2$ = kinetic energy

We have discussed these forms of energy in previous chapters. As a reminder, potential energy represents the energy derived from the position of the matter and kinetic energy is the energy of motion. The latter is readily visualized as a function of wind velocity.

Latent heat is more difficult to visualize. It is the quantity of heat released or absorbed, without any change in temperature, during the transformation of substance from one state to another (e.g. from a solid to liquid). The main source of latent heat in the atmosphere is the water vapour; when this condenses to water it releases heat; when it is formed by the evaporation of water it absorbs heat.

The sensible heat can be thought of as the temperature of the atmosphere. More specifically it is the temperature (T) times the specific heat (C_p)* of the air at a constant pressure. Sensible heat is gained from the ground surface after the absorption of short-wave radiation, or by the release of latent heat through condensation.

The potential energy of the atmosphere is essentially a function of its height above the ground surface (z); gravity (g) is a constant. As air moves in the atmosphere it tends to change its height and alter its energy content. If the air sinks slowly, the potential energy decreases. Normally it is converted to sensible heat, and the air becomes warmer

---

*The specific heat of a substance is the amount of heat required to raise the temperature of 1 g of that substance by 1°C. This is defined at a constant pressure because adding heat normally alters the volume/pressure relationship of the substance. The specific heat of still air at 10°C is 1.010 J g^{-1} °C^{-1}.

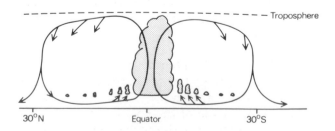

Figure 6.11 Details of the meridional air circulation or Hadley Cell in the inter-tropical zone

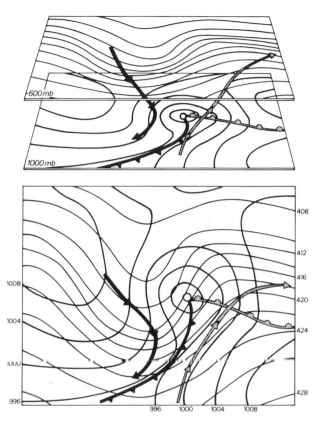

Plate 6.3 A good example of linear cloud development along the Equatorial Trough or Inter-Tropical Convergence Zone off West Africa. More convection is evident over Central Africa. 4 September 1982. Visible waveband. The corresponding infrared image is shown on Plate 3.1

Figure 6.12 Energy exchanges in a mid-latitude depression. The lower diagram shows schematic 1000mb and 600mb contours; upper diagram shows a perspective view of the same with three-dimensional trajectories of air parcels originating in the central part of the cold air and in lower levels in the warm sector. The three portions of each trajectory are for approximately equal time intervals (after Palmen and Newton, 1969)

as it subsides. If the air rises, the temperature tends to decline but the potential energy increases.

Kinetic energy is proportional to the square of the velocity of the wind ($V^2/2$). Therefore strong winds have more kinetic energy than gentle winds, as the damage they cause indicates. In fact, on a global scale, hurricanes and other strong winds are relatively rare, so the quantity of energy in the form of kinetic energy is limited. Even in the regions of strongest winds it probably reaches no more than 0.5 per cent of the total energy content of the atmosphere.

Energy transfer in the atmosphere

The pattern of energy transfer in the atmosphere is complex, and we can only consider here some of the general components of the pattern. As a starting point let us look at a simplified model of what happens in the tropics (Figure 6.11).

The circulation between the tropics consists of two cells. Air blows in towards the low pressure belt of the equator (the equatorial trough) across the sub-tropical seas. As it does so evaporation of water from the ocean utilizes vast quantities of energy so that the sensible heat transfer to the atmosphere is often small. In contrast, over the desert land masses very little evaporation occurs, energy loss is limited, and the incoming radiation heats the ground surface which then heats the atmosphere. Thus much more of the energy is in the form of sensible heat. During the night this energy is reradiated back to space for the dry air is unable to intercept much outgoing long-wave radiation.

Anyone who has camped in desert areas will know the effect is to produce great ranges in temperature from day to night. More specifically, the net surplus of radiation is fairly small. As we have seen, winds approaching the equator rise as they meet the equatorial trough, creating a cloudy zone which can often be seen on satellite photographs (Plate 6.3). The ascent of this air is not a continuous, widespread phenomenon, but occurs mainly in association with localized, often intense and short-lived updraughts such as in thunderstorms. As the air rises and cools, the water vapour condenses and releases latent heat. The increased height of the air also represents an increased potential energy.

The equatorial air then diverges and flows polewards, so the potential energy is exported to higher latitudes. The cycle is completed as radiational cooling causes subsidence of the air. In the process the air dries and warms as the potential energy is converted to sensible heat. It also checks the rise of convection currents in these sub-tropical areas, producing clear, cloudless skies.

In the temperate and polar areas the processes of energy transfer are more complex. There is no general, cellular circulation of air as in the tropics, but instead a complicated pattern in which individual, rotating storms play an important part. Within these storms warm air masses rise, releasing latent heat and gaining potential energy. They then become intermixed with descending cold air and gain sensible heat (Figure 6.12). The rotating storms are moving, so the position of this intermixing changes constantly, although there is a tendency for a concentration in certain zones in the northern

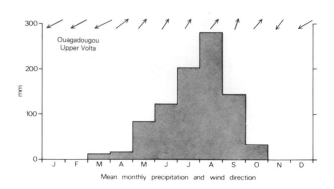

Figure 6.14 Mean monthly precipitation and wind direction at Ouagadougou, Upper Volta. Arrows indicate mean wind direction. Arrow pointing upwards represents a southerly wind

hemisphere. Labrador, Newfoundland and Greenland are associated with these areas of activity, experiencing cool, southward-moving flows of air. Britain and Scandinavia, in contrast, tend to be influenced far more by warm northerly-moving air, a phenomenon that greatly improves their climate.

All these transfers of energy through the atmosphere are highly variable, and major differences in the intensity and character of transfers occur over time. Thus, the flows of energy represent net increments, often produced by individual, temporary processes. It is for this reason that it is difficult to detect the nature of energy transfer directly from the pattern of general circulation.

Wind patterns

The general circulation of the atmosphere reflects the operation of the atmospheric system as a whole. It is clear, however, that the system is composed of many important subsystems and it is these – the main wind belts of the globe – which provide much of the climatic variation and consistency in the world. We have already indicated that the westerly winds dominate the climate of the temperate latitudes; similarly, the equatorwards movement of air in the regular easterly trade winds have a prevailing influence on tropical climate.

Surface winds

Four main surface wind belts can be distinguished. Around the equator, in the low pressure equatorial trough occurs a zone of convergence where the north-easterly trade winds blowing from the Tropic

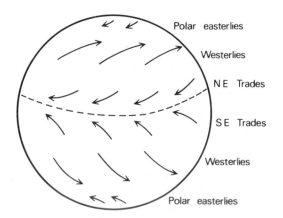

Figure 6.13 The main surface wind belts. The equatorial trough is the confluence zone between the trade wind belts; it is relatively narrow

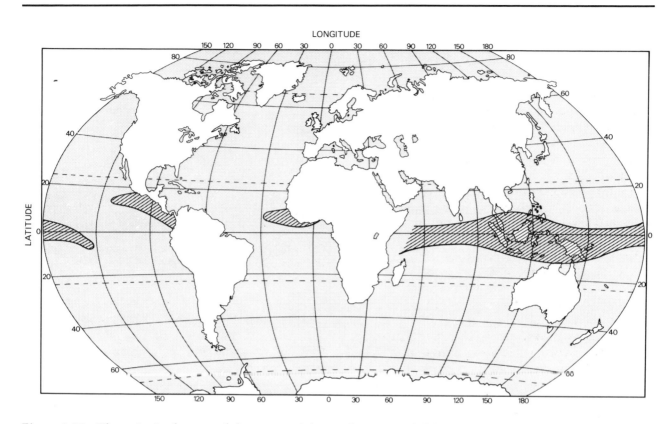

Figure 6.15   The principal areas of the equatorial trough zone or doldrums (after Crowe, 1951)

of Cancer meet the south-easterly trade winds blowing from the Tropic of Capricorn. Either side of the equatorial trough these winds dominate, giving the trade wind belt. Polewards of the tropics, in the temperate latitudes, we find a zone of prevailing westerlies, while around the poles occurs a belt of easterlies (Figure 6.13). We will examine each of these zones separately, but as we do so it is important to remember that, in reality, these wind belts do not operate in isolation. They are closely inter-related.

The equatorial trough

The equatorial trough or Inter-Tropical Convergence Zone is a shallow trough of low pressure generally situated near the equator. Over the oceans it is fairly static, because seasonal temperature changes are small. In the Pacific, for example, its average position varies by no more than 5° of latitude within the course of a single year. The situation is very different over the continents, however. During summer in continental areas the trough sweeps polewards, reaching 30° or even 40°

latitude over eastern China. Behind the trough the winds are predominately westerly and are the main rain-bearing winds to these areas (Figure 6.14). Where they reach into higher latitudes, they are called **monsoons** (an Arabic word meaning season) and they show an almost complete reversal of direction from summer to winter, a change that tends to occur with uncanny regularity about the same dates each year.

With the exception of the monsoon, the winds in the equatorial trough tend to be light and variable and, because sailors often found themselves becalmed there, this area became known as the Doldrums (Figure 6.15).

The trade winds

The trade wind belts lie between the equatorial trough and the sub-tropical highs at 20–30° of latitude. This zone occupies nearly half of the globe, much of it ocean, and within this area the steady easterly trades provide a stable and constant climate (Figure 6.16). At the surface the winds have a component towards the equator being from the

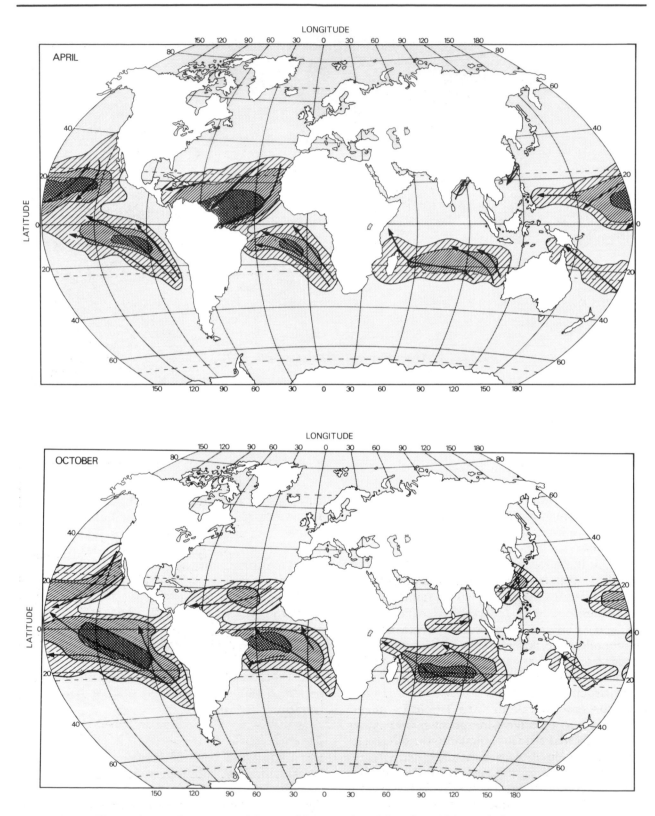

Figure 6.16 The trade wind systems of the world in April and October. The isopleths are in terms of relative constancy of wind direction and enclose shaded areas where 50, 70 and 90 per cent of all winds blow from the predominant quadrant with speeds above 3.3 m s^{-1} (after Crowe, 1971)

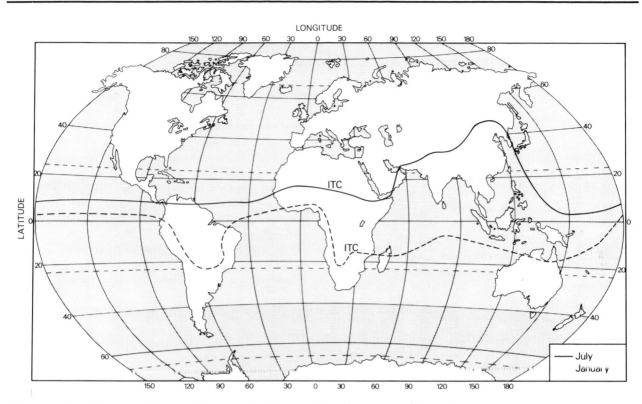

Figure 6.17   Mean positions of the equatorial trough at the times of the extreme seasons

north-east in the northern hemisphere and the south-east in the southern. Above the surface friction layer, the winds become more easterly.

Viewed from the air, the oceanic trade winds contain innumerable, uniform small clouds, all with a similar base and depth (see Plate 7.4, p. 101). These are the visual expression of the transfer of latent heat from the sea surface, through evaporation, before condensing at cooler levels in the atmosphere.

As we have noted, the seasonal movement of the equatorial trough is slight over the oceans, so the oceanic tropical areas are dominated by the trades. On the continents the trades are far more restricted in extent, and the equatorial westerlies and monsoons are more important. The two belts interact closely; it is the convergence of moisture in the trade winds that feeds the equatorial trough. The shift in the position of the trough thus determines the relative extent of the easterlies and westerlies. When the trough is further north, with the overhead sun in July, the trades are restricted in the northern hemisphere, particularly over the land. In January the trough is at its most southerly position and the trades extend to the equator. In the

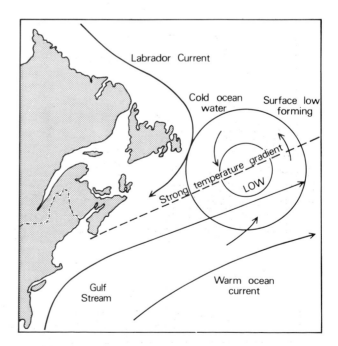

Figure 6.18   Generation of low pressure systems along the steep temperature gradient at the juxtaposition of the Gulf Stream and Labrador Current off north-eastern USA

83

southern hemisphere less marked variations occur, for the predominance of ocean means that the southern limit of the trough remains close to the equator (Figure 6.17).

The westerlies

In comparison to the winds of the tropics, the westerlies of the mid-latitudes seem unreliable and fickle. They are westerlies only on average.

Polewards of the sub-tropical anticyclones, rotating storms are the main mechanism of energy transfer. Unlike hurricanes, these systems cover vast areas and can clearly be seen from space, identified by their characteristic spiral of clouds (Plate 6.1, p. 77). They tend to move north-eastwards, although directions vary from north to south-east. Typically, they follow an evolutionary pattern which we will be examining more closely in the next chapter. The storms are initiated in areas of strong temperature gradients, such as off Newfoundland, where the cold Labrador current and warm Gulf Stream are in close proximity, forming almost circular air masses with low pressure at the centre and a rotational movement of winds around the periphery (Figure 6.18). They are known as lows, cyclones or depressions. As they evolve they become initially more intense, central pressure falling to as little as 960 mb, before filling as the storm declines. On average the location of maximum intensity is in the areas of Iceland in the Atlantic and the Aleutian Islands in the Pacific. Thus meteorologists speak of the Iceland Low and the Aleutian Low. In the southern hemisphere there are no distinct areas for the genesis of storms, so lows form throughout a wide belt.

As a low approaches, winds increase in strength, initially from a southerly direction, then becoming westerly and, finally, as the low moves away, they veer to north-westerly or even northerly. The tracks of the lows reach further poleward in summer than in winter, so the area affected by the storms varies seasonally. Nowhere is this seasonal pattern more clearly seen than over the Mediterranean basin and in California. In winter, when the cyclones follow a more southerly path, they bring rain to more southerly latitudes. In summer, the cyclones move away to be replaced by the sub-tropical anticyclones and dry, hot weather.

The regular march of cyclones and anticyclones through the temperate latitudes produces a majority of winds between north-west and south in the northern hemisphere and south-west and north in the southern hemisphere. This pattern is far from invariable, however, and depending upon the precise tracks taken by the lows, winds from any direction are possible (Figure 6.19). The prevailing westerlies, therefore, are anything but prevailing. Moreover, the strong north–south component of winds in these areas allows a more active transfer of energy between the tropics and polar regions.

Polar easterlies

Around the poles, beyond the main westerly belt, there is some evidence of prevailing easterlies. The winds are variable and linked to the shallow polar anticyclones. In the northern hemisphere, they are often influenced by the circulation around the northern edge of cyclones. As a result they change direction according to the local weather and topography.

In the southern hemisphere the vast Antarctic ice cap controls the atmospheric circulation around the pole. Anticyclones develop frequently over eastern Antarctica, and strong south-easterly winds develop around the margins of the ice plateau with consistencies similar to those of the trades, and occasionally of great strength. The hostility of the climate in these regions can only be imagined. At Byrd (85°S), for example, the mean wind speed in August (mid-winter) is 10.2 m s^{-1} with a constancy of 85 per cent. The daily mean temperature in this month is −37.0°C.

Upper winds

The nature of the upper winds

Looking up at high clouds on a clear day it is not unusual to find that their direction of movement is different from that of the surface winds. As this implies, winds in the upper atmosphere can be affected by forces operating in a different direction to those at the surface and may appear to be part of a different system of circulation to the surface winds. If we were to make an ascent by balloon into these upper wind systems we would find that the change from surface to upper atmosphere conditions was not abrupt but transitional. With increasing height, we would discover that the winds tend to follow a gradually more distinct zonal (east–west) direction and that they become stronger. The main reason for this change is the disappearance of the frictional influence of the ground surface upon the winds. In other words, the flow more nearly approximates to the geostrophic winds that, it will be remembered, result from the interaction of the pressure gradient

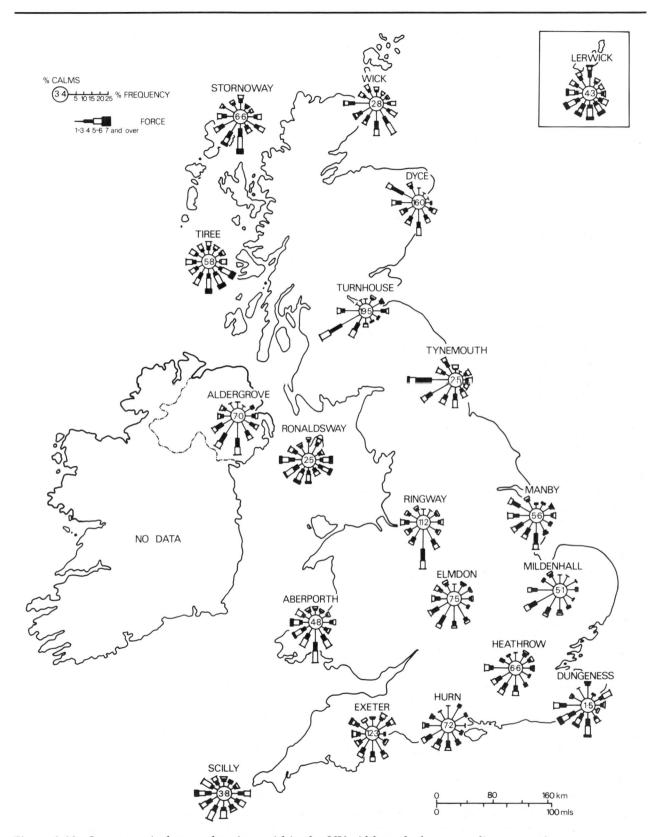

Figure 6.19 January wind roses for sites within the UK. Although the westerlies are at their strongest in January, winds frequently blow from other directions (after Shellard, 1976)

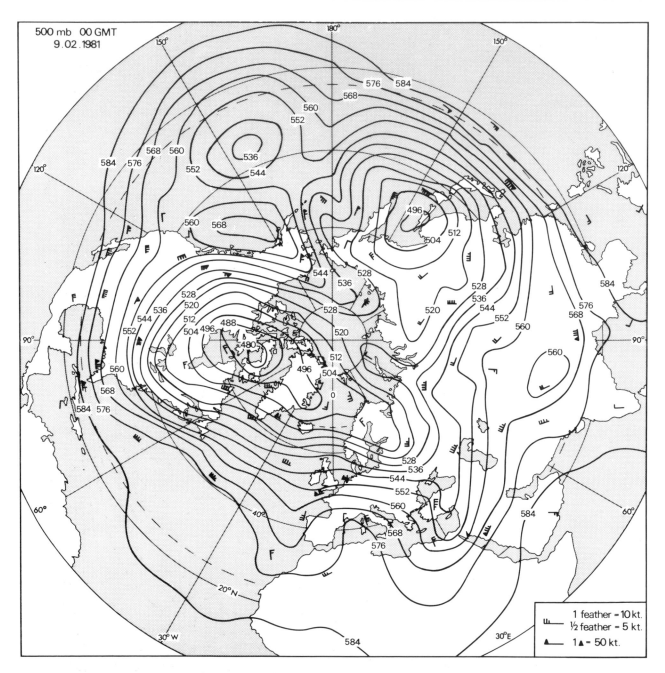

Figure 6.20 The upper westerlies at 500mb on 9 February 1981. The map shows the height of the 500mb pressure level above a fixed datum near sea-level. Winds blow parallel to the contours at a speed proportional to the gradient. Although flow is dominantly westerly, well-marked troughs and ridges can be seen, for example, over the eastern USA and Alaska respectively (reproduced by kind permission of the Deutscher Wetterdienst, Offenbach)

and Coriolis Forces (Figure 6.20).

The zonal flow of the upper winds can be shown on average as a cross-section from north to south (Figure 6.21). In fact variations around this average picture are slight, except in the monsoon areas of Asia. At each season the same basic pattern exists. Between about 30°N and 30°S we have a zone of high-level easterly winds which are relatively

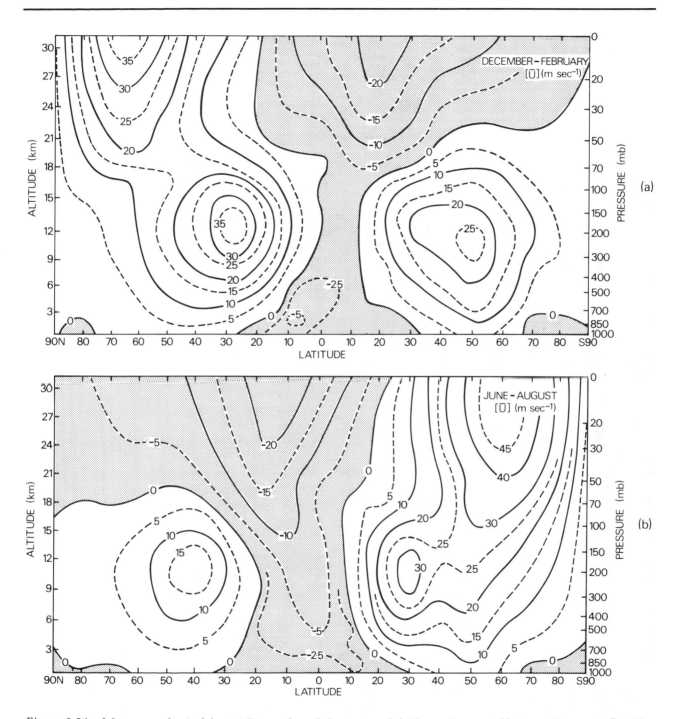

Figure 6.21 Mean zonal wind for (a) December–February and (b) June–August. Units are in m s⁻¹. Positive values denote a westerly wind (after Newell *et al.*, 1969)

weak, reaching a maximum speed of 4 to 5 m s⁻¹ (about 17 km hr⁻¹) at about 3 km. On either side of this belt occurs a ring or vortex of much stronger westerly winds.

The upper westerlies

These high altitude westerly winds are a major feature of our atmosphere. They reach their maximum speed at approximately 12 km between 30°

and 40° latitude. The mean speed is as much as 35 m s⁻¹ (125 km hr⁻¹) and maximum speeds of several hundred kilometres per hour are not uncommon. It is not surprising that aircraft can travel from the USA to Europe more quickly than on the return journey.

Although these wind patterns are generally constant, seasonal variations do take place, especially in the northern hemisphere. The upper westerlies are strongest in the winter when the temperature differences between the tropics and temperate latitudes are at their greatest. From June to August temperatures in the northern hemisphere are relatively warm, even in polar regions, so the pressure gradient is reduced and the upper westerlies decline to speeds of as low as 15 m s⁻¹ (55 km hr⁻¹).

As ever, changes in the southern hemisphere are less pronounced, largely due to the greater thermal stability there. The vast areas of ocean absorb large quantities of heat without any significant increase in temperature. The ice plateau of Antarctica also stays very cool, so the temperature gradients do not change very much from winter to summer.

The position of the boundary between the westerlies and easterlies (of both the upper and surface winds) varies throughout the year. In December to February, the polar vortex of the winter (northern) hemisphere expands, pushing the belts southwards so that, at the surface, the boundaries are at about 30°N and 35°S. As the year progresses, the other polar vortex begins to expand as winter sets in over the southern hemisphere. The boundaries eventually reach about 35°N and 30°S by June to August. The separation between the two systems is not vertical. As a result, we have some parts of the tropics with easterlies in the lower atmosphere and westerlies above. Only over a small area of the globe do easterlies occur at all levels, whereas westerlies extend throughout the atmosphere over a large proportion of the earth.

Rossby Waves and jet streams
The pattern of easterlies and westerlies in the upper atmosphere is only part of the total picture. In addition to the marked zonal flows, there are less apparent but none the less important meridional flows. In the circumpolar areas, for example, there occur wave-like patterns of flow called **Rossby Waves** (after C. G. Rossby, a Swedish meteorologist) that play a vital role in the energy exchange between the temperate and polar areas.

It is not easy to detect these meridional flows

within the pattern of strong zonal circulation by normal methods of depicting winds. These normally show average conditions, so that processes that balance each other, flowing northwards for six months, perhaps, then southwards for the next six months, are lost. Yet this is exactly what happens in the case of the Rossby Waves. At a particular location southerly flows may last for a few days to be followed by more northerly winds as the wave progresses eastwards.

In order to see these waves it is necessary to use a rather different technique of presenting atmospheric circulation. Instead of mapping the actual wind directions or speeds, the height at which a particular pressure surface is reached can be plotted. This may seem a strange way of depicting winds, but as we know the geostrophic winds blow parallel to the isobars, at a speed inversely proportional to the distance between the isobars. Similarly the winds blow parallel to the contours of the pressure surface. Where the contours are close together, the winds are rapid. Irregularities in the pressure surface indicate local patterns of wind movement.

Figure 6.22 shows a pressure surface (500 mb) map for January. The projection of the map may make it difficult to appreciate the direction of flow immediately. What is clear is that the flow is not perfectly circular around the north pole. Areas occur, even on this monthly chart, where the mean flow is northwards, and other areas where it is southwards. Effectively the air is flowing in a series of waves around the pole, carrying warmer air northwards on parts of its track and cold air southwards elsewhere. These are the Rossby Waves. In January the most prominent features of these waves are the pronounced troughs in the pressure surface near 80°W and 140°E, with a weaker trough between 10°E and 60°E. In July (Figure 6.23), the circulation is less intense and the troughs less well marked.

Many experiments have been conducted to determine the reasons for this pattern. Clearly surface features play an important part, even at this height. The presence of the Rocky Mountains and the Himalayas is believed to 'lock' the troughs at 80°W and 140°E respectively. The distribution of land and sea is also thought to be of importance. In the southern hemisphere there are no mountain ranges of comparable size, nor such marked land and sea temperature contrasts. As a result, the mean circulation is much more symmetrical around Antarctica.

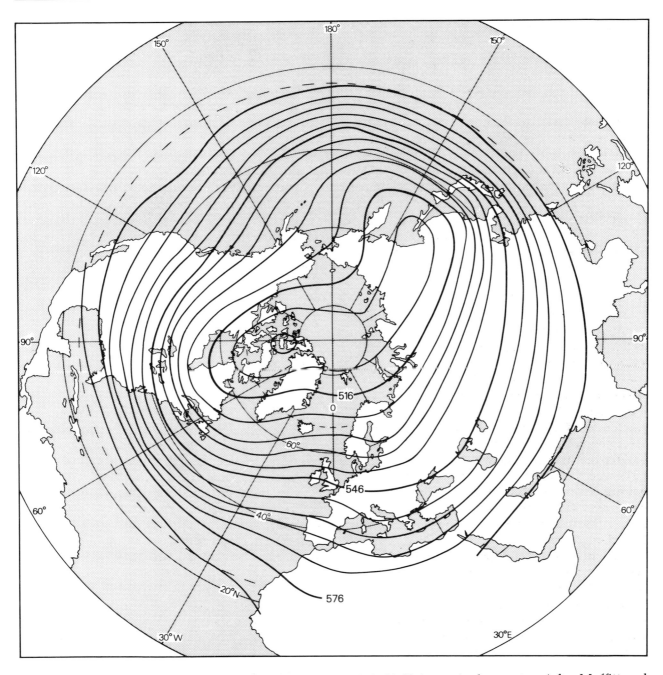

Figure 6.22   Monthly mean 500mb contours for January 1951–66. Units are in decametres (after Moffitt and Ratcliffe, 1972)

Similar wave-like patterns have been formed in models simulating the earth. If a temperature gradient is applied in a rotating fluid, the heat exchange will take place in a wave-like form. This can be shown by using aluminium flakes as markers of fluid flow (Plate 6.2). Because surface features are not usually included in the model, the waves slowly change their position.

Even on a shorter time period, waves may exist in the upper westerlies though their shape is less regular. The smaller waves tend to be associated with an individual depression and move more

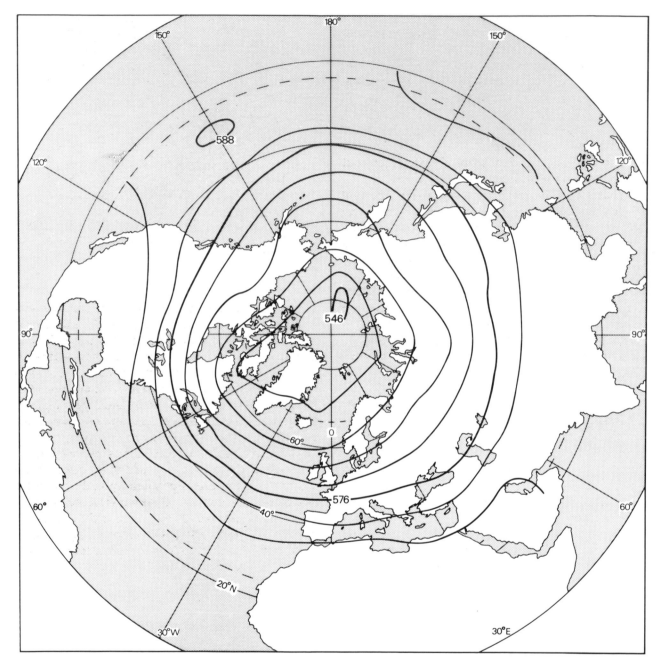

Figure 6.23 Monthly mean 500mb contours for July 1951–66. Units are in decametres (after Moffitt and Ratcliffe, 1972)

rapidly, perhaps up to 15° longitude per day. The longer waves – usually between four and six are apparent – move more slowly and are linked to the major circulation features such as the sub-tropical highs and Icelandic lows. The long wave flow tends to 'steer' the shorter waves, moving them north-

wards when ahead of a trough and southwards when to the rear of a trough.

Links between high-level and low-level flows

Within the Rossby Waves are found bands of especially strong winds. The existence of these winds or

jet streams was not appreciated until the increased use of aircraft during the Second World War. Bombers heading across the Pacific towards Japan reported headwinds so strong that they could hardly advance relative to the ground. More recent investigations have shown that speeds up to 135 m s^{-1} (490 km hr^{-1}) can exist locally in the jet stream maximum. A number of major jets have been found in the troposphere – the polar-front jet, the sub-tropical jet and the tropical easterly jet – but others exist in the stratosphere.

What is a jet? Basically it is a very narrow current of air travelling at great speeds through the lower westerly flow. Jets are formed in regions of rapid temperature gradient and they can lead to intense accelerations (and decelerations) of air in their vicinity. As we shall see later, when air is forced to change its rate of flow, tropospheric vertical motion may be started. In turn this may influence weather events at lower levels. Work in Britain and the USA has shown how surface rainfall amounts could be closely linked to the position of the jet streams (Figure 6.24).

Conclusion

Atmospheric movements, together with oceanic circulation, provide the main processes by which energy is transferred through the global system. They act to maintain a steady state in the system by transporting excess energy from areas which receive high inputs of solar radiation to areas where inputs are small. These movements involve two general patterns of flow; the dominantly zonal flow

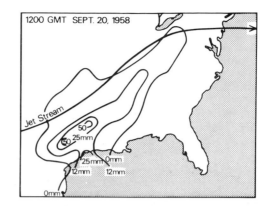

Figure 6.24 The relationship between jet stream maxima and surface precipitation (after Richter and Dahl, 1958)

of air within the main wind belts and the less apparent but equally important meridional transfers. Both circulations are controlled by the pressure gradient force which acts as the driving force for atmospheric motion. The earth's rotation – the so-called Coriolis Force – and friction modify the simple pattern of circulation initiated by the pressure gradient force to give the complex systems we find in the earth's atmosphere.

These atmospheric movements are vital for a number of reasons. Many of the features of the world's climates are dependent upon the character of atmospheric circulation, as we shall see in Chapter 8. Seasonal and daily variations in the circulation affect our life directly, and extreme events may have dramatic implications for man, topics which will be covered in Chapter 7.

7

Weather-forming systems

Weather and weather-forming systems

Not a year goes by without weather events somewhere in the world causing damage or loss of life. Floods, blizzards, tornadoes, hurricanes or even heatwaves can create problems and generate much economic stress over the affected areas. To be prepared for such events, it is vital for man to understand the weather, to be able to predict with accuracy, and preferably well in advance, events such as these. It is important too that man understands the vagaries not only of day-to-day weather conditions, but also longer-term trends. How useful it would be for the farmer to know what the weather over the next few weeks or even the whole growing season will be; then he could plan his sowing or ploughing or harvesting far more successfully. How useful it would be to have a clear idea of the weather in the year ahead so that corn harvests could be predicted, plans for winter snow could be made and contingencies could be taken to deal with droughts. Any such detailed understanding is a long way away. It will come only as we gather far more knowledge about the medium-term processes operating within the atmosphere, and about the myriad of factors that influence these processes.

The key to understanding and predicting the weather lies in understanding what we call weather-forming systems. If we look at a satellite photograph showing half the globe (Meteosat), it is clear that the distribution of clouds is not random (Plate 7.1). In some areas clouds are abundant, sometimes showing certain patterns which make it possible to identify their means of formation. Many areas are devoid of cloud altogether and surface features can be seen. By comparing this photograph with a map of surface pressure we would see that the large spirals of cloud were associated with cyclones in the middle latitudes and the main cloud-free areas with the large anticyclones of the sub-tropics. Between these areas, the cloud patterns are less clear, though over the South Atlantic

Plate 7.1 Meteosat image taken at 12.25 GMT, 24 September 1982 in the visible waveband. The polar, temperate and equatorial regions have high cloud amounts and are separated by the clear skies of the Sahara and southern Africa. There is extensive cloud cover over the cool waters of the South Atlantic Ocean

Ocean the trade winds have produced some interesting forms and, over the cold Benguela current off south-west Africa, there are extensive layers of low cloud. By viewing this instantaneous picture, we can see the way in which different areas of the atmosphere interact, and by using the surface pressure information we can relate these cloud patterns to the weather systems which produce them.

Air masses

An air mass is a large, uniform body of air with no

92

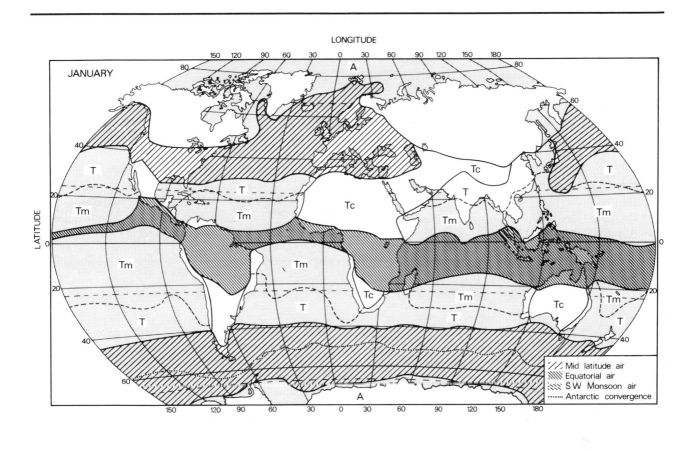

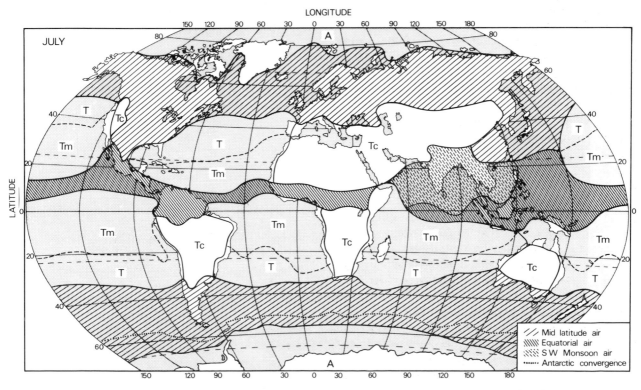

Figure 7.1 Air mass source regions (after Crowe, 1971)

Table 7.1 *Average properties of air masses*

Air mass	Symbol	Properties	Temperature (°C)	Specific humidity (g kg⁻¹)
Continental Arctic	cA	Very cold, dry	−40	0.1
Continental polar	cP	Cold, dry (winter)	−10	1.4
Maritime polar	mP	Cool, moist	4	4.4
Continental tropical	cT	Warm, dry	24	11
Maritime tropical	mT	Warm, moist	24	17
Maritime equatorial	mE	Warm, very moist	27	19

After A. N. Strahler and A. H. Strahler, *Elements of Physical Geography* (1979), New York: Wiley.

major horizontal gradients of temperature, wind or humidity. In the anticyclonic areas of the world, where air movement is slight, the air is in contact with the ground surface and gradually acquires the thermal and moisture properties of the ground. We find that the air then has relatively uniform distributions of temperatures and humidity over large areas – for example, the Canadian Arctic in winter. Whether or not the air will fully reach equilibrium with the surface characteristics will depend upon how long it remains in the source region.

The character of an air mass is dependent upon conditions in the area in which it forms. Because of this it is possible to classify air masses on the basis of their source area. Four main types are recognized: Arctic (or Antarctic), Polar, Tropical and Equatorial, and these are further subdivided into continental (for those forming over large land masses) and maritime (for those forming over the oceans) (Table 7.1, Figure 7.1).

As the air mass moves away from its source area, its character changes due to the influence of the underlying surface. Air moving towards the poles generally comes into contact with cooler surfaces. This causes it to be cooled from below, so that it may become saturated, with the result that low clouds are formed. In addition, the air is made more stable, so rainfall is less likely (Figure 7.2). Conversely, air moving towards the equator becomes warmer as it meets warmer surfaces. As we saw in Chapter 5, warming of the lower layers of the air steepens the lapse rate, making the air less stable and convectional showers more likely.

Changes in air masses in these ways are particularly marked in the mid-latitudes. Here, cyclones draw in air from several sources; the air is modified by the new surfaces it encounters and is gradually mixed as it rises around the cyclone centre.

Weather forming systems of temperate latitudes

Anticyclones

The anticyclone is a mass of relatively high pressure within which the air is subsiding. The major anticyclonic belts are in the sub-tropics, centred about

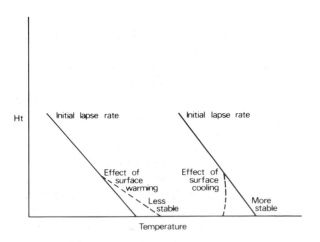

Figure 7.2 The effects of surface warming and cooling on lapse rates

30° from the equator. They represent the descending arm of the Hadley cell circulation of the tropics. As air descends it gets warmer and drier (Figure 7.3), but in these regions its descent is restricted by the layer of cool oceanic air below. This produces a semi-permanent inversion. This combination of circumstances results in very stable atmospheric conditions, reducing the possibility of precipitation. These anticyclonic belts are therefore associated with the main desert areas of the world.

In the middle latitudes, anticyclones often develop as a result of convergence in the upper westerlies, particularly where the waves in those winds are large. The anticyclones then build up within the depression tracks, diverting the cyclones from their usual routes and giving rise to exceptional patterns of weather. **Blocking anticyclones**, as they are called, are most frequent over north-west Europe and the north Pacific. Blocking in the Atlantic caused the drought in north-west Europe in 1975–6 and the severe winter of 1978–9. Unfortunately we do not yet know enough about the causes of blocking anticyclones to predict their behaviour.

Anticyclones are normally associated with dry, calm weather. Clear skies or extensive cloud and very warm or very cold conditions generally occur. Which we get depends upon the time of the year, the degree of moistness, the source of the air and the intensity of the anticyclone. In Europe, in summer, anticyclones often bring hot dry weather, but in winter cold weather is usual with the amount of cloud and position of the anticyclone determining the intensity of the cold.

Cyclones

Wind speed rises, pressure falls and the clouds get thicker; a common sequence of events in the mid-latitudes heralding the approach of yet another cyclone. The **cyclone**, or **depression** or **low** as it is also known, brings with it conditions very different from those associated with anticyclones. Air pressure is relatively low and the air rising. They usually move relatively quickly; in the northern hemisphere this is often towards the north-east. They are smaller in size, but within them air is rising more quickly than it descends in an anticyclone. Pressure and temperature gradients are much steeper so that horizontal winds are strong. In essence, they are the main mobile systems of the middle latitudes and they are responsible for the characteristic climates of these regions. Much of the

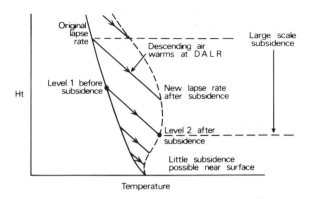

Figure 7.3 The effect of large-scale subsidence on lapse rates

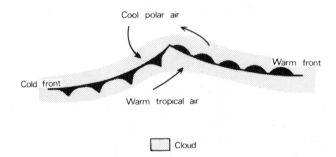

Figure 7.4 A small wave depression developing between polar and tropical air

precipitation here comes from this source. Cyclones are also responsible for the sudden swings in temperature from hot to cold or vice versa.

If we follow a series of cyclones over a period of several days, we find that they conform to a general pattern. Initially, a small wave develops in the front between polar and tropical air masses (Figure 7.4). In some cases, no further development takes place and the wave gradually dies out. More often, the wave begins to amplify and a small low pressure centre forms. Gradually the air pressure within this centre falls, but at the same time, to the observer on the ground, it becomes smaller. Eventually the system starts to fill and the cyclone disappears.

What we see at the surface is only part of the story, however, for the cyclone also extends up into the atmosphere. The low pressure centre represents a column of rising air – one which is often visible on satellite photographs (Plate 7.2). To understand the cyclone more completely, we need to ascend to the top of this column, to the upper atmosphere where we find the ridges and troughs in the upper westerlies. The flow around these

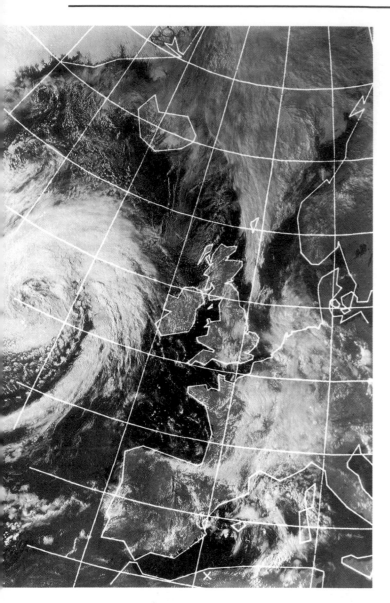

Plate 7.2 An example of the typical large spiral cloud pattern associated with the mid-latitude cyclone. The centre of the spiral marks the position of the lowest pressure with the cold front clearly defined by the band of cloud leading away south-westwards. Visible waveband, 1 September 1983 (copyright, University of Dundee)

waves is not always in equilibrium with the pressure gradients. Where air moves out of a trough it accelerates; as it approaches a trough it slows down (Figure 7.5). The air moving away from the trough draws air from the lower atmosphere, causing a reduction in pressure at the surface. Thus, air is seen to converge at the ground within the cyclone,

rise upwards into the upper atmosphere, and there diverge as it flows away from the trough (Figure 7.6). The relative rates of surface convergence and upper-air divergence control the development of the cyclone. If divergence exceeds convergence the cyclone intensifies as air is drawn out of the system. At this stage we find air pressure at the ground falling. If convergence exceeds divergence, the cyclone fills and air pressure at the surface rises. This is what happens in the final stages of the cyclone.

In the northern hemisphere, the troughs and ridges of the upper westerlies tend to favour certain locations. There is normally a ridge near the Western Cordillera and a trough near the east coast of the USA. This means that the area most favoured for cyclone formation is in the area south of New-

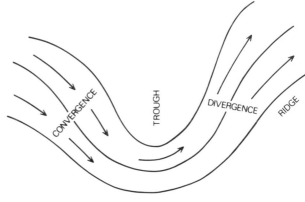

Upper air flow (500 mb)

Figure 7.5 Air flow around a trough in the upper atmosphere. As the air approaches the trough strong convergence builds up. On leaving the trough divergence develops associated with the spreading out of the streamlines

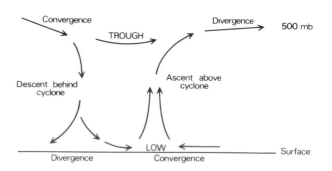

Figure 7.6 Interaction between surface and upper atmospheric flow near an upper trough

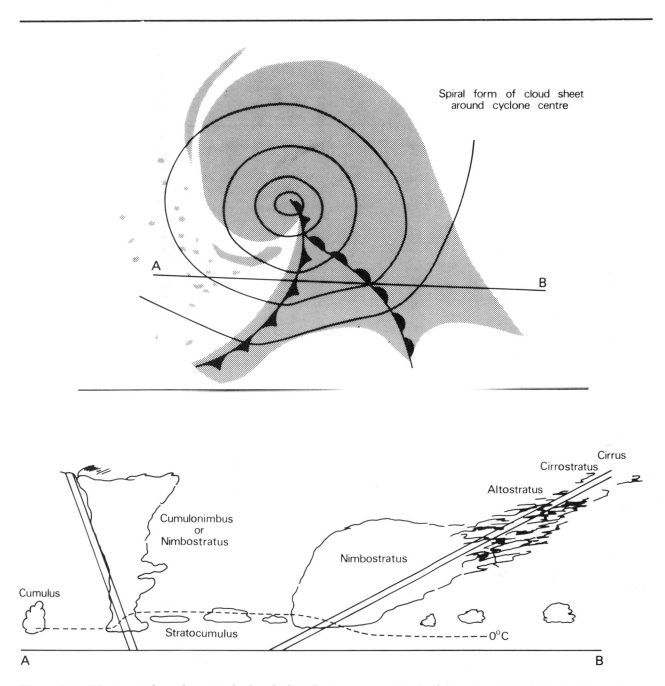

Figure 7.7 Horizontal and vertical cloud distribution associated with a model mid-latitude cyclone. Detailed differences occur in most cyclones. Plates 5.8, 6.1 and 7.2 show horizontal cloud distributions for particular examples

foundland. The depressions intensify, reach their maximum intensity near Iceland, then decay. The average position of the cyclones shown on charts is near Iceland for this reason. It represents the most frequent track of the depressions and where, on average, they reach their lowest pressure.

Because the cyclones are areas of rising air, they are almost always accompanied by extensive cloud and precipitation. The steep pressure gradients and rapid falls of pressure which can occur cause problems to the affected areas.

Figure 7.7 shows the typical vertical and horizontal cloud distribution and temperatures associated with a cyclone in middle latitudes. Details of cloud

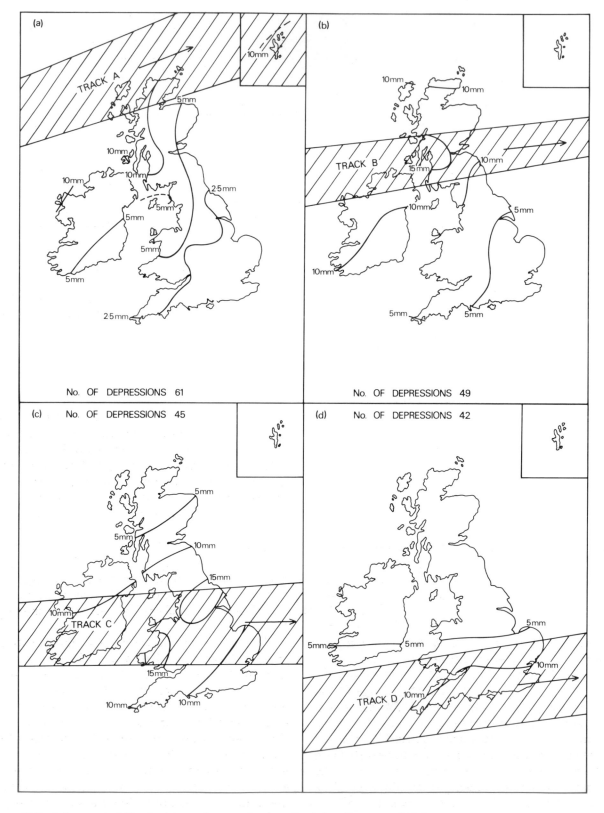

Figure 7.8 Mean rainfall relative to depression tracks (after Sawyer, 1956)

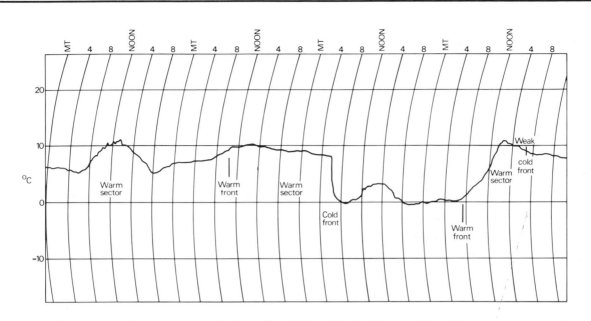

Figure 7.9 Thermograph trace recorded in Sheffield, 20–23 March 1981 illustrating the effects on temperature of the passage of two cyclones. The fall of temperature at the cold front on 22 March was unusually large for the UK

location and thickness will depend upon the nature of the upper atmospheric divergence and temperatures on the time of year and sources of the air. If we look at the surface pattern of precipitation from the cyclone (Figure 7.8), we can see how the areas of highest rainfall tend to be just on the northern side of the depression track with amounts decreasing northwards and southwards. The width affected may stretch for about 1200 km but this will vary between cyclones. The actual track of the cyclone is determined by airflow in the upper atmosphere and the temperature gradient at the ground.

Fronts

In many cyclones, we would find that there is not a gradual change of temperature as the systems pass, but several sudden changes. Figure 7.9 shows the trace of a thermograph during the passage of the cyclone. If it has been cold before the storm approaches, temperatures may rise slightly. This is due to cloud and wind stirring up cold air. If it has been warm, temperatures may fall, because the sun will no longer be shining. Suddenly the temperature starts to rise, perhaps by several °C within a few hours. Temperatures will then remain fairly stable until the cold air in the rear of the

cyclone reaches us. The fall of temperature is usually even more sudden than the earlier rise; falls of 10°C within a few minutes are not unknown.

The sudden change of temperature clearly indicates a change of air mass. The separation surface between air of different origins is called a **front**. Where warm air is replacing cold air, we have a **warm front** and where cold air is replacing warm air we have a **cold front**. The cloud structure of the fronts is shown in Figure 7.7. The clouds show the main areas of rising air produced by divergence in the upper atmosphere. This is why the clouds do not follow the frontal surface as closely as one might expect.

The warm front slopes at a low gradient of about 1 in 300, which means that the first clouds associated with the front can be seen long before the surface front is near. Cirrus clouds are the first indicators of the approach of the front followed by a sequence of a gradually thickening and lowering cloud base. Cirrostratus clouds are followed by altostratus then nimbostratus clouds, by which time rain will be falling. In general, the atmosphere is fairly stable at a warm front, but some convection does occur in the middle levels, producing areas of heavier precipitation. Figure 7.10 shows an example of the

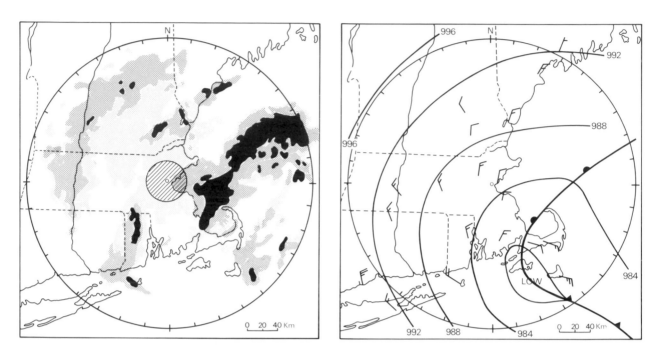

Figure 7.10 Precipitation patterns associated with a cyclone off Massachusetts obtained by radar (after Austin and Houze, 1972). Density of shading represents the following values of equivalent rainfall rate: 2, 4 and 8 (black) mm hr^{-1}

rainfall patterns associated with a warm front.

The slope of the cold front is much steeper at about 1 in 50. Weather activity at the cold front is often much more intense than at the warm front. If the warm air is unstable, the effect of uplift at the front generates thunderstorms and even tornadoes. The line of deep cloud may be seen on satellite photographs (Plate 7.3) as a very distinct band. The cold air descending with the heavy rain can intensify the effect of the fall in temperature.

Where the air in the warm sector between the fronts is rising, cloud development near the fronts follows the pattern described above; this is known as an **ana-front** (from the Greek word meaning up). However, further away from the cyclone, the intensity of uplift decreases and cloud thickness may gradually thin as the front dies out. In this stage of only weakly rising air, the front is termed a **kata-front** and the transition zone of temperature is fairly

Plate 7.3 A band of cloud associated with a stationary front to the north-west of Scotland. The lighter shades indicate high and cold clouds contrasting with the lower and warmer clouds over Ireland and southern Scotland. Infra-red waveband, 30 November 1982 (copyright, University of Dundee)

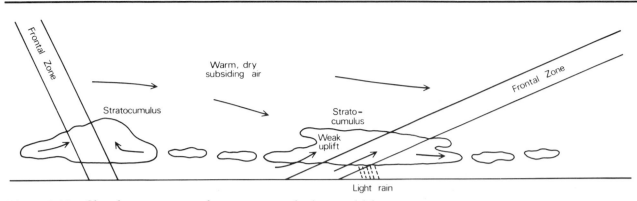

Figure 7.11 Cloud structure at a kata-warm and a kata-cold front

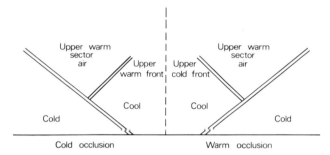

Figure 7.12 Simplified cross-sections through cold and warm occluded fronts

broad (Figure 7.11). Rainfall is slight from kata-fronts as the clouds are not deep enough and the updraughts are weak.

In most cyclones, the cold front moves more rapidly than the warm front. The air of the warm sector is raised above the ground surface as the cold front catches up with the warm front. This is known as the stage of **occlusion**, or the **occluded front**. The nature of the front will now depend upon the relative temperatures of the two cold air masses (Figure 7.12). Where the air behind is colder than that ahead, we will have a structure rather like a cold front. If it is warmer than the air ahead, the structure will resemble a warm front.

The detailed air movements and cloud distribution at an occluded front are complex. As it represents a mixing of air of different origins, humidities, temperatures and stabilities it is not surprising that great variation can occur between fronts or even along the same front.

At one time it was believed that it was the air rising along the frontal surface that caused the development of a cyclone. However, the role of divergence in the upper atmosphere is now

believed to be the most important factor, with the fronts being a consequence of the rotation of air around the cyclone's centre. From being a cause of the depression, the front has been relegated to a consequence. Nevertheless the weather activity associated with fronts is still a very important aspect of the cyclone.

Weather-forming systems of the tropics

Easterly waves
The weather of the trade wind zone normally shows little variety. It is characterized by small convectional clouds drifting across the sky in

Plate 7.4 Typical trade wind cumulus clouds. The clouds provide visible evidence of the continuous evaporation from the warm tropical seas (photo: Peter Smithson)

response to the prevailing winds (Plate 7.4). Showers may develop in the afternoon and they are likely to be heavier and more frequent in the summer season, but otherwise the weather remains remarkably constant throughout the year.

Occasionally disturbances arise to upset this quiet regime. On a dramatic scale there is the tropical cyclone, which is discussed in the next section, but on a smaller scale there is the easterly wave. As its name implies, this represents weather-forming systems related to wave-like structures in the easterly flow of air. They reach their maximum intensity at about the 700 mb level.

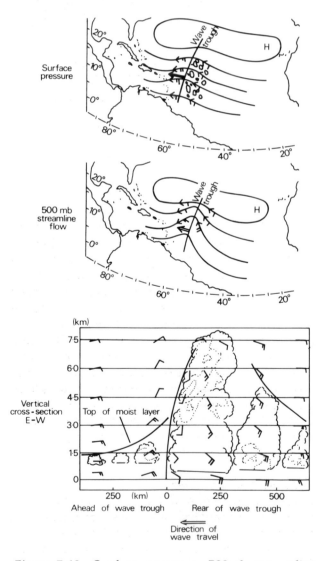

Figure 7.13 Surface pressure, 500mb streamline flow and vertical structure of an easterly wave (after Malkus, 1958)

The wave does not necessarily move at the same speed as the easterly flow, and it may even exceed the average windspeed. Preceding the wave, convectional cloud dies down due to surface divergence and subsidence of the air, while the wind backs towards the north-east (Figure 7.13). As the main axis of the wave approaches, convergence becomes dominant, causing ascent of the air, cloud-formation and precipitation just ahead of the low pressure trough. The wind veers suddenly as the wave passes, to be followed fairly quickly by a clearance of the cloud and a return to undisturbed trade wind flow.

The passage of the wave is not dramatic, therefore, but in areas where the weather hardly changes it does at least provide a little variety. Moreover, in areas such as the Caribbean, where the waves are frequent, they are responsible for a significant proportion of the annual precipitation.

Tropical cyclones

Throughout the tropics one of the other main features of the weather is the tropical cyclone. Unlike its counterparts in the middle latitudes, this is not large; instead it consists of a small, intense, revolving storm. It goes under a variety of names: the hurricane in the Caribbean and the USA, typhoon in the Pacific, or the cyclone in the Bay of Bengal. To qualify as a hurricane, the storm must contain winds reaching over 63 knots (32 m s^{-1}). Less intense storms are called tropical cyclones or tropical storms.

If we look at the parts of the globe affected by these cyclones, it is apparent that they only develop over the warmer parts of the seas (Figure 7.14). For each hemisphere, it is the summer and autumn periods when cyclones are most likely to strike.

The Pacific Ocean has the most hurricanes (Table 7.2). Some affect California, but the majority form

Table 7.2 *Average number of tropical cyclones per year by ocean basins (1958–77)*

North-west Atlantic	8.8
North-east Pacific	13.4
North-west Pacific	26.3
North Indian	6.4
South Indian	8.4
Australian	10.3
South Pacific	5.9

After W. M. Gray, in D. B. Shaw (ed.), *Meteorology of the Tropical Oceans* (1978), London: Royal Meteorological Society.

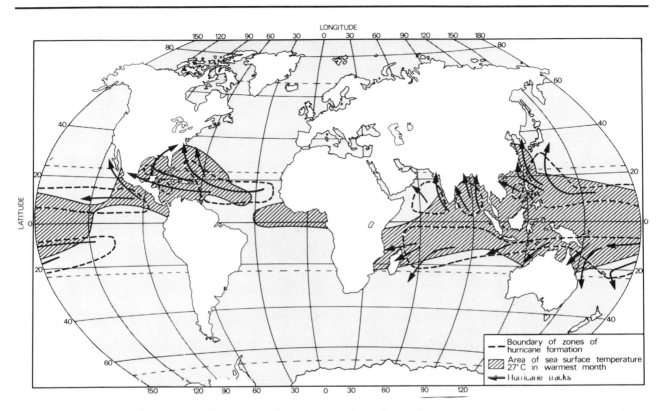

Figure 7.14 Areas of hurricane formation, their principal tracks and mean sea-surface temperatures of the warmest month (after Barry and Chorley, 1982)

further west and cross the Philippines before curving northwards towards Hong Kong, China or Japan. Other areas have fewer storms but they can still cause serious damage in Australia, India, Madagascar and even Arabia.

Effects of hurricanes

To people who have never experienced a hurricane the devastation is unbelievable. The storm consists of a spiral of thick cloud bands from which torrential rain falls (Figure 7.15), centred around an eye of almost clear skies with high temperatures and little wind. But it is the wind which is the most dramatic feature. Around the eye winds may exceed 250 km hr[-1], with mean speeds above 175 km hr[-1], covering an area some 150 km in diameter. Beyond this ring speeds gradually decrease, but gale-force winds (above 60 km hr[-1]) can extend over an area 600 km in diameter. Ahead of the storm, winds blow from one direction, usually northerly in the northern hemisphere. After the eye has passed, the wind resumes with equal ferocity from the opposite direction (Figure 7.16).

When the storms pass on to land they weaken.

Winds decrease and the structure of the storm becomes less clear, but heavy rain may still fall. It has been estimated that hurricane rainfall accounts for nearly one-quarter of the south-eastern USA's annual precipitation. Even in the north-east, heavy falls may occur. In June 1972, Hurricane Agnes gave more than 30 cm of rain over parts of Pennsylvania in a single day during the decaying stage. It is from heavy rain and storm surges along the coast that most hurricane deaths occur. Today's better weather forecasts enable preventive measures to be taken more quickly than formerly and fewer lives are now lost, though storm damage can still be vast. Satellites constantly monitor the skies for signs of the spiralling cloud pattern indicating a hurricane.

The causes of hurricanes

Despite the danger and damage of hurricanes, we know surprisingly little about their origins, except that they all form over the tropical seas where temperatures are above 27°C and that they do not form within about 5° latitude of the equator. Once developed, they move towards the west within the trade winds, gradually increasing in intensity.

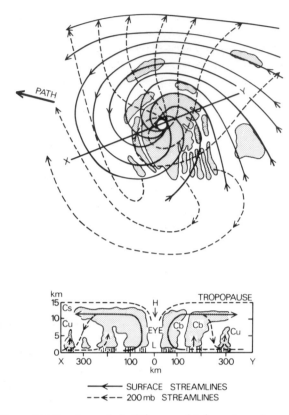

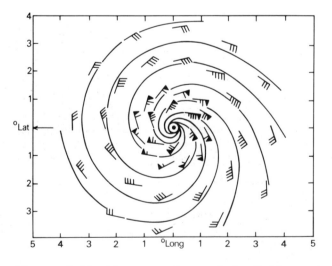

Figure 7.15 A model of the areal (above) and vertical (below) structure of a hurricane (after Barry and Chorley, 1982)

Figure 7.16 Mean vector wind field and streamlines. A solid triangle represents 50 knots (25.8 m s^{-1}), a single feather 10 knots (5.1 m s^{-1}) and a half feather 5 knots (2.6 m s^{-1}) for wind speed (after Riehl, 1979)

Before dying out the storm begins to swing polewards. A few manage to maintain their identity but they gradually decay and acquire the characteristics of a mid-latitude depression. Many September storms and floods in north-west Europe can be traced back to Caribbean hurricanes.

Once started the development of the hurricane is fairly predictable. But what starts it off? In order for the cyclonic wind circulation to develop, we must have air converging, which requires some form of initial disturbance. We do find a variety of small disturbances within the tropics where vertical movements and rotation can be started. As they are small in size and found over the seas, their location is difficult to predict. Once the air begins to rise, it cools and, on reaching saturation, large quantities of latent heat are released. It is this process which is believed to be responsible for giving so much energy to the storm. Once the storm moves over land, the main source of energy is lost and so the storm decays.

Tornadoes

Tornado! The very word brings alarm in areas such as the Mid West and Mississippi Valley of the USA. It conjures up the vision of a darkening sky, the appearance of a pale cloud, the familiar and frightening tornado funnel (Plate 7.5). The funnel may descend from the cloud base, getting larger and darker, until it eventually touches the ground, accompanied by a tremendous roaring wind. Debris is caught in the funnel, and as it moves across the countryside it leaves complete devastation in its wake (Plate 7.6).

The tornado is normally narrow, about $\frac{1}{2}$ km wide and seldom does it move more than 20 km. But exceptions do occur, with some being up to $1\frac{1}{2}$ km wide and travelling 500 km. How fast it moves we cannot tell; no recorder has survived the passage of a tornado. But speeds of over 400 km hr^{-1} are believed to occur.

Tornadoes occur in most parts of the world, but they achieve their greatest strength and frequency over the continental plains of the USA. The reason for this concentration here is the frequent juxtaposition of layers of air with great contrasts in air temperature, moisture and wind. Air ahead of the cold front may be drawn in from the Gulf. Behind it, cold air may be sweeping southwards from the Canadian Arctic. Such a situation is ideal for the development of cumulonimbus clouds needed to spawn tornadoes.

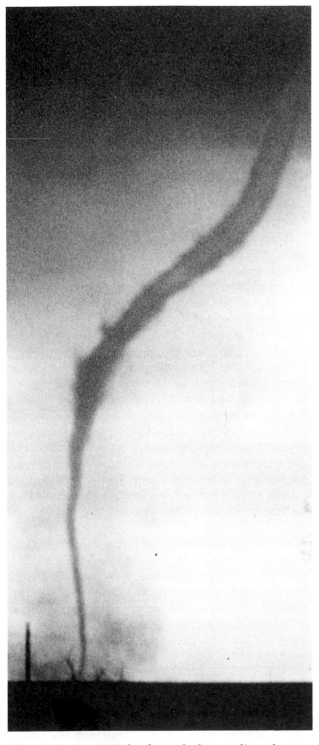

Like hurricanes, the precise mechanism whereby a funnel forms is not understood. It is probable that tornadoes are produced by thermal and mechanical effects acting in the cloud. But why some clouds generate tornadoes and others do not is a mystery. Nevertheless, favourable conditions are recognized and tornado warnings are issued by the local weather services.

Over the sea, similar funnels are found and termed waterspouts. As convection over the sea tends to be less intense than over land, the waterspout is much weaker than the tornado but may cause some damage to small boats or to light buildings if it passes over land.

Weather prediction

We all know from experience how much daily and seasonal variations in weather influence our lives. Clearly it is useful to have an idea of the weather which is in store for us. But how is it possible to foretell the weather? In the past we relied heavily on folklore. 'Red sky at night, shepherd's delight; red sky in the morning shepherd's warning', says one country adage. 'When there's sheep-backs

Plate 7.5 A tornado funnel descending from a large cumulonimbus cloud. The funnel becomes visible when moist air moves into the area of sharpley lowered pressure and condenses. Debris raised from the ground can add to effect (photo: Barnaby Picture Library)

Plate 7.6 Tornado damage to a high school, Virginia, USA (photo: J. Allan Cash Ltd)

(cumulus clouds) in the sky, not long wet, not long dry' goes another. Clearly these sayings sometimes contain a grain of truth – that is presumably why they have survived – but not enough for them to be reliable.

During the course of this century, as man has started to understand atmospheric processes in more detail, methods of forecasting have become more sophisticated. There are a number of different approaches. One of the main methods is that of extrapolation. Using experience gained from years of analysing weather patterns, the forecaster predicts the behaviour of the weather system from a knowledge of its recent history. This approach depends upon the principle that weather, like history, repeats itself; thus weather analogies can be drawn from the past to use as a basis for understanding the near future.

The trouble with this approach is mainly the fact that the weather does not always follow the expected course; similar conditions may result in very different weather. Moreover, it is difficult to obtain all the necessary information. A second approach is therefore used more widely. This involves understanding the basic processes of weather formation and using these physical laws to predict events. Unfortunately, although we know many of the basic physical laws, and can express them mathematically, the equations we end up with are difficult to solve. Only recently, with the development of computers, has it been possible to tackle this mind-stretching task; the first attempt to forecast weather in this way, without computers, in 1921, took several months! Today, however, numerical approaches to forecasting are widely used. They involve integrating information on many different atmospheric properties.

To forecast values of pressure and wind, the northern hemisphere is subdivided into a grid consisting of about 4000 squares each with an area of 90,000 km^2. For each point, the upper atmosphere is subdivided into fifteen levels and values of the critical atmospheric properties determined, mainly from satellite information. The physical equations of motion, continuity and thermodynamics are applied to each grid point at each level to predict the new value a short time period ahead. The new data set then provides the starting point for the next set of predictions and so on until the 24 hour or 48 hour forecast is available. Clearly there is a vast amount of calculation required, though, with the speed of present computers, six-day forecasts only take about 15 minutes of computing time. Realistic results are produced in this way and most meteorological services now use computer methods to predict future weather patterns.

We might get the impression from this technique that we can forecast the weather for the distant future but this is not true. It appears that small deviations can seriously affect the development of weather-forming systems. New predictions have to be made on a daily basis to incorporate small-scale changes which could become very important. The problem is that we do not have enough information (nor large enough computers) to solve the equations accurately.

Efforts are being made to improve our techniques of long-period forecasting but success has been very limited. In the UK, monthly forecasts were prepared based on previous weather analogies. For example, if the atmospheric circulation in, say June 1985, were very similar to that of June 1927, then it would be assumed that the weather of July 1985 should be the same as that of July 1927. Other controlling factors such as sea temperatures and ice cover would also be assessed to improve the accuracy of the forecast. However, even if the basic circulation is correctly predicted, slight shifts in cyclone or anticyclone tracks can produce markedly different weather. Due to lack of success, therefore, the experiment has been terminated. Numerical methods have also been used for long-period forecasting but for predictions beyond about four days the results are not encouraging.

Weather predictions and hazards

It is possible to predict many of the weather hazards discussed earlier, but we have to distinguish between large-scale and small-scale hazards. Tropical and temperate latitude cyclones can be predicted reasonably well so that we know approximately the areas they are likely to affect with their strong winds and heavy precipitation. At the smaller scale, the warnings of tornado formation are announced for a large area, but precisely where the funnel clouds will touch down on the surface is not known. It is probably impossible to forecast such conditions for more than a few minutes ahead. Flash floods from a single thunderstorm are in a similar category. We have to accept them as one of the microscale features of our atmosphere that occasionally may cause devastation over a small area. The chance of any one site being affected by them is very small.

8

Climates of the world

The nature of climate

On a July evening in New York the residents are sweltering. Temperatures have touched 29°C during the day, a little above average, but relative humidity is over 60 per cent and this makes it feel very uncomfortable. Meanwhile, at about the same latitude on the west coast, San Francisco is much cooler at 18°C. Even though the levels of humidity are slightly higher than in New York the air feels more pleasant; discomfort is less. Why are these climates so different at the same distance from the equator?

If we take an example of a difference in latitude rather than in longitude, the contrast becomes even more pronounced. In northern Alaska, there is Barrow Point near the newly discovered oilfields of the Arctic. The average July daytime maximum temperature is now only 7°C and night-time temperatures often fall below freezing point. If we headed southwards to Maracaibo on the Venezuelan coast at the same longitude as New York but at latitude 10°N, we would find average temperatures in July to be 34°C; a large jump from Barrow Point, or even New York.

Repeating the experience in January would not find these cities in the same order of warmth (Table 8.1). At Maracaibo, daytime maximum temperatures at 32°C would be only slightly cooler than in July. It is San Francisco which comes second warmest at 13°C; New York would be shivering at about 3°C but at Barrow, north of the Arctic Circle, there would be little daylight and with average *maximum* temperatures of −23°C it would be decidedly chilly.

Almost all the figures we have quoted have been averages. Clearly if we measured temperatures on any single day, we might find a very different pattern, due to the influence of different weather systems, or winds from different directions. What we are dealing with here is not the weather – the short-term, transient conditions of the atmosphere

– but the **regional climate**, the average, long-term pattern.

What we get is a static picture, a statistical summary of the climate, described by graphs and tables. It is a useful picture, for it helps us to compare different areas or to define briefly the character of a region. It allows us to see the way in which general atmospheric conditions vary across the globe. Later in this chapter we will look at the nature of this regional variation in climate, and in Chapter 30 we will discover how these climatic patterns relate to patterns of vegetation and soil and landscape. First, however, we need to ask two more fundamental questions. What are the characteristics that make up what we call 'climate'? And what are the factors which influence climate at a regional scale?

The elements of climate

The first of these two questions refers to what we generally call the **elements of climate**; the properties that we need to measure if we are to describe the climate. The main elements are the temperature, the rainfall, the nature of the winds and the degree of humidity. But each of these elements can be measured in a variety of different ways.

We can determine the mean annual temperature, for example, and this gives us some indication of average conditions throughout the year. Yet other aspects of temperature are also important. The

Table 8.1 *Latitudinal temperature transect of the Americas*

City	July (mean max)	January (mean max)
Maracaibo (10°N)	34°C	32°C
San Francisco (38°N)	18°C	3°C
New York (41°N)	29°C	13°C
Barrow (71°N)	7°C	−23°C

monthly pattern of temperature is useful, because it shows the seasonal variations. The temperature range (the difference between the temperature of the hottest and coldest month) is also helpful. All three may be used to describe and classify the regional climate.

Similarly, in measuring precipitation, it is important to know not just the annual average, but also the monthly or seasonal pattern. In some cases the reliability or variability of the rainfall is significant, since this reflects the likelihood of excessive precipitation or drought.

These two elements – temperature and rainfall – possibly illustrate the main characteristics of the climate. But if we are to get a more complete picture we also need information on windspeed and direction and frequency, on average humidity and perhaps on the rates of evaporation throughout the year. As we will see in later sections, all these elements may be used in climatic classifications.

Factors affecting the climate

The climatic elements are controlled on a day-to-day basis by the passage of the sun, by the nature of the weather systems and by local atmospheric factors, such as local winds and air movements. In the longer term, the climate is determined by the relationship of the area to the sun and by its position relative to major atmospheric features such as the permanent centres of high or low pressure, or the main components of the circulation. Thus it is possible to distinguish seven main factors which control the climate:

1 latitude;
2 altitude;
3 distribution of land and sea;
4 the nature of ocean currents;
5 distribution of mountain barriers;
6 the pattern of prevailing winds;
7 location of the main centres of high or low pressure.

Latitude affects the seasonal pattern of temperature (Figure 8.1), for as we have seen, the inputs of solar radiation decline markedly away from the equator. Altitude also affects temperature, for the atmosphere becomes cooler with height at an average rate of about 6.4°C for every 1000 m (the average environmental lapse rate). The distribution of land and sea has a variety of complex effects. We will mention one of the more important implications in Chapter 9, for the land gains and loses heat more rapidly than the sea; thus the temperature range tends to be greater over the continents than over the oceans. The nature of the ocean currents, the distribution of mountain barriers, the pattern of prevailing winds and the location of the main pressure centres all influence the weather systems which affect the area. They therefore control the 'average' weather which, in the long term, characterizes the climate.

It is these factors which determine the regional pattern of the world's climate.

Climatic classification

Classifying climates

If we compared the climatic record of all the observing sites in the world, it is highly unlikely that we should find any two that were identical. Each location possesses an individuality that is unique. On the other hand some of these differences are so small as to be insignificant. Many similarities may be seen between, say, the levels and pattern of mean temperature during the year, or the monthly distribution of precipitation. It is this aspect of similarity running through individual records that has encouraged climatologists to group together areas which appear to have similar climates.

The Greeks made one of the earliest classifications, dividing the earth into three climatic zones. In low latitudes there was the winterless zone with high temperatures throughout the year. In high latitudes there was the summerless zone with generally low temperatures. To the Greeks, the UK and much of the USA would have seemed without summer at all and have fallen into this second category. Third, there was an intermediate or middle latitude zone where one season was hot and the other cool or cold.

Such a scheme has at least the merit of great simplicity. The system works if you only want to distinguish areas of the world in terms of temperature. But it fails to take into account other aspects of the climate such as precipitation, nor does it provide any information about the range of values or their variability.

After the Greeks, few attempts were made to classify climates until our century when interest revived. Numerous classifications have been proposed based upon, for example, vegetation, water needs, atmospheric circulation and human physiological responses. Some of the best known are by

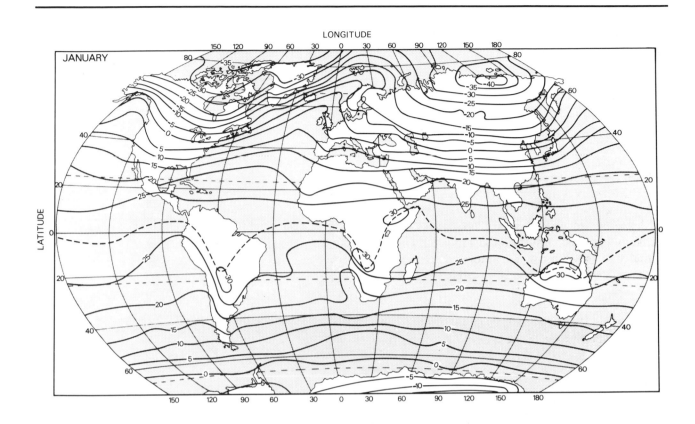

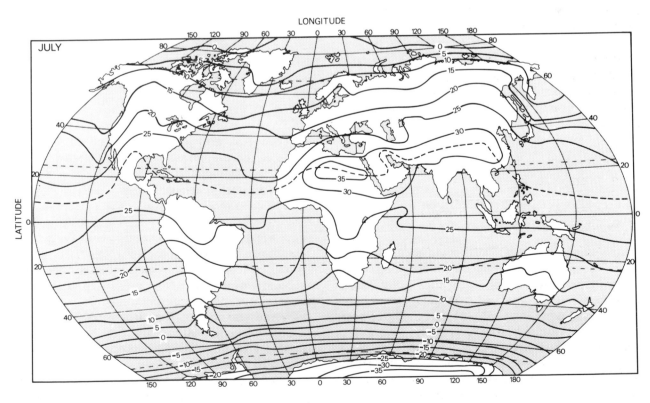

Figure 8.1 Mean sea-level temperatures in January (above) and July (below) (°C). The approximate positions of the thermal equator are shown by the dashed line (after Barry and Chorley, 1982)

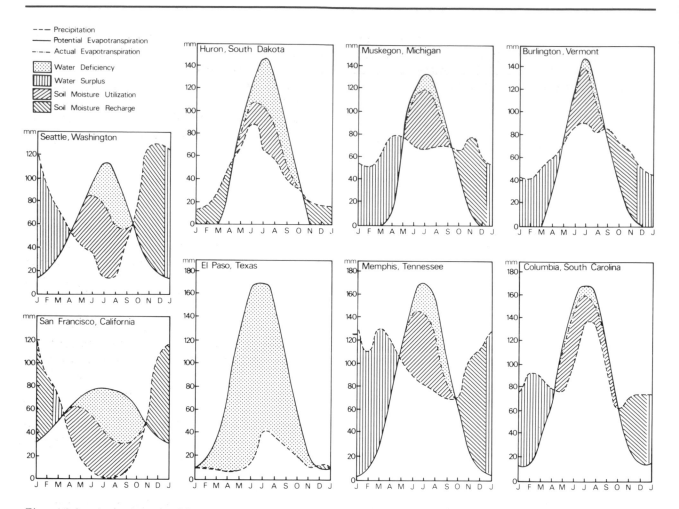

Figure 8.2 Average monthly water budgets at selected stations in the USA (after Mather, 1974)

Miller (1951), de Martonne (1948), Köppen (pronounced Kerp'n) (1918) and a derivation by Trewartha (1954), and Thornthwaite (1933 and 1948). The Köppen and Thornthwaite schemes were noteworthy in that they incorporated quantitative information to define the boundaries between climatic categories.

The Köppen system

The Köppen system has become most popular as a teaching aid, especially with the modifications made by Trewartha or Geiger. Details of the scheme are given in Appendix I. Briefly, the classification is based on annual and monthly means of temperature and precipitation. Five of the six major categories are recognized on the basis of monthly mean temperature:

A Tropical rainy climate. Coldest month > 18°C.
B Dry climates.

C Warm temperate rain climates. Coldest month between −3°C and 18°C; warmest month > 10°C.
D Cold boreal forest climates. Coldest month < −3°C; warmest month > 10°C.
E Ice climate. Warmest month < 10°C.

These temperature divisions were chosen because they appeared to correspond with vegetation boundaries. Plants were assumed to provide a good overall indication of climatic conditions since they respond not only to mean conditions of temperature and precipitation but also to the variability and seasonality. Not surprisingly a close correspondence is found between many climatic regions and vegetation. (Compare the maps in Appendix I and Chapter 30.)

If these had been the only zones identified the classification would have been very limited. In addition, further descriptions are provided to yield more information about precipitation and temperature conditions. From these combinations, twelve major climatic types were designated. Details of the meaning of the letters and their distribution are given in Appendix I.

Af	Tropical rain forest climate		Steppe climate
		BS	
Am	Monsoon variety of Af		Desert climate
		BW	
Aw	Tropical savanna climate		
Cf	Mild humid no dry season		
Cw	Mild humid dry winter		
Cs	Mild humid dry summer		
Df	Snowy forest climate, moist winter		Tundra
		ET	
Dw	Snowy forest climate, dry winter		Perpetual frost
		EF	

Thornthwaite classification

A development away from vegetationally derived classifications came with the Thornthwaite scheme in 1948. He based his work on the concept of potential evapotranspiration (PE) and the soil moisture budget. Potential evapotranspiration is defined as 'the water loss which will occur if at no time is there a deficiency of water in the soil for use of vegetation'. Potential evapotranspiration is essentially a function of climate alone and so can be used as a classification system. The most important influences are solar radiation and temperature, atmospheric moisture and wind. In general, PE increases as sunshine, temperature and wind increase, and as the humidity decreases.

In the classification the boundaries are drawn by comparing precipitation with potential evapotranspiration. If we compare the monthly variation of precipitation and PE in a graphical form (Figure 8.2) we can easily see whether or not precipitation exceeds PE. The graph for Seattle illustrates the classic case of rainy winters and dry summers. When the rainfall is greater than PE the excess is assumed to be absorbed by the soil until an equivalent of 100 mm of precipitation has been added. As rainfall is increasing rapidly in the autumn, this recharge of soil moisture is achieved by November and until PE becomes greater than rainfall in late April we have a water surplus amounting to 391 mm. Most of this will reach the river systems as runoff. During early summer, the 100 mm of moisture in the soil is utilized, then

further PE leads to a water deficit amounting to 208 mm which lasts until October. Distinct break-points in the curves can be seen.

Thornthwaite's system has a certain scientific rigour but it is difficult to use in large areas of the world because of the absence of the necessary climatic data. As a result it is not as popular as the Köppen system.

Regional climates

Whatever might be argued for or against different methods of climatic classification, the main application for climatic data is descriptive; interest is in the temperature levels, rainfall totals, their seasonality and extremes. If some understanding can also be given of why the climates are as they are, then so much the better. It is clearly impossible (and unnecessary) to describe all individual examples of climate, so some scheme must be adopted to identify groups. In this book, we shall make a basic subdivision into tropical, temperate and polar climates. First a discussion of the main features of each atmospheric circulation will be made, followed by examples of different climatic regions which exist within that area.

Tropical climates

The tropics have been described as the firebox of our atmospheric engine. Most of the sun's energy is absorbed here; energy which is transferred eventually into the cooler, energy-poor latitudes. This transfer is brought about by wind flows and ocean currents.

A simple approach to climate in the tropics is to distinguish four main areas: (a) the equatorial trough zone (Inter-Tropical Convergence Zone); (b) the sub-tropical highs; (c) the trade wind areas; and (d) the monsoons (Figure 8.3). The monsoon area is really a modification of the trade wind zone brought about by the effects of the continents. Nevertheless, as it is such an important part of the atmospheric circulation, it will be treated separately. The main features of the tropical circulation were outlined in Chapter 6 to which reference may be made.

Equatorial trough
It is the equatorial trough area that most closely meets people's idea of a tropical climate. During the day, clouds build up into massive cumulonimbus displays. Rainfall is frequent and abundant,

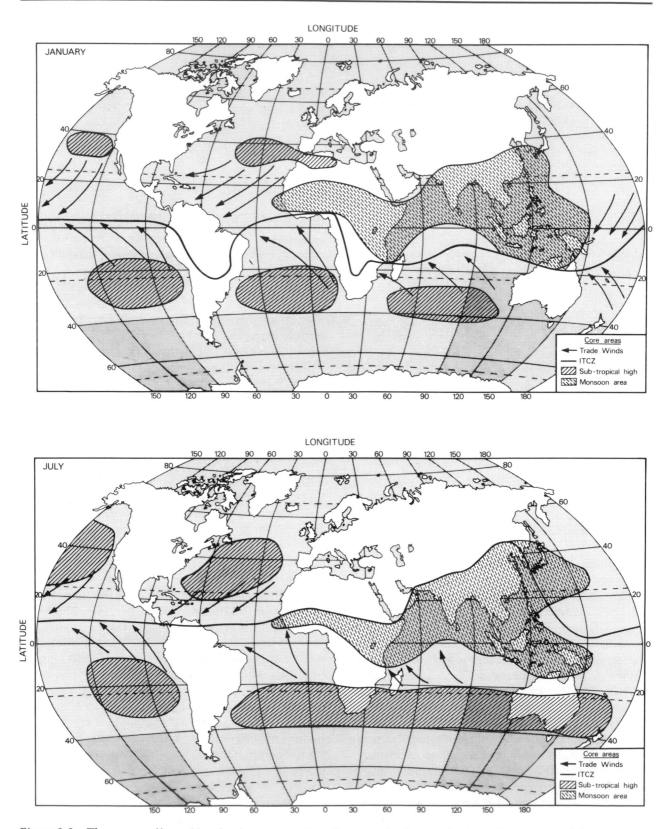

Figure 8.3 The areas affected by the four main climatic zones in the tropics in (a) January and (b) July. The boundaries are really transition zones and may show considerable shifts in position from year to year

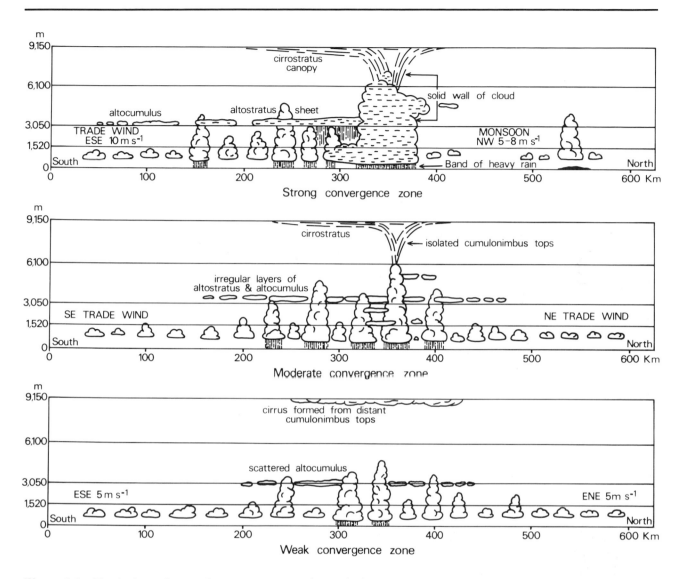

Figure 8.4 Typical north–south cross-sections through the equatorial trough zone. The cloud structure is related to the degree of surface convergence. With strong convergence it is difficult to portray the cloud field because both stratiform and cumuliform clouds are intermeshed (after Gabites, 1943 and Steiner, 1980)

temperatures and humidity are high, acting together to give us the tropical rain-forests. The structure of the atmosphere, though, is not so simple as this model may suggest. The multitude of names which have been used for the area give some idea of its variety – the doldrums, the intertropical front, intertropical convergence zone, intertropical trough, equatorial trough or intertropical confluence zone. For simplicity we shall refer to it as the equatorial trough although it does extend towards the sub-tropics, and it is quite variable in character.

The equatorial trough has many different forms. It represents the area of low pressure somewhere near the equator towards which the trade winds blow. The precise form it takes will depend upon the stability of the trades, their moisture content and the degree of convergence and uplift. Figure 8.4 shows some of the variability which can develop. Much of the trough is over the oceans and it is only recently that satellite photographs have shown us more about the detail of cloud forms and the structure of the equatorial trough.

One great surprise that these photographs showed was that over the oceans the trough could have a double structure (Figure 8.5). The reasons for this are not fully understood. Even where the

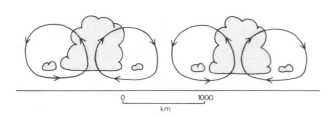

Figure 8.5 The double cell structure of the equatorial trough which is often found over the tropical oceans

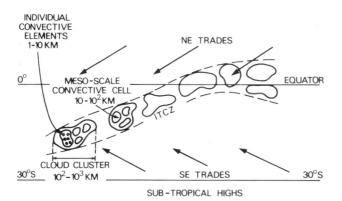

Figure 8.6 Schematic representation of the Inter-Tropical Convergence Zone or equatorial trough showing the organization of associated convective cloud systems in plan view (after Mason, 1970)

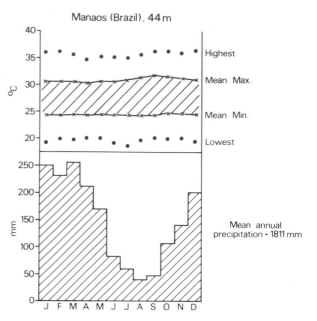

Figure 8.7 Climatic graphs for Manaos, Brazil in the equatorial trough zone

trough is single, its structure is variable (Figure 8.6), and careful scrutiny can show that a hierarchy of cloud is present. The larger element is a cloud cluster, perhaps 100 to 1000 km in length. Within the cluster there are convective cells and embedded in the cells are individual convective elements which can give the heavy rain characteristic of the equatorial trough.

What is the climate of the equatorial trough like? Figure 8.7 gives an example of monthly rainfall and temperatures at Manaos in Brazilian Amazonia. The mean monthly maximum temperature varies by 2.8°C over the year and the mean monthly minimum by only 0.6°C. Extremes are rare and insignificant by temperate latitude standards. Mean annual rainfall is high, with 1811 mm, though even in this zone there is a drier season when raindays are fewer. This is true of most of the equatorial trough zone, though the intensity and duration of the dry season vary. Only a few areas have no dry season, as in Indonesia, where Padang in Sumatra (7 m above sea-level) receives an average rainfall of 4427 mm and only one month has less than 250 mm. The driest season occurs when the trough moves polewards in response to continental heating in the summer hemisphere. As one moves further away from the equatorial trough zone so the dry season lengthens and we reach the monsoon or trade wind areas.

Sub-tropical highs
The **sub-tropical high pressure** zones act as the meteorological boundary between the tropical and temperate latitudes. The dominant air movement is away from the highs; the circulation is maintained by the subsiding air from the Hadley cell. Because the air is subsiding it tends to be warm and dry. An inversion develops in the lower atmosphere (Figure 8.8) and so these sub-tropical highs are generally cloud-free and deficient in rain. Where the highs remain fairly constant in position we find the main desert areas of the world – the Sahara, the Kalahari and the Great Australian Desert.

If we look at a map of surface pressure, the high pressure centres that we would expect over the desert areas may be absent (Figure 8.9). On the contrary, there is often a weak low pressure area. These lows are the result of intense heating of the

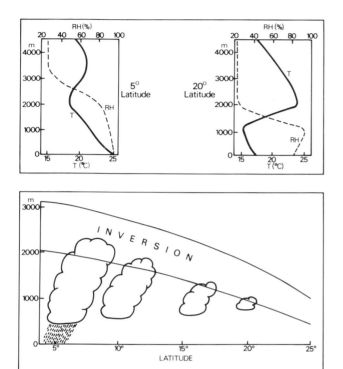

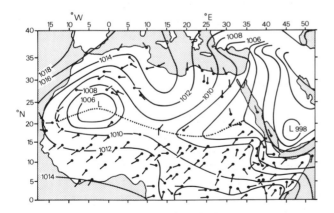

Figure 8.8 A generalized meridional cross-section in the trade wind zone and (above) vertical temperature and humidity profiles at 5° and 20°N (after Nieuwolt, 1977)

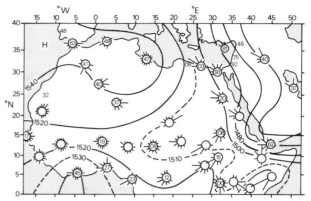

Figure 8.10 Mean 850 mb pressure surface and wind rose over northern Africa in July. The figures inside the wind rose denote the mean height of the pressure surface, i.e. 37 indicates 1537 metres (after Thompson 1965)

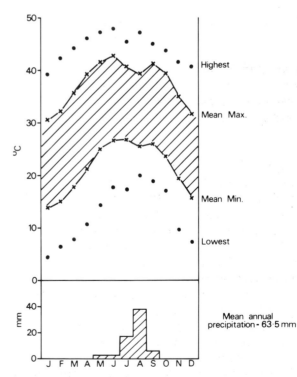

Figure 8.9 Mean surface pressure and wind direction over northern Africa in July. The dotted line indicates the mean position of the equatorial trough (after Thompson, 1965)

Figure 8.11 Climatic graphs for Atbara, Sudan, dominated by the sub-tropical anticyclones

ground surface during the cloudless days, taking temperatures above 40°C in summer. The air becomes less dense and thermal lows form. As they are a consequence of surface heating they tend to be fairly shallow and are replaced by high pressure by the 850 mb level (Figure 8.10).

From what has been said, we would expect the climate of these zones to be characterized by little rain and extremes of temperature. The data for Atbara (355 m) (Figure 8.11) confirm this point. In mid-summer, mean maximum temperatures are 42°C but in winter the mean minimum temperatures are only 8°C and frost can occur occasionally. The very dry atmosphere helps by allowing long-wave radiation from the ground to escape to space with little counter-radiation from water vapour or clouds. Even on the coast at Bahrain night-time temperatures are cool in winter, though frost is very rare. Precipitation is negligible. Rain falls on about ten days per year giving a total of about 75 mm. Most of this falls in winter and spring when temperate latitude depressions extend their effects far south and do give occasional rain.

Desert rainfall is notoriously variable. No rain may fall for several years to be followed by several heavy showers giving a few centimetres. It is this variability which makes the average rainfall figures for desert areas almost meaningless. Annual rainfall totals at Phoenix, Arizona for a 100-year period, for example, show a long-term mean rainfall of 186.9 mm, but most years (55 per cent) have less than this and a few years have very much more. It has been said that for deserts average rainfall is the total which never falls.

In some of the sub-tropical high pressure belts

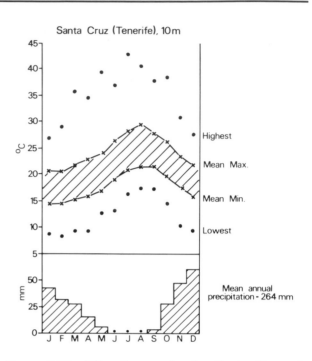

Figure 8.13 Climatic graphs for Santa Cruz de Tenerife, Canary Isles, displaying a maritime trade wind climate

additional factors reduce the likelihood of rain. On the west coast of the Sahara, the Kalahari and the Atacama Deserts cold ocean currents flow offshore. They cool the air and make it even more stable. Mist and fog may be frequent but rain is rare. One of the driest places in the world is Arica in the Atacama Desert of Chile (Figure 8.12). Years have elapsed between rainstorms and even then only a few millimetres may fall. Conditions here are similar to other coastal deserts near a cold ocean current but in addition the prevailing winds blow from the southeast. To reach the Chilean coast they must descend the main mountain barrier of the Andes, some 5000 m, which further emphasizes stability and dryness. The result of all these factors acting against the mechanisms of rainfall generation is to produce one of the driest parts of the earth.

Trade winds

Blowing away from the sub-tropical anticyclones of each hemisphere are the **trade winds**; northeasterlies in the northern hemisphere and southeasterlies in the southern hemisphere. The trades can be some of the most constant and reliable winds of the world. If we measure constancy by the persistence of winds from the same general direc-

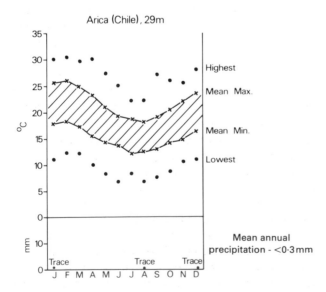

Figure 8.12 Climatic graphs for Arica, Chile in the sub-tropical high pressure belt. Global and local factors ensure that there are few occasions favouring precipitation

tion (± 45° of the mean direction), then over 80 per cent of the time the winds will be from this direction (Figure 6.16, p. 82).

Around the tropics the trade winds are very stable, being greatly affected by subsidence so the moist-layer near the surface is thin (Figure 8.8). The sudden rise of temperature and drying of the air at about the 900 mb pressure surface is known as the **trade wind inversion**. In the north-east of the Atlantic and Pacific Oceans, it may be only a few hundred metres above the ground, effectively preventing rainfall over the oceans. When islands, such as the Canary Islands, rise through the inversion, the lower windward slopes may be moist due to cloud and some rain, but above the mean level of the inversion and on the leeward slopes we have deserts. Figure 8.13 shows the climatic conditions at Santa Cruz de Tenerife in the Canary Islands. The cool Canary current helps to keep temperatures moderate in summer.

The trade winds gradually pick up moisture as they blow away from their source areas – the anticyclones. We can see this in the shape of the trade wind cumulus clouds. They are the visible sign that moisture is being evaporated from the seas and partly condensing as clouds. With more moisture being added and the influence of the anticyclones weakening, the intensity of the trade wind inversion weakens and it gets higher. Rainfall is now more likely to occur and if we look at the situation on the western side of the Atlantic Ocean we have a much moister climate, although with a distinct wet and dry season. In the Caribbean the wet season is usually summer and autumn when rainfall from the moist trades is supplemented by torrential rain from tropical storms. Their effects will be discussed in more detail later.

Monsoons

In some parts of the world, the wind systems appear to experience a seasonal reversal. In one season they may be blowing from the south-west; in the other season they are from the north-east. There has been much discussion among learned authorities about the precise limits of the **monsoon**, but Figure 8.3 summarizes these ideas. Without a doubt a large area of the tropics is affected by a seasonal reversal in areas where we might expect the trade winds to be dominant.

The reason for the reversal is the positions of the continents in the northern hemisphere. During the northern hemisphere summer, surface heating of

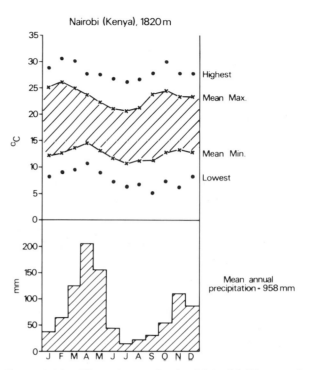

Figure 8.14 Climatic graphs for Nairobi, Kenya. A bimodal rainfall distribution related to lulls in the main monsoonal flows unlike that of other monsoon areas where the heaviest rains are associated with the strongest monsoon flows (see Figure 8.15)

the continental land masses is intense. A shallow surface low pressure centre forms over the Sahara, over India and over central Asia. The equatorial trough moves northwards allowing an inblowing of moist south-west to south winds to give the wet season in West Africa, India, China and most of south-east Asia. In winter the continents cool down, high pressure becomes established at the surface and winds between north-east and north predominate. This is the cool dry season for the monsoon areas of the northern hemisphere.

In the southern hemisphere, land masses are smaller and only Australia develops the semblance of a monsoon. Even then its influence does not extend very far inland. Over East Africa, set astride the equator, a seasonal reversal does occur, but the winds tend to be blowing parallel to the coast. As a result, the rainy season is between the main monsoon flows, rather than during one of them as in most of the other regions. Nairobi (Kenya) has its rainy season from March to May and again in November and December (Figure 8.14), in the lulls between the main north-east and south-east monsoon.

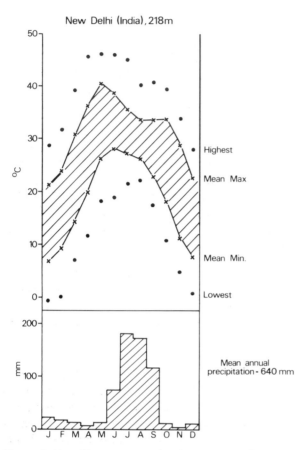

Figure 8.15 Climatic graphs for New Delhi, India showing the typical unimodal precipitation peak for most monsoon areas. Note the summer temperature peak occurs prior to the onset of the main rains

Freetown (Sierra Leone) illustrates the classic monsoon climate of West Africa. Temperatures are high throughout the year; there is no source of cold air to affect the area in winter, though away from the influence of the sea temperatures fall as low as 15°C. From January to March, rainfall is unusual. In April and May, thundery activity begins to develop, often associated with westward moving storms or disturbance lines in the north-easterlies, or the Harmattan as they are known locally. Towards the end of May, the monsoon rains start and rain falls on most days. Large totals accumulate; water is flowing everywhere and crops are growing rapidly. The air is very humid even during the day. During October, the steady rains cease, to be replaced briefly by thundery activity of the disturbance lines again. By December the dry season has returned and the land dries out.

This sequence of events – dry season, disturb-ance line thunderstorms, monsoon rain – has been used to distinguish the climates of West Africa. The zones are aligned parallel to the equatorial trough and move northwards and southwards with the season. To the north of the area called the Sahel, the monsoon rains may not penetrate in some years and rainfall will be from the disturbance lines only. When the trough is at its northernmost extent a short dry season appears in the southernmost parts of the region on the coast. The effects of the South Atlantic sub-tropical high are believed to cause this mid-summer dry season.

The best known area of the monsoon is India. Variability does exist across the vast sub-continent, but for much of the area there is a summer wet season and a winter dry season. New Delhi (Figure 8.15) illustrates typical conditions. The wet season is shorter and not so rainy as Freetown; most of the rain falls in the three months July to September with little at other times. The cloudiness associated with the rain reduces sunshine and tempers the heat. As a result, it is May and June which are the hottest months of the year with mean maximum temperatures well above 40°C. The monsoonal cloud clears in late September and October so the temperatures remain high, the lack of cloud com-pensating for the reduced intensity of the sun. In winter, when the winds blow from the continental interiors of Kashmir, temperatures fall and even frost may occur. Occasional mid-latitude depres-sions enter India from the north-west and give light rain to the north-western areas of India including Delhi. Further north-west into northern Pakistan, the summer monsoon rainfall is weak and the winter depressions make December and January the main rainy season.

In detail, the onset, retreat and origins of the monsoon are more complex. They depend upon an interaction between the massive land and sea breeze system generated by heating and cooling of the continents, the winds of the upper atmosphere and, in Asia, by a modification by the Tibetan Plateau whose mean altitude reaches 3660 m – almost half-way through the atmosphere. Nor-mally the south-westerly monsoons are overlain by easterly winds. Waves may develop in the upper easterly flow and, when they favour ascending motion by divergence, rainfall develops in the moist monsoonal air (Figure 8.16). Northwards the layer of moist air gets thinner and thus rain frequency and intensity decreases as shown in West Africa.

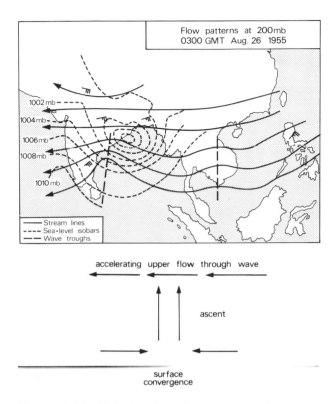

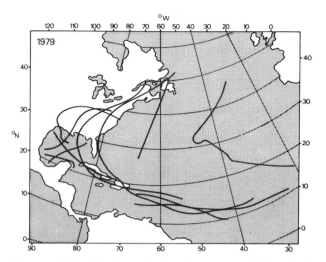

Figure 8.17 Tracks of hurricanes and tropical storms during the 1979 Atlantic hurricane season. Not all tropical storms reach hurricane intensity (> 33 m s⁻¹) but exhibit similar properties (after Hebert, 1980)

Figure 8.16 Relationship between surface and upper atmospheric flow during the Indian summer monsoon. The upper diagram shows the flow patterns at 200mb on 26 August 1955 (after Koteswaram and George, 1958)

Tropical cyclones

The above sections describe the main flows of air of the tropics. Within these airflows, disturbances can develop which alter their characteristics and are often the source of extensive precipitation.

The major disturbance of tropical latitudes is the cyclone. Main features have already been mentioned in Chapter 7, so here we summarize their regional distribution and effects on climates.

If we look at the parts of the globe affected by tropical cyclones, it is apparent that they only develop over the warmer parts of the oceans (Figure 7.14). In each hemisphere, cyclones are most likely to strike in the summer and autumn. Along the Atlantic and Gulf coasts of the USA, the normal hurricane season is from June to November. Early in the season, storms develop in the Gulf of Mexico or the Caribbean but there is a progressive eastward movement of their starting points until September when they may start as far east as the Cape Verde Islands off West Africa. After September the area of

origin shifts back towards the Gulf of Mexico. Another seasonal change affects the **zone of recurvature**. Most storms initially move westwards but at some stage many begin a curving track towards the north and then the north-east (Figure 8.17). The average latitude of recurvature is at its northernmost position in August and furthest south in November, linking with the change in position of the upper westerlies which help to steer the hurricanes.

The Pacific Ocean has most hurricanes but the figures are difficult to compare. Until satellite photography became available, many storms could reach hurricane intensity and decay without being recorded by the global observing network. It was only in the Atlantic that shipping was sufficiently dense to record most of the storms. As a result, in all areas except the Atlantic, the number of reported hurricanes has increased markedly, especially in the eastern Pacific and the south Indian Ocean. Changes in the frequencies of the hurricanes do occur. Figure 8.18 shows these variations in the Atlantic where records are most reliable. The number of storms is closely linked to the surface temperatures of the sea and the degree of cyclonic curvature in the trade winds; warmer seas and more cyclonicity mean more storms.

Hurricanes affecting the USA have been classified according to their intensity (Tables 8.2 and 8.3). Few recent storms have been category 5, though

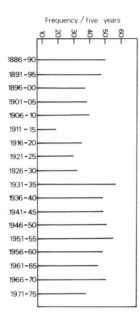

Frequency / five years

Figure 8.18 Atlantic hurricane frequency 1886–1975 (after Riehl, 1979)

Camille (1969) and David and Frederick (1979) reached this level. There has been a trend towards fewer deaths due to storms but an increased cost of damage. Improved forecasts of storm tracks have enabled preparations to be made to reduce risk to life but with greater material prosperity plus inflation, the cost of a storm inevitably rises, especially if it crosses an urbanized area (Plate 8.1).

The mean rainfall totals in summer and autumn for areas affected by tropical cyclones reflect the vast amounts of water which a hurricane can release. Data for Manila (Philippines) illustrate this point (Figure 8.19). In some 24-hour periods, the monthly average fall has occurred. Such a large daily total is likely to have been caused by a tropical cyclone.

Not all tropical areas are affected by tropical storms or easterly waves, but other, less organized disturbances give appreciable precipitation, as in Mozambique and parts of Brazil. In India, monsoon depressions develop in the moist south-easterly flow up the Ganges valley and bring extensive rains. A few areas miss major disturbances altogether. Anomalous dry zones occur in north-east Brazil where annual rainfall of less than 500 mm is found. In Somalia in Africa, mean annual precipitation is below 250 mm and aridity prevails (Figure 8.20), although the area is only just north of the equator.

Table 8.2 *Hurricane scale of intensity*

Scale number	Central pressure (mb)	Winds (ms^{-1})	Surge (m)	Damage
1	980	33–42	1–1.5	Minimal
2	965–979	43–49	1.5–2.5	Moderate
3	945–964	50–58	2.5–3.7	Extensive
4	920–944	59–69	3.7–5.5	Extreme
5	<920	>69	>5.5	Catastrophic

After   P. J. Hebert and G. Taylor, 'Hurricanes', *Weatherwise*, **32** (1979), pp. 60–7.

Table 8.3 *Number of hurricanes (direct hits) affecting United States and individual states, 1900–78*

Area	Categories					All	Major hurricanes
	1	2	3	4	5		
United States	47	29	38	13	2	129	53
Texas	9	9	7	6	0	31	13
Louisiana	4	6	6	3	1	20	10
Mississippi	1	1	2	0	1	5	3
Alabama	3	1	3	0	0	7	3
Florida	18	11	15	5	1	50	21
Georgia	1	3	0	0	0	4	0
South Carolina	4	3	2	1	0	10	3
North Carolina	9	3	6	1	0	19	7
Virginia	1	1	1	0	0	3	1
Maryland	0	1	0	0	0	1	0
Delaware	0	0	0	0	0	0	0
New Jersey	1	0	0	0	0	0	0
New York	3	0	4	0	0	7	4
Connecticut	2	1	3	0	0	6	3
Rhode Island	0	1	3	0	0	4	3
Massachussetts	2	1	2	0	0	5	2
New Hampshire	1	0	0	0	0	1	0
Maine	4	0	0	0	0	4	0

After   P. J. Hebert and G. Taylor, 'Hurricanes', *Weatherwise*, **32** (1979), pp. 100–7.

Temperate climates

Oceanic areas

Temperate latitudes are dominated by westerly winds. At the surface, winds can be very variable depending upon the movement of depressions and anticyclones. In the upper atmosphere the westerlies really become dominant forming a series of waves (Figure 6.22) which affect the movement and development of the main surface pressure systems. There is a clear seasonal movement of the main centres of activity. In the summer, the depressions tend to be less intense and follow more

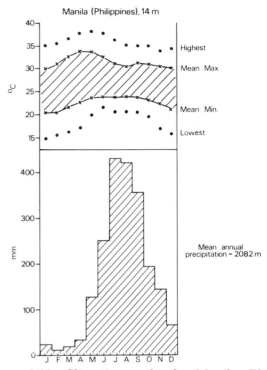

Plate 8.1 Hurricane damage in Darwin caused by Hurricane Tracy, 26 December 1974 (photo: Frank Lane Picture Agency Ltd)

Figure 8.19 Climatic graphs for Manila, Philippines, a maritime tropical climate affected by typhoons

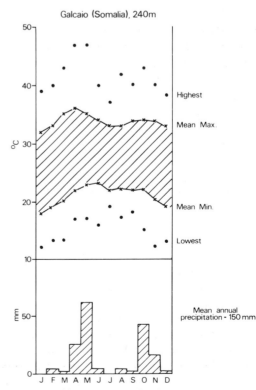

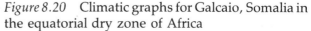

poleward tracks; the anticyclone centres also have a more northerly location (Figure 8.21). During winter the main centres of the sub-tropical highs move southwards and depressions become more frequent in lower latitudes. This type of climate, with mild wet winters and hot dry summers (when anticyclones are dominant), has been called a **Mediterranean climate**.

As winter is the only rainy season, the frequency and intensity of the actual depression tracks will determine how much rain falls. Consequently, variability is often large with very dry years interspersed with sudden flooding or prolonged heavy

Figure 8.20 Climatic graphs for Galcaio, Somalia in the equatorial dry zone of Africa

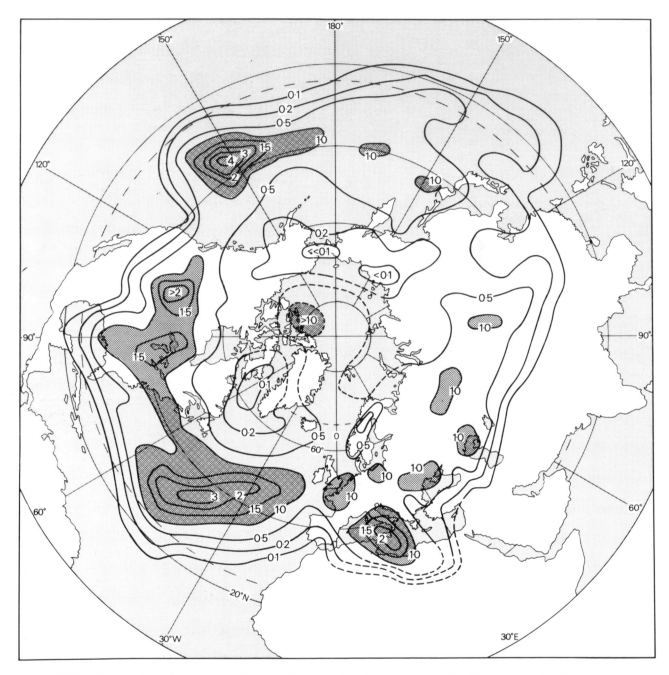

Figure 8.21a   Percentage frequency of anticyclone centres in summer (after Petterssen, 1950)

rain. Perth, in Western Australia, has a similar type of climate (Figure 8.22), although winter in the southern hemisphere is June to September.

In the higher latitudes beyond the Mediterranean zone, the influence of depressions extends throughout the year and so the rainfall is more evenly distributed in time as well as space. Depres-

sions normally move from west to east with the consequence that, on the western side of the continents, winds blow onshore. The sea is cooler than the land in summer so air blowing off the sea tends to temper the continental effect, reducing summer temperatures. In winter, the sea is warmer, maintaining higher temperatures in the cool season.

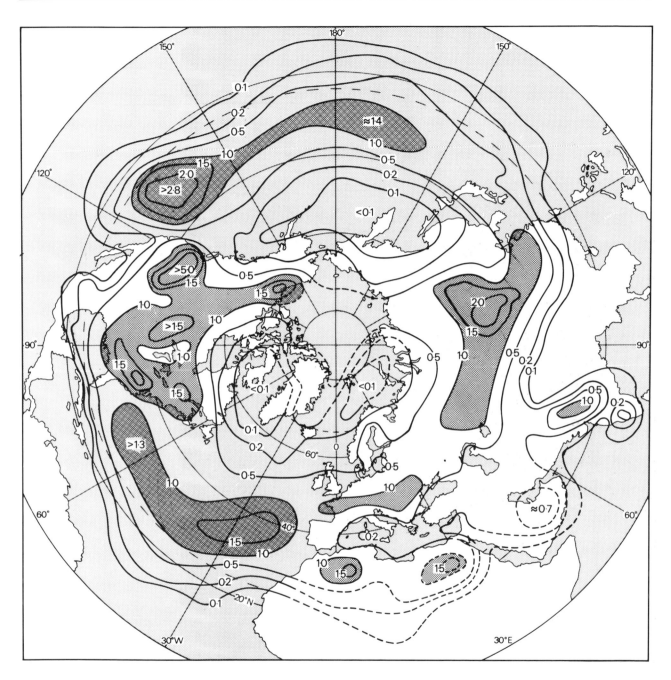

Figure 8.21b Percentage frequency of anticyclone centres in winter (after Petterssen 1950)

Vancouver (Figure 8.23) is an excellent example of this type of climate, with abundant rainfall at all seasons and its minimum in summer. Temperatures are equable throughout the year, uncomfortable extremes being a rare event. Much of north-west Europe experiences a similar climate and, because there is not a comparable range of mountains to the Rockies to prevent mild air penetrating the continent, this climatic type extends much further into the continent of Europe than it does in America. The effect of prevailing westerly winds across the Pacific and Atlantic Oceans drives warm water north-eastwards in the Kuroshio and Gulf Stream/North Atlantic Drift.

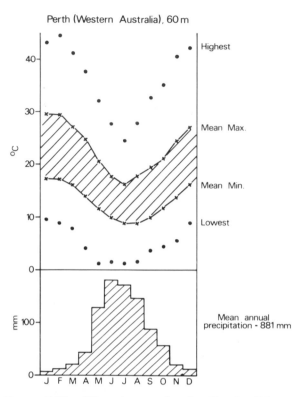

Figure 8.22 Climatic graphs for Perth, Western Australia showing a Mediterranean-type climate. Being in the southern hemisphere, the May to September rainy season is during winter

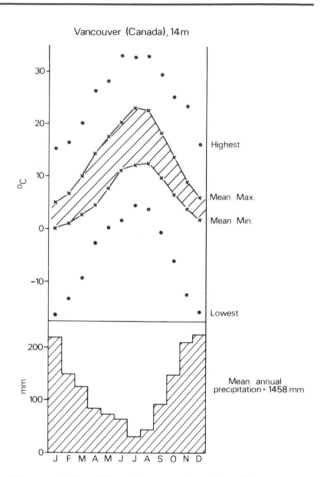

Figure 8.23 Climatic graphs for Vancouver, Canada with a moist, temperate climate

Temperatures are maintained at a higher level than would be expected on the western side of the continents by this oceanic energy transport.

Continental areas
On the eastern side of the continents (e.g. north-eastern USA and eastern China and USSR) the prevailing airflow is offshore so oceanic influence is reduced. In summer, air will be blowing off the heated landmass, but in winter the land cools down and even on the coast very low temperatures can be found. New York is a good example of this climatic region (Figure 8.24) with hot summers, cool winters and precipitation evenly distributed throughout the year. Comparison with Eureka (Figure 8.25) at an identical latitude stresses this contrast between the windward and leeward coasts. The cool California current makes summer temperatures particularly low in coastal northern California.

In the southern hemisphere, only South America and New Zealand extend so far southwards, but continental areas at these latitudes are small. Hence the capacity for major heating or cooling of the air is low. The record for Dunedin (Figure 8.26) shows this with a small annual temperature range. The frequency of depressions means that rainfall is abundant (161 days with rain per year), evenly distributed, and of low intensity giving a mean annual total of 937 mm.

Away from the coasts, rainfall gradually decreases, and temperatures become more extreme, with cold winters and hot summers. For example, at Winnipeg in Manitoba, conditions are extreme. In winter, beyond all warming influences and with the sun low in the sky at midday, average daily maximum temperatures in January are −13°C. Average minimum temperatures drop to as low as −48°C. Once more radiation starts to be received from the sun, temperatures start to rise rapidly. Highest temperatures occur in July, soon after the radiation maximum which is characteristic of continental areas, and up to 42°C has been observed.

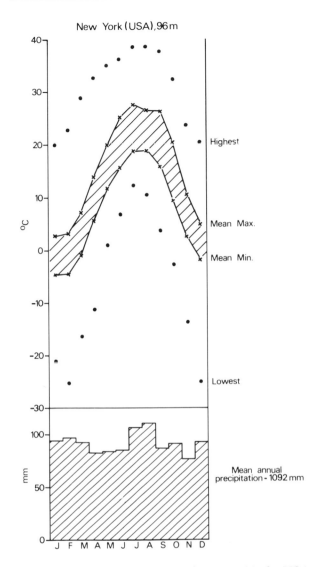

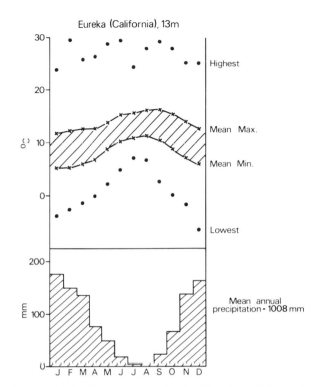

Figure 8.25 Climatic graphs for Eureka, California, USA. The rainfall regime is similar to that of Perth (Figure 8.22) in the southern hemisphere

Figure 8.24 Climatic graphs for New York, USA. Note the contrast with Eureka (Figure 8.25) on the west coast of the American continent at the same latitude as New York

The moisture content of the atmosphere in winter is very small. Summer is the rainy season, convectional showers giving short spells of heavy rain, although depressions do give some light snowfall in winter. Mean annual totals remain small.

In continental USSR, where depressions penetrate less frequently and moisture sources are distant, rainfall can fall to very low values even in the westerly belt. Kazalinsk (Figure 8.27) has a mere 124 mm on average which, with the hot summers, gives a desert climate. Even near the coast, mountain barriers can dry out the air. When this dry air descends into inter-montane basins or on the leeward side of the barrier, the likelihood of rain is small.

In the central areas of the USA, the climate is renowned for its extremes. Although this area is still within the main westerly zone, a spell of prolonged northerly or southerly winds can draw air from the Arctic or the Gulf with few relief features to provide shelter. Thus at Sioux City (Iowa) (Figure 8.28) the mean temperature in January is −7°C, but the average of the highest temperature is 11°C with 20°C being recorded once; the mean lowest temperature is −26°C with an extreme of −37°C. So the mean figure of −7°C is a little misleading unless this is borne in mind. In July, extremes from 44°C to 5°C have occurred. Some of the waves of cold air from the north sweep southwards and occasionally reach the Gulf coast, injuring frost-sensitive crops. These waves usually occur to the rear of a cold front when high pressure builds up to the west, strengthening the west–east pressure gradient.

The northern extremes

In more northern parts of the oceanic temperate

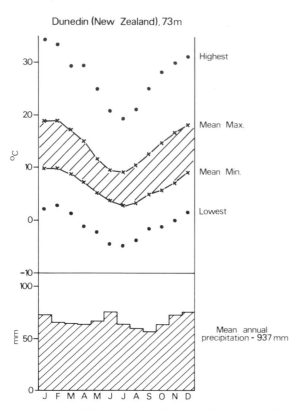

Figure 8.26 Climatic graphs for Dunedin, New Zealand in a maritime temperate climate

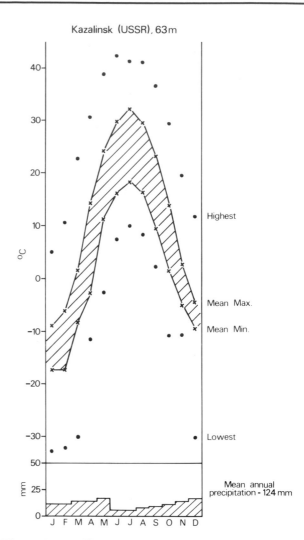

Figure 8.27 Climatic graphs for Kazalinsk, USSR (45.7°N, 61°E) with a continental climatic regime

latitudes, the effects of the proximity to the pole become apparent. These areas are directly in line with depression centres or even to the north of them. As a result, they experience cold easterly winds when the depression centres pass to the south rather than the mild air of the warm sector (Chapter 7). Because of this, the climate is cool, moist and cloudy with frequent gales. Reykjavik (Iceland) shows the extreme oceanicity of many of these areas, being exceptionally mild for its latitude (64°N) in winter but cool in summer (Figure 8.29). Highest rainfall is in winter, when some will be in the form of snowfall, and least in late spring and early summer. This area is surrounded by oceans, so extremes are rare. Nearer the continents, colder or warmer air may be drawn in occasionally. At Tromsö (Norway) at latitude 70°N, the same as Barrow Point in Alaska, the mean annual temperature is 2.5°C compared with −12°C at Barrow. Extremes range from 28°C to −18°C. The reason for this tongue of warmth which extends so far polewards is the direction of the depression tracks in the North Atlantic. The majority start in the Newfoundland area, reach their greatest inten-

sity near Iceland then decay off northern Norway, bringing with them air which originated in more southerly latitudes. Ocean currents follow a similar trend to maintain ice-free waters so far north.

Polar climates

The polar regions have low energy inputs. Even though the sun may be above the horizon for 24 hours in midsummer the intensity of its radiation is low. Also, so much energy is reflected from the ground or used in melting ice that the potential for heating is limited. As a result temperatures are low. Only when warm air is rapidly advected polewards can even relatively warm temperatures result.

Under polar climates we can distinguish three regimes:

1 In northern Canada and USSR we have the tundra zone. The winter snows do melt in summer to give a brief growing season but temperatures soon fall and winters are bitter.

2 The Arctic Ocean which is covered by ice throughout the year, though summer melting can give much water on the surface and at this time it is a very cloudy area.

3 Antarctica, which again is an ice-covered surface, is separated from any source of warm air by the cold southern oceans and has a mean altitude of over 3000 m. As a result little melting occurs, except on the coast, and summer days are much sunnier than in the Arctic.

Let us look at these three areas in a little more detail.

Tundra zone

The tundra zone is based on a vegetation type rather than a climatic region. It is the area of lichens, mosses, sedges, grasses and a few birch trees. Much of the ground is permanently frozen though the surface layers thaw out in the summer. This presents great problems for human development. If the permafrost thaws, perhaps as a result of an influx of heat from a building, the meltwater will flow away from the site and subsidence takes place. Exploration for oil in the Arctic has experienced many problems because of this. We are dealing with a very sensitive environment where small changes to the system can generate major consequences.

Depressions with their fronts and warm air rarely penetrate into the tundra zone so the climate is determined by more local energy inputs and outputs. In winter there is little input and so temperatures fall dramatically as shown by Yakutz (USSR) (Figure 8.30). Average daily maximum temperatures are only −43°C in January; the mean minimum is −47°C and even −64°C has been reached. The clear skies, dry atmosphere, lack of wind and thin snow cover of the continental interior present ideal conditions for long-wave radiational cooling. At some sites, like Verkoyansk, the radiational cooling is emphasized by cold air drainage into the valley bottoms (Chapter 9) and −68°C has been observed here, the coldest site on earth outside Antarctica (with climatological observations!). Because of the cold and lack of depressions, winter precipitation is scanty.

As the sun gets higher in the sky during spring, the land responds quickly to heating once the snow

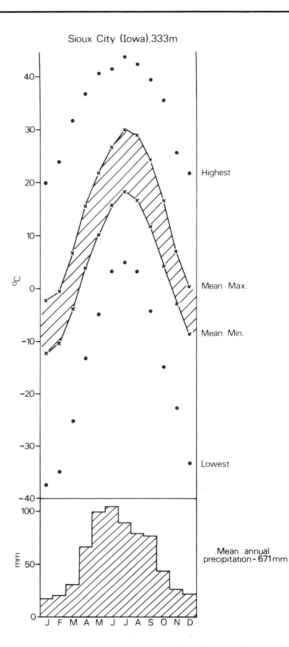

Figure 8.28 Climatic graphs for Sioux City, Iowa, USA. The city is well away from the moderating influences of the oceans and so has an extreme climate with cold winters and hot summers but is much wetter than Kazalinsk (Figure 8.27) at a similar latitude in the continental interior of the USSR

cover melts. Because the snow is thin, this does not take long. Temperatures increase rapidly so that by midsummer mean temperatures reach 17°C and even temperatures over 30°C are not unknown. The heating can generate convectional showers so

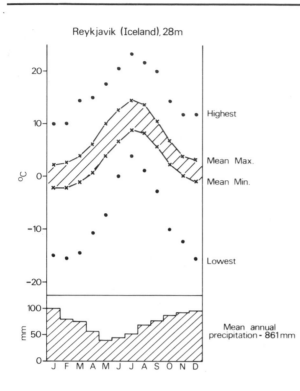

Figure 8.29 Climatic graphs for Reykjavik, Iceland. Despite being 20° latitude north of Sioux City, the oceanic location of Reykjavik ensures much warmer winter temperatures

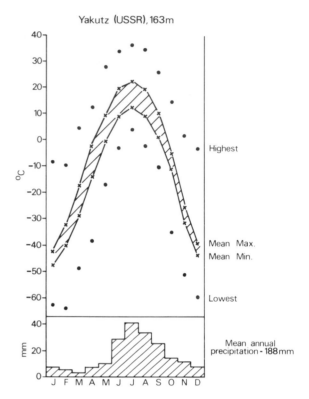

Figure 8.30 Climatic graphs for Yakutz, USSR (62.1°N, 130°E) in the tundra zone with brief warm summers and intensely cold winters

summer is the wetter season. Even then amounts are small and if it were not for the low evaporation these areas would be like deserts. The mean annual total of 188 mm for Yakutz is the same as Phoenix (Arizona).

Arctic Ocean Basin

The Arctic ice cap covers a large area, but compared with Antarctica it is relatively uniform. Climatic stations are sparse and of short duration. The reason for this is that, being a surface of sea ice, the ice moves taking the observers with it until it eventually melts.

In winter the shallow polar anticyclone dominates the scene though its centres are often over Siberia or Canada. Mid-latitude depressions approach the margins of the ocean, especially between Norway and Spitzbergen, and between Greenland and Baffin Land. With few depressions, skies are fairly clear with a mean value of cloudiness of about 40 per cent. Precipitation is very slight and

temperatures fall to a daily mean value of about −35°C. In summer, pressure gradients are weak over the polar ocean with a low pressure centre near the North Pole and a high near Alaska. The warmer temperatures associated with constant daylight result in the surface ice melting. The process of melting and evaporation both require energy. Almost all available energy during summer goes into melting or evaporation leaving virtually nothing for heating the air. Once away from the land surfaces temperatures rarely rise above 5°C.

With so much moisture around from melting ice and relatively low temperatures the air soon reaches saturation and cloud is abundant. The mean cloudiness for much of the Arctic in July is 90 per cent so, in this part of the Land of the Midnight Sun, the sun is rarely seen. With conditions favouring rising air being infrequent, precipitation tends to be small. Absolute amounts are difficult to measure because of drifting snow but 100–200 mm per year is a realistic estimate.

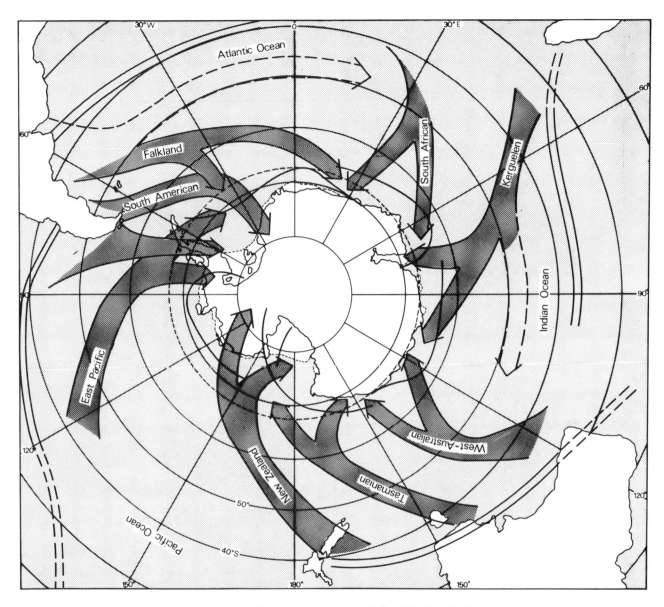

Figure 8.31 Paths of cyclones during winter in Antarctica (after Rusin, 1964)

Antarctica

Conditions in the Antarctic continent are very different. It is isolated from other continents by the vast expanse of the southern oceans. More than 55 per cent of its surface lies at an elevation of more than 2000 m and about 25 per cent at more than 3000 m above sea-level. The effects of latitude, altitude and isolation result in the vast majority of the continent being permanently ice covered. Only about 3 per cent of its area is believed to be free of a permanent ice sheet. This occurs in the coastal zone or in some of the arid valleys near McMurdo Sound.

Recording sea-level pressure is difficult when much of the continent is so high. The lowest standard pressure surface which is above land level is the 500 mb surface. In both winter and summer low pressure prevails but this is due to the cold dense air rather than the area being a centre of active depression formation. In fact depressions rarely penetrate on to the ice cap, especially the higher ice plateau of eastern Antarctica (Figure 8.31). We can visualize the continent as an area of relative calm surrounded by the mid-latitude storms carrying with them clouds and blizzards. As

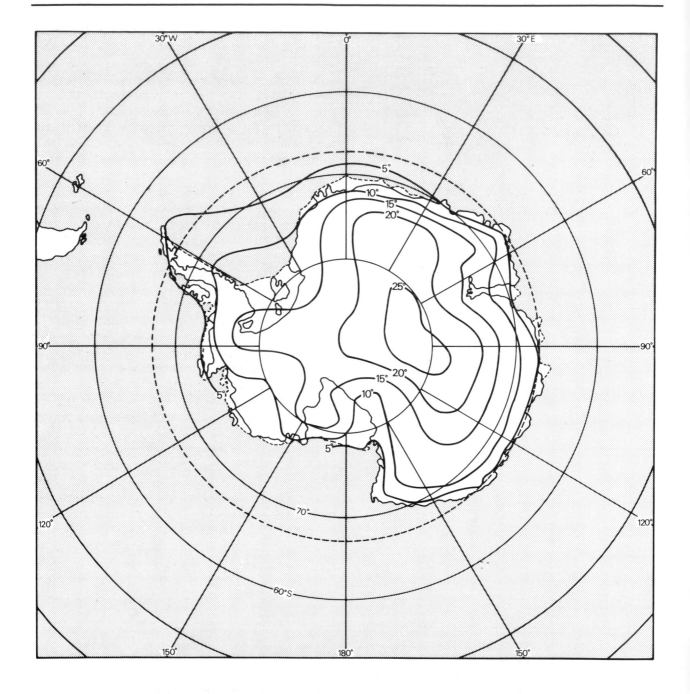

Figure 8.32 Isolines of the mean temperature difference (°C) between surface and 20m over Antarctica in winter (June–August) (after Phillpot and Zillman, 1969)

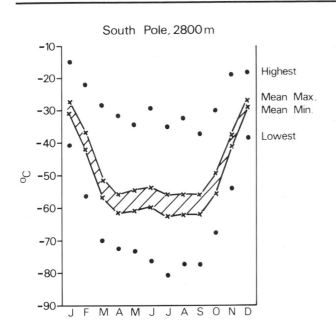

Figure 8.33 Climatic graphs for South Pole. Altitude greatly affects the temperature levels. Precipitation is believed to be low but is very difficult to measure because of drifting snow

sources of warm air are so distant, even warm sectors of the depressions are cool and warm fronts much less significant in the climatology of the Antarctic than in the northern hemisphere.

The relative absence of depressions over much of the continent means that cloudiness is low. This affects the radiation exchanges. In summer radiation input is high because there is little cloud to reflect the insolation. In winter the absence of cloud and sunshine means extensive long-wave radiational losses and a large fall of temperature. From this we might expect hot summer and cold winters, but much of the surface is ice with a high albedo (>80 per cent). Hence most of the incoming radiation is reflected and lost to the surface. The high emissivity of snow and ice assists the loss of long-wave radiation and with little counterradiation the surface has a large radiational deficit. Coupled with the high altitude, therefore, this means that temperatures rarely rise above freezing and the melting, such a feature of Arctic summers, does not take place.

As most of the long-wave losses take place from the surface and advection is limited, so the surface usually has the lowest temperatures producing a marked inversion. This is a conspicuous feature of the Antarctic climate. Except for the two summer months it is ever-present over the high ice plateau and frequent elsewhere. Figure 8.32 shows the average intensity of the inversion in winter. The presence of the inversion means that temperature statistics for Antarctica are much lower than that in the free air aloft. Nevertheless, it is at this level that human activity takes place. Figure 8.33 illustrates some of the features of the climate. At South Pole there is an annual temperature range of 30°C but temperatures never rise above freezing. Precipitation is impossible to measure because of drifting snow and so is omitted from the graph. On the coast at McMurdo Sound, daytime summer temperatures do rise above freezing but the warmth is short-lived and night-time temperatures fall below freezing on average.

Another distinctive feature of the Antarctic climate is the wind. Figure 8.34 shows the average pattern of surface wind flow. In essence it resembles a vast katabatic wind (one blowing down a slope) flowing off the ice cap. Because of the Coriolis Force the usual direction is south-easterly rather than southerly. Average speeds are strong: 6.5 m s⁻¹ at McMurdo and 6.2 m s⁻¹ at South Pole. Near the coast, the winds can be funnelled by glacier valleys descending steeply to the coast to give much stronger than average flows. They are strongest when the katabatic wind is supplemented by the pressure gradient wind. At one particularly severe site, Cape Denison, the annual mean wind speed is 19.4 m s⁻¹, with the highest monthly mean speed being 24.9 m s⁻¹. The combination of strong winds, low temperatures and driving snow can produce the worst possible weather conditions on earth.

Conclusions

The regional variation of climate across the earth's surface may appear complex and illogical at first sight, but it is controlled by a few dominant factors: the distribution of solar radiation, the global circulation of the atmosphere and the general pattern of continents and oceans. What we experience as we move from one climatic zone to another are changes in all the interacting components of the climate in response to variations in these factors.

The pattern of the world's climates is not only important in relation to atmospheric processes;

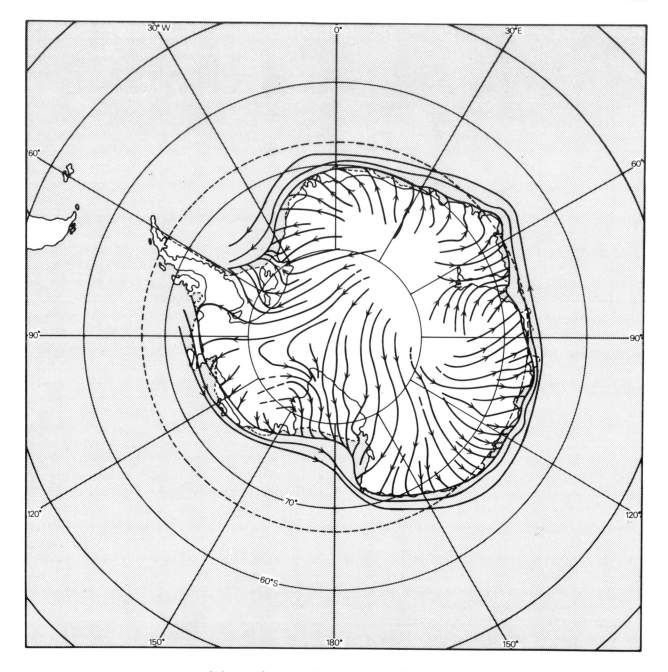

Figure 8.34 Average pattern of the surface wind flow, inferred from predominant wind frequencies at stations and from traverse records (after Mather and Miller, 1967)

many other aspects of the global system show a similar variation, for it is the climate that governs processes of landscape formation, soil development and vegetation evolution. All three vary in relation to climate. Thus, knowledge of the climatic regions of the world will provide an important starting point for looking at other aspects of the environment.

9

Micro and local climates

The climate near the ground

Living as we do in the lowest few metres of the atmosphere, we should have special interest in the climate of this zone. Unfortunately, it turns out to be a very diverse and complicated zone. Climatic differences equivalent to a change in latitude of several degrees can occur in a matter of a few metres. These are examples of microclimatic conditions in the atmosphere.

When the sun is shining, for instance, the ground may become too hot to walk on barefooted, as on a dry sunny beach in midsummer or a black road surface. At the ground surface the temperature may reach above 65°C although at head height it may only be 30°C, and in the shade, where most temperature observations are made, it might be as low as 20°C. Similar variations can be found in wind and humidity. So, we may ask, what is it about this layer near the ground which produces such major differences; differences that are not repeated anywhere in the free atmosphere?

The main reason for this variability is that we are dealing with the main exchange or activity zone between the ground surface and the atmosphere. Energy is reaching this zone from both the sun and, to a much lesser extent, the atmosphere. It is absorbed and then returned to the atmosphere in a different form, or is stored in the soil as heat. This absorption process is very sensitive to the nature of the ground surface. Conditions such as surface colour, wetness, vegetation, topography and aspect all affect the interaction between the ground and the atmosphere. We can see these effects clearly in snowy weather. Clean snow reflects solar radiation and so the surface remains cool and the snow fails to melt. But where the snow is dirty, it absorbs the radiation, heats up and melts. Vegetation, too, may protect the snow from the heat of the sun, while even late in the spring snow may be preserved in sheltered hollows or on hillslopes facing away from the sun.

Let us look at the causes of these differences in more detail. We will start by considering the simplest possible conditions of a bare soil surface.

Microclimate over a bare soil

Many different properties of the soil influence conditions in the thin layer of atmosphere above it. Soils vary in colour, and darker soils, such as those rich in organic matter, absorb radiant energy more efficiently than do light coloured soils. Moisture in the soil is also important. Wet soils are normally dark, but water has a large **heat capacity**, that is to say, it requires a great deal of energy to raise its temperature. A moist soil, therefore, warms up more slowly than a dry one (Figure 9.1).

A further complication in heat transfer into soils is that air is a poor conductor of heat. If there is a large amount of air between the soil particles, heat transfer into the soil is slow. This means that the heat is trapped in the upper layers, so the surface layers warm up more rapidly. Because of this, dry sandy soils get very hot on sunny days. Water conducts heat more easily than air, so soils which contain some moisture are able to transmit warmth away from the surface more easily than dry soils.

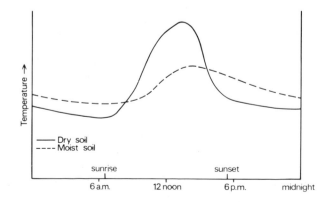

Figure 9.1 Diurnal temperature changes over a dry soil and over a moist soil surface

However, if the soil contains a lot of water, the large heat capacity of the water will prevent the soil warming despite heat being conducted from the surface. For most agricultural crops a balance is needed so that soils warm up fairly quickly at depth and are neither too wet nor too dry. This is achieved when the moisture content of the soil is about 20 per cent.

The nature of heat transfer from the soil

As the ground surface gets hotter through absorbing the sun's energy, the layer of air in contact with the ground becomes warm by conduction. If this was the only mechanism of heat transfer, it would take a very long time before even the lowest 1 m of air was warmed. The daytime maximum temperature at this height would not occur until about 9 p.m. Clearly this cannot be the only process transferring heat, although it is the most important in the lowest few millimetres where temperature gradients are extreme. Above this level, the effect of heating the air causes it to become less dense than its surroundings and so it rises, carrying heat with it. Cooler air then moves in to take its place. This air is heated in turn. Consequently we have **convection currents** rapidly transferring heat to the cooler layers of the lower atmosphere. If there is a strong wind blowing, mixing of heat is encouraged and the temperature profile in the lower atmosphere becomes less steep (Figure 9.2).

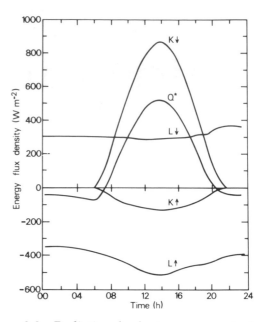

Figure 9.3 Radiation budget components for 30 July 1971 at Matador, Saskatchewan (50°N) over a 0.2 m stand of native grass. Inputs to the surface have been plotted as positive and outputs negative to assist interpretation. K is short-wave radiation, L is long-wave radiation and Q* is net radiation (after Oke, 1978)

The radiation balance at night

At night the radiation balance at the ground surface is negative (Figure 9.3). More long-wave radiation is lost from the ground than is returned as counter-radiation from the atmosphere. This is especially true with clear skies and dry air which allow the radiation to escape to space more easily. In cloudy or humid conditions heat loss is less effective, for water vapour readily absorbs long-wave radiation. During the night, therefore, the surface gets cooler although heat may flow up from lower levels in the soil to maintain surface temperatures. In a sandy soil, with its large air spaces, this process is limited so the surface becomes particularly cool, as anyone who has slept outdoors on beach sand will know. The air in contact with the ground also gets cooler, making the air denser and preventing any thermals of warmer air rising and mixing with the air above.

The night-time profile of temperature during calm conditions is shown in Figure 9.4 and this illustrates the major cooling at the surface. If windy, the cooler air will be mixed with the warmer air above to give a smaller increase of temperature with height. Clouds are efficient emitters or radiators of

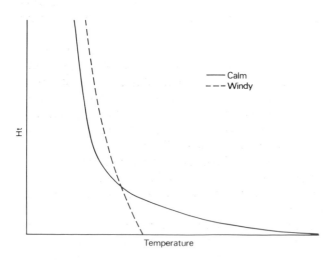

Figure 9.2 Temperature profiles above a heated ground surface on calm and windy days

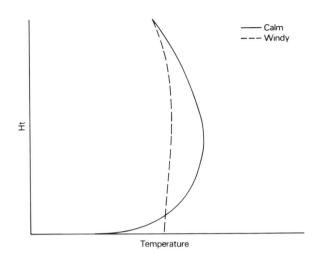

Figure 9.4 Night-time temperature profiles on calm and windy nights with clear skies

long-wave radiation, so low clouds encourage counter-radiation to the surface and the net loss of energy from the surface is reduced. The factors most favouring low surface temperatures at night are consequently clear skies, dry air, no wind and sandy or peaty soil. If these conditions occur at the beginning of the growing season in most temperate latitudes then damage is likely to frost-sensitive crops.

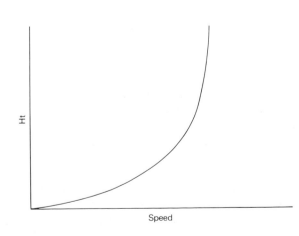

Figure 9.5 Wind speed profile near the ground. The precise form of the profile will depend upon the roughness of the surface as well as upon any buoyancy

Wind near the ground

The profile of wind speed near the ground is somewhat similar to that of temperature (Figure 9.5). As we approach the ground, wind speed decreases very rapidly to almost zero in contact with the soil surface. This is largely due to the frictional drag exerted on the air by the underlying rigid surface; the rougher the surface is, the more it slows down the air (see Chapter 24). Over a soil surface, the effect on the wind is fairly simple, but when we are dealing with a vegetation layer or an urban area, interference is much greater as we shall see later.

The microclimate at a soil surface represents the simplest case of modification of the energy balance by ground conditions. Both the inputs and outputs of radiation are changed and this alters the way the energy is used in terms of sensible and latent heat, and heat flow or storage into the soil. This is illustrated in Figure 9.6 where the energy balances over a wet and a dry soil are contrasted and in Table 9.1 where the thermal differences above the our faces are demonstrated.

Microclimate above a vegetated surface

The nature of microclimatic conditions and processes becomes far more complex when a vegetation cover is present, for not all the energy is absorbed at a single surface. Some is absorbed by the top of the vegetation, some penetrates into the plants, and some may even reach the soil surface. The amount that gets through to the soil depends upon the height of the crop, the density of the leaves and the angle of the sun's rays. As the size of the plants increases, so does the degree of microclimate modification.

Let us look at some of the detailed effects of plants on the microclimate by considering conditions around a single leaf. The amount of short-wave radiation absorbed by a leaf depends upon the quantity of radiation reaching its upper surface, the angle between the leaf and the sun's rays, and the colour of the leaf. Through absorption the temperature of the leaf rises and, consequently, the amount of long-wave radiation emitted also increases. Some radiation is transferred downwards towards the soil, and some flows upwards. With a large number of leaves, the sun's rays are increasingly obstructed by the leaves so the amount of sunlight reaching the ground may be small. The actual quantity depends upon the type and amount of leaves (or **leaf area index**) and the crop height.

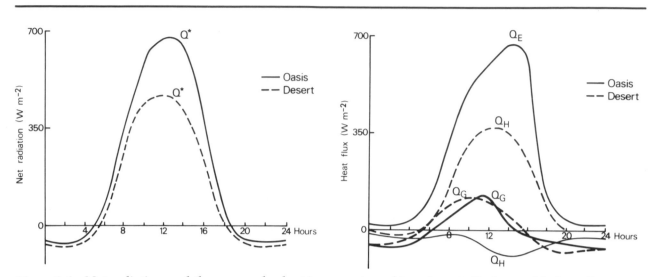

Figure 9.6 Net radiation and the energy budget in an oasis and in a desert. Q_H is sensible heat flux, Q_E is latent heat flux and Q_G is soil heat flux

Table 9.1 *Twenty-four hour diurnal air temperature variation in July*

Height above surface (m)	1	5	7	9	11	13	15	17	19	21
					Irrigated oasis					
2	21.4	18.9	20.7	25.4	30.5	33.2	33.9	33.7	30.0	26.4
25	23.8	20.8	21.8	25.3	30.2	33.0	33.9	34.3	30.9	29.7
50	26.2	22.6	22.5	25.5	30.0	33.1	33.5	34.5	31.6	32.7
100	28.6	25.9	23.8	25.9	29.9	33.0	33.3	34.0	31.9	34.2
					Semidesert					
2	23.0	19.9	23.1	28.4	33.5	37.0	36.7	37.8	33.9	29.4
25	24.5	21.4	23.5	28.1	33.4	35.6	35.3	36.5	33.5	32.2
50	26.2	22.6	23.9	28.2	33.4	35.0	34.7	36.3	33.0	32.9
100	28.6	23.9	25.1	27.8	32.9	34.6	33.9	35.8	32.8	33.1

After I. A. Goltsberg, *Microclimates of the U.S.S.R.* (1969), Jerusalem: Israel Program for Scientific Translation.

Because of its agricultural importance there have been numerous studies of the climate within crops. Agronomists and plant physiologists use this information in order to increase yields from plants best suited to the micro- and macroclimate in which they grow. It is now possible to determine the type of plants growing and to check their health by aircraft or satellite photography. The nature of the radiation reflected and emitted from leaves varies from one species to another and from healthy to unhealthy plants due to alterations in the distribution of pigments in the leaves.

Temperatures in the vegetated layer
If we look at mean profiles of wind speed, temperature and humidity within a plant crop, there is some similarity with those found above a bare soil surface (Figure 9.7). In this instance, the main heat exchange zone is found slightly below the canopy top rather than at the soil surface. As a result, daytime temperatures reach their maximum values within the canopy. The actual location represents a balance between the reduction in sunlight intensity as it penetrates into the crop and the decrease in wind speed and turbulence which would help to remove the heated air. At night under clear skies, long-wave radiation continues to flow from the leaf surfaces, but only that from the upper leaves is able to escape from the plant system. At lower levels in the crop, radiation is

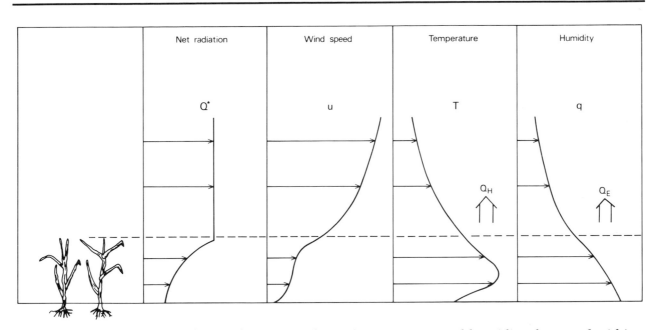

Figure 9.7 Typical profiles of net radiation, wind speed, temperature and humidity above and within a plant canopy (after Saugier, 1977)

trapped and re-emitted, maintaining warmer temperatures. Thus the temperature profile has a minimum value just below the canopy top and gets warmer both into the atmosphere and down to the soil surface. If the crop has a low density, with large gaps between plants, the air cooled by contact with the radiating leaves becomes denser and sinks to the ground to give the minimum temperatures there. In the soil, temperature changes are much smaller because surface heating and cooling are greatly reduced through shading by the leaves.

Wind in the vegetated layer

The wind speed profile is also more complex due to the presence of the crop. Its precise form depends upon the nature of the crop and the prevailing wind speeds. By day, there is normally a progressive decrease in speed as far as the middle canopy. Below this level, most crops have fewer leaves, enabling the wind to blow through the crop. So we get a slightly windier zone before the final decrease towards the soil surface. This effect can be felt behind a hedgerow or wind-break where the stems are not so effective at reducing wind speed as the leafier branches at higher levels.

Moisture in the vegetated layer

Daytime humidity levels usually show a progressive decrease from the soil surface, through the crop, into the atmosphere. Moisture is evaporated from the soil and transpired by the plant leaves, so that the main moisture sources are within the crop. As wind speeds are low, much of the moisture remains within the vegetation, but that in the upper layers may be carried away by convection to mix with the drier air above. At night the shape of the humidity profile is more complicated. Cooling may give rise to dewfall on the upper leaves, producing an inverted profile for a short distance, but normally humidity differences are relatively weak throughout the crop.

Within a plant canopy, moisture exchanges are extensive and of vital importance to the well-being of the crop. In reality these processes are highly complex, but we can get an idea of the exchanges by constructing a simple model of the water balance.

Figure 9.8 shows the inputs and outputs of moisture we might expect for an ideal crop. The major input for most climatic regimes is precipitation either in the form of snow or rain, but hail, dewfall, frost and fog can add small amounts. Some of this moisture is intercepted by the leaves. Depending upon the intensity and duration of precipitation and the nature of the leaf, the water may drip off the leaves or be directly evaporated without ever having reached the ground surface. This effect is greatest when the rainfall is light and the leaf density high – it is much more sensible to

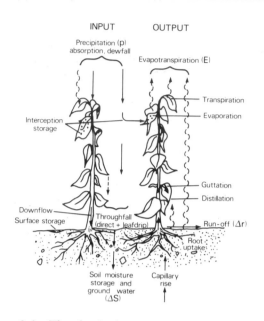

Figure 9.8 The hydrological cascade in a soil–plant–atmosphere system (after Oke, 1978)

shelter from the rain under a yew or beech than a tree of low leaf density such as the ash or Scots pine. Small quantities of moisture may flow down the stems of the plants, but with heavy or prolonged rain some droplets will fall right through the crop to moisten the soil surface, eventually reaching the plant roots (see Chapter 14).

The output from the system is primarily through transpiration from the leaves and evaporation from both soil and leaves. Moisture is extracted from the soil to maintain transpiration, but if it becomes too dry during droughts, the plants may wilt or even die. During periods of rain, input is usually far higher than evapotranspiration alone. This surplus then goes to recharge moisture in the soil or it becomes runoff – the horizontal flow of moisture on the soil – which eventually forms part of the river system. We will examine some of these processes in more detail in Chapter 14.

The microclimate of woodland

So far we have been dealing with the microclimate either on, or at least very close to, the ground surface. From this, we have been able to illustrate the processes controlling the climate at that level. As the crop or vegetation gets larger, so the degree

of modification increases and the active zone extends from the higher canopy down to the soil surface. The extreme example of this effect is seen in the mature forest. So much has been written about the microclimate within a forest that the term **forest climate** is frequently used to indicate the wide variety of conditions that can be experienced.

On a hot summer's day it is noticeable that temperatures in a forest are much lower than outside, providing a respite from the strong glare and baking heat of the sun. Air movement is weak. It feels humid, and the impression is quickly gained of an entirely different climate. This affects plant and animal life as well as man. Quite different ecosystems develop because of the climatic environment produced by the forest. Because of the differences in scale, the microclimates within a forest are more distinct than in grasslands or with low crops.

Radiation exchanges in woodland

It is apparent on entering a forest that the forest canopy cuts out much of the incoming radiation. Most of the energy is absorbed by the tree canopy. A significant proportion is reflected; about 5–15 per cent on average, although in some cases reflection may reach 20 per cent. Only a very small proportion reaches the ground directly, and this is normally in the form of small patches of light called **sun-flecks**. The remainder penetrates the vegetation indirectly; it is reflected, absorbed and re-emitted and arrives as diffuse radiation. During the progress of radiation through the forest vegetation, considerable changes in its spectral composition take place, as specific wavelengths are filtered out or scattered by the canopy. The shorter wavelengths (i.e. blue light) are preferentially removed by the leaves, while amounts of longer-wave red and infra-red radiation increase. This change in the composition of the light is responsible for the characteristic colours that we encounter in woodlands. It also makes the light less suitable for plant growth. As a result, the range of plants that can survive on the forest floor is limited.

The woodland affects not only the inputs of radiation; it similarly affects outputs. The manner of this modification is far more complex, for out-going long-wave radiation comes from a wide range of sources – from the atmosphere, the top of the canopy, from the leaves and branches of the trees, from the undergrowth and from the soil surface. There is inevitably a great deal of interception, absorption and re-emission of the long-wave radiation, so that little escapes directly to space.

Variations over time

These patterns of microclimate are only averages. Considerable variations occur over time due to changes in the inputs of solar radiation and to changes in the woodland itself. If we measured short-wave inputs of radiation throughout the day we would find that levels remained low with the exception of brief periods associated with the development of sunflecks; the peak intensity would occur about midday (Figure 9.9). During the night, cooling is slow for the vegetation traps and returns much of the outgoing long-wave radiation.

This pattern also changes seasonally. In winter the inputs of radiation are low and the effect of the forest on the microclimate diminishes. Moreover, in deciduous woodlands, the trees lose their leaves so that there is much less interception and absorption. If we compare woodland temperatures with those in open land, therefore, we find a much smaller difference in winter. The effect of the woodland is at a maximum when the trees are in full leaf and radiation inputs are high (Figure 9.10).

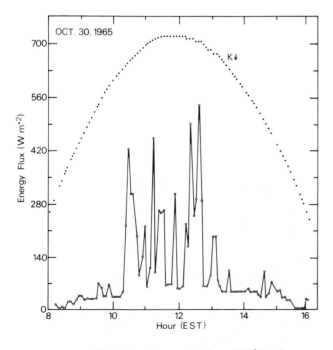

Figure 9.9 Global solar radiation (K↓) above a pine plantation and at one point on the floor (after Gay *et al.*, 1971)

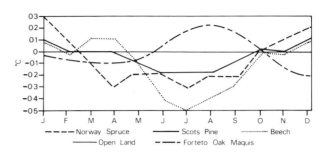

Figure 9.10 Mean monthly forest temperatures compared with thermal conditions in the open. The woodlands are normally cooler in summer and slightly warmer in winter. Anomalies are beech in summer because of the late opening of the leaves and the very dense oak maquis which has little transpiration in summer (after Smith, 1975)

The effects of woodland type

The microclimate of woodland depends very much upon the type of woodland we are dealing with. Deciduous trees show a much more marked seasonal change than conifers, which continue to intercept radiation throughout the winter. But considerable variations occur between different species of deciduous trees. Birch leaves, for example, are smaller and have a lower density than beech or oak, so that even when they are in full leaf, birch trees allow more light to reach the ground. As a result more plants grow on the woodland floor. Similarly, pine trees give a less dense canopy than do spruce; the dark, unvegetated floor of plantations of Sitka spruce contrast with the much lighter conditions in pine woodland.

It is not only the amount of radiation absorbed by the trees that varies with tree species. So too does the amount of reflection from the canopy. Coniferous trees have albedos of 5 to 15 per cent, while deciduous trees may reflect as much as 20 per cent (although the average is about 12 per cent; see Table 9.2). In addition, of course, the nature of the understorey is important. An open canopy allows the development of one or more layers of understorey plants and these, too, intercept both incoming and outgoing radiation. The extreme example is shown by the tropical rain forest. Although radiation inputs are high, the successive layers of trees and bushes and shrubs intercept so much radiation that only about 1 per cent reaches a height 2 m above the ground. Less than 0.1 per cent may reach the forest floor.

Winds in woodland

Patterns of wind in woodland are similar to those in grassland, although the zone of modified flow extends to a much greater height. Above the canopy wind speed may be increased slightly, but, as the canopy is approached, velocity falls rapidly. Lowest windspeeds are often found within the leafy canopy and, where the undergrowth is also dense, velocities may remain low. In most cases, however, the main trunk zone is more open, so there is less interference to airflow and windspeeds increase again. Near the ground, friction and the effect of low-growing plants cause velocity to fall once more. Complex patterns of flow often develop in the forest, with local funnelling and deflection of the wind. We can often see the nature of these flow patterns in the distribution of dead leaves on the woodland floor. Sheltered areas trap deep layers of leaves, while more exposed zones are swept clear by the wind.

Moisture in woodlands

In general, vapour pressure is slightly higher in a forest or woodland than outside it. This is mainly due to the large amount of leaf area in a forest, transpiring moisture into the atmosphere which is not easily dispersed because of the lighter winds. On the other hand, interception of moisture by vegetation reduces the amount of water available at the forest floor so the net effect on humidity levels is small. As daytime temperatures are cooler than those outside, the **relative humidity** of the air should also be greater even if the forest atmosphere contained the same absolute amount of water vapour. Experiments suggest values about 5 per cent above those outside though the precise differences depend upon the type of woodland as well as the time of year and weather conditions (Table 9.3).

Table 9.2 *Tree albedos (per cent)*

Aleppo pine	17	Eucalyptus	19
Monterey pine	10	Sitka spruce	12
Loblolly pine	11	Norway spruce	12
Lodgepole pine	9	Birch and aspen	
Scots pine	9	late winter	25
Oak		Orange trees	32
summer	15	Tropical rain	
spring	12	forest	13
		Cocoa	16

Table 9.3 *Difference of relative humidity (per cent) between the inside and outside of a forest (Positive values indicate that inside the forest is more humid.)*

Forest	January	April	July	October	Year
Deciduous broad-leaf	3.4	3.2	−0.8	1.1	2.2
Needle tree (conifer)	4.8	4.8	6.5	9.5	6.8
Japanese cedar	1.6	−1.1	1.5	0.5	0.8

After M. M. Yoshino, *Climate in a small area* (1975), Tokyo: University of Tokyo Press.

Urban climates

The climate modifications found in woodlands are small compared to what man achieves when he builds cities. Instead of a mixture of soil and vegetation, he covers the earth with a mosaic of concrete, glass, brick, bitumen and stone surfaces ranging to heights of several hundred metres. Among this, he scatters grass surfaces and a few trees to variegate the 'concrete jungle'. The building materials have vastly different physical properties to soil and plants. For example, the warmth of concrete in the evening is due to its high heat capacity. This means that if large quantities of heat are added to the material while the sun is shining it is slowly released during the night, adding warmth to the urban atmosphere. In this way the city temperatures are kept relatively high. We notice the effect most in the evening when we travel from the cool of the open countryside to the heat of the city (Figure 9.11). It is an effect called the **urban heat island**. Early blooming of flowers and decreased snowfall and frost are both indicators of this effect.

The urban heat island

We can illustrate the different responses of the city and rural areas by comparing their heat budgets as shown in Figure 9.12. It is the change of the heat budget by the urban surface which helps to produce the distinctive urban climate, so let us look in more detail at the way changes are produced. By day, both rural and urban areas experience a radiation surplus. Smoky urban atmospheres may reduce the size of this surplus slightly, but as the quality of urban air improves because of pollution controls, the differences in inputs have become slight.

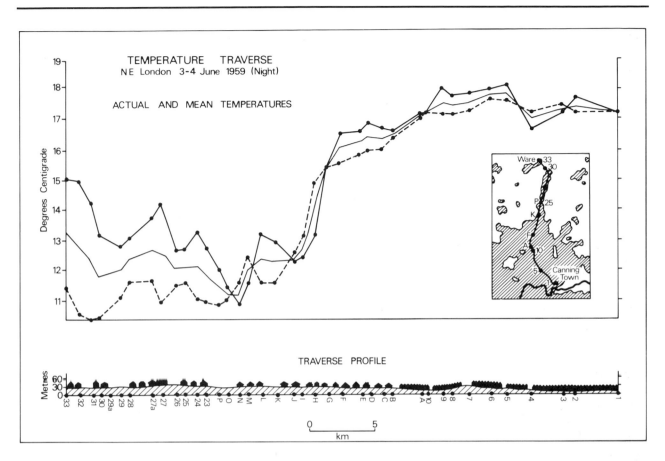

Figure 9.11 Temperature traverse from rural Hertfordshire to central London. The solid line shows the inward journey, the dashed line the outward journey and the thin line is the mean value. The lower diagram depicts relief and building density (after Chandler, 1965)

At a smaller scale, differences are more significant. Trees and crops allow a certain amount of radiation to pass through them to the ground surface. They transpire moisture and have a low heat capacity. As we saw earlier, this results in cooler temperatures beneath the canopy. In the city, the building materials of concrete, brick and stone all have high heat capacities enabling them to store large amounts of heat. Shadowing can be important but there are still numerous surfaces exposing large, dry areas to the sun's rays. When the angle between the receptive surface and the sun's rays approaches 90° the heat input will reach its maximum. This effect is likely to occur much more frequently in an urban area, with its vertical walls, than in a rural area.

Of the energy which is available as net radiation, some is used to heat the air, some in evaporation and the remainder is absorbed by the soil or buildings and other artificial surfaces. This is where the main contrasts arise. In a city, sewers and drainage systems lead to a rapid removal of water. Surfaces soon become dry once rain has stopped, so the use of energy for evaporation is small. This means that more is available for heating the air and the buildings rather than being used for evaporation which is 'non-productive' in terms of heating. Conversely, when rain falls in a rural area, the leaves intercept some and the soil soaks up much of the rest so that there is plenty of moisture around for evaporation. The energy used in evaporation cannot be available for heating the air or soil and so temperatures stay cooler. A final factor is significant in the city. Large amounts of fuel are used there to heat buildings, in industrial processes and in car travel. Even human activity generates appreciable amounts of heat where population density is high and all this heat is eventually released into the urban atmosphere (F in Figure 9.12). On Manhattan Island, research has shown that, in the average January, the amount of

(a) Day

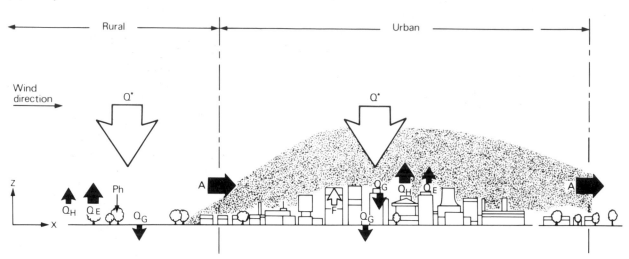

(b) Night

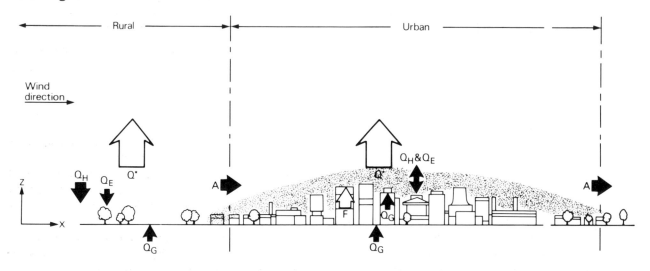

Figure 9.12 Two-dimensional schematic diagrams of the heat balances of urban and rural surfaces by day (top) and by night (bottom). Ph is energy use in photosynthesis and A is net advected energy. See Figure 9.6 for meaning of other symbols (after Fuggle and Oke, 1970)

heat produced from combustion alone is greater than the amount of energy from the sun by a factor of 2.5. In summer this ratio is only about 0.15.

At night, the ground surface loses energy resulting in cooling. In the rural areas, the ground becomes cooler than the air above giving an inversion of temperature. There is then a weak transfer of heat to the surface from the soil and from the atmosphere but these additions do not compensate

for the radiational losses and so temperatures fall. In a hot summer this may feel refreshing compared with the humid warmth of the city. There the buildings continue to give off heat which they absorbed and stored during the day (Q_G in Figure 9.12) and, coupled with the heat of combustion (F in Figure 9.12), this reduces the rate of cooling. We can see this in Figure 9.13 where the cooling rates in Montreal are compared with a nearby rural site.

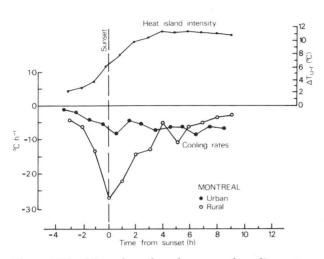

Figure 9.13 Mean hourly urban–rural cooling rates and heat island intensities based on data from summer nights with clear and calm conditions in Montreal (after Oke and Maxwell, 1975)

This relative warmth prevents the development of an inversion so heat transfer and evaporation still take place. Dewfall or condensation is much less frequent than in rural areas. It is this urban heat which many city dwellers find so uncomfortable in the summer; why they long for the coolness of the countryside; and why, irritated by the conditions, they may tend to react violently.

The effect of winds

If winds were strong, all this surplus heat would be rapidly removed from the city to be mixed with the cooler air around, and there would be no such thing as the urban climate. When there is little wind and clear skies, we find the greatest differences between urban and rural areas (Figure 9.14). The pattern of night-time minimum temperatures usually shows highest values near the city centre, fairly uniform levels in the low-density suburbs and then a sharp boundary into the cooler rural areas. This is seen most clearly in cities with few relief features. Valleys, hills and parkland within the urban area can produce major changes. Slopes can give rise to areas of greater heating by day but, by night, cold air may move downslope and accumulate in the valley bottom counteracting any urban effect. The parklands have different heat capacities, albedos and emission temperatures from surrounding buildings giving slightly lower day- and night-time temperatures. The advantages of these 'urban lungs' extend well beyond their aesthetic appeal as

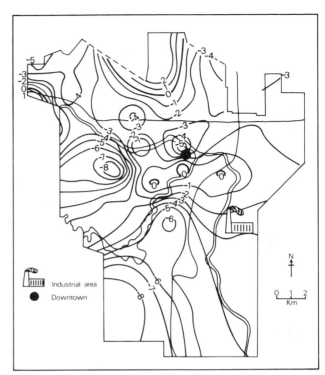

Figure 9.14 Night-time temperature field in Calgary, Canada (Temperatures in °C) (after Nkemdirim, 1976)

shown by the Londoner's love of Hyde Park, especially during hot summer weather.

Even when winds are not light, the presence of the urban structure tends to slow down the air movement. Wind records from city centre sites show lower average speeds than suburban or rural locations nearby, although the degree of gustiness may be higher, especially in summer. As the air flows over the very irregular surface of a city, friction with the buildings retards the wind in the lowest layers (Figure 9.15). The presence of skyscrapers, however, produces eddies (Figure 9.16) which can cause strong local winds. At street-level these can become quite unpleasant, raising dust, perhaps even rubbish, and making walking difficult for pedestrians. Quite a few shopping precincts have been unpopular with shoppers until the architects realized that such winds could be a problem and took measures to prevent them forming.

Cloud and precipitation in cities

Most of the climatic changes brought about by urbanization have been well documented. These are summarized in Table 9.4. Some of the changes are appreciable though the decrease in use of coal as a fuel and energy source should lead to smaller modifications in insolation, contaminants and fogs. The increase in cloud and precipitation over cities was one aspect which took some time to prove. It is only recent work, particularly in St Louis, USA, which has confirmed conclusively the urban effect. There appear to be multiple causes for the increases in cloud cover and precipitation. Added heating by air crossing the city, increases in contaminants, the mechanical effects on airflow and altered moisture all appear to have a role.

The confluence zones induced by these urban effects may even lead to preferential development of clouds and rain. Which factor becomes dominant in a particular storm varies depending upon the nature of the air circulation over the city on that day. As the effects are less noticeable in winter than in summer, it follows that it is the natural, not artificial, heating effects, which are most important. Observations have shown that in some cities precipitation is higher from Monday to Friday when industrial activity and pollution levels are greater than at the weekend. However, contradictory results have also been found so that the role of industrial activity on precipitation is still under investigation.

As the degree of urbanization has increased so an ever greater number of people are affected by an urban climate. Apart from the more obvious effects of pollution, wind and summer heat, few people may realize that their urban area has changed other aspects of the pre-existing climate. The nature of the urban area represents an extreme example of the way in which human modification can change the climate near the ground.

The microclimate of slopes

So far, all examples quoted have assumed that the ground surface is almost flat. In reality there are few areas of the world that are so level that the effect of topography can be ignored. The reason we need to know more about the topography is that slopes modify how much short-wave radiation reaches the surface. We saw earlier that the maximum intensity

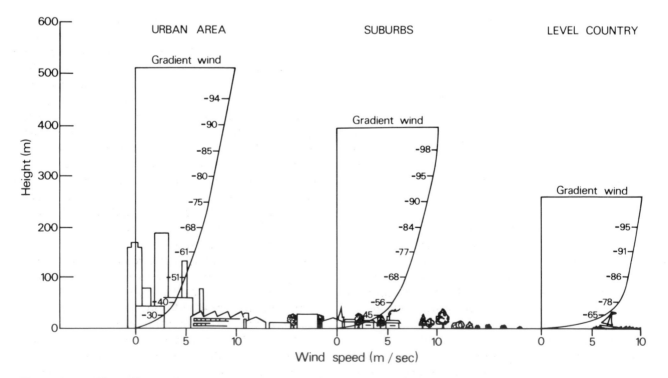

Figure 9.15 The effects of terrain roughness on the wind speed profile. With decreasing roughness the depth of the affected layer becomes shallower and the profile steeper (after Davenport, 1965)

Table 9.4 *Effects of urbanization on climate (Average urban climatic differences expressed as a per cent* of rural conditions)*

	Annual	Cold season	Warm season
Pollution	+1000	+2000	+500
Solar radiation	−22	−34	−20
Temperature (°C)	+2	+3	+1
Humidity	−6	−2	−8
Visibility	−26	−34	−17
Fog	+60	+100	+30
Wind speed	−25	−20	−30
Cloudiness	+8	+5	+10
Rainfall	+14	+13	+15
Thunderstorms	+15	+5	+30

Note
*   Except for temperature.

Source: S. A. Changnon, 'Inadvertent weather modification', *Water Resources Bulletin*, **12** (1976), pp. 695–718.

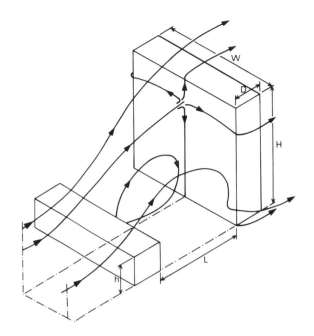

Figure 9.16 Typical flow pattern on the windward face of a building (after Wioc, 1971)

of radiation is received when the angle between the surface and the sun's rays is 90°. If a horizontal surface is tilted so it is at right angles to the rays then the amount of radiation received increases. This factor is exploited by sunbathers who can tilt the angle of their reclining seats to achieve a maximum heat input. If this was the only factor, calculating the new input for a slope would be easy. However, while the slope remains constant in its angle and direction of slope, the sun is continuously changing its position in the sky throughout the day and throughout the year. Slopes, unlike sunbathers, cannot adjust their position. Consequently a slope that receives maximum intensity at one time on a certain day of the year may be in shadow at other times.

Effects on the radiation balance

As the movement of the sun across the sky is known, it is possible to calculate the intensity of short-wave radiation falling on a slope of any combination of gradient and orientation (**azimuth**) for clear skies (Figure 9.17). More frequently we are interested in the total radiation rather than the intensity but even this problem has been overcome using computers. A computer program can be devised to calculate the intensity of radiation on the surface for any particular time and slope. So, for the start of the program, radiation intensity is cal-

culated for sunrise. Then the computer determines the sun's position in the sky, say 10 minutes later, works out the new radiation intensity and then adds this value to the previous total. This is continued until sunset (Figure 9.18). We then have the daily total of short-wave radiation based on intensity values every 10 minutes. The contribution from diffuse radiation is assumed to be constant throughout the day and so does not add to the spatial variability of solar receipt at the surface. None the less it is vitally important for slopes with a northerly aspect which would otherwise receive very little short-wave radiation. Moon explorers are able to see this, for with no atmosphere there is no diffuse radiation and any surface that is not directly in sunlight appears almost black.

These effects of slopes upon radiation inputs mean that the radiation balance varies locally with topography. In the northern hemisphere, slopes with a southerly aspect receive a greater input of radiation than northerly ones, resulting in larger exchanges in sensible heat and higher temperatures (Table 9.5). In high latitudes this additional energy may be an advantage in sunshine-starved areas, but in more arid countries the increased radiation will evaporate moisture more quickly and may produce moisture stresses in cultivated plants.

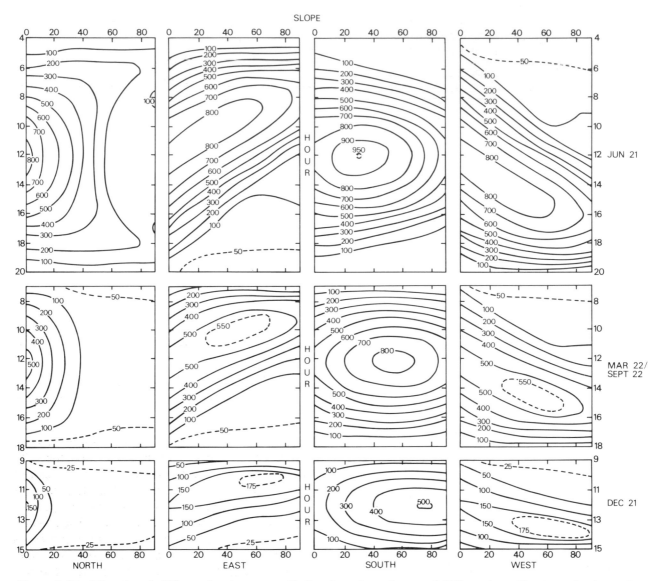

Figure 9.17 Direct and diffuse short-wave radiation input on slopes of different gradients and orientation at the solstices and the equinoxes under clear skies at 53°N. Units in W m⁻². Apart from minor differences due to atmospheric moisture content the values will be the same for 22 March and 22 September. Based on a model developed by the Department of Building Science, University of Sheffield

Slopes at night

At night when there is no input of short-wave radiation, the effect of a sloping ground surface on the energy budget is less pronounced. Figure 9.19 shows the exchanges taking place. For slopes between 0° and 30°, emission of long-wave radiation follows the **cosine law** ($E_{sl} = E_{horiz.} \cos \theta$); at higher angles more radiation is emitted than would be predicted. The only effect of slope direction is in influencing surface heating during the day which, through heat storage, may affect night-time temperatures and hence emission rates. If the sky is obstructed by trees, other valley slopes or even buildings, much of the long-wave emission is absorbed and reradiated back to the ground. This reduces the rate of cooling from the ground and is one of the factors important in keeping city centre temperatures high at night. It can sometimes be seen in frosty weather when open grassy surfaces are white, but, beneath trees or near buildings where counter-radiation has been greater, there is no sign of frost on the ground (Plate 9.1).

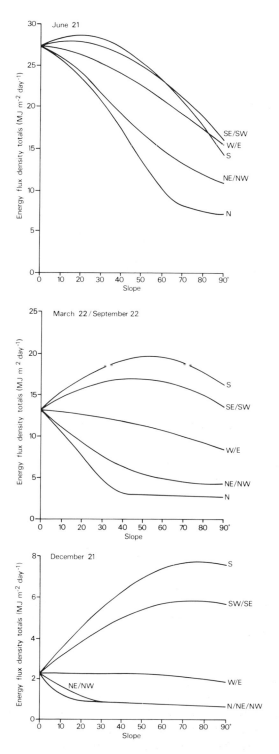

Table 9.5 *Influence of slope orientation on microclimate*

		After five dry days	After two days with rain
Maximum temperature (°C)	N	−1.9*	−1.5
	E	−1.3	0.0
	S	2.6	1.4
	W	0.5	0.2
Minimum temperature (°C)	N	−0.3	−0.4
	E	−0.1	−0.4
	S	0.4	0.3
	W	0.0	0.5
Daily mean temperature (°C)	N	−0.9	−0.4
	E	0.1	−0.3
	S	1.1	0.6
	W	−0.4	0.2
Relative humidity at 13.00	N	8%	1%
	E	3%	5%
	S	−13%	−3%
	W	6%	−4%

Note
* Figures are relative to a horizontal surface nearby.
Source: Translated from Fuh Baw-Puh, 'Influence of slope orientation on the microclimate', *Acta Meteorologica Sinica*, **32** (1962), pp. 71–86.

Of much greater importance at night is what happens to the air as it cools through contact with the ground surface. As the air becomes cooler, it gets denser. If the surface is flat, the cold air remains at ground level. However, on a slope, cool air may move downslope as a **katabatic wind**, increasing in strength and volume until it meets a physical barrier, such as a fence, wall or embankment, or until it is no longer colder than the surrounding air. Once the cold air stops moving it continues to cool through radiation emission and may eventually reach very low temperatures. This microclimatological effect can be very pronounced on clear, calm nights which allow radiational cooling to continue at a high rate.

One result of this process is the formation of **frost hollows**. Farmers should always take care that frost-sensitive crops are not grown where cold air is likely to accumulate and give ground or even air frosts. It is for this reason that in frost-susceptible areas, fruit orchards are cultivated on valley slopes, allowing the cold air to drain through the trees without accumulating. A classic example of a frost hollow was found in the European Alps. A limestone sink-hole with a steep back-wall facing northeast allowed cold air to become stagnant. Figure

Figure 9.18 Total daily direct and diffuse solar radiation incident upon slopes of differing angle and aspect at 53°N for 21 December, 22 March and 21 June. Note different scale for December. Based on a model developed by the Department of Building Science, University of Sheffield

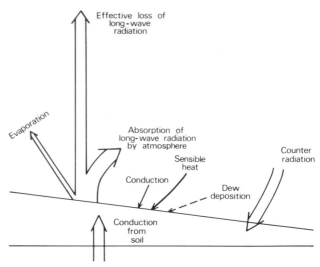

Plate 9.1 Effects of trees on night-time temperatures. The area beneath the tree is free of frost (photo: Peter Smithson)

Figure 9.19 Night-time energy exchanges. The magnitude of the components will vary greatly depending upon weather conditions such as cloud amounts, wind speed and humidity

9.20 shows temperatures at different levels on one particular night. Towards the WSW the sink-hole is intersected by a col which allows the stagnant cold air to remain in the lowest 50 m of the hollow. Temperatures as low as −51°C have been recorded when the ground was snow covered. Even coastal Antarctica is usually much warmer than this!

Valley-breeze systems

If the katabatic winds, described above, are not prevented from flowing then we find they begin to form an organized system of cold air drainage downslope and down-valley. Speeds are low, perhaps 1 m per second or less, and the movement tends to pulsate with intermittent surges – like that which can be seen in water running off a sloping road surface. The downslope flows eventually combine into a down-valley flow, known as a **mountain wind** where they emerge on to the lowlands.

By day, this **cold air drainage** does not occur, except where snow and ice surface maintain cooling. Instead, it is replaced by **anabatic winds** upslope. These are produced by heating on the slope which forces the warm air to rise upslope. Cool air from the valley floor flows in to replace the warm air and a **valley breeze** is generated (Figure 9.21). These valley breeze systems could not last long unless a continuity of the flow was maintained. This is usually found as a counter-wind at

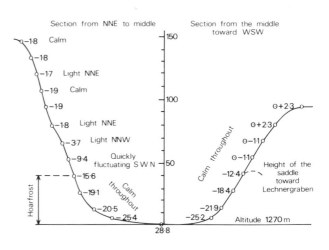

Figure 9.20 Temperature distribution in the Gstettneralm sinkhole near Lunz, Austria, 21 January 1930 (after Schmidt, 1930)

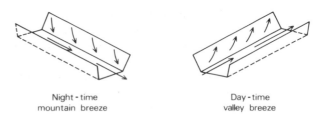

Night-time mountain breeze

Day-time valley breeze

Figure 9.21 Katabatic (left) and anabatic (right) airflow in a valley–mountain breeze system

148

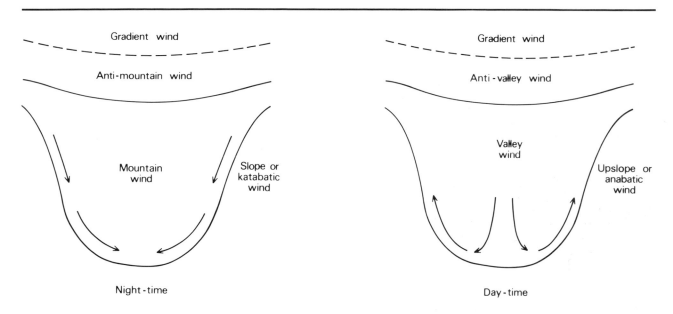

Figure 9.22 Valley cross-section to illustrate the vertical relationships between surface and upper airflow in a valley–mountain breeze system

higher levels (Figure 9.22). If the pressure gradient wind is strong this increases local mixing so that major temperature differences are prevented. No cold air is available to sink downslope nor warm air to rise upslope, so the formation of the breeze is stopped. Like so many microclimatological phenomena, valley and mountain breezes require clear skies and light winds for their operation.

Sea breezes

The driving force for the valley and mountain breezes is a temperature gradient. Temperature contrasts develop between slopes and valley floors, between uplands and lowlands, so that the nature and strength of the wind depends upon the precise form of the gradient. This thermal control of winds occurs at all scales from the general circulation of the atmosphere (Chapter 6) right down to the smallest eddy of heat rising from the ground. We have already referred to one wind system which forms at the local scale, but an even more widespread thermally-driven wind at this scale is the **sea breeze**.

Sea breezes are formed by the different responses to heating of water and land. If we have a bright, sunny morning with little wind, the ground surface warms rapidly as it absorbs short-wave radiation. Most of this heat is retained at the surface although some will be transferred through the soil at a rate

dependent upon the amount of air and water in the soil. As a result of this, the temperature of the ground surface increases and some of this heat warms the air above. When the sun sets, the surface starts to cool rapidly as there is little store of heat in the soil. Thus we find that land surfaces are characterized by high day (and summer) temperatures and low night (and winter) temperatures.

The response of the sea is very different. First, sunshine can penetrate through the water to about 30 m, as any skin diver knows. Second, water has a large heat capacity, so a lot of sunshine has to be absorbed to raise its temperature. In addition, the warming surface water will be mixed with cooler deeper water through wave action and convection. Instead of a thin active layer which we have in a soil, the top 20 m or so of water provides the active layer; consequently temperature changes are slow. Slight warming occurs during the day and slight cooling at night. This means that the sea is normally cooler than the land by day and warmer by night. (On a longer time scale, the sea is cooler relative to the land in summer and warmer in winter unless there are unusual currents offshore.)

The higher temperature over the land by day generates a weak low pressure area. As this intensifies during daytime heating, a flow of air from the cool sea develops, gradually changing in strength and direction during the day. At night the reverse

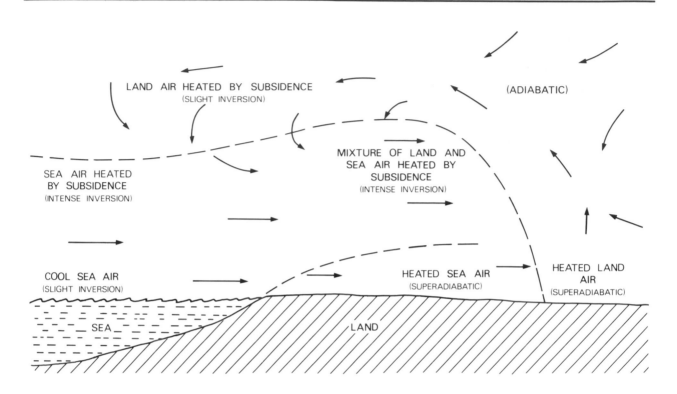

Figure 9.23 Schematic representation of a sea-breeze when the geostrophic wind is light (after Munn, 1966)

circulation evolves, with a flow of air from the cooler land to the warmer sea, though as the temperature difference is usually less and the atmosphere stable, the land breeze is weak. At higher levels, we find a flow in the opposite direction (Figure 9.23) compensating for the surface land or sea breeze. Even large lakes show a breeze system of this nature (Figure 9.24), though in winter the lakes may freeze up so that temperature differences do not develop.

In tropical areas, the strength and reliability of the sea breeze brings a welcome freshness to the climate along the coast, and its effect can extend up to a couple of hundred miles inland.

Conclusions

The atmospheric processes of radiation, convection, evaporation and advection interact between themselves and with the variable nature of the ground to produce a mosaic of microclimates. Distinctive effects can be found at a wide variety of scales in increasing size from the microclimate of a single leaf, through crops, forest, valley slopes, urban areas and sea–land breezes. In most cases, there is not a firm boundary between scales, the micro- and local climates form part of a continuum or spectrum from smallest to largest. Certainly within the larger scales like urban climates there would be innumerable microclimates resulting from surface modifications. This diversity makes their investigation fascinating. Equally it presents problems of explanation and interpretation as it is physically impossible to measure the wide variety of possible microclimates and it is easy for so-called understanding to degenerate into a series of case studies. A final understanding (if there is such a thing!) will only come when we appreciate the interactions and links between the myriad of atmospheric processes and surface conditions.

The importance of microclimate modifications goes far beyond the study of climate, however. It is at this scale that we can see the relationships between climatic processes, landscape and ecosystems. Landforms and vegetation modify the microclimate; the microclimate in turn controls many of the processes involved in landscape and

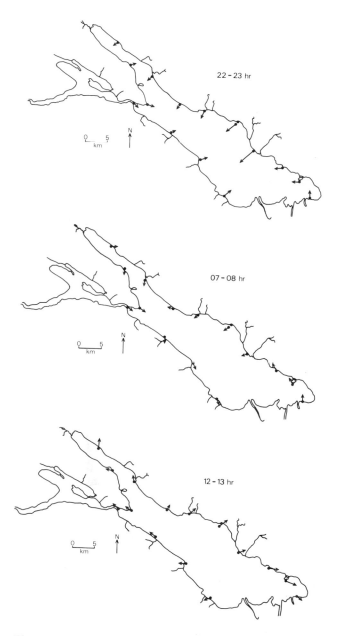

Figure 9.24 The land-lake breeze over Lake Constance, S. Germany. (a) 22–23 hr – and-breeze phase; (b) 07–08 hr – intermediate phase; (c) 12–13 hr – lake breeze phase (after Huss and Stranz, 1970)

soil development and plant growth. Here, as in so many cases, we need to remember that the world does not fall as conveniently into compartments as students (and authors of textbooks) would sometimes like! This may make the study of geography rather complicated, but it also makes it intriguing.

10

Climatic change

When climatology was still in its infancy in the nineteenth century, it was a common belief that climate was a non-changing feature of our environment. By averaging climatic data over a sufficiently long period of time, perhaps thirty or fifty years, it was assumed that the true climate would be determined. As more has been learnt about former climates it is clear that this idea was nonsensical; climate does fluctuate all the time. The evidence of these climatic fluctuations is undeniable, but it comes from a variety of different sources. During the seventeenth century in Britain, for example, instrumental records tell us that winters were much colder and harvest failures not uncommon. Before that, during the twelfth century, historical records show that there was a period of great warmth; droughts were common and famine often resulted. Switch the clock back another 10,000 years and we would be in an age of almost polar coldness. 100,000 years before that there were hippopotami and elephants where London now stands. Thus, we have the proof of historical records and a wealth of geological and biological materials demonstrating frequent, and often drastic variations in climate. The subject of former climates and the evidence used to interpret climatic variations is now so vast that this chapter must be only a brief and relatively simple introduction. Let us start by examining what types of evidence are used and how they can help us to interpret the climatic record.

The evidence

Geological and biological evidence

A wide range of evidence may be used to provide clues about climate, although interpreting it is not easy. By comparing certain landforms and sediments with the type of environment in which they are forming today, it may be possible to infer the climate during their formation. The relationship between landforms and climate is not simple but some landforms can give relatively precise information about former climates (Table 10.1). Permafrost features such as **pingos** only develop where mean annual temperatures are below about −5°C.

Table 10.1 *Some geomorphological indicators of climate*

Landform	Type of climatic information
Pingos, palsen, ice-wedge casts, giant polygons	Permafrost, and through that, mean annual temperatures below −5°C
Cirques	Temperature through their relationship to snowlines
Closed lake basins	Precipitation levels associated with ancient shoreline formation
Fossil dunes of continental interiors	Former wind directions and precipitation levels
Angular screes	Frost action with the presence of some moisture
Misfit valley meanders	Higher discharge levels which can be determined from meander geometry
Aeolian-fluted bedrock	Aridity and wind direction
Tufa mounds	Higher groundwater levels, wetter conditions

After: A. S. Goudie, *Environmental Change* (1983), Oxford University Press.

Similarly, **cirques** can be used to determine former snowlines. In arid areas, we find that sand dunes form only where precipitation totals fall below 200 mm; the prevailing wind direction during their formation can be found from the alignment of the dunes. In the Orinoco Basin of South America, fossil sand dunes can be found indicating that at some time in the past dry, wind-blown material was available in sufficient quantities to form dunes in an area which is now quite wet. Some glacial and fluvial sediments are sufficiently distinctive and related to climate to be used as indicators of former climate. It was from the recognition of fossil **tills** that evidence of early glaciations in southern Africa was first accepted. Another characteristic sediment is **varve clay**, a layered deposit of clay and silt which accumulates in still, lacustrine waters. Varves are a series of alternating coarse and fine layers which are believed to represent an annual input. The relative thickness of the layers is related to the amount of sediment entering the lake, and therefore to the rainfall and ice melt during that year. During the winter period the lake freezes over so that the finest clay sediment is deposited and no coarser material is brought in. The cycle begins again in spring when the ice melts. Over time, the changing frequency of thick and thin layers can give some idea of climatic fluctuations.

More frequently it is the biological material contained in the sediments which is more useful in making interpretations about climate. Of particular use are the remains of pollen from plants. Seed-bearing plants release large quantities of pollen into the atmosphere during their flowering period. The pollen returns to the earth's surface by settling or wash-out in rain and may be preserved if trapped in an anaerobic environment such as lake muds or peat. If we take a series of samples through the deposit, extract the pollen by chemical methods and identify them under a microscope (Plate 10.1), we can build up a picture of the vegetation of the area at the time of deposition (Figure 10.1). The different composition of the preserved pollen for different time periods at our site indicates something about the vegetational history of the area. On the assumption that the plants producing the pollen responded to climatic conditions in the past as they do now, we can draw conclusions about the nature of the climate. The method is not without difficulties. Not all plants produce the same amount of pollen so some species will be over-represented relative to others even if they occupy a similar proportion of the land surface. Not all the pollen will be locally derived, some may have been blown in from other locations where different plant species grow. Suitable sites for preservation are not uniformly distributed so vegetation growing round marshes or lakes is more likely to be preserved. We are also making the assumption that the vegetation is in equilibrium with the prevailing climate which may not have been true for all occasions. Nevertheless, despite its limitations, pollen analysis can tell

us about broad changes in vegetation and, indirectly, in climate as long as the limitations are recognized and understood.

Numerous other fossil remains provide similar evidence. The fragmentary remnants of beetles, snails, plant remains (including seeds, nuts and leaves as well as logs), marine and freshwater molluscs such as foraminifera, ostracods and vertebrates all provide varying levels of information about former climatic conditions.

Tree rings

One of the most widely used and seemingly accurate sources of information on past climates is **tree rings**. Most temperate-latitude trees have an annual growth ring (Plate 10.2), the width of which is influenced by weather conditions during the growing season. By comparing the pattern of rings with climatic data from periods for which instrumental records are available, the relationship between ring width and climate can be determined. This relationship can then be used to ascertain the general nature of climatic conditions for earlier periods.

Tree rings respond to different climatic factors in different areas, of course. In arid areas moisture is often limited, so growth relates to rainfall conditions. In other cases, such as the mountains of California, temperature is more important. More often the ring-widths depend upon a combination of temperature and precipitation but sophisticated statistical techniques have enabled scientists to reconstruct former climates, for example in maritime Europe, on the basis of the changes in ring-widths of oak trees.

The dating problem

One of the problems of using geological evidence of past climatic conditions is that we cannot always date the materials accurately. This is a severe limitation, for there is little use in knowing that, at one site, the climate was once cooler than at present if we cannot say when, nor compare it with other areas. It is sometimes possible to date events relative to each other without any reference to their absolute age; for example, where organic material is found above and below a distinctive inorganic horizon, it can be assumed that the material above the layer is younger than that below the layer but there is no indication of the time period involved.

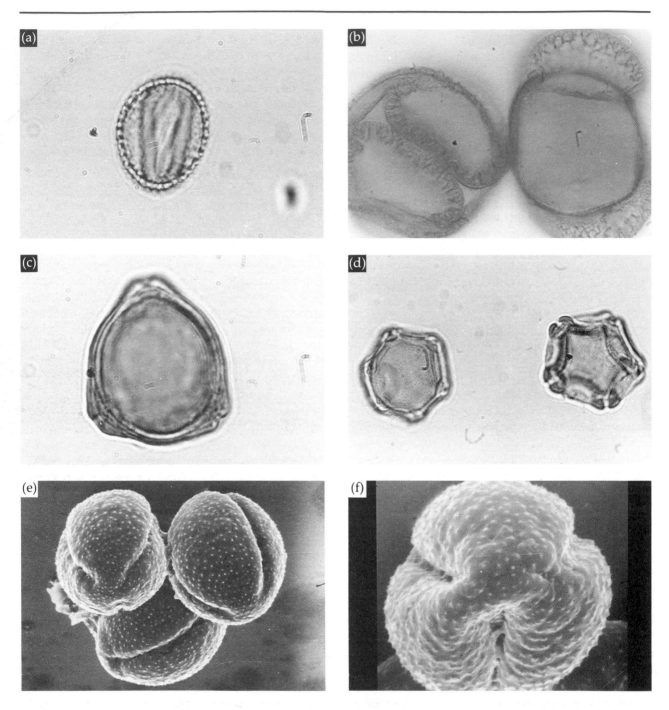

Plate 10.1   Examples of pollen: (a) *Salix* (willow); (b) *Pinus* (pine); (c) *Betula* (birch); (d) *Alnus* (alder) – photographed under a microscope; (e) and (f) *Ranunculus* (buttercup species) – taken by a scanning electron microscope. The length of the pine pollen grain is about 0.003 mm and about 0.001 mm for the willow, birch and alder, and 0.004 mm for *Ranunculus* (photography by courtesy of Department of Botany, University of Sheffield)

Archaeologists have recognized different cultural levels by their artefacts and given them a relative age through the evolution of styles, but this is not very satisfactory or reliable.

Of much greater use is absolute dating where we can give a definite age to material. Two techniques

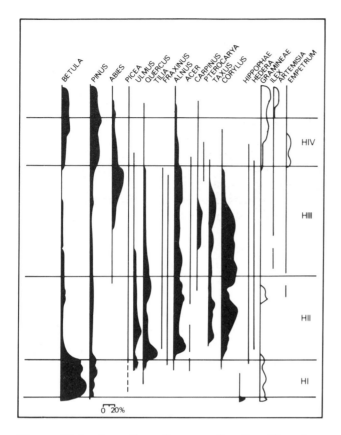

Figure 10.1 A pollen diagram for the Hoxnian Interglacial from a clay pit at Mark's Tey, Essex (after Turner, 1970)

Plate 10.2 Annual growth-rings of a 40-year-old Corsican Pine tree from Bawtry Forest, South Yorkshire (photo: A. Wickramasinghe)

already mentioned have provided methods of absolute dating. The alternation of a coarse and a fine layer in a varve is believed to represent an annual input. Some years, perhaps as a result of more meltwater or rainfall than usual, produce particularly thick bands. Varve chronology depends upon the correlation of these distinctive bands in different areas so that older and older bands can be recognized and counted. A similar technique has been used with tree rings. The science of **dendrochronology** involves comparing the pattern of rings from trees of different ages in order to seek an overlapping sequence which takes us further into the past (Figure 10.2). In this way, using Bristlecone Pine trees from California and their preserved branches, experts have been able to trace and date climatic conditions back more than 8000 years.

This method is time consuming. Fortunately, radioisotope methods have been developed and now provide the main method of dating. Radio-

active substances decay at fixed and known rates. For example, carbon-14, an isotope of carbon, decays slowly so that in 5730 years half the radioactive atoms will have broken down into carbon-12. This method of measuring the amount of radioactive carbon can be used on substances containing suitable carbon atoms such as charcoal, plant material or shells for ages less than about 60,000 years. For older samples, there is too little radioactivity left to be measured.

A number of other radioisotopes can be used based on the uranium series of elements, on potassium or on rubidium. The reversals of the earth's magnetic field also provide a reference series for the last 4 million years. Table 10.2 shows the known reversals dated by the potassium method. This catalogue is useful as the reversals are a global event. Oceanic sediments from different places will have had different rates of deposition but their identification is helped by the additional detail provided by the reversals (Figure 10.3).

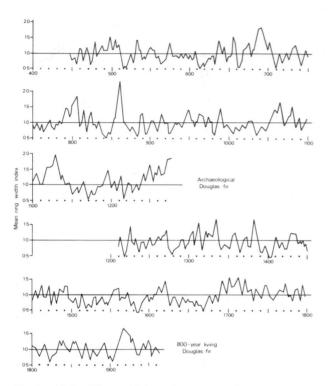

Figure 10.2 Ringwidth index series from Douglas fir trees on the south-western Colorado plateau, 2000 to 2500 m above sea level, from AD 440 to 1964: five year means plotted at the middle year. Ring widths are directly related to precipitation and inversely related to temperature over about thirteen months from June to June (after Fritts *et al.*, 1965)

Table 10.2 *Major palaeomagnetic reversals dated by potassium-argon method*

Brunhes normal epoch	Present to 690,000 years BP
Matuyama reversed epoch	690,000 to 2.43 million years BP
Gauss normal epoch	2.43 to 3.32 million years BP
Gilbert reversed epoch	3.32 to over 4.55 million years BP

Documentary and historical evidence

Weather often makes the headlines in our heavily-populated world, perhaps as a result of excessive rains, severe cold, droughts or heat-waves in some part of the globe. Similar events occurred in the past but our knowledge of them often depends upon the chance factor of documentary records. Many documents have survived from historic times, perhaps the best known being the Bible, and some of these provide contemporary reports of the weather and climate. Unfortunately, many records give a distorted view of what really happened and it is very difficult to discern objective reality for any length of time. Some attempts have been made, such as the analysis of wheat prices in England and wine harvest records in France, but weather is not the only factor determining price or harvest times. Lamb, the well-known climatologist, who has spent much time devoted to this problem, believes that there is only limited scope for deriving climatic information from human history.

The most recent period of our climatic history has the advantage of introducing instruments which provide us with a direct method of measuring climate. By AD 1700, most of the standard meteorological instruments had been invented though the quality of manufacture and the siting of the equipment left a lot to be desired. Much time has been spent trying to improve the quality of early data as rarity enhances their value. As a result of this work we now have average temperatures for central England by season and for the whole year going back to AD 1659 (Figure 10.4). Precipitation measurements dating back to 1677 have survived in a form suitable for analysis, but it was not until 1725 that sufficient data became available for it to be possible to derive monthly values of rainfall that are reasonably accurate, and even then it was for only a few parts of the country. Over the intervening years, instruments and their siting have become standardized, at least within a particular country, so now it is possible to see directly how our climate has varied. Unfortunately, our understanding of the causes of these variations is far less complete.

Nature of changes during the Quaternary

So far in this chapter we have indicated how former climates may be interpreted from a wide range of different types of evidence. In some instances there is abundant information about the type of climate which must have existed and for other periods it is rather sparse, but wherever possible all lines of evidence should be used to reinforce the conclusions. Occasionally evidence may be contradictory

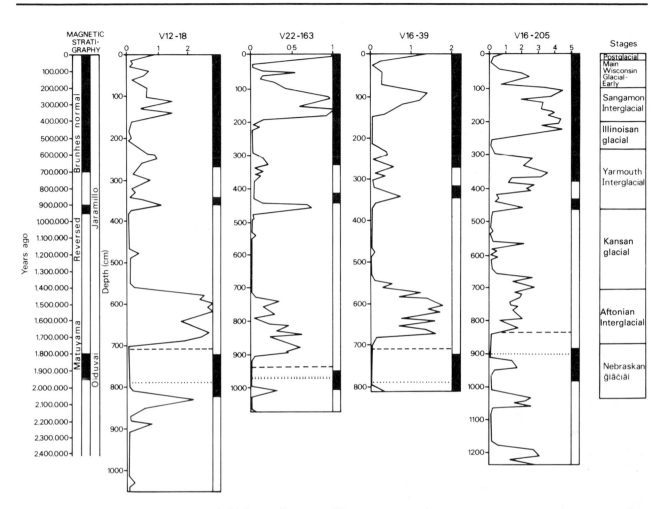

Figure 10.3 Frequency curves of *Globorotalia menardii* species in relation to magnetic and stratigraphic timezones for four cores in the tropical South Atlantic off South America. Scales at the top of the column are ratios of the number of shells of the *Globorotalia menardii* complex to the total number of foraminifera in the sample. Dashed and dotted lines indicate the levels at which certain species become extinct in each core (after Ericson and Wollin, 1968)

but with the large amounts of information about climatic history it is now possible to outline the changes which have taken place.

During the last 3 million years, large parts of the globe have experienced much colder conditions than at present. At times ice sheets have advanced equatorwards over areas which now have a temperate climate. This period of cooling is known as the **Pleistocene Epoch** and together with the recent period since the last glaciation (the **Holocene**) forms the Quaternary Period. Recent studies have indicated that our original ideas on the glaciations of the Pleistocene Epoch were too simplistic. For a long time it has been argued that there were four distinct **glacial** episodes in this period, separated by long,

warmer **interglacial** stages. The names given to these episodes in Britain and their assumed correlations with similar events in other parts of the world are shown in Table 10.3.

It now seems certain that these events were far more complex than this implies. During the glacial phases there occurred relatively warm periods (**interstadials**) when the ice retreated, and cold periods (**stadials**) when it advanced. The interglacials were not long, uninterrupted periods of warmth, but undoubtedly contained cooler phases (Figure 10.5). It is probable that there were many more than four periods in Britain deserving the name glacial, and more than three interglacials. Unfortunately, the evidence of the Pleistocene is far

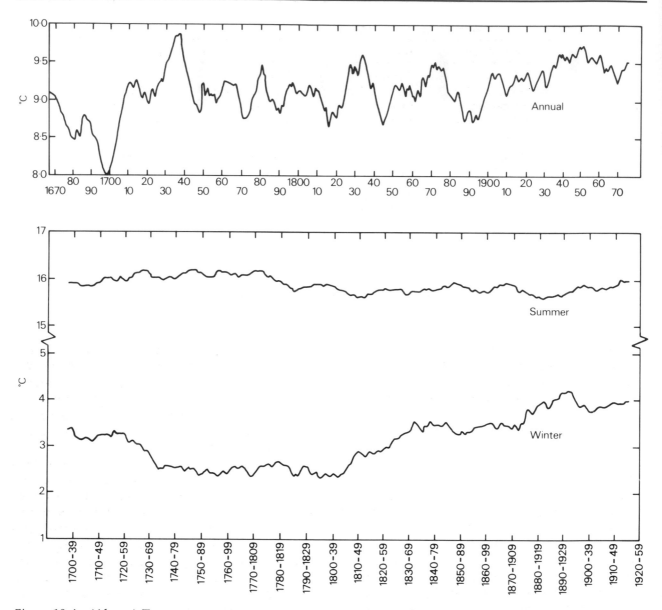

Figure 10.4 (Above) Ten year running means of Central England temperatures from 1650 to 1975 (after Mason, 1976); (Below) forty year running means for summer and winter temperatures over Central England (after Lamb, 1977)

from complete. The deposits laid down by the various ice sheets and glaciers have been eroded, rearranged, often removed; as a result they are difficult to interpret. In addition, deposits from interglacial periods are rare.

Some of the best evidence for events during the Pleistocene comes from deep sea sediments. These show that the early Pleistocene probably involved a long period of irregular cooling during which the polar and mountain ice sheets expanded. As they did so, water was taken out of the hydrological

cycle; less reached the sea and sea level fell. Eventually, possibly about 1–1½ million years ago, the ice sheets had grown large enough to affect the climate significantly, and more rapid expansion occurred as the ice triggered off further cooling. Interestingly, it seems that as the ice caps grew a gradual westward migration of the main centres of ice accumulation took place, as snow was concentrated on the windward side of the ice sheets.

Quite suddenly (in geological terms) this glacial advance turned into a retreat, and the ice melted;

Table 10.3 *World Quaternary nomenclature*

		North America	Britain	N. European Plain	Alps
Upper		WISCONSIN	Flandrian	Flandrian	
		WISCONSIN	DEVENSIAN	WEICHSELIAN	WÜRM
		Sangamonian	Ipswichian	Eemian	Riss-Würm
		ILLINOISAN	WOLSTONIAN	SAALIAN	RISS
Middle		Yarmouthian	Hoxnian	Holsteinian	Mindel-Riss
		KANSAN	ANGLIAN	ELSTERIAN	MINDEL
		Aftonian	Cromerian	Cromerian	Gunz-Mindel
Lower			BEESTONIAN		
			Pastonian		
		NEBRASKAN	BAVENTIAN	MENAPIAN	GUNZ
			Antian	Waalian	Donau-Gunz
			THURNIAN	EBURONIAN	DONAU
			Ludhamian	Tiglian	Biber-Donau
			WALTONIAN	PRAETIGLIAN	BIBER

Note: Glacial periods are printed in capitals; interglacials in lower case. The relationships shown here are constantly under review through further research

sea levels rose; land, which had been depressed by the weight of the ice, rebounded through isostatic recovery. The world entered an interglacial period. But this, too, was temporary. The climate cooled again, and the processes were repeated.

Exactly how many times this happened we cannot tell. At their maximum extent the ice sheets covered some 34 per cent of the globe, extending across most of northern Europe as far south as southern Britain (Figure 10.6) and in America reaching the present Ohio and Missouri rivers. Ice also spread out from the mountain ranges; from the Alps, from the Himalayas, from the Andes and even from the mountain areas of Australia and Africa. The exact relationship between these events in the northern and southern hemispheres remains unclear. It is not certain, although it seems likely, that the glacial advances coincided in both hemispheres. As can be seen, there is still much work to be done in deciphering the glacial history of the Pleistocene.

Glacials

During glacial phases, we can usually tell how far the ice sheets extended by moraines or outwash material. Few organic remains are preserved near the ice front to indicate temperature levels, where the close proximity of ice would ensure that summer air temperatures would be unable to rise much above freezing and in winter very low temperatures would be experienced. Further away from the ice, the glacial climate would remain cold but some

plants grew to give a vegetation assemblage resembling that of the Arctic tundra today. On a global scale, there appears to have been an equatorward shift of the vegetation zones, though compression of these zones also occurred as most of the present desert areas were very arid during the last glacial phase (see Figure 28.10, p. 463).

Temperatures may be assessed by five main lines of evidence: the nature of floral and faunal remains, the extent of permafrost, the limits of frost-affected sediments, the level of cirques and the snowline and isotopic measurements. Although some inconsistencies arise, all indicators confirm a major drop in temperature during the glacials of between 5 and 10°C with the change being even greater in ice-covered areas. A much clearer record is now available for the oceans, where the fossils preserved in ocean sediment cores have enabled scientists to reconstruct the probable sea surface temperatures during the maximum of the last glaciation (Fig. 10.6).

Not all the glacial periods consisted of massive ice advances. During the interstadials ice-covered areas were considerably less extensive although the climate retained tundra characteristics in much of Britain. At times warm summers may have lasted for a long enough period to affect the flora and fauna. For example, about 60,000 years ago, boreal forest existed in the English Midlands and about 40,000 years ago beetle remains suggest a summer temperature at least 5° higher than those in the following glaciation. It is possible that average July temperatures became slightly warmer than those of

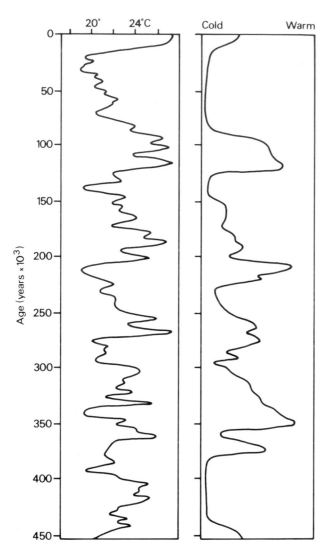

Figure 10.5 (Left) A temperature curve derived from oxygen isotope ratios in the Caribbean showing fluctuations in temperature during the interglacials; (Right) a foraminifera curve from the Caribbean showing less detail of the temperature fluctuations and poorer correspondence with the isotope method for older sediments (after Pittock *et al.*, 1978)

today, though winter temperatures were much colder. The climate would appear to have been less affected by oceanic influences than that of today. A summary of temperature changes during this interstadial is shown in Figure 10.7.

Interglacials

During the interglacial periods, temperatures reached were similar to those of today and in some cases slightly higher. The change from glacial to interglacial appears to have been sudden in most parts of the world, so sudden that vegetational changes do not always accord with the temperature levels indicated by beetles which are able to respond rapidly to climatic changes. During the warm interglacials, vegetational succession in Europe saw a change from tundra through birch and coniferous woodland to mixed oak forest, though each interglacial differed in detail over the timing of when species appeared and which species became dominant. In more northerly latitudes the climate did not always become warm enough for mixed oak forest to grow and coniferous forest or even birch woodland may represent the climax vegetation. In earlier interglacials, the oak forest eventually gave way to coniferous forest before the gradual return to tundra and then glacial conditions.

Rainfall is much more difficult to determine except on the very general scale of dry, moist and wet. In temperate latitudes rainfall does not seem to have differed greatly from that of today, but in the tropics periods of greater moisture have alternated with drier times. Most of the evidence is relatively recent and little is known about rainfall or lake levels in previous interglacials.

In the past, the duration of full interglacial temperatures has been from 10,000 to 15,000 years interspersed with more prolonged cold periods during which the ice sheets expanded. We should not forget that, as our present interglacial has lasted already for about 10,000 years, we may be close to its natural end.

Pleistocene climatic history

In the space available, it would be impossible to cover this topic in detail, so in this section an outline is given of how the British climate has changed during the Pleistocene with most emphasis on the most recent period. The main feature of the Pleistocene has been the oscillation of cooler and warmer climates identified as glacials and interglacials in Table 10.3. Ice sheets did not reach the British Isles during the earlier cold periods; but coolness is indicated by vegetation and pollen types. Later, in the Beestonian, evidence of permafrost is found in sand and gravels. It is not until the Anglian period that we have definite evidence of ice sheets spreading across the British Isles, eventually reaching just north of London. Ice advances also occurred during the Wolstonian and the Devensian though they did not extend so far south. Between the glacial

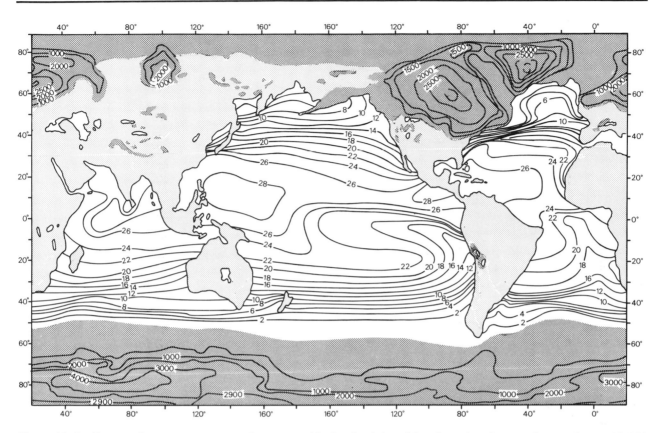

Figure 10.6   Sea-surface temperatures, ice extent (dark stipple) and ice elevation for southern winter, 18,000 years ago with sea-level at −85 m (after CLIMAP Project Members, 1976)

periods, there were brief warm interglacials when mixed oak forests spread across much of Europe. The Cromerian, Hoxnian and Ipswichian periods do have slightly different vegetational assemblages which enable distinctions to be made but they do not appear to have differed significantly from the present interglacial. Many minor fluctuations were superimposed on this basic cycle of glacial and interglacial but evidence from land surfaces is rarely complete; the picture is fragmented. To obtain a more complete picture, marine sediments have to be investigated (Figure 10.3) and related to what happened on land. As we have seen in Chapter 5, the climate over the ocean can be very different from that over land so it is not always possible to use this approach. Not much is known about the earlier glacial and interglacials, and they do not appear to have been greatly different from the most recent glacial and the current interglacial, so we will examine the latter in a little more detail.

At the beginning of the Devensian glaciation about 80,000 years ago, there was a sudden fall in temperatures in most parts of the world. Few types of evidence can be equated directly with temperature but estimates of the actual temperature change indicate a decrease of about 10°C. Organic deposits from Cheshire show a very continental climate in the United Kingdom by 60,000 years ago with moderate summer temperatures but with mean January temperatures in the range −10 to −15°C compared with about +4°C today. There are no sedimentary records yielding continuous information about our climate for this period but material dated by radiocarbon methods has shown other similar spells of summer warmth during the Devensian glacial. Figure 10.7 shows the main peak at about 43,000 years ago when mosses, pollen and insect evidence indicates summer temperatures close to those of today. In some deposits assemblages of both Arctic and more southern species occur together which appears to indicate climatic conditions which do not exist on the earth at the present time. Understanding of this long sequence of strange climate in north-west Europe, spanning the time from 50,000 to about 25,000 years ago, is still a long way off.

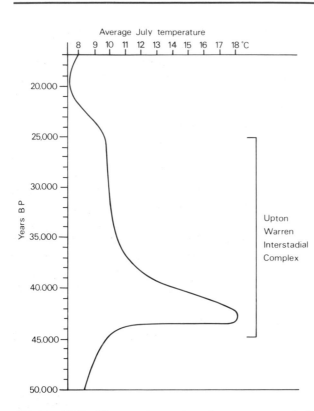

Figure 10.7 Fluctuations in the average July temperature in lowland England during the Upton Warren interstadial complex (after Coope, 1975)

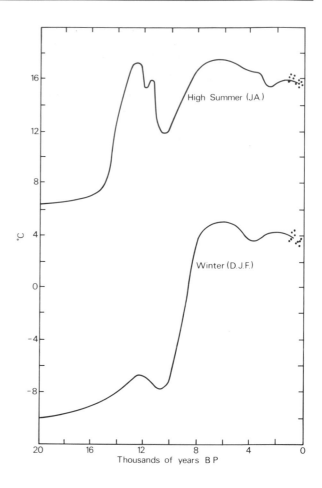

Figure 10.8 Believed course (500–1000 year averages) of the temperatures prevailing in central England over the last 20,000 years as derived largely from pollen analysis. JA = July and August; DJF = December, January and February. Dots show the summer and winter temperatures derived for the last twelve centuries (after Lamb, 1982)

Further cooling after 25,000 years ago led to a general ice advance which reached its culmination about 18,000 years ago. By then, ice extended as a lobe as far south as north Norfolk and more generally to the Humber on the eastern side of the country and Wolverhampton in the West Midlands. The ice margin in Wales is still subject to much debate. The duration of ice at this position was brief. In East Yorkshire, organic material beneath the glacial sediments has been dated to 18,240 ± 250 years BP (before present). The mosses found in this deposit now live in glacial environments so presumably were living there immediately prior to the ice advance. Lake muds above the glacial tills have been dated to 13,000 years old. In the southern Lake District by 14,500 years ago, organic mud was being deposited in some lakes, indicating that the ice sheet was beginning to thin rapidly or had disappeared from this area already.

The change from a glacial to our present interglacial state took place very rapidly though not without fluctuations. Fortunately there is much evidence available about the changes in flora and fauna which accompanied these fluctuations while radiocarbon methods allow us to date these events fairly accurately. The warming 15,000 years ago was so sudden that in many cases the vegetation was out of equilibrium with the climate as deduced from insect evidence. It takes years for trees to spread from their refuge areas, but animals and insects can move more quickly in response to a change in climate as long as their food supply moves too. Insect evidence indicates that in this Late Glacial or Allerød warm period summer temperatures were as high as those of today (Figure 10.8). But the warmth did not last. About 11,000 years ago, temperatures dropped dramatically (Figure 10.8).

Cirque glaciers became re-established in many of the British uplands and in western Scotland ice readvanced towards the lowlands near Glasgow. Mean July temperatures fell below 10°C and trees temporarily disappeared from Britain. This brief period of about 800 years is called the Loch Lomond stadial. It was too short for extensive glacier growth but the piles of gravelly debris in many mountain cirques show the erosion which took place (Plate 10.3).

The Post Glacial Period

About 10,000 years ago after the cold phase of the Loch Lomond stadial, the climate began to improve. Temperatures began to increase which allowed a further immigration of plant species and trees into the British Isles. Space does not permit a detailed examination of the environmental changes which took place, but from the pollen record in lakes and ponds, we can see that alder, elm, oak and lime trees were becoming widespread by 8000 years ago. As these trees grow abundantly throughout much of southern Britain now, it appears likely that the climate of the Post Glacial had improved rapidly to approach that of the present day.

Even this geologically brief period has shown climatic variations, though the size of the swings has been less marked. Figure 10.8 shows some of the fluctuations in summer and winter temperatures in central England based on pollen evidence. Highest temperatures appear to have occurred about 5000 to 7000 years ago. This period of maximum warmth has been termed the **Climatic Optimum**. It appears to have been a worldwide event. Trees extended further polewards and to higher altitudes than now. Some wetter and drier periods have been identified by changes in the character of the vegetation.

Following the Climatic Optimum, temperatures

Plate 10.3 Mounds of glacial moraine deposited during the Loch Lomond Stadial; Cwm Idwal, north Wales (photo: R. Hartnup)

in general have decreased slightly with occasional colder and warmer fluctuations. About 1000 years ago the climate was in a warm, dry phase with few winter storms in the North Atlantic. The Vikings took advantage of this quieter period to colonize Iceland and Greenland and probably visited North America. By AD 1200 cooling set in with increased storminess. Large parts of Friesland in north Holland were flooded after storm damage in 1216. At the same time, drought was starting to affect Indian settlements in Iowa and South Dakota. An increase in strength of the westerly circulation in the northern hemisphere often has this effect on the leeward side of the Western Cordillera of the United States.

The Little Ice Age

After a partial return to more favourable conditions from 1400 to 1550, the climate grew colder again and for 300 years Europe experienced a cold spell. This was known as the **Little Ice Age**. In central England, the mean annual temperature for the 1690s was only 8.1°–about 1.5° below the current figure. Agriculture in upland areas became more difficult as the growing season shortened, leading to an abandonment of many farms whose land often reverted to moorland or rough grazing. An added problem during this period of cooler temperatures appears to have been enhanced variability of the temperature level. This was not merely a swing from one year to another, but a period of several successive years with similar temperatures and precipitation before a change to a period of markedly different character. Other parts of Europe were also affected. Glaciers advanced in the Alps, farms had to be abandoned in Iceland and Scandinavia; in upland Languedoc in southern France there were food shortages and famines associated with severe winters and wet summers. In Spain, agricultural difficulties arose through increased aridity and temperature variability. Elsewhere the extent of snow and ice on land and sea seems to have reached its highest levels since the Late Glacial though the timing of its culmination varies. It appears to have been earlier over the United States and China than in Europe and the southern hemisphere. Much of the evidence for cooling is based on ice advances. However, it is the combination of temperature and precipitation as well as the 'response time' of the glacier which determines advances rather than temperature alone.

Towards the present

By the middle of the nineteenth century, the Little Ice Age was waning and we begin the steady warming to the peak reached in the 1930s and 1940s. During this warm phase the westerly winds were much more frequent over the UK and the strength of the northern hemisphere circum-polar westerly winds was high. This pattern of winds did not bring benefits everywhere; it also coincided with drought in the Great Plains of the USA. Since this peak, there has been a decline in mean northern hemisphere temperatures to the early 1970s, followed by a sudden and rapid warming into the 1980s (Figure 10.9). The impact of a hemispheric mean temperature change of a few tenths of a degree may seem very small but change is not uniform. Some areas experience more significant increases or decreases of temperature, while rainfall patterns are notoriously variable. For example, in the Sahel area of Africa, we can see from the rainfall record that at certain times there have been sequences of higher than average rainfall followed by periods with lower than average rainfall (Figure 10.10). These changes may have significant human impacts. The most recent drought has had serious consequences for the population which had been increasing rapidly during the previous wet spell. An estimated 100,000 people died, largely from starvation, as a result of the drought, and it still continues into the 1980s.

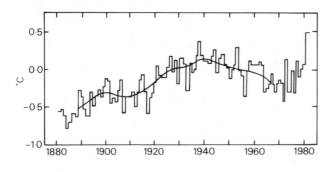

Figure 10.9 Annual means of surface air temperature for the northern hemisphere and for the Arctic (expressed as departures in °C from the 1946–60 period). The smooth curve shows the main trend (after Jones and Kelly, 1983)

Causes of climatic change

One question we need to ask is, Why does the climate vary so much? It is a question which may

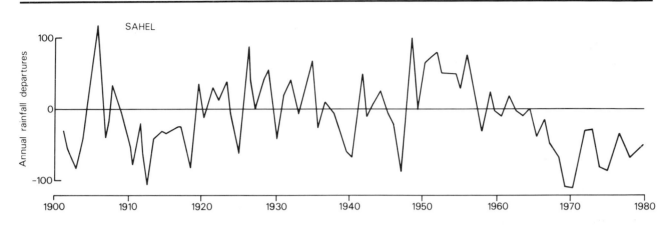

Figure 10.10 Annual rainfall departures for the Sahel, 1900–80. The plotted values are approximately equivalent to a regionally-averaged 'per cent of standard departure' (after Nicholson, 1983)

not have only one answer. There are at least four different time scales which require explanation: glacial/interglacial, stadial/interstadial, post glacial oscillations and fluctuations over the last 100 years. In looking for possible causes we can think of influences external to our planet, of purely internal factors between the atmosphere and the surface, or a combination of both. Let us look at these in turn.

Variations in energy inputs

As the sun is the main source of energy, is there any indication that its output has changed with a consequent effect on our climate? Modern observations of the sun's output have not shown any significant variations and the only feature of note is the periodic absence of sunspots from 1645 to 1715, 1460 to 1550 and 1100 to 1250. In any case, one of these coincides with the peak of the Little Ice Age and another with a warm period. It seems unlikely that we should expect variations of more than 1 per cent in solar output and simple calculations of the earth's radiative balance suggest that even a 1 per cent difference would only lead to a change of 0.6°C in mean annual temperature. Nevertheless, this small figure could be important in climatically marginal areas.

An indisputable cause of variation in solar energy input to the earth's surface is the astronomical relationship between the sun's rays and the earth. The orbit of the earth around the sun is approximately elliptical. The nearest point of this orbit to the centre of the orbit is known as the **perihelion** (Greek *peri* = near + *helios* = sun), and is about 14.71×10^7 km from the sun. The furthest point is known as the **aphelion** (greek *ap* = far + *helios*)

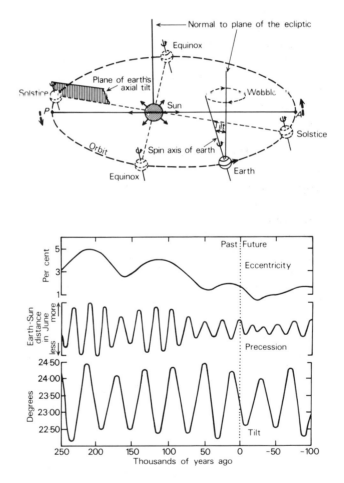

Figure 10.11 (Above) Geometry of the sun–earth system showing the factors causing variation in radiation receipt by the earth (after Pittock *et al.*, 1976); (Below) changes in eccentricity, tilt and precession for the last 250,000 years and the next 100,000 years (after Imbrie and Imbrie, 1979)

and is approximately 15.2×10^7 km from the sun. At present the perihelion occurs on 3 January while the aphelion is on 4 July, but over time the dates change gradually. The eccentricity produces a minor fluctuation in the input of solar radiation, with a maximum of 1400 W m^{-2} at the perihelion and a minimum of about 1311 W m^{-2} at the aphelion. Thus, solar inputs at the extremes of the orbit vary by about 7 per cent. The time of year at which the earth is nearest to the sun changes through time in a cycle which takes about 21,000 years to complete. Its effect is referred to as the **precession of the equinoxes** (Figure 10.11).

An even longer cycle of variation occurs due to slight changes in the shape of the orbit. Over a period of about 90,000 years this varies from almost circular to strongly elliptical, and as a consequence the seasonal variation in solar inputs shows a marked fluctuation (Figure 10.11). Its effect is referred to as the **eccentricity of the orbit**.

The final source of variation in the distribution of solar inputs occurs due to the changes in the tilt of the earth's axis of rotation. Although, at present, the tilt is about 23.5°, the limits vary from about 22° to 24.4°. This means, in essence, that the Tropics of Cancer and Capricorn shift slightly. When the axis has a greater tilt, the position of the overhead sun at midday at the solstices is further polewards by about 2.5° than when the tilt is at its least. Again the change is regular with the full cycle taking about 40,000 years. The variation is sometimes referred to as the **obliquity of the ecliptic** or more simply as the variation in tilt (Figure 10.11).

The variations in the solar radiation at different latitudes of the earth's surface due to these orbital changes were first calculated by **Milankovitch** in 1930. Apart from a few modifications, the basic pattern has remained unchanged and is shown in Figure 10.12. In high latitudes it is the 40,000-year cycle which dominates but at lower latitudes the 21,000-year cycle is more important.

Calculations have been made of the amount of heat that would be available to the different latitudes based on the Milankovitch variations with allowances being made for the amounts of ice existing at each period of time. These show that the orbital variations did have the correct timing and size to start the succession of major advances and retreats of the ice sheets during the last 300,000 years. This is seen most clearly in some of the ocean cores where undisturbed sediments have accumulated over many thousands of years. Fluc-

tuations in temperature as determined from their fossil and carbonate contents do tie in closely with the Milankovitch cycles (Figure 10.13).

The orbital changes take place only slowly. However, sudden changes in climate can occur. For example, in one deposit near Birmingham, UK a typical northern assemblage of beetles was found dated to 10,025 ± 100 years BP. 10 cm higher no Arctic fauna survived at an age of 9970 ± 110 years BP. Conversely the rapid cooling at about 10,900 BP brought a catastrophic readvance of the ice which destroyed fully grown forests, and caused desiccation in Colombia and a marked cooling in Antarctica within a time span of only 200–300 years. It

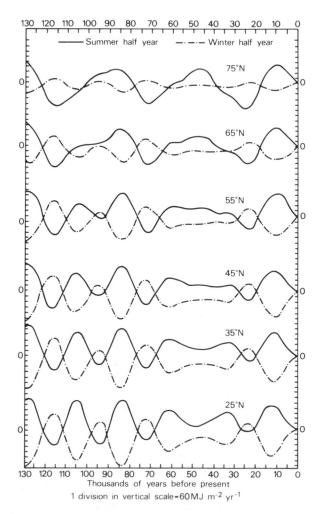

Figure 10.12 The variation over the last 130,000 years of the radiative flux between latitudes 25° and 75°N in the summer and winter half years based on calculations by Milankovitch (after Mason, 1976)

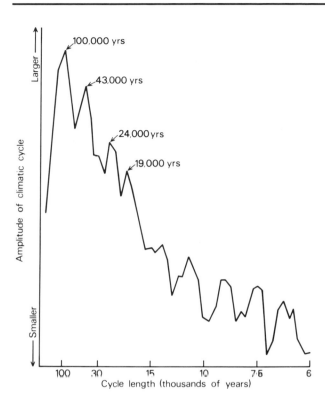

Figure 10.13 Spectrum of climatic variation over the past half million years. This graph – showing the relative importance of different climatic cycles in the isotope record of two Indian Ocean cores – confirmed many predictions of the Milankovitch theory (after Imbrie and Imbrie, 1979)

seems highly unlikely that the orbital variations could be responsible for such sharp climatic fluctuations as described here. For these changes we must look to other mechanisms.

Variations in internal conditions

Feedback processes

Our climatic system consists of several subsystems, such as the atmosphere, the oceans, the ice sheets and the land surfaces. These are all closely related and changes in one may affect the others. Moreover, changes within one of these components may act as positive or negative feedbacks ultimately influencing inputs of solar radiation to the earth. These feedback mechanisms may be responsible for many of the more rapid fluctuations in climate that have occurred throughout the earth's history.

Positive feedback leads to more dramatic and far-reaching changes. The initial effect is magnified, so that quite small changes in the environment produce major adjustments in the system. Perhaps this is why the climate sometimes changes abruptly without any evidence of a clear change in external conditions. Figure 10.14 shows such an effect which has been proposed as a cause of the ice ages. A quite small cooling of temperatures at the poles of only 1–2°C delays the summer melting of the Arctic ice cap. Because the ice survives for longer, the albedo of the surface is high for longer. More incoming short-wave radiation is reflected back to space. Reduced heating of the surface therefore occurs, allowing the ice caps to survive even longer, which increases reflection further which lowers temperatures further. . . . The cycle is self-perpetuating. Once they have been initiated, positive feedback processes magnify the effect of the initial change and cause major adjustments in the system; possibly even an ice age.

Mountain building and volcanic activity

Mountain building and volcanic activity are also responsible for climatic changes. Uplift of mountains is generally slow – too slow to account for the more rapid variations in climate. But its effect can be marked (Figure 10.15). As the surface is raised above the snowline, the extent of snow-covered land increases, altering the earth's albedo. In addition, as we have already seen, major mountain ridges interfere with the circulation of the atmo-

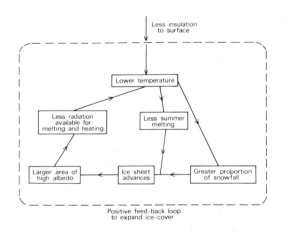

Figure 10.14 A positive feedback loop demonstrating how a decrease in insolation and lower surface temperatures may generate further cooling and perhaps even an ice age

sphere and consequently may affect the climate over a wide area. The changing configuration of land and sea as a result of plate movements may affect ocean currents with similar extensive modifications of climate being possible.

The impact of volcanoes is more immediate. Volcanic activity appears to follow a fairly random pattern which means that occasionally there could be an unusually frequent set of eruptions as, for example, in 1750–70 or 1810–35, and periods when there are few, such as 1913–62 and 1857–77. A volcano can eject large quantities of material into the atmosphere. The particles will absorb and back-scatter part of the solar beam, decreasing the amount of direct radiation reaching the earth's surface, although increasing diffuse radiation. Spectacular sunsets are often seen at these times due to this scattering effect. If the particles reach the polar stratosphere they may remain there for three to five years under average conditions. Several eruptions may lead to a prolonged dust layer in the stratosphere above the ice caps. Some of the dust will be deposited on the ice reducing its albedo, and more long-wave radiation will be emitted by the dust-laden stratosphere, both effects increasing the rate of melting of the ice and influencing weather conditions throughout the polar region.

The effects of man

Air pollution

The period for which accurate instrumental records are available is the period within which man's impact upon climate could have been most marked. Superimposed upon the natural changes of the atmosphere, therefore, are the effects of human activity. Air pollution in particular may have caused variations in temperature.

Smoke and carbon dioxide are two of the most widespread pollutants. They are released by the burning of a wide range of products, especially fossil fuels such as coal and oil. For over 200 years, coal was the main source of energy, and throughout the industrialized world large quantities of smoke and carbon dioxide, as well as heat, were emitted into the atmosphere. Less obvious pollutants are sulphur dioxide, nitrogen oxides and hydro-carbons, from sources such as electricity power stations and vehicle exhausts.

The impact of these pollutants upon the local climate is often straightforward. The particles of

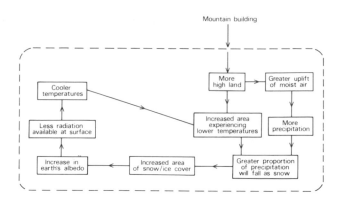

Figure 10.15 A positive feedback loop showing some of the effects of mountain building on surface climate

smoke and dust act as nuclei for condensation, encouraging the formation of fog. In Los Angeles photochemical fogs form as a result of the interaction between strong sunlight and hydrocarbons emitted by automobiles. Instead of being dispersed by winds and mixed with cleaner air, the choking pollutants hang in the still, sheltered air, causing problems for transport and health. Similar dense fogs used to be common in London due to coal smoke. In 1952, the deaths of 5000 people were blamed on a particularly bad fog. Following this, legislation was introduced to restrict the use of coal, and since then the problem has almost disappeared.

Solid atmospheric pollutants also influence the receipt of solar radiation by the surface. Direct radiation is reflected and scattered, so the amount of energy reaching the earth is reduced. We might expect that this would cause a decline in surface temperatures but in reality the opposite seems to happen. The reason is apparently that the pollutants intercept, absorb and reradiate outgoing long-wave radiation, so that more energy is trapped within the lower layers of the atmosphere.

It is not only solids in the atmosphere that affect climate. Gaseous pollutants may also interfere with the inputs of short-wave solar radiation and the outputs of long-wave radiation. Carbon dioxide intercepts little short-wave radiation, but does absorb outgoing infra-red radiation. This again traps energy in the lower atmosphere and raises temperatures. In the period from 1880 to 1940 there was a progressive rise in the mean temperature of the northern hemisphere (Figure 10.9), and this was thought to be a result of the increase in levels of

Table 10.4 *Energy consumption for selected countries*

	Area (10³ km²)	Population (m)	Energy consumption (MW × 10⁶)	Energy consumption density (W m⁻²)	Energy consumption per capita (kW)
West Germany	246	62	0.336	1.36	5.4
East Germany	108	17	0.150	0.91	5.8
USA	7,760	196	1.586	0.24	9.3
USSR	22,400	233	1.380	0.05	4.4
Japan	366	99	0.263	0.71	2.7
UK	242	55	0.295	1.21	5.4
France	573	50	0.188	0.32	3.8
Italy	299	53	0.160	0.53	3.0

Table 10.5 *Changes of heat budget after conversion from forest to agricultural use*

	Albedo	Bowen Ratio (Q_H/Q_E)	Q^* (W m⁻²)	Q_H (W m⁻²)	Q_E (W m⁻²)	Evapotranspiration (mm month⁻¹)
Coniferous forest	0.12	0.50	60	20	40	41
Deciduous forest	0.18	0.33	53	13	39	40
Arable land, wet	0.20	0.19	50	8	42	43
Arable land, dry	0.20	0.41	50	15	35	36
Grassland	0.20	0.67	50	20	30	31

carbon dioxide. Since then, the trend has reversed twice with decreasing temperatures until the 1970s with a subsequent rapid increase. Clearly the explanation is more complex than once believed.

Direct warming of the atmosphere by waste heat also raises temperatures. Estimates of the global energy production have indicated that 8×10^6 MW are generated annually, most of it in densely populated urban and industrial areas (Table 10.4). Long-period temperature records at city-centre sites usually show an increase of temperature through time because of this effect coupled with storage of heat by buildings.

Changes in the surface vegetation

Outside urban areas man's effect on climate has been less marked, but extensive changes in vegetation and land use will have an effect at the microscale and may be important at larger scales. Forest clearance, tillage of grassland and desertification due to soil erosion all increase the albedo of the surface and may alter the earth's radiation balance slightly (Table 10.5). A change in land use from forest to farmland is believed to lead to an increase in the CO_2 content of the atmosphere which again would affect the radiation balance. In tropical areas, where most rainfall is the result of

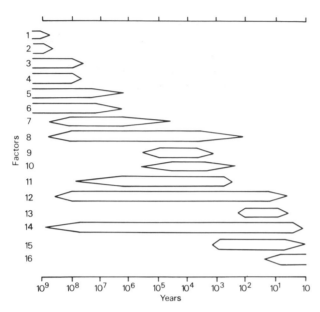

Figure 10.16 Potential causative factors in climatic change and the probable range of time-scale of change attributable to each (after Mitchell, 1968). Factors: (1) evolution of the sun; (2) gravitational waves in the Universe; (3) galactic dust; (4) mass and composition of the air (except CO_2, H_2O and O_3); (5) polar wandering; (6) plate tectonic movement; (7) orogenic and continental uplift; (8) CO_2 in air; (9) earth-orbital variations; (10) air–sea–ice cap feedback; (11) abyssal oceanic circulation; (12) solar variability; (13) CO_2 additions by fossil fuel combustion; (14) volcanic dust in the stratosphere; (15) ocean–atmosphere autovariation; (16) atmospheric autovariation

convection, the reduced radiation absorbed at the surface would result in less heating and so less convection which in turn means less rainfall. We are in the positive feedback situation. Recent work on the prolonged drought in the Sahel region of West Africa has suggested that changes in surface vegetation through overgrazing and land mismanagement are probably crucial factors in the major decrease of rainfall which has occurred (Figure 10.10).

Conclusions

It is clear that in the medium to long term, over the time-scale of tens to thousands of years, our climate varies, not randomly, but systematically. Broad, consistent fluctuations occur, giving periods of rela-

tive warmth and periods of coldness, years of aridity and years of wetness. The reasons for these fluctuations are not clear; variations in the earth's orbit and rotation, changes in solar output, internal adjustments in the vegetation, topography and atmosphere may all be contributory factors. The time scale of possible causative factors is shown in Figure 10.16 but we must not forget that most of the factors are interactive; we cannot isolate a single process and describe its consequences with much confidence. In recent centuries, man too has had an impact on climate.

Two questions remain. What is the effect of these climatic fluctuations? And where are we going now?

Some of the effects are all too apparent to us. In those areas which are marginal for agriculture and human habitation, minor changes in the climate may have an appalling consequence, bringing crop failure, soil erosion and famine. Some of the effects are more subtle, but none the less significant for man. As the pattern of climate changes man tends to move; new areas become favourable for habitation, others become unfavourable. It has been suggested that the stimulus for the Viking invasions and settlement of Iceland, Greenland and Britain was climatic deterioration in Scandinavia. Nomadic tribes today respond to similar stimuli.

The effects of climatic change are not confined to agriculture and man. As we will see in later chapters these fluctuations also influence landscape processes. Throughout the temperate regions of the world, the imprint of past climatic change is clear within the landscape. Glacial landforms lie hundreds of kilometres beyond the limits of the present ice caps; lakes which were at one stage huge inland seas are now small pools by comparison; river valleys that once carried vast torrents of water are now occupied by small, placid streams. Effects of a similar magnitude can be detected in the vegetation. In many areas the range of plants that we find today is a result of migration and mixing of vegetation in response to climatic change. The global system, as we have noted before, is intricately interrelated. Changes in one part affect others, and the effect is nowhere more apparent than in the influence of climatic change.

So to our last question. What will be the climate of the future? Numerous predictions have been made. It has been suggested that the temperate latitudes are entering a phase of cooler climate, possibly another Little Ice Age. It is also argued that we are at present within an interglacial; at some stage in the future we will move into another glacial period. The problem with such predictions is that at the time-scale of our measurements changes in climate are almost imperceptible. Moreover, climatic change involves numerous different trends, superimposed upon one another and each operating on a different periodicity. Consequently, it is almost impossible to tell how long any trend we identify will persist. We can only guess what the future holds.

The global water balance

The hydrological cycle

Raindrops splattering against the window panes or drumming on the roof are such a common occurrence that we rarely give them a second thought. But what happens to all this water suddenly descending from the skies, and how did it get there in the first place? Rainfalls of several centimetres in a day are not uncommon, but can this amount of moisture be held in the atmosphere or does it have to be constantly replenished from elsewhere? Do the vast quantities of water which flow down our rivers each day come directly from precipitation alone, or is there some underground source which

provides a steady flow of water? Questions like this make us think more about the movement of water between the atmosphere, the lithosphere and the oceans. Clearly they must be interlinked, but how does this movement take place and why? In this chapter we shall describe the main features of the hydrological cycle and outline at a general level the main processes and components which comprise the cycle.

The whole system of water movement has been termed the **world hydrological cycle**. In Figure 11.1 we can see the main components of the system, with inputs, outputs, flow regulators and storages, rather like the solar energy cascade which we

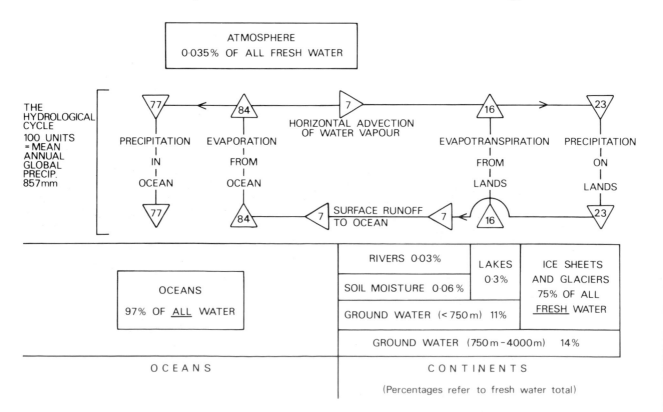

Figure 11.1   The global hydrological cycle (after More, 1967)

looked at in Chapter 3. The main differences here are that the storage components in the system are much larger relative to the inputs, outputs and flows than they were for solar energy. The vast amounts of water stored in the oceans represent an enormous stabilizing influence on the cycle, as do the rather smaller volumes stored as ice in the ice sheets of Antarctica, Greenland and the Arctic Ocean.

The main inputs to the surface hydrological cycle are from precipitation. The main outputs are by evaporation and transpiration. Inputs and outputs are linked by flows in the atmosphere, in the oceans and as rivers on the continents. Storage occurs in the oceans, the cryosphere (ice-covered areas of the world) and the groundwater. Over time the hydrological cycle is more variable than the solar energy cascade, for changes can take place in the amounts of water stored in the ice and oceans, as the ice ages of the recent geological past have indicated. However, on a shorter time-scale, there is an approximate balance between the various components, and no sudden, major changes in ocean content, ice cover or groundwater storage take place.

Inputs to the surface hydrological cycle

Precipitation

Precipitation is by far the most important input to the surface hydrological system. The distribution of this input across the world shows a marked relationship to the distribution of factors influencing precipitation, in particular, the incidence of storms, the atmospheric moisture content and the oceans. Unfortunately it is difficult to be precise about the amount of precipitation occurring in different parts of the world, for data on rainfall are not available everywhere. Over the continents, there is a satisfactory network of recording stations only in the developed countries. Over the oceans, measurements are sparse. No rain-gauges are maintained at sea, so precipitation has to be estimated by indirect

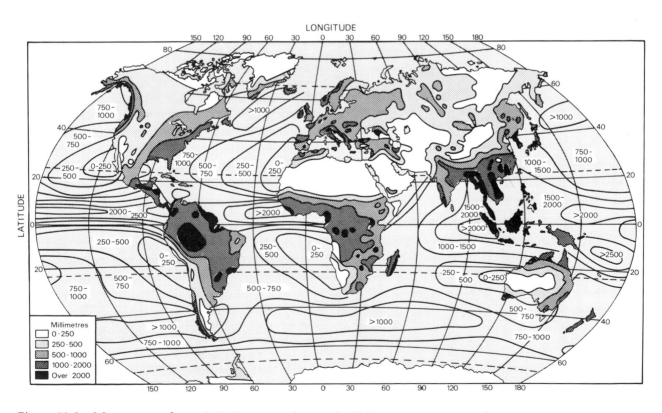

Figure 11.2 Mean annual precipitation over the earth. Values over the oceans are approximations as no rain-gauges exist apart from on islands. Satellite cloud information allows some assessment to be made (partly after Critchfield, 1983)

methods. Formerly, ships' reports were used to give an approximate measure of precipitation, but more recently estimates have been made from satellite images of the clouds. As yet, however, these results have not been incorporated into global figures, so maps such as Figure 11.2 are little more than intelligent guesses for many parts of the world.

The latitudinal distribution of precipitation

Putting the continental and oceanic data together we can obtain a general assessment of the precipitation inputs across the world. Figure 11.3 shows the latitudinal distribution of precipitation. The curves are approximately symmetrical, with peaks of rainfall at 40–50° N and S, and a further peak at 0–10° N. These correspond to the mid-latitude cyclone belts and the equatorial trough zone respectively. Superimposed upon this general pattern, however, is a smaller-scale pattern which reflects the distribution of oceans in the world. Prolonged precipitation is only possible if the moisture in the atmosphere is constantly replenished, and the main source of renewal is the oceans. Thus, rainfall tends to be greater in association with areas of extensive ocean. We can see this by comparing the patterns of precipitation in the two hemispheres; both in the dry

zone between 20 and 30°, and in the wet zone from 40 to 50°, the southern hemisphere receives significantly higher amounts of precipitation. Towards the poles precipitation declines dramatically in both hemispheres, totals becoming almost zero over the South Pole.

This description indicates only the average latitudinal pattern of precipitation inputs. What we must not forget is that precipitation is highly variable in both quantity and character. Especially within the sub-tropics, the inputs of rainfall are very unreliable, and long dry periods may result in extensive drought, interspersed by periods when excessive rainfall occurs, causing flooding and soil erosion. In contrast, many parts of the middle and high latitudes, which receive much less rainfall in total, are blessed with very consistent and reliable patterns of rainfall. From both the human and the geomorphological point of view, variability may be as important as quantity, as we will see in Chapter 12.

Precipitation occurs in a variety of forms. Throughout most of the world the major input is in the form of rainfall, but in the high latitudes, and in many mountain areas, much of it falls as snow. Moreover, significant inputs may occur in the form of fog, dew or rime (frozen fog). Again, the character of the precipitation has an important influence on what happens to the water after it has reached the ground. So, too, does the intensity of the precipitation. We will see in the next chapter that rainfall may take place as gentle, prolonged drizzle or as sudden, intense storms; snow may fall as light flurries or as blizzards. The effect upon the hydrological cycle, upon geomorphological processes, and, above all, upon man are almost always greater when precipitation is intense.

Outputs from the surface hydrological system

Evaporation

Evaporation is a much less dramatic process than precipitation. The only visible evidence we have of it taking place is the gradual drying-up of the surface – rapidly in the case of roads, rather more slowly in the case of the soil. Despite this, evaporation is one of the major outputs from the earth's surface (Figure 11.4).

The rate of evaporation depends upon a number of factors. Most important is the supply of energy,

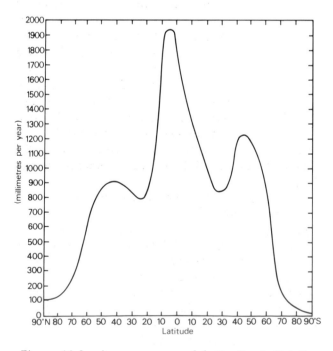

Figure 11.3 Average annual latitudinal distribution of precipitation (after Sellers, 1965)

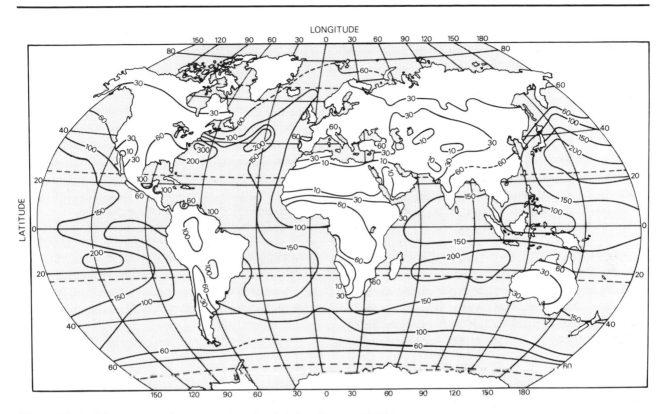

Figure 11.4 Mean annual evaporation (cm) (after Barry, 1969b)

for evaporation involves the conversion of water to water vapour and this requires considerable inputs of energy. Another important factor is the availability of moisture at the surface. As the surface dries out and moisture becomes less available, rates of evaporation tend to decline. Thus, all else being equal, wet surfaces, such as oceans or lakes, result in higher rates of evaporation than do relatively dry surfaces. In addition, evaporation is favoured by a moisture gradient between the surface and the air above, and thus rates decline when the atmosphere is moist. Finally, wind plays an important part by removing the moist air and maintaining a moisture gradient.

The global pattern of evaporation reflects the interaction of all these factors. As we have seen, energy inputs are greatest in the tropical and equatorial areas, and we might expect that these would be the zones of greatest evaporation. But it is in these areas that the atmosphere is often most humid; clouds build up and intercept the incoming radiation, restricting the amount of evaporation. In many of the sub-tropical continental areas, energy inputs are also high and humidity low, but so is rainfall with the result that there is insufficient

moisture to maintain high levels of evaporation. As a consequence, we find the greatest rates of evaporation in the sub-tropical ocean areas; here there is an open water surface, plenty of sunshine because of the low cloudiness, and the trade winds to remove the moisture from near the sea surface. Some of the moisture evaporated in these areas can be seen in the characteristic trade wind cumulus clouds, but much of the remainder is not visible. It stays in the form of vapour until it reaches the equatorial trough zone where it contributes to the cloudiness and precipitation that typifies the humid tropics.

We might expect that negligible amounts of evaporation occur from the final component of the hydrological system – the cryosphere – yet this is not entirely true. One of the processes by which the ice sheets decay is direct evaporation of ice, a process known as **sublimation**. This takes place mainly at the margins of the ice sheets, where high seasonal inputs of solar radiation supply sufficient energy to convert the ice crystals to water vapour. Today this process accounts for no more than 2 per cent of the moisture return to the atmosphere, but in the past it was considerably more important.

During the latter part of the Ice Age, when the extensive ice sheets that had invaded the temperate latitudes were stagnating and decaying over an area of thousands of square kilometres, sublimation must have been much greater.

The details of the precise distribution of evaporation across the world remain obscure. The data from which the precipitation maps were drawn may have been unreliable over the oceans, but in the case of evaporation they are no more than intelligent guesses almost everywhere. Unlike precipitation, evaporation is not something we can easily measure; we cannot use an inverted rain-gauge and hope to catch the moisture being evaporated! Because of this, evaporation values are usually calculated on the basis of the meteorological factors which control evaporation – net radiation, wind flow and the moisture gradient. These estimate the **potential evaporation**. The rates of actual evaporation are almost impossible to monitor for they depend upon the state of the ground and the amount of moisture in it. Data on such detailed conditions are rarely available.

Transpiration

Over vegetated land, much of the moisture is returned to the atmosphere not by evaporation from the ground surface but by **transpiration** from the plants. In reality it is often difficult to separate

Plate 11.1 Mist forming over the tropical forest of Sarawak as a result of moisture transpired from the trees condensing in the surrounding humid air. Visibility is poor because of the high humidity (photo: J. Rose)

the two processes; they are both similar in mechanism and they often act simultaneously. By transpiration we mean evaporation from the leaves of plants and, where the surface is fully covered by vegetation, such as in a forest of a well-grassed pasture, this is the main process by which water is lost. We can sometimes see the effects of the process. On cold spring or autumn mornings, temperatures may fall low enough for the air in immediate contact with the leaves to be close to saturation. As moisture is liberated from the plant by transpiration it is not evaporated, for the air is too humid. Instead, it forms a small bead of water on the leaf, called a guttation drop. Similarly, on humid days, we may see clouds of water vapour rising from trees as moisture, released by transpiration, condenses in the moist air (Plate 11.1).

Nevertheless, it is often more convenient to consider evaporation and transpiration as a single process. We then combine them to give the composite term **evapotranspiration**.

Internal transfers

Between the input of water in the form of precipitation and its output in the form of evaporation or transpiration, a great deal may happen. Extensive movements of water take place, in the atmosphere, on land and in the oceans. In the atmosphere these horizontal transfers are known as **advection**. On land they involve the flow of water in rivers and ice sheets. In the oceans they occur as part of the oceanic circulation. Together they act to redistribute the water, so that inputs and outputs are kept in balance.

The routes that the water reaching the ground may take before being returned to the atmosphere are shown in Figure 11.5. The pathway followed in any particular instance depends to a great extent upon the area we are dealing with and the form in which the precipitation occurs. In the higher latitudes and in many mountain areas, for example, the main input is in the form of snow. This may accumulate over time, slowly being compressed into ice. As ice it may slowly move under gravitational forces as a glacier or ice sheet. Eventually the water is released by melting, to form streams and rivers which then flow to the sea. In addition, many large ice sheets in the Arctic and Antarctic terminate in the sea and release their water directly into the oceans as huge icebergs (Plate 11.2).

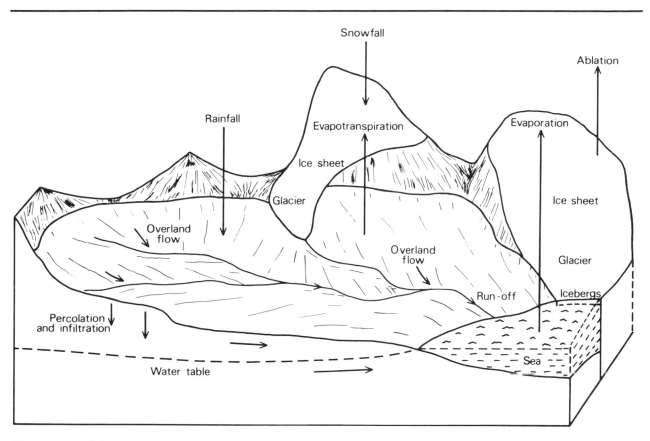

Figure 11.5   Schematic diagram illustrating the main components of the hydrological cycle

Plate 11.2   Icebergs calving off the Antarctic peninsula. The iceberg summit is about 5 m above water level (photo: J. D. Hansom)

In more temperate areas the route taken by the water is more complex, but the processes of transfer are much more rapid. Most of the precipitation occurs as rainfall and this either collects on the ground surface or soaks into the soil. The water which remains on the surface tends to flow over the land as runoff. This rapidly reaches the streams and is then carried to the sea. On the whole, channel flow is rapid. It takes water about three days to travel from the source to the mouth of the Thames, a distance of some 350 km. It probably takes about eight weeks for it to pass the 6300 km length of the Mississippi. Thus the streams represent one of the most dynamic, important – and also sensitive – routes by which these transfers of water from land to the oceans occur.

The water which soaks into the ground travels more slowly. Some of it flows gradually through the soil and ultimately emerges as seepage water in springs or in stream banks. Some of it drains downwards into the bedrock and there enters the almost immobile groundwaters, where it may be stored for many years before emerging again at the surface.

Flows in the ocean are less apparent and less well understood. On a global scale we can see them only as general patterns of movement in the three-dimensional oceanic circulation. But they also operate on a smaller scale. The action of waves and currents in coastal areas involves a constant transfer of water towards the land with a return flow seawards. Often, between the two, there is a lateral flow, giving a cell-like circulation of water. The nature of these local flows is determined by the direction and strength of the winds and the morphology of the coastline. The character of the broader oceanic circulation is related to the pattern of atmospheric circulation, the distribution of solar radiation inputs and the shape of the oceans (see Figure 15.13, p. 240).

Runoff

The most obvious process of water movement in the hydrological system is surface runoff: the flow of water over the ground in distinct channels. The process is easy to follow during a period of heavy and prolonged rainfall. Initially the soil may be dry and most of the rain may soak into the soil, but as the surface becomes wetter or where it is impermeable, signs of running water start to appear. Small depressions fill with water, spilling over as the water flows downslope. Tiny channels called **rills** are cut in the soil and these are gradually

Plate 11.3   A Landsat image of Lake Eyre, Australia taken on 15 February 1973 after heavy rains had almost filled the lake. Darker areas show water; light areas are salt deposits (photograph by courtesy of CSIRO, Canberra)

enlarged as they merge to form more permanent gullies. If we follow the gullies we find them entering stream channels, and if we follow these we see them joined by other streams, forming rivers that flow to the sea.

Not all rivers reach the sea however. On the way the water is evaporated, and in drier parts of the world the streams may dry up before they get to the sea. Often they gather in the wetter mountain areas and spread out on the flat arid plains at the mountain edge where much of the water soaks into the ground. In other cases, the waters flow into large inland depressions where they are trapped, producing permanent inland lakes such as Lake Eyre in Australia, Lake Chad in Africa or the Caspian Sea in the USSR (Plate 11.3).

The global distribution of runoff
Looked at on a global scale, the distribution of runoff is seen to be very unequal (Table 11.1). Some of the continents receive large amounts of rainfall

Table 11.1 *Continental runoff (mm yr⁻¹)*

Europe	300
Asia	286
Africa	139
N. and C. America	265
S. America	445
Australia, New Zealand and New Guinea	218
Antarctica and Greenland	164

After M. I. Lvovitch, *Symposium on World Water Balance* (IASH–Unesco, 1972).

and experience plenty of runoff; others are relatively dry and surface runoff is rare. The areas having most runoff are those with high rates of precipitation and low rates of evaporation. Thus, South America has the largest value, mainly because much of the continent lies in the humid tropics. The temperate latitude continents of Europe, Asia and North America all have similar amounts of runoff. Africa has a low ratio of runoff to evaporation because of the large area of desert across the Sahara and the smaller proportion of the surface with tropical rain forest. The driest continent is Australia, where large permanent rivers are scarce. Antarctica is rather an unknown quantity. Evaporation must occur but is very small. Presumably the snow which accumulates on the continent is eventually returned to the oceans through the calving of icebergs. As this process is very slow, the *annual* runoff rate is likely to be small.

Infiltration and percolation

We only have to watch what happens during a rainstorm to appreciate that a considerable proportion of the precipitation does not flow across the surface as runoff, but soaks into the ground. This process is known as **infiltration**, and in humid temperate areas of the world as much as 95 per cent of the precipitation may be absorbed by the soil in this way. Once in the soil, the water comes under the influence of a variety of forces, including gravity and the suction forces applied by the small soil particles. Under the influence of gravity the water drains downwards into the soil and downslope. Downslope movement carries the water through the soil into the river valleys, a process referred to as throughflow or interflow. In the valley, the water may emerge to contribute to the stream runoff (Figure 11.5). Drainage through the soil is called

percolation, and much of the percolating water makes its way to the bedrock where it meets a zone of almost permanent saturation. The top of this saturated layer is marked by the water table; beneath the water table lies the groundwater (Figure 11.6). This is often almost immobile, and it may lie dormant in the bedrocks for thousands of years. In some cases, as in well-fissured limestone rocks, the groundwater is more active, however, and it flows through narrow cracks, tunnels and caves to re-emerge as major springs, often only a few hours after it entered the ground. Again, once it is at the surface, the water joins with the channel runoff to flow to the sea. From man's point of view, the water stored in groundwater is highly important, for it provides him with an enormous store of fresh water; about 25 per cent of the world's fresh water supply is held in the ground. Moreover, release of this water is slow and almost independent of rainfall events, so the flow of water from springs and seepages contributes a very steady and reliable supply to the streams. It is largely because of this supply that most streams go on flowing even during long periods without rain.

Movement in the cryosphere

Another important store of fresh water is the cryosphere – the vast ice-covered regions of the world. This contains about 26 million cubic kilometres of water, about 75 per cent of the world's fresh water, or enough to maintain the Mississippi for about 40,000 years! The water enters this huge store largely as snowfall and it makes its way to the sea at imperceptibly slow rates, often only a metre or less per year. As a result, snow falling on the central

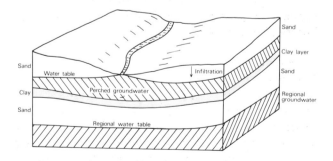

Figure 11.6 The relationship between topography, lithology and the water table. No recharge of the regional groundwater will take place in this situation because the clay layer will prevent percolation to the lower level

parts of the Arctic or Antarctic ice caps may be trapped there for many thousands of years; dating of ice in Antarctica, for example, has shown that some of it collected over 100,000 years ago. Even in the fastest moving glaciers, ice may take several centuries to flow from the source area to the snout, and once released the water may still be trapped in the almost permanently frozen ground which surrounds many of the glacial areas.

Oceanic flows and storage

It is apparent if we examine the volumes of water entering the oceans from surface runoff (Table 11.1) that the oceans do not receive equal amounts of runoff. The Atlantic Ocean collects vast quantities of fresh water from the Amazon, the Mississippi and the Congo. The Pacific is supplied by relatively few major rivers, mainly because of the mountain ranges near the coasts in the American continents, from the Andes right through to Alaska. Only from China does the Pacific get an appreciable contribution. Despite its size, it is the Arctic Ocean which receives the greatest inflow, not from the ice cap but from the northward flowing rivers of the Ob, Yenisei and Lena in Russia and from the Mackenzie in Canada. The vast proportion of this inflow occurs during the summer months, for during the winter most of these rivers are frozen.

If we consider also the input of water through precipitation over the oceans (Figure 11.2) and the losses by evaporation (Figure 11.4), it is clear that some oceans receive far more water than they return to the atmosphere. As the oceans are all interlinked and must keep the same mean level, it is clear that a balance is maintained by the lateral flow of water from areas of excess to areas of deficit. We cannot be precise about the magnitude of these flows, for measurements of ocean water movement are sparse, but as Table 11.2 indicates, large quantities of water must be involved.

Both the Atlantic and Indian oceans, for example, are 'dry' oceans, in that they lose more water by evaporation than they gain by precipitation and runoff. Conversely, the Pacific and Arctic oceans have a moisture surplus, precipitation and runoff exceeding evaporation. To maintain a balance, water therefore flows from the Arctic to the Atlantic and from the Pacific to the Indian Ocean. The net flow into the Atlantic is estimated to be about 0.14×10^6 m^3 s^{-1}, while that into the Indian Ocean is about 0.70×10^6 m^3 s^{-1} (about 100 times the flow of the Mississippi). Of course these are only net

Table 11.2 *Oceanic flows of water*

Atlantic Ocean	$+0.14 \times 10^6$ m^3 s^{-1}
Indian Ocean	$+0.70 \times 10^6$ m^3 s^{-1}
Pacific Ocean	-0.68×10^6 m^3 s^{-1}
Arctic Ocean	-0.16×10^6 m^3 s^{-1}

Notes:
Positive values indicate an inflow.
Negative values indicate an outflow.

For comparison the average Gulf Stream total transport is approximately 70×10^6 m^3 s^{-1}.

flows. In reality water is constantly moving from one ocean to another. For example, there is a considerable surface flow across the Bering Straits and near Iceland into the Arctic Ocean, but this is more than compensated by deeper flows back into the Atlantic and Pacific. Similarly, a two-way flow occurs between the Pacific and Indian Oceans. In the long term, however, these flows maintain equilibrium in the oceanic water budget.

This flow of water within the oceans cannot disguise the fact that the world's ocean basins represent a vast trap for water in the hydrological cycle. The oceans contain about 1,350,400,000 km^3 of water, or about 97 per cent of the total global supply. Not a lot is known about how long the water stays in these immense reservoirs, but it is clear that much of it remains there for thousands of years before being evaporated. It is also clear that the size of this store may vary markedly over geological time. During the glacial periods, for example, release of water from the continents in the form of runoff was slowed down as more and more water became locked up in the expanding ice sheets. The water budget became imbalanced, therefore, and the oceans lost more water than they gained. Sea-levels throughout the world fell; at their extreme they were possibly 100 metres lower than they are today. Imagine the effects of this. The Bering Straits would have become land; Newfoundland would have no longer been an island. The British Isles would have become linked to the mainland of Europe as the North Sea dried up. Similarly, Borneo, Sumatra, and Java would have been joined by land to Malaysia. The whole oceanic circulation would have changed as a consequence. Indeed, the world's hydrological system must have been very different from what it is now.

Atmospheric movement and storage

We started off by talking of precipitation as the main input of moisture to the hydrological system, and this might imply that the atmosphere contains large quantities of moisture waiting to be released. This is far from the truth, however; it has been estimated that no more than 0.035 per cent of the world's fresh water – a meaninglessly small proportion of the total water volume of the world – is contained in the atmosphere. Precipitation, and the continued operation of the hydrological system is only possible if the atmospheric moisture is constantly being replenished by evaporation.

The inputs of moisture to the atmosphere are not everywhere in balance with outputs by precipitation. Over the oceans, for example, evaporation is high and precipitation relatively low; the atmosphere gains more moisture than it loses. Over the continents, even Australia, evaporation is less than precipitation, and the atmospheric moisture budget is negative. As with the oceans, therefore, horizontal flows of moisture (advection) must occur to maintain equilibrium.

We can get some idea of these flows (Figure 11.7) by comparing evaporation and precipitation levels throughout the world. Unlike other diagrams of heat and water budgets this does not show maximum values over the equator. Nor does it show flows in the same direction throughout each hemisphere. What we find is that in the northern hemisphere there is a net transfer of moisture towards the equator in the tropical and sub-tropical areas. To the north of the sub-tropical high, flow is towards the pole. A similar pattern exists in the southern hemisphere.

The sub-tropics, therefore, constitute the main source of moisture for the atmosphere; as we saw earlier it is in this zone that evaporation is at a maximum. The water from this area is carried both equatorwards and polewards. In the mid-latitudes, at about 35°N and S, the quantity of advected moisture is at its greatest. This moisture is then carried by the mid-latitude cyclones into the temperate areas of the world, where it falls as cyclonic rain.

Conclusions

The movement of water through the hydrological cycle is one of the vital processes operating within the global system. A major component of this cycle

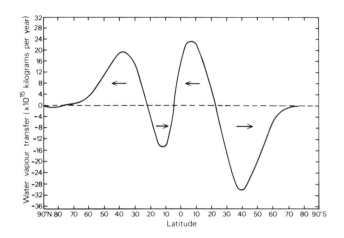

Figure 11.7   The latitudinal distribution of the average annual meridional transfer of water vapour in the atmosphere (after Sellers, 1965)

is the oceans, which cover some 70 per cent of the earth's surface and contain 97 per cent of the world's water. Because the oceans store so much of the total water volume, they act as a major regulator upon the flow of water through the system, influencing in particular evaporation and the return of water to the atmosphere, and thereby controlling to a great extent flows of moisture in the atmosphere and rates of precipitation. Another major control upon the hydrological cycle is provided by the cryosphere. As the events of past Ice Ages have testified, changes in the volume of water stored in the ice caps have far-reaching effects upon the rest of the hydrological system.

Increasingly in recent decades, however, a further regulator of the hydrological system is emerging – man. Man depends upon water for life; he uses it not only for drinking, but in his industry, to feed his crops and to dispose of his waste products. Because he depends upon the hydrological cycle, he also tries to manage and control it. How successfully he does this depends upon how well he understands the system he is dealing with and especially the interactions with other natural systems such as the atmosphere, the lithosphere and the biosphere. For our own sake, therefore, it is vital that we begin to comprehend the delicate processes and interrelationships of the hydrological system.

12

Precipitation

The nature of precipitation

For sixteen months during 1975 and 1976 much of northern Europe experienced prolonged and uninterrupted periods of drought. Total precipitation for this period was only 489 mm at Bournemouth in Hampshire, about 48 per cent of the average. The effects were widespread. Crops and vegetation suffered and many trees died, extensive fires occurred in the dry moors and woodlands, streams dried up, the level of water in reservoirs fell (Plate 12.1) and water use was restricted. In places, drying of the soil led to shrinkage of the ground and buildings and bridges were undermined.

For those of us who live in the humid regions of the world, precipitation is normally so common that we take it for granted. During times of drought the importance of precipitation and its role in feeding the hydrological system become all too apparent. For precipitation represents the input of water to the surface hydrological system. It is the

nature of this input – the character and distribution of precipitation – which we will consider in this chapter.

Types of precipitation

To most people three types of precipitation come immediately to mind: rain, snow and hail. As we would see if we intercepted these and looked at the raindrops, snowflakes or hailstones more closely, the distinction between them is not always clear, and the terms mean different things in different areas. Moreover, they are not the only forms in which moisture inputs occur. Dew, fog-drip and rime all transfer water from the atmosphere to the ground (Table 12.1). Their contribution, however, is often small.

Rain

We have already discussed the main processes of rainfall generation in Chapter 5; we are concerned here with the nature of the rainfall after it has fallen from the cloud.

Typically, rainfall consists of water droplets that vary in size. Where the rain is produced by thin, low-level stratus clouds, droplets tend to be small, with a majority in the range from 0.2 to 0.5 mm in diameter. Where the clouds are thicker, strong updraughts hold the droplets in the atmosphere longer so the number of collisions increases and the rain is composed of larger droplets, often several millimetres in diameter. Droplet diameter is also affected by events during its fall through the atmosphere. Further collision and coalescence cause them to grow in size, while evaporation makes them smaller, and may cause the total loss of very tiny droplets. In general, however, the droplets reaching the ground show a logarithmic size distribution, with a large number of small droplets and a much smaller number of large drops. The details of the size distribution vary according to the rainfall intensity, as Figure 12.1 indicates.

The size of the droplets has considerable signifi-

Plate 12.1 Nant-y-moch Reservoir, mid-Wales towards the end of the 1975–6 drought (photo: Peter Smithson)

Table 12.1 *Types of precipitation*

Type	Characteristics	Typical amounts
Dew	Deposited on surfaces, especially vegetation; hoar frost when frozen	0.1 to 1.0 mm per night
Fog-drip	Deposited on vegetation and other obstacles from fog; rime when frozen	Up to 4 mm per night
Drizzle	Droplets <0.5 mm in diameter	0.1–0.5 mm hr^{-1}
Rain	Drops >0.5 mm in diameter, usually 1–2 mm	Light <2 mm hr^{-1} Heavy >7 mm hr^{-1}
Hail	Roughly spherical lumps of ice 5–>50 mm in diameter, often showing a layered structure of opaque and clear ice in cross-section	Highly variable
Snowflakes	Clusters of ice crystals up to several cm across	
Granular snow	Very small, flat opaque grains of ice; solid equivalent of drizzle	
Snow pellets (graupel or soft hail)	Opaque pellets of ice 2–5 mm in diameter falling in showers	
Ice pellets	Clear ice encasing a snowflake or snow pellet	
Sleet (UK)	Mixture of partly melted snow and rain	
(USA)	Frozen rain or drizzle drops	

cance for, together with the strength of updraughts in the air, it controls the fall velocity of the rain. In still air, the fall speed of a droplet 0.2 mm in diameter is about 70 cm s^{-1}; for a drop of 2 mm diameter it is about 650 cm s^{-1}. The momentum of the droplet when it reaches the ground is known as the **terminal velocity**, and, with the mass of the drop, determines its kinetic energy.

The total kinetic energy of a storm depends upon the number of raindrops reaching the ground. This is a measure of the rainfall intensity and in Figure 12.2 the relationship between the intensity and total kinetic energy of rainfall is illustrated. Rainfall intensity varies considerably both within an individual storm and between storms. Rainfall from thick cumulus type clouds is particularly variable due to spatial differences in cloud thickness and updraught strength, but intensities may be as high as 200 mm hr^{-1} or more. Precipitation from stratiform clouds is less variable and intensities are often low – less than 10 mm hr^{-1}.

Snow

In most areas of the world, rainfall is by far the most important input to the surface hydrological system. Snow occurs mainly in winter, and, despite its thickness and persistence, the quantities of moisture involved are relatively small (Table 12.2). In general, 12 cm of freshly fallen snow produces only about 1 cm of water. Where snow is formed in very cold, dry air the moisture equivalent is even smaller, and it may take as much as a metre of snow to produce 1 cm of water. In high mountain and polar regions where temperatures are low throughout the year, the majority of precipitation falls as snow. Even so, because of the low temperatures preventing the atmosphere holding much moisture, many of these areas are, in fact, quite arid. There are no real data to provide accurate estimates, but it seems likely that on a world basis no more than 1 per cent of the total annual precipitation occurs as snow.

Snowfall usually starts in the atmosphere as tiny

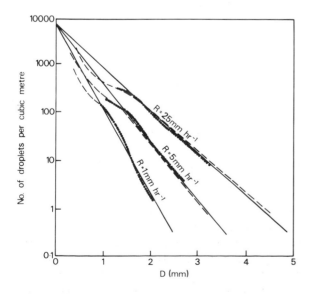

Figure 12.1 Measured drop-size distributions (dotted lines) compared with best-fit exponential curves (straight lines) (after Marshall and Palmer, 1948)

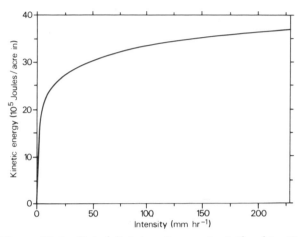

Figure 12.2 Rainfall intensity against the kinetic energy released by rainstorms (after Bennett and Chorley, 1971)

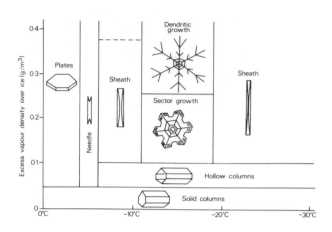

Figure 12.3 Ice crystal forms in relation to temperature of formation and excess vapour density over ice compared with water

Table 12.2 *Water input during a snowstorm in California*

Period	Duration (hr)	Depth (mm)	Intensity (mm hr^{-1})
Heaviest clock hour	1	5	5
Heaviest 3-hour period	3	11	3.7
Heaviest 6-hour period	6	17	2.8
Heaviest 12-hour period	12	26	2.2
Heaviest 24-hour period	24	38	1.6
Total storm	42	49	1.2

After: D. H. Miller, *Water at the Surface of the Earth* (Academic Press 1977).

ice crystals produced at temperatures well below freezing. As the crystals fall they tend to aggregate, particularly where there is sufficient moisture in the air to bind the crystals together. This mainly occurs where temperatures are close to freezing point, and in these conditions large snowflakes may be formed (Plate 12.2). At lower temperatures, however, moisture is lacking and the crystals do not aggregate (Figure 12.3).

As with rain, the fall velocity of the snowflakes depends on size and, all else being equal, large flakes fall more rapidly than small ones, with maximum speeds of about 100 cm s^{-1}. As we all know, however, snowflakes vary considerably in shape and this too may influence fall speeds. Moreover, the density of snowflakes is very low; large flakes often have a density of as little as 100 kg m^{-3} (compared with approximately 1000 kg m^{-3} for raindrops). Consequently, for their size, snowflakes are

light and they are readily blown by the wind. For this reason the distribution of snowfall during a storm is greatly influenced by surface wind conditions, and even after reaching the ground the snow may be redistributed to form deep drifts and snow-free areas.

Another important feature of moisture inputs in the form of snow is that it is often many weeks or, in polar areas, even years before the water is actually released. Thus, in mountain areas, winter snowfall may survive into the spring and the snow represents a temporary store of water which is released only by melting. In Arctic and Antarctic regions the snow accumulates for centuries, then moves with imperceptible slowness in the ice sheets and glaciers before melting perhaps thousands of years later. Unlike rainfall, therefore, snow is not always a direct input to the hydrological system.

Hail

The word hail can strike fear into the heart of farmers in many parts of the world. Damage to crops can be severe though normally the devastation is localized. The hail storms usually produce a swath of stones as the parent cloud moves across the country. Because of the limited areal extent of the storms and their relative infrequence, their contribution to water inputs is generally small.

Hailstone sizes vary considerably, but are usually less than 1 cm in diameter (Figure 12.4). Stones of this size can cause damage, but it is the larger stones, possessing considerable kinetic energy,

Plate 12.2 Large snowflakes falling on an April evening in SE England with temperatures at −1°C. A small disturbance in an Arctic airstream deposited 15–20 cm of snow (photo: C. J. Richards)

that produce spectacular effects, such as damage to cars, greenhouses and vegetation.

Data on the frequency of hailstorms are not entirely reliable. Standard statistics probably underestimate the true frequency of hail because many storms pass between observing stations. For example, in South Africa, where hailstorms are prevalent, the standard network of recording stations gave an average of five storms per year. When the network was increased to one observer per 10 km², eighty days with hail were recorded.

Dew, fog-drip and rime

Walking through the fields on a cold autumn morning after a clear night, we would almost certainly be conscious that the ground surface was wet with dew. Similarly, in a dense forest with lots of mist or cloud we might see water dripping from the leaves, as fog-drip. In these two cases we are dealing with some of the smallest contributors to the precipitation input, although locally they may have some importance. Dew forms on cold surfaces at night when the air is close to saturation. Under these conditions, of course, the air can hold little moisture and, as the atmosphere loses moisture to the ground in the form of dew, it dries out further. Consequently, the total amount of dewfall that can occur in a single night is normally limited, rarely more than 0.6 mm. As evaporation rates are high once the sun rises, such small quantities of moisture are soon returned to the atmosphere, so the contribution of dew to the local water budget is likely to be small.

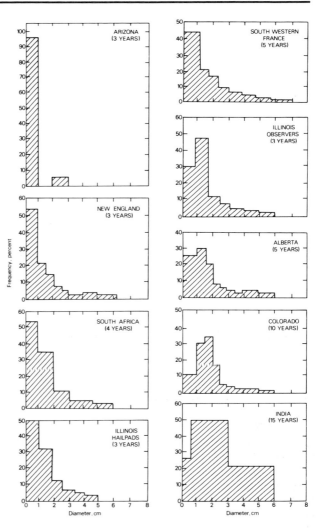

Figure 12.4 Frequency distributions of maximum hailstone sizes for various locations (after Changnon, 1971)

Where cloud droplets are blowing continuously across a rough surface such as a forest we get fog-drip. The process results from the deposition of small water droplets moving horizontally by contact with the vegetative surface. Eventually the droplets combine to form larger drops which fall to the ground. The effect is accentuated if the trees increase the turbulent motion of the air as it moves over the canopy. The vertical motion brings the cloud droplets downwards and impacts them on to the leaves.

Fog-drip is most important in areas where forested mountain ridges extend into persistent cloud sheets. Examples include California, Hawaii, the Canary Islands, Japan and western Germany. Studies in all these areas indicate that there is a significant increase in moisture input at the ground, which would not otherwise occur. At Berkeley, California, as much as 200–300 kg of water per square metre of surface is found during the summer when little rain falls. On Hawaii, trees planted at 800 m altitude catch trade-wind cloud droplets at a rate of about 4 mm per day; over the year this represents an input of about 750 mm to the island's water budget. Without this additional input it is unlikely that the forests would be able to survive.

If temperatures at the ground are below freezing point, then the drifting cloud droplets freeze on the vegetation to give rime. Although this may occasionally produce spectacular scenes (Plate 12.3), the weight of rime can be damaging to trees. The contribution to water input and to tree damage has been investigated in West Germany where coniferous tree growth can be severely hampered by damage to the growth points.

Plate 12.3 Rime accumulation in the Westerwald, West Germany. The direction of air movement can be determined as rime will grow into the wind when super-cooled droplets freeze on contact with the frozen surface (photo: Peter Smithson)

Measurement of precipitation

Rain

As every farmer knows, it is easy to measure rainfall. Any water-tight container sited well away from buildings and trees will act as a rain-gauge. How much it collects will depend not only upon the amount of rain, but also on the gauge diameter and its height above the ground (Table 12.3). Because of this, rain-gauges in the UK have a standard diameter of 12.7 cm and are set a fixed distance of 30 cm above the ground. Unfortunately the standard varies from country to country, so that in Canada the diameter is 9 cm at a height of 30 cm above the ground, while in the USA the gauges are 20 cm wide and 78 cm high (Figure 12.5). Comparisons of rainfall totals between countries are therefore more difficult than might be expected.

The main reason for these differences in gauge height is snow. We have already mentioned that snowfall is difficult to measure because of its lightness and tendency to drift. In the USA the same gauge is used to measure both snowfall and rain, so it has to be well above the level of drifting snow. In Canada, separate gauges are used, while in Britain snowfall is a relatively small component of the annual precipitation.

In normal operation, the amount of rainfall collected in a gauge is measured once a day. An appropriately calibrated stick is used to measure the depth of water which has accumulated in the gauge to obtain the quantity of rainfall. In Canada and the UK the rain water in the gauge is poured into a glass measuring cylinder, where the rainfall equivalent can be read directly. A standard rain-gauge will only record the total rain which has fallen between readings; in many cases it is important to know when the rain fell and at what intensity. For this purpose recording rain-gauges are used (Figure 12.6).

Snow

In some countries, the water equivalent of snowfall is found by melting the snow which has accumulated in the gauge. Clearly this is not very accurate, especially during heavy snowfall when a low gauge may be totally covered. In the USA, the tall gauge prevents this happening but the gauge tends to underestimate the amount of snow reaching the ground. In Canada and the USSR, separate snow gauges are used. Recently, experiments have also been made to measure snow depth

Table 12.3 *Variation of rainfall catch with height*

Height of gauge mouth above ground (cm)	5	10	15	20	30	46	76	152	610
Catch as percentage of that at 30 cm	105	103	102	101	100	99.2	97.7	95.0	90.0

After: J. P. Bruce, and R. H. Clarke, *Introduction to Hydrometeorology* (1966), Pergamon Press.

photogrammetrically, either with aerial or satellite photography. Where the snowfall is substantial, the depths can be obtained fairly accurately, but without ground observations the water-equivalent of the snow is unknown.

Whichever approach is used, measurements of the water-equivalent of snowfall always present problems and probable inaccuracies. We just have to accept that, apart from a few areas of intensive observations, we do not know the precise input of water to the ground surface by snow.

Hail

Hail measurement is even more difficult. Hail stones possess considerable kinetic energy and many will bounce out of a conventional gauge, causing underestimation of the total fall. The size distribution of hail stones can be obtained from a hail pad which measures the degree of impaction made by the stones. If pads are left out for known times, the amount of ice and water-equivalent can be found. Fortunately hail is normally insignificant as a precipitation input to the hydrological cycle, so it is normally recorded separately in terms of the number of days with hail.

Fog-drip and dewfall

The water content of fog-drip and dew is small, so special measurement techniques have to be used. Fog-drip falls to the surface after contact with leaves or trees so trough-shaped rain-gauges have been designed to increase the sampling area and make measurements more accurate. In principle they work like an ordinary gauge.

The most commonly used instrument for dewfall is an accurate weighing device. The dewdrops collect on hygroscopic plates which are attached to a balancing system to weigh the amount of water collected. All methods suffer from the basic uncertainty of how accurately the gauges collect dew compared to natural surfaces. Fortunately, water quantities are minute so that even large errors are insignificant in relation to the total precipitation input.

Temporal variations of precipitation

Variability in the short term

The variations of rainfall over time are of vital importance to hydrologists. Decisions about bridge size, storm sewer construction, culvert dimensions and even flood protection measures must be taken on the basis of the expected inputs of precipitation. For this type of decision a single total is not very informative. We need to know not only how much rainfall is likely to fall, but over what period of time. 25 mm of rainfall in a day may not be significant, but if that amount of rainfall fell in an hour, or even less, then there could be drastic consequences. Surface runoff may occur, soil erosion might be initiated, streams might start to swell and flooding might result. Clearly, the precipitation intensity is extremely important.

If we monitored a storm, we would normally find that precipitation intensity – that is, the amount of rainfall per unit time – varied considerably. Heavy bursts of rain are normally seen to alternate with relatively quiet periods. All types of rainfall show these variations; there is rarely such a thing as steady rain. In convectional storms, the variations are often associated with the passage of the main convection zones across the land. Where the updraughts are strong, the raindrops are held in the cloud and prevented from falling, but as the updraughts weaken the drops fall more easily to the ground, giving periods of higher intensity (Figure 12.7). Cyclonic rain, too, shows considerable variability, often associated with temporary zones of instability in the cyclone (Figure 12.8).

In fact, it is only when the source of precipitation is held stationary that we get anything like steady

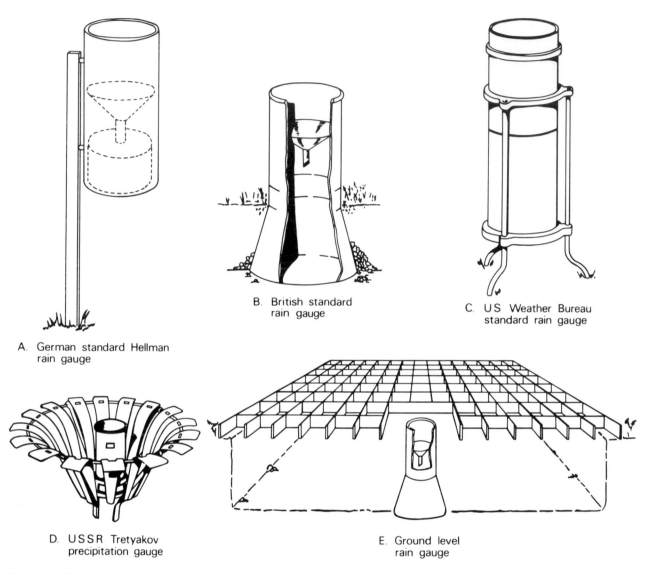

A. German standard Hellman
rain gauge

B. British standard
rain gauge

C. US Weather Bureau
standard rain gauge

D. USSR Tretyakov
precipitation gauge

E. Ground level
rain gauge

Figure 12.5   Types of standard rain-gauges (after Rodda, 1969)

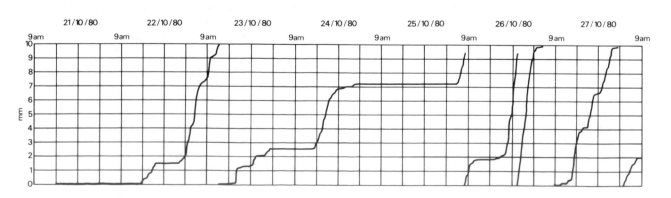

Figure 12.6   Chart from a recording rain-gauge

rainfall. One of the most common situations in which this occurs is where moist air is forced to rise over a mountain barrier. If the moist air is blowing from the sea at a constant speed, the air will be fairly uniform and the conversion of vapour to water droplets will proceed at a constant rate. Rainfall then is often prolonged and steady (Figure 12.9).

The short-term variability of rainfall differs greatly from one area to another. It tends to be greatest in the tropics, and at Djakarta (the capital of Indonesia), for example, the annual rainfall of 1800 mm falls in only 360 hours on average. By contrast, the average rainfall in London is only 600 mm, yet this takes 500 hours to fall. Variability in precipitation is often most important, however, in the more arid parts of the world, for here even quite small storms may be a rare event (Table 12.4); channels that have been dry for months or even years may fill with water, and the baked clay used to make houses may crumble and be washed away. Within a matter of hours the rainfall may have ceased and the water almost vanished; within weeks the vegetation will have disappeared again.

Seasonal variability

In many climates there is a predictable and consistent cycle of rainfall during the course of the year related to the latitudinal migration of the wind and pressure systems. Precipitation areas associated with areas of convergence and uplift tend to shift polewards in summer and equatorwards in winter. Some areas, like the British Isles, remain within the same pressure system throughout the year and so seasonal variations are subdued. This is also true in the equatorial trough zone, where rainfall can occur at any time throughout the year (Figure 12.10) and in deserts, where rainfall is almost negligible. The brief, rare storms which do occur can come at any time, so monthly rainfall, averaged over the long term, shows little variation (Figure 12.11). Even within the same pressure system, some seasonal pattern may be evident. In the mid-latitudes, where

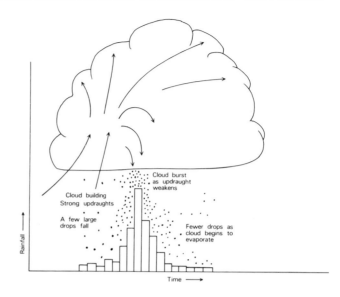

Figure 12.7 Surface rainfall rates relative to cloud development. In the early stages of cloud build-up, strong updraughts prevent most drops reaching the ground. Later, when the cloud drifts downwind, or as the updraught weakens, a period of much heavier rain may ensue

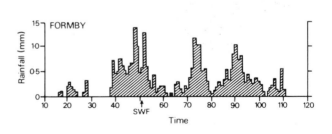

Figure 12.8 Variations in rates of precipitation ahead of, during and after the passage of a surface warm front. The numbers on the horizontal axis represent intervals of time after an arbitrary starting point. Each interval is approximately 15 minutes (after Atkinson and Smithson, 1972)

Table 12.4 *Rain days and amounts at Timimoun (Algeria), 29°50'N 0°30'E, 1926–50*

Days with no rain	9007	(98.8 per cent)												
Days with rain	118	(1.2 per cent)												
Rain amount (mm)	>1	>2	>3	>4	>5	>6	>7	>8	>9	>10	>15	>20	>25	>30
Number of days in 25 years	63	40	30	18	13	12	9	9	7	5	3	2	1	1

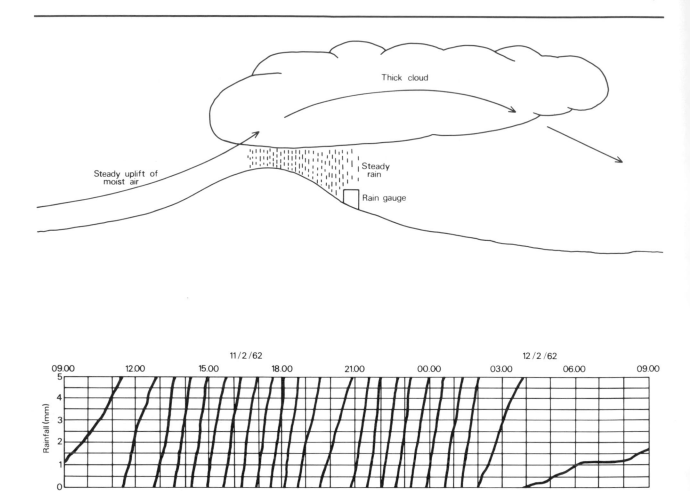

Figure 12.9   Diagrammatic representation of orographic rainfall. The rainfall chart is for a site in the Scottish Highlands experiencing these atmospheric conditions throughout a 24-hour period

rainfall is associated with the activity of the rain-bearing cyclones, the winter and autumn are relatively wet, for it is at these periods that the westerlies bring the most intense storms.

In the tropics and sub-tropics, where convectional rainfall is more important, precipitation tends to be more abundant during the summer months (Figure 12.12). The magnitude of these seasonal variations is even more marked in the monsoonal areas of the world where the year can be subdivided into a wet and a dry season. At Tilembeya in the Sahel region of Mali, for example, rainfall during August is over 200 mm; from December to February it is almost zero. Similarly in the monsoon areas of Asia and northern Australia (Figure 8.15, p. 118) seasonal differences are great so that hydrological conditions vary considerably throughout the year. During the dry season there is practically no surface runoff to be followed by a wet season when runoff is extensive. Vegetation, geomorphological processes and man all respond to these changes.

Precipitation is more abundant during winter months in those parts of the world which experience the mid-latitude depressions during winter only. The Mediterranean Sea area is the best known example of this type of precipitation regime, although other areas do experience similar rainfall patterns. In some areas, the seasonal patterns may be more complex. There may be more than one peak in rainfall totals, as found in many areas of supposedly Mediterranean climate (Figure 12.13) and in the tropics where the seasonal migration of the equatorial trough produces two maxima (Figure

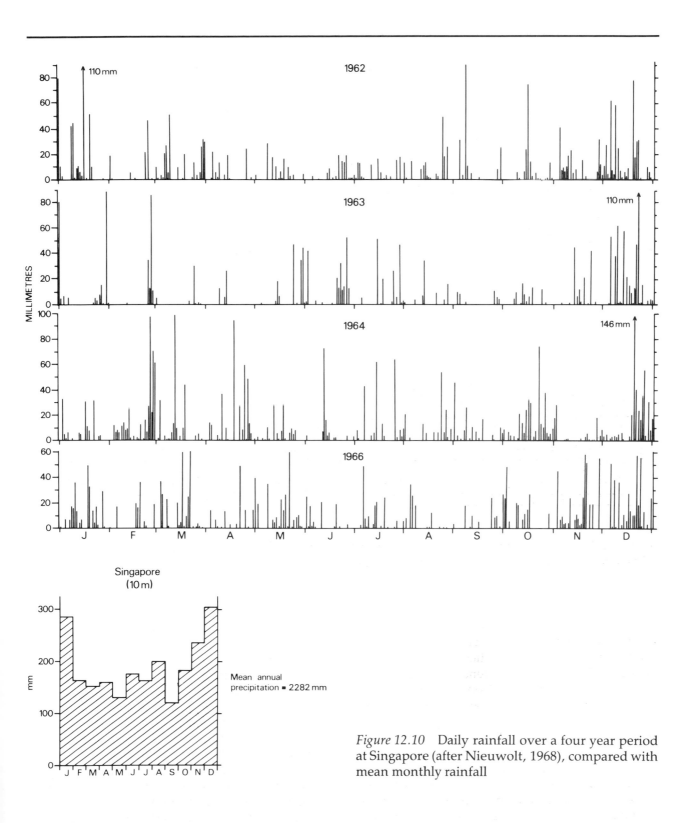

Figure 12.10 Daily rainfall over a four year period at Singapore (after Nieuwolt, 1968), compared with mean monthly rainfall

12.14). In parts of East Africa, a similar pattern occurs (Figure 12.15) but not as a result of the passage of the trough; the rainfall occurs when the monsoonal flow is at its weakest.

Rainfall frequencies

In view of the important consequences of extreme variations in rainfall, it is useful to have some measure of the reliability of precipitation. This may

191

be expressed in a number of different ways. One of the most common is to plot graphs of what are called rainfall recurrence intervals. Using data from a long time period, say fifty years, it is possible to estimate the frequency with which storms of a particular size or intensity are exceeded. In general, small storms occur most commonly and very heavy storms only rarely. Thus, a graph like that in Figure 12.16 is obtained. From this, it is possible to tell how frequently a storm giving, for example, 50 mm or less in a day will occur; or how many years it will be on average between storms of 100 mm or more per day. Such information can be very useful in planning bridges, drains or canals, when the aim is normally to produce something that will cope with all but the most extreme events. It is important to remember, however, that the figures are only probabilities, derived from average conditions. It is quite possible for two storms with an average recurrence interval of fifty years to occur on successive days!

Another way of expressing information on rainfall variation is to plot annual rainfall totals on

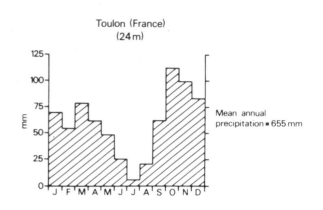

Figure 12.13 Mean monthly precipitation at Toulon, France

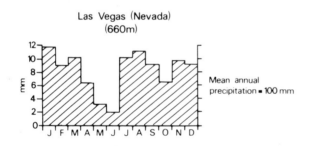

Figure 12.11 Mean monthly precipitation at Las Vegas, USA

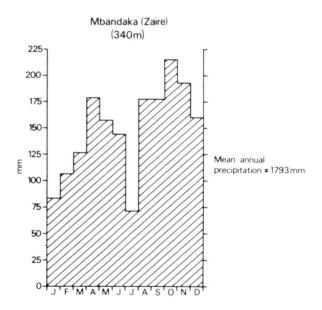

Figure 12.14 Mean monthly precipitation at Mbandaka, Zaire

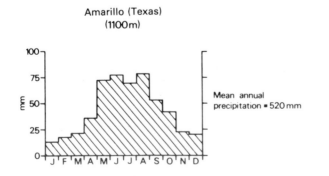

Figure 12.12 Mean monthly precipitation at Amarillo (Texas) USA

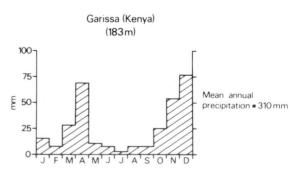

Figure 12.15 Mean monthly precipitation at Garissa, Kenya

similar graphs. Thus, in Figure 12.17 the frequencies of annual rainfall for Sheffield and Timimoun are shown. We can see that there is a 50 per cent probability of at least 760 mm of rainfall occurring in any year at Sheffield, while at Timimoun the equivalent total is 14 mm. It is also apparent from the graphs that the variability at Sheffield is fairly small compared to that at Timimoun, although the latter is, on average, much drier.

Again this type of data may be very useful. It may be known, for example, that a particular crop may only grow satisfactorily if the annual rainfall exceeds 600 mm. From information on annual rainfall frequencies it is possible to determine the likelihood of receiving this amount of precipitation. If the probability is, say, 90 per cent the farmer may well think it worthwhile to grow the crop; if it is only 20 per cent, then it is unlikely to be worth the risk. Similarly, it is possible to determine in the same way how often, on average, it will be necessary to irrigate crops.

Rainfall variability may also be expressed statistically by the **coefficient of variation**. This is calculated from the formula

$$CV = \frac{s}{\bar{x}} \times 100 \text{ per cent}$$

where $\bar{x}$ is the average rainfall and s is the **standard deviation**.* This defines the variability relative to the mean. With a standard deviation of 200 mm and a mean annual precipitation of 1600 mm, the coefficient of variation would be 12.5 per cent, but with the same standard deviation and a mean of only 400 mm, the coefficient of variation would rise to 50 per cent. This is a useful measure since it gives an indication of the importance of the variability. The pattern of variability, measured in this way, is shown for the UK in Figure 12.18.

Spatial variations in precipitation

Atmospheric causes of variation

We all know that annual rainfall totals vary from one part of the world to another, even when altitude is allowed for. Locally, however, it seems likely that annual totals will be fairly consistent. It is also clear that, in the short term, quite marked differences in rainfall may occur within short distances, depending upon the route taken by a particular storm or cyclone; indeed, it is sometimes possible to see it raining on one side of the street and dry on the other.

In order to study spatial variations at a local scale, we need a dense network of recording stations, for otherwise individual storms may be missed as they pass between the rain-gauges. One such investigation was carried out in Illinois, where fifty recording rain-gauges were set up in an area of 1400 km² of flat rural land. The experiment was maintained for five years measuring individual storms, and for thirteen years for monthly and seasonal analyses. Comparisons were made by correlating rainfall at a gauge at the centre of the area with all other gauges. Correlation is a statistical measure which provides an index of the strength of the relationship between two variables; a value of +1.0 indicates a perfect positive linear relationship, a value of −1.0 shows a perfect negative linear relationship, and a value of 0.0 shows no relationship (Figure 12.19).

For the shortest time period studied (one minute) the degree of correlation fell rapidly with distance

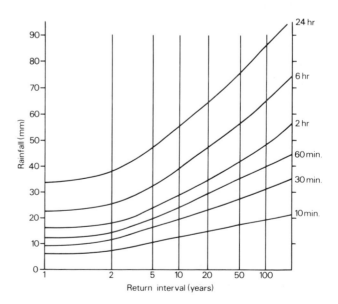

Figure 12.16 The return period for precipitation totals at Sheffield (based on methods in Flood Studies Report, NERC, 1976)

---

*The standard deviation is a measure of dispersion of the data. It is calculated from the formula $S = \sqrt{\frac{\Sigma(x - \bar{x})^2}{n}}$ where n is the number of observations in the data set and Σ means 'sum of'.

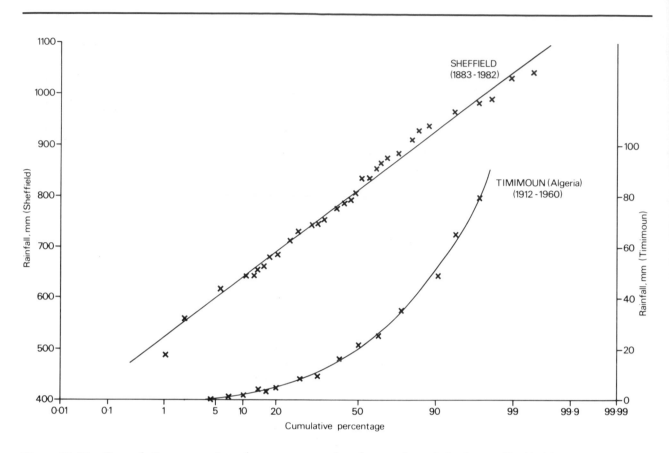

Figure 12.17 Cumulative percentage frequency graphs of annual precipitation at Sheffield (1883–1982) and Timimoun, Algeria (1912–60)

from the central gauge (Figure 12.20). Thus, at a distance of only 8 km from the central gauge, the rainfall pattern is different; in many cases it would have been raining at the central gauge but not 8 km away. This is what we would expect if rainfall was produced by localized summer convection storms, each affecting an area of only a few square kilometres.

At a longer time-scale, the degree of correlation is better. Taking rainfall totals for whole storms (Figure 12.21a), it is apparent that the gauges close to the central station are quite strongly correlated. Nevertheless, at a distance of about 16 km the degree of correlation is low. Again, this is probably due to the effect of local variability caused by the passage of small summer convection storms. If, instead, we look at frontal storms or storms associated with low pressure systems, we get a different picture (Figure 12.21b). Now, most of the area shows a close correlation with the central gauge; indeed, almost a perfect correlation – indicating that these more general storms affected the whole

area equally.

These results indicate some of the atmospheric factors controlling rainfall variability. **Convectional storms** give high levels of spatial variation, while **cyclonic rainfall** is spatially much more uniform. In the tropics, where a great proportion of the rainfall comes from convectional storms, the spatial variation is particularly marked. Table 12.5 illustrates this point. It gives rainfall totals at rain-gauges only 3.2 km apart in a level area of Dar-es-Salaam in Tanzania. It is clear that the totals are very different. The reason is that most of the rainfall is derived from individual cumulonimbus clouds which produce intense precipitation over an area of about 2 to 60 km². The storms often build up without any significant movement so areas just beyond the limits of the cloud may receive no rainfall at all. Sometimes the storms develop over a wider area, perhaps 500 km², but even so they do not give rain everywhere. Using the correlation method, therefore, we find that the relationships between rain-gauge totals falls to zero within 100 km and is

Figure 12.18 Coefficient of variation of annual precipitation, 1901–30 (after Gregory, 1968)

Table 12.5 *Daily rainfall at two stations, 3.2 km apart near Dar-es-Salaam, Tanzania, April 1967(mm)*

	6th	7–8th	10th	12th	13th	24th	25th	Month
Station A	54.9	33.3	2.3	13.5	13.5	14.2	21.1	315.2
Station B	100.3	3.3	31.8	64.8	5.1	8.4	0.8	437.6

After: I. J. Jackson, *Climate, Agriculture and Water in the Tropics* (1977), Longman.

negative beyond. In other words, if rainfall were high for a particular period in one area, it would tend to be low beyond 100 km. Over the short term these differences might be considerable, but in the long term we would expect them to balance out.

Surface modifications of precipitation

So far we have considered rainfall variability over essentially flat terrain. Few areas of the world are extensively flat, however, and surface irregularities interfere with atmospheric processes to give even more complex spatial patterns of variation in rainfall. Even relatively small hills can have a marked effect. The Chiltern Hills in southern Britain, for example, only rise some 90 m above the surrounding land but they receive appreciably more rainfall (Figure 12.23).

Within any climatic region, the relationship between rainfall and altitude is generally quite consistent. In most cases, precipitation increases with increasing altitude, even in relatively arid areas. At Grand Canyon, for example, average annual precipitation increases from less than 250 mm on the

canyon floor at 760 m, to 400 mm on the southern rim of the canyon at 2100 m. On the forested northern rim, 2600 m above sea-level, rainfall totals over 600 mm.

Nevertheless, the progressive increase in rainfall with altitude does not always extend to the summits of the mountains. The Sierra Nevada in California are no wetter on the summit than they are 1200 m lower (Figure 12.22a). In Hawaii, the peaks of Mauna Loa and Mauna Kea receive far less rain than the windward slopes, even close to sea-level. It is also apparent that the relationship between altitude and precipitation varies from one part of the world to another. In the tropics, much of the precipitation is produced by warm clouds whose upper limit is only 3000 m above the ground; thus

Figure 12.20 Correlation patterns associated with one minute rainfall rates in warm season storms in Goose Creek, Central Illinois, USA (after Huff and Shipp, 1969)

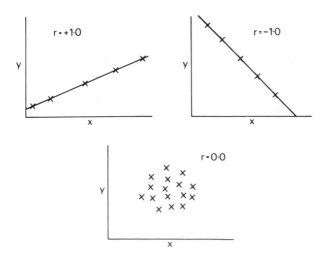

Figure 12.19 Scatter diagrams demonstrating different correlation coefficients

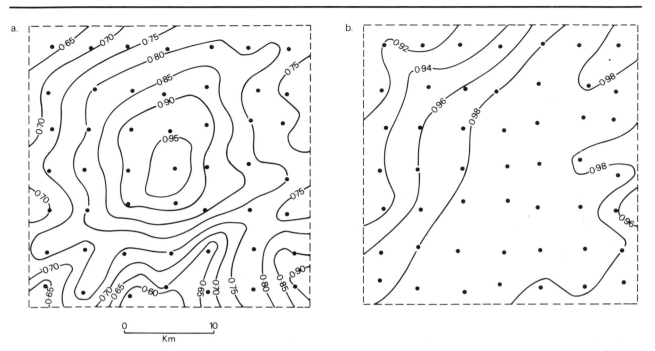

Figure 12.21 Correlation patterns associated with (a) air mass storms and (b) low pressure centres, during May–September in Illinois, USA (after Huff and Shipp, 1969)

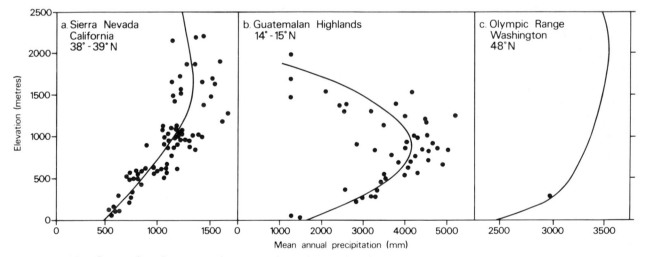

Figure 12.22 Generalized curves showing the relationship between elevation and mean annual precipitation for west-facing mountain slopes in North and Central America (after Barry and Chorley, 1982)

the effect of altitude is subdued (Figure 12.22b). In contrast, in temperate areas, a large proportion of the rainfall comes from deep stratiform clouds that extend through a considerable part of the troposphere. Here the effect of altitude on rainfall is more marked (Figure 12.22c). Comparisons are difficult, however, because much of the precipitation on the mountains in temperate areas falls as snow and, as we have seen, this is impossible to measure accurately.

The importance of surface topography on precipitation is indicated at a general scale for the British Isles by Figure 12.23. As can be seen, the general pattern of rainfall is appreciably modified by the Welsh Mountains. This gives rise to higher totals on the westerly slopes, and a marked rain-shadow effect on the east. In contrast, the Cairngorms have a relatively minor influence. This is because the prevailing winds have lost much of their moisture by the time the eastern side of the

Figure 12.23   Mean annual precipitation over the British Isles in millimetres, 1931–60 (after Atkinson and Smithson, 1976)

country is reached. Although precipitation does increase with altitude the gradient is not so steep as in the west.

Conclusions

The precipitation input to the hydrological cycle is probably one of the most important regulators of the system, for it determines the intensity and distribution of many of the processes operating within the system. As we will see in the next chapter, it is closely related to the rate of evapotranspiration, and as we will find in later chapters it also influences the pathways of runoff and underground flow and the magnitude of streamflow. Through these processes, and through the direct effects of the impact of rainfall on the ground, it also takes part in many geomorphologi-

cal processes; it causes rainsplash and soil erosion and it plays a vital role in weathering and rock breakdown. The distribution of rainfall across the globe therefore controls to a large degree the operation of the landscape system. Precipitation is similarly a vital input to the ecosystem, and the distribution of vegetation, fauna and man owes much to the pattern of rainfall.

For these reasons, and because of their ultimate importance for man, a great deal of attention is given to measuring, mapping and predicting precipitation. As we have seen, scarcity of data, particularly in less accessible parts of the world, limits our ability to gain a complete picture of precipitation inputs. On the whole, however, rainfall is one of the easiest components of the hydrological cycle to measure, and for that reason if no other it provides an ideal opportunity for coming to grips with real-world hydrological processes.

13

Evapotranspiration

Evapotranspiration processes

Evaporation and transpiration form the major flows of moisture away from the earth's surface. Because we can rarely see the processes taking place, it is easy to neglect this component of the hydrological cycle, but it is an extremely important one. It returns moisture to the air, replenishing that lost by precipitation, and it also takes part in the global transfer of energy. In this chapter, we will examine the processes involved in evaporation and transpiration and show their significance at a global and local scale.

Evaporation

Evaporation can be defined as the process by which a liquid is converted into a gaseous state. It involves the movement of individual water molecules from the surface of the earth into the atmosphere, a process occurring whenever there is a water-pressure gradient from the surface to the air. Thus, evaporation requires that the humidity of the atmosphere be less than that of the ground. The process also requires energy: 2.48×10^6 J to evaporate each kilogram of water at 10°C. This energy is normally derived from the sun, although sensible heat from the atmosphere or from the ground may also be significant. However, when the air reaches saturation (100 per cent relative humidity) evaporation cannot take place.

Transpiration

Transpiration is a related process involving water loss from plants. It occurs mainly by day when small pores, called **stomata**, on the leaves of the plants open up under the influence of sunlight. These expose the moisture in the leaves to the atmosphere and, if the vapour pressure of the air is less than that in the leaf cells, the water is transpired. As a result of this transpiration, the leaf becomes relatively dry and a moisture gradient is set up between the leaf and the base of the plant.

This draws moisture up through the plant and from the soil into the roots (Figure 13.1).

As far as the plant is concerned, this is a passive process; it is controlled largely by atmospheric and soil conditions and the plant has little influence over it. Consequently, transpiration results in far more water passing through the plant than is needed for growth. Only 1 per cent or so is used directly in the growth process. Nevertheless, the excessive movement of moisture through the plant is of great importance, for the water acts as a solvent, transporting vital nutrients from the soil into the roots and carrying them through the cells of the plant. Without this process, plants would die.

Evapotranspiration

In reality, it is often difficult to distinguish between evaporation and transpiration. Wherever vegetation is present, both processes tend to be operating together, so the two are normally combined to give the composite term evapotranspiration.

Evapotranspiration is governed mainly by atmospheric conditions. Energy is needed to power the

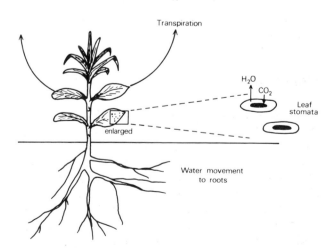

Figure 13.1 Schematic diagram showing exchanges of water and gases by transpiration in plants

process, and wind is necessary to mix the water molecules with the air and transport them away from the surface. In addition, the state of the surface plays an important part, for evaporation can only continue so long as there is a vapour pressure gradient between the ground and the air. Thus, as the soil dries out, the rate of evapotranspiration declines. Lack of moisture at the surface often acts as a limiting factor on the process.

We can therefore distinguish between two aspects of evapotranspiration. **Potential evapotranspiration** (PE) is a measure of the ability of the atmosphere to remove water from the surface assuming no limitation of water supply. **Actual evapotranspiration** (AE) is the amount of water that is actually removed. Except where the surface is continuously moist, actual evapotranspiration is significantly lower than PE.

Potential evapotranspiration

Energy inputs

The main variable determining potential evapotranspiration is the input of energy from the sun, and it has been estimated that this accounts for about 80 per cent of the variation in PE.

The amount of radiant energy available for evapotranspiration depends upon a number of factors, including latitude (and hence the angle of the sun's rays), day-length, cloudiness and the amount of atmospheric pollution. Thus PE is at a maximum under the clear skies and long hot days of tropical areas, and at a minimum in the cold, cloudy polar regions. In the short term, however, rates of potential evapotranspiration may vary considerably at any single place. Daily variations in radiation inputs cause marked fluctuations in PE, so that very little evapotranspiration occurs at night. Even subjectively we can get some idea of this by noting how long the ground stays wet after a shower of rain during the night, yet how quickly it dries out during the day. Similar patterns occur seasonally. Potential evapotranspiration reaches a peak during the summer months and declines markedly during the winter (Figure 13.2).

Wind

The second important factor is the wind. The wind enables the water molecules to be removed from the ground surface by a process known as **eddy diffusion**. This maintains the vapour pressure gradient

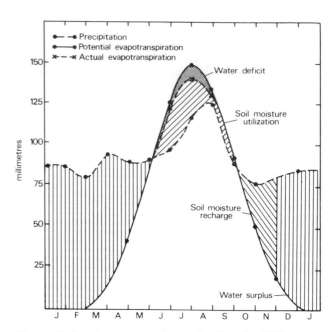

Figure 13.2 Average moisture budget for Wilmington, Delaware, USA. The difference between potential and actual evapotranspiration is small because precipitation is distributed relatively uniformly throughout the year (after Mather, 1974)

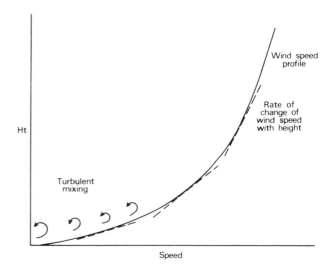

Figure 13.3 Wind speed profile and its relationship with the rate of change of wind speed with height

above the surface. Wind speed is obviously one of the variables determining the efficiency of the wind in removing the water vapour, but it is not the only one. The rate of mixing is also important, and this depends upon the turbulence of the air and the rate of change of windspeed with height (Figure 13.3).

Vapour pressure gradient

Third, evapotranspiration is related to the gradient of vapour pressure between the surface and the air. Unfortunately, the vapour pressure gradient has proved very difficult to measure precisely in the layer immediately above the surface, so wherever possible, methods of calculating PE use measurements of vapour pressure at one level only.

Actual evapotranspiration

Actual evapotranspiration only equals PE if there is a constant and adequate supply of water to meet the atmospheric demand. This situation exists over moist, vegetated surfaces and it is also approximated over water surfaces such as the open sea or large lakes, but most land surfaces experience significant periods when water supply is limited. As a result, actual evapotranspiration falls below PE. We can get some idea of the importance of surface conditions by considering evapotranspiration in a variety of situations. Let us start by examining evapotranspiration from an open water surface.

Evaporation from water surfaces

Because there is an unlimited supply of water to maintain evaporation, and because there is no vegetation to complicate the process, the surfaces of oceans or large lakes provide the simplest situation in which to study evapotranspiration. In this situation, of course, transpiration does not occur and water loss is entirely by evaporation. The main factors determining water loss are therefore the atmospheric conditions, and there is generally a close relationship between actual evaporation and PE.

Nevertheless, the relationship is not perfect and the main reason is that the water is able to absorb a large amount of energy which is not used in evaporation. This energy is expended in heating the water, and much of it is recirculated through the water body by advection.

Evaporation is greatest when the sea is warm in comparison to the air. In general this is the case as air temperatures are slightly below those of the sea over much of the globe for much of the time. Where upwelling of cold water from the sea bottom occurs, however, the surface temperatures are greatly reduced and the difference between sea and air temperatures becomes small; in some cases the sea may even be cooler than the atmosphere. An exam-

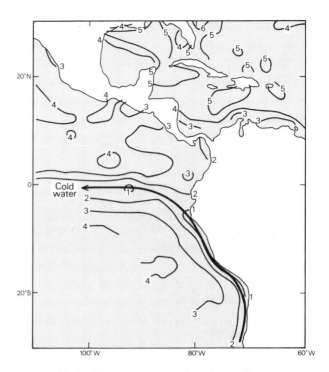

Figure 13.4 Mean evaporation in millimetres per day from the tropical Atlantic and eastern Pacific Ocean (after Hastenrath and Lamb, 1978)

ple of this phenomenon occurs off the coast of Peru where the cold Humboldt current brings bottom waters to the surface. As a result, the air is warm relative to the sea, it retains moisture and so the humidity gradient above the surface is low. This greatly reduces the rate of evaporation, and, as Figure 13.4 shows, the effect continues some way out into the Pacific.

At the oceanic scale, there have been few studies of evaporation. Smaller, freshwater bodies, such as Lake Ontario, have been investigated more fully, however, and Table 13.1 lists the water and energy budget components for the lake on a monthly basis. During the spring, as air temperatures are rising, the water surface remains cold and evaporation becomes negative (i.e. there is a net condensation of water on the lake). All the available radiation and heat from the air goes into warming up the deeper water of the lake. During the cooling phase of the cycle, from late August to about March, the situation is reversed and heat is released from the depths of the lake. Evaporation can continue, maintained by the sub-surface heat.

This discussion of the energy budget of the lake may sound unimportant to all but the fish of the

Table 13.1 *Water and energy budget of Lake Ontario*

	J	F	M	A	M	J	J	A	S	O	N	D
Net income of energy (W m⁻²)	+2	+35	+85	+140	+185	+200	+200	+160	+110	+50	+10	−5
Exchange of heat between lake surface and deep water (W m⁻²)	+250	+160	−15	−165	−245	−190	−130	−70	0	+65	+145	+200
Energy available at lake surface (W m⁻²)	+246	+192	+67	−25	−59	+10	+70	+89	+109	+115	+153	+190
Surface temperature (°C)	+3	2	2	2	4	10	19	20	17	11	7	5
Evaporation (mm day⁻¹)	−3.2	−2.6	−1.4	+0.6	−0.3	−0.1	−1.6	−2.6	−3.6	−3.0	−3.0	−2.9
Latent heat flux (W m⁻²)	−90	−75	−40	+15	−10	−5	−50	−75	−105	−85	−85	−85

After D. H. Miller, *International Geophysics*, no. 21, New York, Academic Press, 1977.

lake. However, it has vital implications for the climate of the surrounding areas. The very high rates of evaporation in the autumn and early winter mean that water is being added to the atmosphere in large quantities. With low temperatures, much of the moisture is returned to the ground as snow on the south-eastern shores of the lakes (Figure. 13.5). In spring, the low water temperatures of the lakes keep air temperatures low along the lake shores and delay blossoming of the orchards. This reduces the chances of severe frosts damaging the fruiting of the trees, one of the main reasons why this is

an important fruit-producing area. The low temperatures throughout the summer reduce convection over the lakes to such an extent that annual totals of rainfall have been estimated to be 6 per cent (80 mm) less over Lake Michigan than over the surrounding land.

Where the water body is shallow, heat storage becomes less important and may only be apparent on the diurnal scale. Temperatures are higher than in deep water bodies for there is less water to heat up, and the thermal properties of the lake approach that of a moist ground surface – as alligators well know! As the proportion of soil to water increases, so the amount of evaporation decreases, very little stored heat being available for evaporation. Measurements in Australia, comparing evaporation from a water surface and a wet soil, showed that the rate for the soil was only 86 per cent of that from the water surface.

Evapotranspiration from land surfaces

Because of the importance of the energy and water balance to growing crops, there have been a large number of studies of evapotranspiration from vegetated surfaces. The presence of a vegetated surface complicates the energy exchanges taking place, however, for the plants intercept radiation and rainfall inputs, they affect the temperature and wind profiles near the ground and they also modify humidity. The degree of these effects varies with the character of the vegetation, so evapotranspiration from a vegetated surface often differs markedly from PE (Figure 13.6).

Within a mature crop we can identify three layers: the upper layer or canopy, the main stem

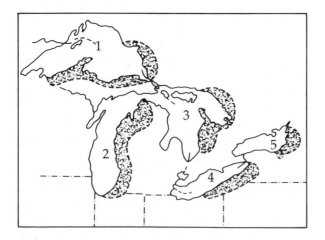

Figure 13.5 Snowbelts of the Great Lakes (after Eichenlaub, 1970). 1 – Lake Superior; 2 – Lake Michigan; 3 – Lake Huron; 4 – Lake Erie; 5 – Lake Ontario

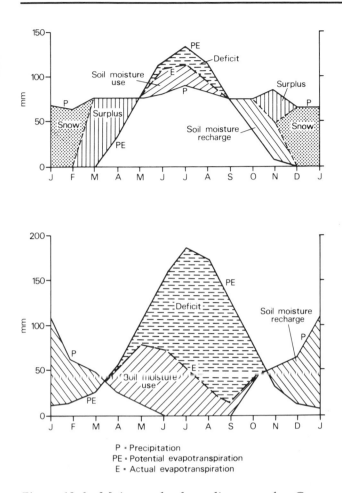

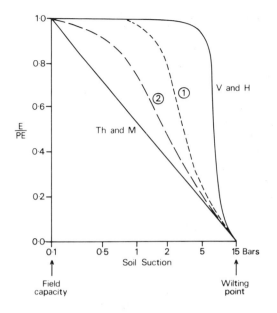

Figure 13.7 The relationship between the ratio of actual to potential evapotranspiration $\frac{E}{PE}$ and soil moisture. V and H = Veihmeyer and Hendrickson; Th and M = Thornthwaite and Mather; 1 and 2 = schematic curves for a vegetation-covered clay loam under low evaporation stress and a vegetation-covered sandy soil under high evaporation stress (after Barry, 1969)

P = Precipitation
PE = Potential evapotranspiration
E = Actual evapotranspiration

Figure 13.6 Moisture budget diagrams for Concord, New Hampshire, USA and Aleppo, Syria. The method assumes that 50 per cent of the soil water surplus runs off in the first month, 50 per cent of the remainder in the next and so on, unless additional surplus forms (after Barry, 1969)

zone and the ground surface. During the day most of the incoming radiation is absorbed by the canopy. The air space between the canopy and the ground acts as an insulator so that it is the top of the vegetation rather than the soil that acts as the active surface. Consequently, transpiration rather than evaporation takes place.

So long as moisture is available in the soil, the plants are able to transpire at or very close to the potential rate. Thus, in a moist soil evapotranspiration proceeds unhindered, water being drawn up the plant from the soil to replace that lost from the leaves. As the soil dries out, however, the plants experience increasing difficulty in extracting moisture and the rate of transpiration cannot be maintained. Several changes take place. The plants start to suffer from moisture stress and nutrient deficiencies, and in some cases the stomata in the leaves may close, reducing transpiration further. But the drain upon the soil moisture store continues, so the moisture stress gets worse. Progressively, the rate of actual evapotranspiration falls below PE (Figure 13.7).

It now seems clear that the effect of declining moisture availability depends upon a variety of conditions, including vegetation type, rooting depth and density, and soil type. In a heavy clay soil, for example, it seems that evapotranspiration rates fall only slightly as the soil dries out until the point is reached where no more water is available to plants. Evapotranspiration then falls rapidly. Conversely, in a sandy soil, the decline in actual evapotranspiration rates is much more regular (Figure 13.7) as the sandy soil's capacity to retain moisture is less than that of a clay.

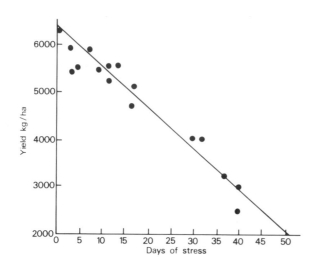

Figure 13.8 Relationship between grain yields of maize and the number of days of critical soil-moisture tension (soil moisture below 12 per cent on a loess soil) (after Rubin and Bielorai, 1957)

Plant responses to moisture stress

The reduction in evapotranspiration as the soil dries out has a number of implications. Eventually, of course, the plants experience severe nutrient deficiencies and yields are reduced. Thus we often see a close relationship between the degree of moisture stress and crop yields (Figure 13.8). In addition, the moisture in the plants helps to control its temperature; the energy used in transpiration cannot heat the plant. As transpiration declines, more energy is available for heating and the leaves get warmer. Initially, this may encourage growth, but ultimately, if high temperatures are reached, it may damage the plants. The soil, too, is heated more effectively so surface temperatures rise. In one study in Wisconsin, with an air temperature of 28°C, the surface of a dry sandy soil was 44°C, while the same soil kept moist reached only 32°C. The difference was due to the fact that more energy was used in evaporation from the wet soil.

The ultimate effect of continued drying of the soil is that plants can no longer obtain any water and transpiration ceases. Since the water in the plant cells is important in keeping the plants rigid (turgid), as the plants dry out they become limp and wilt. At this stage, therefore, the soil is said to be at **wilting point** (see Chapter 14).

Of course, evapotranspiration does not always continue until wilting point is reached. Instead, renewed rainfall generally wets the soil and rejuvenates water uptake by the plants and increases transpiration. Thus the soil acts as an important store of moisture, as we will see in the next chapter, supplying water for transpiration between each storm.

Where the intervals between each period of precipitation are long, the ability of the soil to supply water may be stretched to the limit and moisture stress may be a common occurrence. In these circumstances the vegetation often adapts to the hydrological conditions, by developing deeper roots, or by regulating water use in a variety of ways. For example, the plants may have a dormant period during the dry season, completing their growth cycle during the brief wet period. The sight of deserts blooming after a storm is a most remarkable one. Other plants adapt by controlling their stomata, closing them during periods of dryness in order to reduce water loss. Others are able to alter the orientation of their leaves so that the stomata are more sheltered from the hot sun. Yet others, like the mesquite of south-west USA, have thick waxy leaves which protect them from the radiation and slow down water losses.

Evapotranspiration and irrigation

One way of combating the effects of excessive evapotranspiration is through the use of irrigation. Man may try to make good the water deficit during the dry months by importing water from elsewhere. In many parts of the world agriculture is dependent upon irrigation; the fertile farmlands of southern Australia, for example, would be little more than desert without man's ability to supply water to the crops in this way.

A wide variety of irrigation methods have been developed over the centuries, some supplied by groundwater (see Chapter 14) and some using surface waters. Underground waters often need to be raised to the surface, and in the more primitive systems buckets, winches and levers may be used. In other cases, however, man has developed ingenious methods of exploiting the topography to feed the water to his fields by gravity. The **qanats** of Chile and Iran (see Chapter 14) and the **galeria** found in the Tehuacan Valley of Mexico (Figure 13.9) are examples. More sophisticated systems use pumps and hydraulic rams. Once at the surface, the water has to be carried to the fields, either in canals

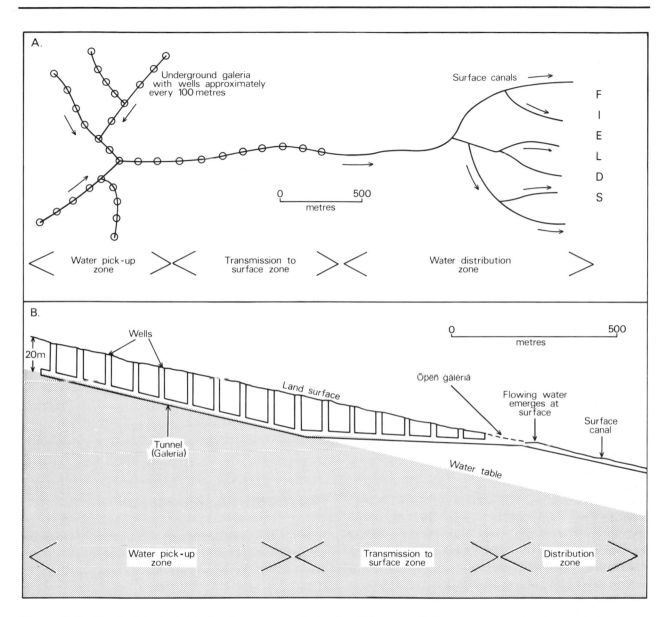

Figure 13.9 Typical 'galeria' irrigation system from the Tehuacan Valley, Mexico; (a) plan and (b) cross-section (after Kirkby, 1969)

or pipes. It is then distributed by a variety of methods to the plants. Common distribution systems are illustrated in Figure 13.10.

Irrigation of crops is a highly skilled task if it is to be totally successful. Not only must the farmer be able to predict when the crops will need water to counteract the effects of excessive transpiration, but he also needs to apply the right amount of water and control its quality. Serious problems may arise if he fails to do so.

Applying water when it is not needed, or in too

great a quantity, is a waste of what, in arid areas, is a valuable and scarce commodity. It can also damage the soil and the crop. It may compact the surface and lead to saturation of the soil. This encourages shallow rooting of the plants which are able to obtain all the water they need from the surface. When irrigation ceases, the water level in the soil falls and the plants start to suffer from moisture stress. In addition, surplus water may leach salts from the soil. To some extent this is beneficial, for there often exist excessive quantities of sodium and

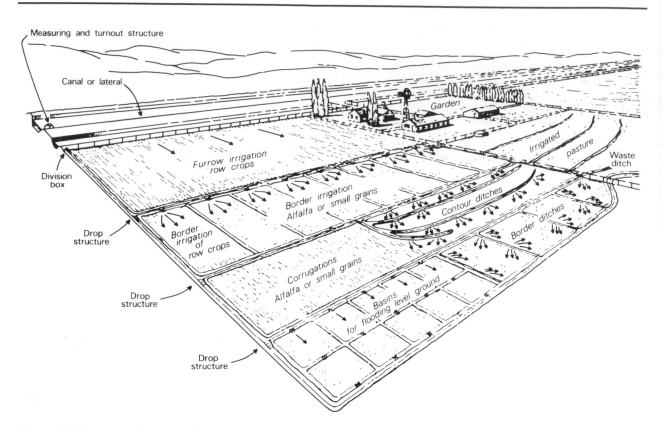

Figure 13.10 Irrigation techniques: various methods of applying irrigation water to field crops (after Israelson and Hansen, 1962)

calcium in the surface layers; irrigation with large amounts of water washes them into the lower layers, away from the roots. But it is not only the harmful salts that are removed; vital plant nutrients may also be washed away, reducing soil fertility.

The application of too little water may also cause problems. Evaporation quickly dries out the surface soil, drawing water to the surface and causing accumulation of salts in the root zone. The plants suffer from high salinity (salt toxicity) and in severe conditions they wilt as the sodium draws the sap from the roots (a process known as **plasmolysis**). In severe conditions, a salt pan may even form at the surface (Plate 13.1).

Perhaps the most serious problem of irrigation is the supply of water of a suitable quality. In arid areas, where excess evapotranspiration is most common, water resources are scarce and may suffer from high concentrations of soluble salts. During conveyance of the water to the fields considerable evaporation takes place, and this too increases the salt concentration. Application of saline water to the crop rapidly leads to a variety of problems.

Measuring evapotranspiration

One of the main needs of the farmer or irrigation engineer is to be able to predict when the plants will suffer from moisture stress and how much water must be applied. This involves being able to measure or calculate the rate of evapotranspiration. Knowledge of evapotranspiration losses is also required by the hydrologist who wishes to plan water management policies; he needs to know what proportion of the precipitation will be available to replenish groundwater or runoff into streams. The measurement of evapotranspiration is therefore important. Unfortunately it is also difficult. Several systems of measurement have been developed, including:

1 direct measurement (e.g. with evaporation pans and lysimeters);
2 meteorological formulae;
3 moisture budget methods.

Plate 13.1 Effects of salinization of the soil in Upper Murray District, South Australia. Most of the vines in the foreground have been killed. Encrustations of salt are visible nearby (photo: P. Crabb)

Plate 13.2 An evaporation tank and gauge (photo: E. M. Rollin)

Direct measurement

Possibly the most widely used method of direct measurement is the **evaporation pan**. This consists of a shallow pan filled with water. The rate at which the water is lost through evaporation is measured with a gauge (Plate 13.2). This procedure measures only potential evaporation, for it does not allow for limitations of moisture supply, nor does it directly determine transpiration losses. In addition, the results seem to vary according to the size, depth, colour, composition and position of the pan, so it is not always easy to compare results from different sites.

An alternative system of direct measurement is the **lysimeter** (Figure 13.11). This may be employed to measure either potential or actual evapotranspiration. To measure PE the column of soil is kept constantly moist so that water deficiencies do not occur. To measure actual evapotranspiration the column is allowed to respond naturally to atmospheric conditions. Regular weighing allows the moisture content to be determined. If the amount of precipitation is known, the moisture loss through evapotranspiration can be calculated (Table 13.2). Commercial lysimeters may weigh up to 60 tonnes, and for really accurate results large columns and

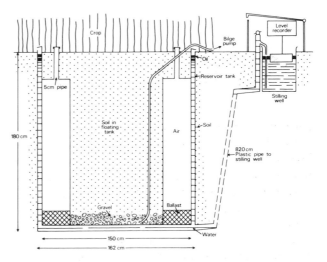

Figure 13.11 Lysimeter installation at Hancock, Wisconsin. The soil block floats in a tank of water. Changes of water level are recorded instead of weighing the block (after Barry, 1969)

precise yet robust weighing instruments are required. But it is quite easy to construct a simple and effective lysimeter from an old tin can and a piece of nylon mesh (Figure 13.12).

Meteorological formulae

Because evapotranspiration is greatly dependent upon atmospheric conditions, it is possible to

Table 13.2 *Calculation of evapotranspiration through lysimeter moisture measurements*

1	2	3	4	5	6	7	8
Date	Precipitation (cm)	Weight of precipitation (g)	Weight of lysimeter (g)	Previous weight of lysimeter (g)	Change in weight (g) (4–5)	Weight transpired and evaporated (g) (3–6)	Water transpired and evaporated expressed in cm (7÷surface area)
1.8.81	0.24	75.36	9110.35	9062.75	+47.60	27.80	0.09
2.8.81	–	–	9097.21	9110.35	−13.14	13.14	0.04
3.8.81	–	–	9042.94	9097.21	−54.27	54.27	0.17
4.8.81	–	–	8986.32	9042.94	−56.62	56.62	0.18
5.8.81	0.51	160.14	9124.67	8986.32	+138.35	21.79	0.07

Note: Surface area of lysimeter – 314 cm².

derive good estimates of PE from data on meteorological conditions. A wide range of formulae have been developed to do this, some of them so complex that it is almost impossible to use them under normal circumstances; the necessary data just do not exist.

This problem is illustrated by what at first seems to be a very simple approach. As we saw in Chapter 4, the energy budget can be expressed as follows:

$$Q^* = Q_H + Q_E + Q_G$$

where Q^* is the net radiation, Q_H is the sensible heat flow, Q_E is the heat usage through evaporation and Q_G is the heat flow into the ground. If we could determine all the other components of the equation we could find Q_E by difference, and this would tell us how much evaporation was occurring. Unfortunately, we rarely know the value of the other components; Q_H, in particular, is difficult to determine. Consequently, to solve this equation, we would have to use further indirect methods to allow us to estimate Q_H.

Because of this problem of obtaining data, a number of simpler, more empirical formulae have been produced. These are much easier to use; they are, however, based not on physical principles but on the observed relationships between evapotranspiration and one or more climatological variables. The relationships have usually been obtained under one particular climatic regime and they may not be applicable elsewhere.

Probably the best known of these empirical equations is the one developed by Thornthwaite to assess PE. We have already encountered this when discussing climatic classification in Chapter 8. In simple terms PE is calculated from the formula:

$$E = 1.6 \, (10T/I)^a$$

where E is the unadjusted value of PE, T is the mean monthly temperature in °C, I is an annual heat index* at the measuring station and a is a constant that varies in relation to I. The value of E has to be adjusted to allow for day-length to give PE.

In this formula, Thornthwaite is using temperature as a substitute for radiation, and it therefore works reasonably well where the two are closely correlated. In the tropics, however, the

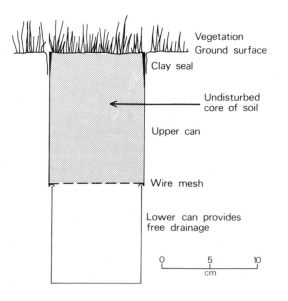

Figure 13.12 Field weighing lysimeter. The can is removed and weighed in the field (after Atkinson, 1971)

---

*$I = \Sigma(T/5)^{1.514}$ where Σ means 'sum of' for all months.

equation underestimates PE because temperatures lag behind radiation inputs. In addition, the method takes no account of wind, although this may be locally important. Nevertheless, the relative simplicity of the method makes it popular, and despite its shortcomings and its inevitable inaccuracies, it is one of the more widely used methods of assessing PE.

In the UK, the **Penman Formula** is the main method for calculating PE. It is less empirically based than the Thornthwaite method, combining the energy budget and the aerodynamic approaches to the estimation of evaporation but requires much more meteorological data. Different versions of the basic formula have been developed. A simple form is:

$$E = \frac{\Delta\,(Q^* - Q_G) + \gamma E_a}{\Delta + \gamma}$$

In this equation the terms are as follows:

Δ is the slope of the saturation vapour pressure curve at air temperature (Figure 5.7, p. 58);
Q^* is the net radiation flux;
Q_G is the soil heat flux;
γ is a constant; and
E_a is a measure of the aerodynamic evaporation.

E_a cannot be measured directly and is obtained from the formula:
$E_a = 0.35(e_a - e_d)\,(1 + U/100)$
where e_a is the saturation vapour pressure expressed in mb at the mean air temperature;
    e_d is the actual vapour pressure expressed in mb; and
    U is the run of wind in km day^{-1} at 2m.

If Q^*, the net radiation flux, is not measured, it can be obtained from the expression:

$Q^* = 0.75Q_a(0.18 + 0.55n/N) -$
$0.95\sigma T_a^4\,(0.10 + 0.90n/N)\,(0.56 - 0.092\sqrt{e_d})$

where Q_a is the amount of short-wave radiation reaching the outside of the earth's atmosphere at the latitude of the site expressed in mm water equivalent;
    n/N is the ratio of observed hours of sunshine to the maximum possible number of hours;
    T_a^4 is the theoretical black body radiation at mean air temperature T_a (in degrees Kelvin); and
    e_d is actual vapour pressure in mb.

One advantage of Penman's method, compared with some others, is that it requires meteorological measurements to be made at one level only. Unfortunately net radiation is rarely measured directly and has to be derived by the method outlined above. Penman's method appears to be the most appropriate approach for conditions in Britain. The pattern of PE across England and Wales calculated by this method is shown in Figure 13.13.

Moisture budget methods

An alternative method of obtaining actual evapotranspiration is by the **moisture balance equation**. At a site, the moisture balance can be expressed as:

$$P + I = E + R + D + \Delta S$$

where P is precipitation, I is irrigation, E is evapotranspiration, R is runoff, D is drainage to bedrock and ΔS is the change in soil moisture content. As before, if we knew all the other elements in this equation we could calculate E by difference. Several of the components present little problem, for precipitation and irrigation inputs can easily be measured, as can runoff. But drainage to bedrock (D) and changes in soil moisture content (ΔS) are rarely known. As a result, the method can only be used on a large-scale, long-term (e.g. annual) basis where it can be assumed that drainage to bedrock is balanced by release from spring seepage, and where changes in soil moisture content are negligible.

It is clear from what we have said that evapotranspiration remains one of the most difficult aspects of the hydrological system to measure. For this reason, if no other, our knowledge of the processes involved remains uncertain. For the same reason, it is difficult to give precise figures to the global pattern of evapotranspiration. None the less, it is useful to consider the role of evapotranspiration within the global hydrological cycle.

Evapotranspiration in the hydrological cycle

As we have noted, evapotranspiration provides the main output of moisture from the surface hydrological system, returning water to the atmosphere. So far we have discussed some of the processes involved, but it is important to appreciate that evapotranspiration occurs in many different stages of the hydrological cycle. Thus, losses of water to the atmosphere may take place at any point within the system (Figure 13.14).

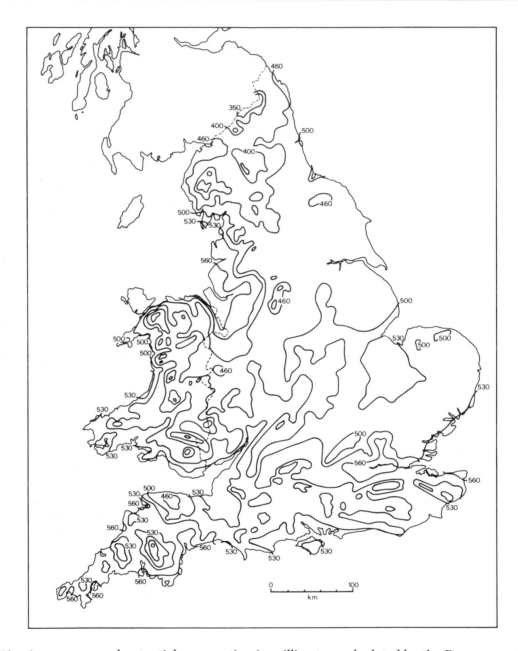

Figure 13.13   Average annual potential evaporation in millimetres calculated by the Penman method for a surface with an albedo of 0.25. Isopleths at 350, 400, 460, 500, 530 and 560 mm intervals (after Grindley, 1972)

One of the major losses, for example, occurs during precipitation. Considerable amounts of moisture may be evaporated during rainfall and the small droplets in particular may be totally evaporated before they reach the ground. Similarly, moisture which is intercepted by the vegetation is also susceptible to direct return to the atmosphere by evaporation. The amount of moisture retained on the vegetation during a storm varies according to

the character of the storm, the species of plant, the leaf density (and therefore the time of year) and, of course, the vegetation density. In the case of woodlands as much as 50 per cent of the incoming water may be retained in the canopy and returned as evaporation. In the case of more low-lying vegetation, such as grass, the amount of interception is not known with such certainty, partly because of the difficulty of measuring interception in such

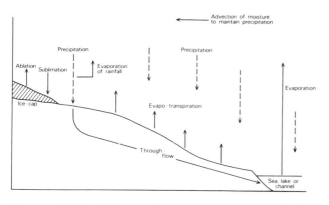

Figure 13.14 Schematic representation of the hydrological cycle showing the main losses of water to the atmosphere

crops. Nevertheless, again it seems likely that interception may reach 20–30 per cent; we only have to walk across a grass field following rain to appreciate the quantity of water trapped in the vegetation. Once more, most of this is lost through evaporation.

A proportion of the water which reaches the soil is also returned by evaporation, for in heavy storms rainfall often collects in surface depressions and these are gradually dried by the sun. Similarly, some of the rainfall flows across the surface as runoff, and further evaporation losses may occur at this stage. Rates are generally low, however, for turbulence mixes the water and disperses the heat from solar radiation through the water body. Thus much greater inputs of energy are needed to heat the water and less is available to carry out evaporation. Nevertheless, losses at this stage are often important for man. Open canals and reservoirs may lose considerable quantities of water through evaporation (Table 13.3), and, in arid areas especially, this may represent an irretrievable loss of an important resource.

In vegetated areas, the major process of moisture return is by transpiration. Rates of transpiration vary according to the character of the vegetation and therefore change over both space and time (Figure 13.15). They are at a maximum when the

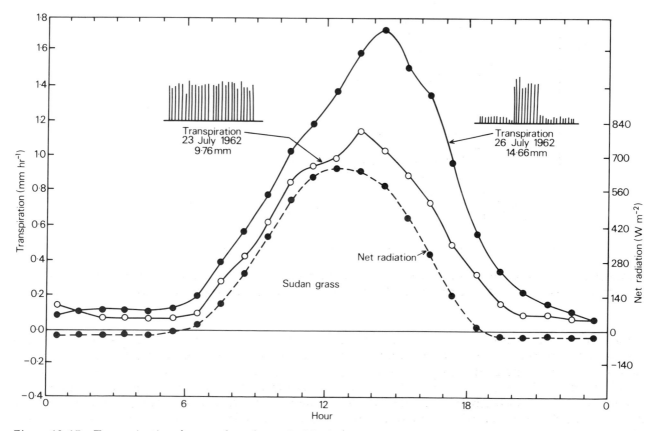

Figure 13.15 Transpiration from a closed stand of Sudan grass at Phoenix, Arizona on 23 July 1962 and from an isolated 1 m² plot of the same grass three days later. Average net radiation for the two days, which differed by only 5 per cent, is also shown (after Van Bavel *et al.*, 1963)

Table 13.3 *Mean daily evaporation from an open water surface (mm day⁻¹), Molato, Italy*

J	F	M	A	M	J	J	A	S	O	N	D
–	–	1.8	2.7	3.0	6.2	7.9	5.7	4.3	4.1	2.3	1.3

Source: H. L. Penman, 'Natural evaporation from open water, bare soil and grass', *Proceedings of the Royal Society*, **193** (1948), pp. 120–45.

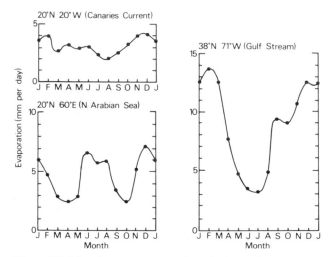

Figure 13.16 Average annual variation of evaporation at selected oceanic locations (after Sellers, 1965)

vegetation is in full leaf and the soil is moist; they decline as the plants lose their leaves or the soil dries out. During the course of a single year, therefore, transpiration losses may show complex fluctuations in response to prevailing conditions.

Without doubt, the major evaporative losses occur from the sea – possibly 85 per cent of the global return to the atmosphere is from the oceans. The reason is not only the great extent of the oceans – some 70 per cent of the world surface – but also the fact that evaporation can continue at or near the potential rate. Unlike evapotranspiration from the land, the process is unhindered by water shortages. Even so, seasonal and regional differences in evaporation can be seen due to the effect of changing meteorological conditions (Figure 13.16).

Finally, evaporation may occur from the other main storage component of the hydrological cycle – the cryosphere. In general losses are small, for it requires large amounts of energy to convert ice to water vapour – a process known as sublimation; some 2.83×10^6 J are needed to evaporate 1 kg of ice at 0°C. Sublimation does occur in the marginal areas of glaciers and ice sheets, however, where seasonal inputs of solar radiation may be high, and perhaps 2 per cent of the moisture is returned to the atmosphere each year in this way. Moreover, in the past the process was much more important. During the latter parts of the glacial periods, for example, as the ice sheets that had spread into the mid-latitudes began to retreat, sublimation must have been one of the main processes of stagnation and decay. Warm, turbulent and often relatively dry air masses then moved across the ice margins, drawing vast quantities of moisture from the ice sheets and causing them to retreat over areas of thousands of square kilometres.

Conclusions

The importance of evapotranspiration to the hydrological cycle is apparent. It provides a vital link between the surface waters and the atmosphere, a link operating at almost every stage of the hydrological cycle (Figure 13.14). In the process of returning water to the atmosphere, evapotranspiration involves a major exchange of energy; thus it is also a vital part of the energy cycle of the atmosphere. In addition, evapotranspiration has an overriding influence upon vegetation growth. Transpiration provides the means by which nutrients are absorbed by plants, and by which plant temperatures and turgidity are maintained. When evapotranspiration is prevented, as for example when the plant is kept in a totally saturated atmosphere or when there is no available moisture in the soil, plant growth ceases.

Through the control evapotranspiration exerts on the hydrological cycle, and through its effects on plants, it also has a fundamental influence upon man. In many parts of the world high rates of evapotranspiration constrain agriculture, and man has been forced to adapt to the conditions by using irrigation or by growing drought-resistant crops. Excessive evapotranspiration, like lack of rainfall, is also responsible for the hardship and famine so apparent in many arid areas. Unfortunately, as we have seen, evapotranspiration remains one of the most problematic parts of the hydrological cycle to study. Much more research is needed before we fully understand the processes and their implications, or can do much to plan for their consequences.

14

Runoff and storage

The hydrological crossroads

King Alfred the Great was not just a great king. He was also something of a philosopher and scientist. None the less, he had what would today be regarded as a curious notion about the working of the hydrological cycle. In a book called *Boethius*, he claimed that all rivers flowed to the sea, but that from there the waters seeped into the rocks and made their way back through unknown passages to the same stream from which they emerged! We all know now, of course, that this is wrong; that streams obtain their water from rainfall. But what happens between the rainfall and the stream? How does the water reach the channel, and what happens to it on the way? These are some of the questions we will be answering in this chapter.

They are questions which have considerable importance, for what happens to the rainfall after it has reached the ground influences the way in which streams behave, the supply of water to man and many processes of erosion in the landscape.

Where, then, does the water go? There are four possible routes (Figure 14.1). It may go straight on at the surface, infiltrating into the soil and percolating downwards towards the bedrocks. It may be deflected laterally, flowing either in the soil or on top of it downslope towards stream channels and thence to the sea. Or it may be trapped and then returned whence it came, by evaporation (or transpiration) to the atmosphere.

Thus the ground surface acts like a huge hydrological crossroads, diverting the rainfall four ways. The relative proportions going in each direction are extremely significant, for man obtains much of his water from groundwaters; these need to be replenished by percolation (route 1 in Figure 14.1). Similarly, stream flow is maintained in part by the water which moves laterally in or on the soil (routes 2 and 3); if this supply of water ceases then the streams may dry up. Plants obtain their moisture

from the soil and if sufficient water is not retained close to the surface then the vegetation cannot continue to transpire and the plants may wilt and die.

It is not only the amount of water that is significant. The rate of movement is also critical. In particular, the speed at which the water flows laterally into rivers and streams affects the rate at which these respond to rainfall, and their liability to flooding. In many of the more arid parts of the world the extremely rapid runoff of water into streams during rainfall results in an almost instantaneous torrent of water developing in the channels. People have been drowned as a result; campers pitched on what were once dry river beds have been swept away by the flash flood before they could escape.

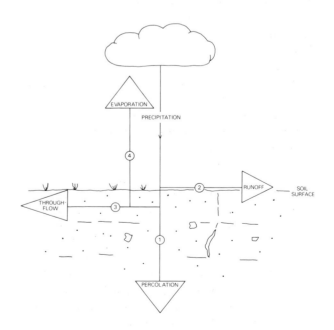

Figure 14.1   The hydrological crossroads

213

Overland flow

The generation of overland flow

The streets of our local town are soon awash during a heavy storm, as water collects in the gutters and pours into the drains. Off the streets, in the parks and gardens, the effect of the storm is less dramatic; very much more of the rainfall disappears. The difference is because in the street the rainfall cannot immediately soak into the ground. The road surface is almost impermeable and much of the water gathers on the surface, then runs off as **overland flow**. The soil, however, is far more permeable, much of the rainfall infiltrates and relatively little remains on the surface. In many cases it is only during the heaviest of storms, or the most prolonged periods of rain, that overland flow occurs on the soil.

These observations provide us with the key to understanding the processes by which overland flow is generated. Two main groups of factors govern the process: the permeability or, more strictly, the **infiltration capacity** of the surface, and the intensity and duration of the storm. Overland flow occurs when the amount of rain reaching the surface is greater than the ability of that surface to absorb the water.

The reasons for this imbalance between rainfall and infiltration vary. Two main theories have been developed to explain it, and as we may imagine there has often been conflict between geographers who support the two views. We will see, however, that both are probably valid under different circumstances; indeed, there is less difference between them than is often implied.

Hortonian overland flow

The traditional concept of overland flow generation is attributed to R. E. Horton, and runoff which seems to fit his theory is often referred to as **Hortonian overland flow**. Horton worked mainly in the south and west of the United States, and as we will see this almost certainly influenced his ideas of what happened during a storm. In simple terms, he argued that overland flow occurred when the intensity of the storm was so great that the soil could not absorb the water quickly enough. In other words, when rainfall intensity exceeded infiltration capacity, overland flow was produced.

Because the infiltration capacity of many soils is high – that is to say, the soil can rapidly absorb large quantities of water – Hortonian overland flow

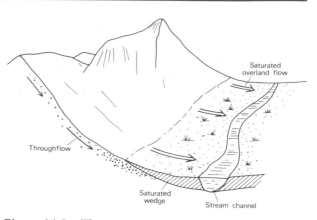

Figure 14.2 The generation of saturated overland flow

occurs mainly in areas which experience very intense and sudden storms, such as the semi-arid areas of south-west USA.

Saturated overland flow

Elsewhere, Hortonian overland flow is less common. Instead, much of the water infiltrates the soil, then passes through the soil layers as throughflow, or percolates to the groundwater (Figure 14.2). Nevertheless, even in these areas, surface runoff does occur, but it is explained by rather different processes. Overland flow in these instances is known as **saturated overland flow** and it occurs as follows.

During the early part of the storm, rainwater infiltrates into the soil and drains slowly downwards. The rate of percolation is often slow, however, especially at depth where the soil is more compact and where channels created by roots and soil animals (e.g. earthworms or termites) are absent. At the same time, inflow of water may come from adjacent areas so that water enters the soil more quickly than it can escape. Ultimately, the surface layers of the soil become saturated. No more water can infiltrate and overland flow occurs.

Rainfall intensity and duration

In both cases, one of the critical factors influencing the generation of overland flow is the amount of rainfall. Hortonian overland flow is produced mainly during very intense storms. Fortunately these are relatively rare and short-lived. As we saw in Chapter 12 they occur mainly in association with unstable or cyclonic air and are often localized. Saturated overland flow tends to occur following more prolonged rainfall. It develops mainly in slope-foot or bottom-land sites where slope angles are low, the soil is naturally wet and high inputs of water from upslope take place (Figure 14.2). Although the intensity of precipitation may be low, therefore, gradual saturation of the soil occurs. Ultimately the soil becomes waterlogged and any additional input of water, either from upslope or direct rainfall, spills across the surface.

Both forms of overland flow are often localized for rainfall intensity varies from one place to another. Often this variation is topographically controlled, and exposed slopes catch much of the rain, while the sheltered areas are left in a slight rain shadow. In addition, the vegetation modifies the rainfall so that the intensity and distribution of rain at the ground surface are not the same as that above the vegetation canopy. We can see the effects of **interception** most clearly in woodlands. The tree canopy may intercept as much as 30 per cent of the rainfall in heavy storms, and much more in light showers (Figure 14.3). Much of this may be trapped on the leaves until returned to the atmosphere by evaporation. Much of the remainder is diverted down the tree trunk, where it is known as **stemflow** (Figure 14.4). Often 10 per cent or more of the precipitation may reach the ground as stemflow, and it is clear that this water is concentrated into a very small area around the base of the trunk. There is also a tendency for part of the rainfall to be shed at the edge of the canopy (canopy drip), giving a zone of high intensity around the fringe of the tree (Figure 14.4). As anyone who has sheltered from a storm knows, relatively little rainfall passes through the canopy by **throughfall**. At least during the early parts of the storm the canopy acts as a most efficient umbrella. As the leaves become saturated, however, more and more of the rainfall makes its way through the canopy as **leaf drip** and this umbrella effect is reduced.

The ability of trees to intercept rainfall in this way varies according to species and time of year. Clearly far less is intercepted when the trees are leafless. Other types of vegetation also intercept rainfall in a similar way, and thus during a single storm inputs of rainfall beneath a vegetation cover are almost always less than in the open land (Table 14.1). In addition, the distribution and intensity of the rainfall beneath the canopy is markedly irregular, with the result that overland flow may be produced locally (e.g. around the tree trunk) even in relatively gentle storms. This, in fact, is one reason why the soil is often bare and eroded next to the trunk.

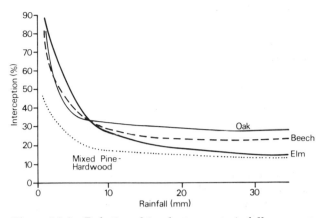

Figure 14.3 Relationships between rainfall amount and interception for different tree species (data from Kittredge, 1973)

Infiltration capacity

The second main factor influencing the generation of overland flow is the infiltration capacity of the soil – the ability of the soil surface to absorb water. This depends upon a variety of conditions, some of them, such as the texture and structure of the soil, relatively permanent, and others, such as the existing moisture content, relatively transient.

The effect of soil texture and structure

Water enters the soil by flowing into the small holes and cracks (the **pores** and **fissures**) in the surface. The pores occur mainly between the individual

Table 14.1 *Rainfall interception by different types of vegetation*

Vegetation cover	Condition	Interception (per cent)
Douglas fir	25 years old	43
Spruce-fir	Mature	37
Lodgepole pine	Mature	32
Hemlock	Mature	30
Heather	Different ages	35–66
Mixed grass sward	Full cover	26
Clover	Full cover	40
Maize	During growing season	16
Maize	Full cover	40–50
Soybean	Full cover	35
Oats	During growing season	7
Oats	Full cover	23

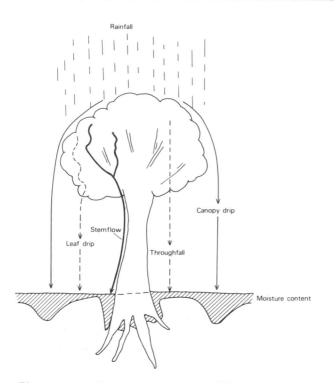

Figure 14.4   The routing of rainfall by trees

by organisms and plants roots which burrow or push their way into the soil. The soil particles, however, are grouped together into aggregates or peds separated by thin fissures and planes of weakness. These provide extensive, large voids within the soil into which water can flow. Clearly the extent, number and size of pore spaces (which together define the porosity of the soil) vary considerably. Many factors affect these properties. The size of the particles making up the soil (soil texture) are important. In general, fine-grained soils are more compact and have smaller voids than coarse-grained soils. As a result, water infiltrates less rapidly. Soil management is also important. Ploughing breaks up and overturns the surface soil, increasing the porosity and encouraging infiltration. Compaction, by machinery (e.g. in the ruts caused by tractors) or by animals, reduces porosity and inhibits infiltration. Vegetation has a further effect. A dense network of roots increases the infiltration capacity by providing channels in the soil, while the debris from the plants may provide a loose, porous surface layer into which the water can readily infiltrate.

Antecedent moisture

The other main factor of importance is the existing moisture content of the soil (the **antecedent moisture**). So long as the pores and fissures in the

particles of the soil and they exist because the particles do not fit tightly together (Plate 14.1). In addition, large pores or micro-channels are created

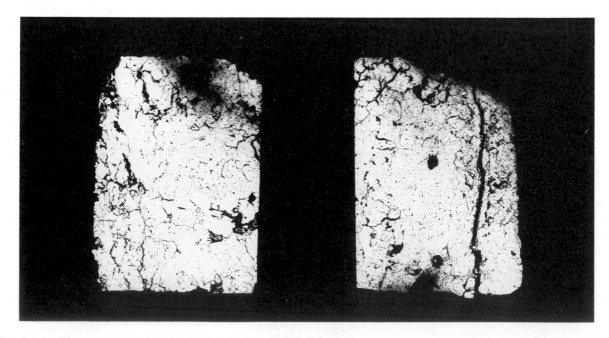

Plate 14.1   Thin section of a block of soil showing pores and fissures (black) within the soil mass (photo: John Owen)

216

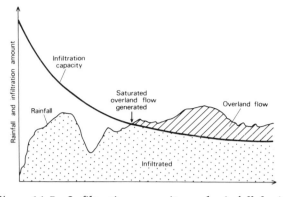

Figure 14.5 Infiltration capacity and rainfall during a storm and the generation of saturated overland flow

soil are free of water they can absorb the rainfall and infiltration can continue. But what happens when the pores become filled with water? The result is that infiltration can no longer occur and, as we have seen, saturated overland flow takes place. Thus, as the soil becomes saturated and the water infiltrating the soil builds up in the upper layers, the infiltration capacity falls until it equals the rate of downward percolation through the soil. At the point where the infiltration capacity falls below the rainfall intensity, overland flow is produced (Figure 14.5).

One factor complicates this picture. During rainfall, as the soil becomes wetter, changes in soil structure tend to occur which act to reduce infiltration capacity even more. The impact of raindrops on the surface may detach particles which lodge in the pores and seal the soil, producing a surface crust. In some cases this **crusting** is very severe and it leads to rapid reduction in infiltration. Clayey soils may also swell as they become wet; water becomes bound up within some of the complex clay particles in the soil, causing them to expand. As a result, the pores are closed up and infiltration capacity falls. When the soil dries again the clay shrinks and cracks may develop. Clearly, in this state, the soil has a very high infiltration capacity.

Measurement
It is apparent that the infiltration capacity of the soil varies markedly over both time and space. Variations in infiltration capacity also account for localized development of overland flow. For this reason, infiltration capacity is an important regulator within the hydrological system, and a vital property to measure when we are trying to monitor the system. In fact, measurement is fairly easy and

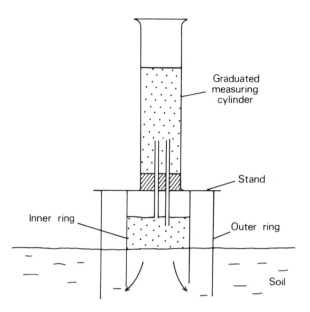

Figure 14.6 The ring infiltrometer for measurement of infiltration capacity. Measurement is normally made in mm hr^{-1}

the simple equipment shown in Figure 14.6 gives reasonably good estimates of infiltration capacity.

Rill and channel flow
When the infiltration capacity is exceeded and water gathers on the soil surface, the first effect is normally for very localized flow to occur as water moves into small surface depressions and gathers there. It may be stored here for some time as **depression storage**, and if rainfall ceases very little overland flow may occur. Afterwards, the water slowly infiltrates into the soil as the underlying layers drain, or it may be evaporated back to the atmosphere. If rainfall continues, however, the depressions become full and the water overflows to make its way downslope. It is at this stage that extensive overland flow becomes visible.

Initially, flow is often widespread and the water forms a thin sheet over the surface. This is known as **sheetwash,** and it is most commonly seen in areas where the soil surface is relatively smooth and on the rounded crests of slopes. Sheetwash, however, rarely persists for any distance and it soon gives way to **rill** or **gully flow**. Rill development may start around stones or the low points of depressions where the flow is concentrated, or where surface unevenness or the effect of raindrops cause the sheetwash to become turbulent. In both cases localized erosion occurs and micro-channels develop which further concentrate the water. Eventually, the rills may become quite large and persistent at which stage they are known as gullies. We will examine the processes involved in gullying in Chapter 22.

Both sheetwash and gully flow are relatively rapid, and rates of flow of as much as a metre per second are common on steep slopes. Flow is most rapid in channels, however, because the water then is deeper so the effect of friction with the bed is reduced. Even so, it is clear that flow rates vary markedly according to slope and surface conditions. Velocities tend to increase with slope angle so that flow is most rapid on steep hillsides. Friction increases with surface roughness, so flow is reduced on ploughed soils or where there is a dense cover of vegetation.

Throughflow

The water that enters the soil does not all drain downwards into the bedrocks. Much of it flows laterally, eventually escaping back to the surface as seepage and contributing to stream flow. This is known as **throughflow** and it takes place mainly when the soil is fully saturated.

Compared to overland flow, movement in the soil is slow and often irregular. The reason is that much of the flow occurs in very small pores and fissures, and here, close to the soil particles, the water is affected by forces which hinder movement. We will examine these forces a little later.

Rates of movement reach a maximum in soils on steep slopes and in large channels or pore spaces. Much of the flow occurs through the coarser pore spaces in the soil (those greater than about 1 mm in diameter) but flow is most rapid in the tunnels created by rodents and other soil animals, and the moving water is often able to erode these tunnels to produce extensive systems of **pipes** within the soil. Rates of flow similar to those on land may be achieved in these pipes. In contrast, rates of movement through the fine pores in a heavy clay soil may be almost imperceptible; perhaps no more than 1 mm per day. Thus, the rates of throughflow are highly dependent upon soil conditions and vegetation as well as topography.

Water moving through the soil tends to gather at the slope foot so that the soil there becomes totally saturated (Figure 14.2). As it does so, the water starts to emerge along seepage lines and produce overland flow. In addition seepage may occur directly into the stream through the bed and banks. In this way, most of the throughflow reappears to join with the water moving by sheetwash and in rills and gullies and enter streams. It contributes to the next stage in the runoff process: **streamflow**.

Streamflow

Streams represent permanent and relatively large features of the landscape, as we will see in Chapter 22. They also play a vital role in the hydrological system for they conduct water across the land to the sea.

Streamflow varies markedly over time in response to the inflow of water from the surrounding land. Thus, if we monitored streamflow we would find that the volume of water passing through the channel – the stream discharge – fluctuated from day to day and even from hour to hour (Figure 14.7).

Measurement of streamflow

Stream discharge represents the volume of water passing through the channel in a given period of time. It is the product of the width and depth of the channel and the velocity of the water:

$$Q = w \times d \times v$$

where Q is the discharge, w is width, d is depth and v is velocity. Normally discharge is measured in cubic metres per second (sometimes called cumecs).

Because width, depth and velocity all tend to vary together as discharge increases, measurement of discharge is complex. One-off measurements

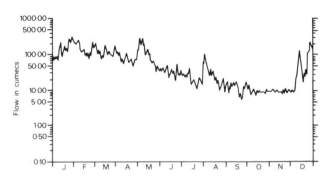

Figure 14.7   A stream hydrograph: River Thames at Teddington, 1978 (from DoE Water Data Unit, 1983)

218

Plate 14.2 A current meter: the speed of rotation of the propeller when it is immersed in the water gives a measure of flow velocity (photo: John Owen)

Plate 14.3 A shallow V Crump weir at Slippery Stones on the Derwent River, Derbyshire. The discharge is determined by monitoring the height of the water surface above the weir crest, using a float attached to a recorder (photo: Rick Cryer)

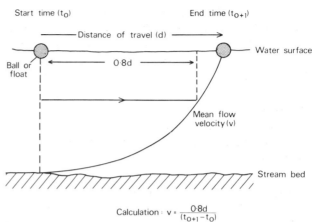

Calculation: $v = \dfrac{0 \cdot 8d}{(t_{o+1} - t_o)}$

Figure 14.8 Measurement of stream velocity by the use of floats. The time taken for the float to pass over a measured distance is determined and the surface velocity computed. This is then converted to an estimate of mean flow velocity by multiplying by 0.8 (the ratio of mean flow to surface flow velocity)

may be made by computing the depth and width of the flow at a cross-section, and then measuring the velocity either with a current meter (Plate 14.2) or by assessing the time taken for floats to pass over a measured distance (Figure 14.8). When we need to obtain a more continuous record of streamflow, however, this approach is clearly too time-consuming, so permanent, automated measuring sites are set up. These involve two main types of structure: weirs and flumes (Plate 14.3).

In general, the relationship between width, depth and velocity of the water flowing over these structures is calculated at different discharges. Then, only the depth of flow needs to be measured; the other parameters can be calculated from these relationships. Commonly, a flow recorder is installed, consisting of a float attached to a pen. As the depth of the water varies the float moves the pen up or down and the pen traces the changes on a rotating chart (Plate 14.4, Figure 14.7). In this way a continuous record of streamflow is gained.

When this record is expressed as stream discharge it allows us to see the change in streamflow during an individual storm. The graph of flow is then called the **storm hydrograph**.

The storm hydrograph
Typically, the storm hydrograph consists of a curve

Plate 14.4 A stage recorder: the pen is tracing a hydrograph on to the rotating chart (photo: John Owen)

something like that in Figure 14.9. Soon after the onset of rainfall, the hydrograph rises rapidly, showing an increase in discharge. The time between the start of rain and the rise in the hydrograph (or the rainfall and streamflow peaks) is known as the **lag**; it represents the time it takes for

the water to move overland and through the soil into the stream channel. Following this steep **rising limb** of the hydrograph the curve peaks, then starts to decline as discharge falls. This **falling limb** of the hydrograph tends to be less steep than the rising limb, showing that the decline in discharge occurs more gradually. Ultimately, the hydrograph settles at a low level and it characteristically continues at this level until the next storm. This is known as the **base-flow stage**.

Contributions to the storm hydrograph

The shape of the storm hydrograph reflects two important factors: the character of the surrounding land (the catchment area of the stream) and the nature of the storm. If we compare hydrographs for different storms or from different streams, therefore, we can detect subtle variations in the shape of the hydrograph. Let us look at the processes involved in creating the storm hydrograph by looking at events during a single storm. We will assume that there has been no rain for several days so the stream is at its base flow stage. We will also assume that the storm which then occurs is sudden and very intense.

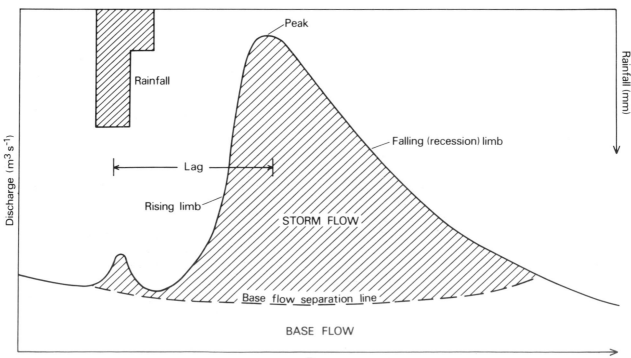

Figure 14.9 The storm hydrograph

The first evidence we see of the storm within the hydrograph is a small 'blip' (Figure 14.9). What causes it? The answer is rain falling directly on to the stream – **channel precipitation** as it is known. Clearly this involves only a very small quantity of water, but in an intense storm it leaves its mark on the hydrograph.

As we have seen, during the early part of the storm much of the rain seeps into the ground, so very little overland flow occurs. That which does often ends up trapped in depressions on the surface and does not reach the channel. For some time, therefore, the stream does not receive any water from the surrounding land. This is what creates the lag. Only after it has been raining for a while is overland flow generated, and there is then a short delay while this water moves to the stream channel. Flow is rapid on the steeper slopes of the valley side, but on the flat land beside the channel flow is slow, so often it is an hour or so before this water reaches the stream. As water starts to enter the channel, the storm hydrograph starts to rise. It goes on rising as the pulse of water from around the stream flows into the channel. When the rate of inflow is at a maximum, the peak of the hydrograph is reached.

By this stage some of the water which had soaked into the soil may also have reached the channel. In particular, that which flows more rapidly through pipes and large pores in the soil may emerge in the stream bank as the soil around the channel becomes saturated. This contributes to the rising limb of the hydrograph.

The inflow of water flowing overland continues for a short time after rainfall ceases, but very soon all this water has washed into the channel and the input of surface water declines. If this were the only input of water then the hydrograph would fall as quickly as it had risen. Indeed, this tends to happen where Hortonian overland flow occurs. Where throughflow is occurring, however, the fall in the hydrograph is slowed down, for water continues to seep into the channel as the flow within the soil makes its way downslope to the stream. It may take many hours or even days for this flow to cease. When it does, the hydrograph returns to its base flow stage.

As we have seen, the interplay of these processes depends very much on the characteristics of the storm and the catchment area of the stream. For this reason the pattern we have identified gives only a general picture of the contributions to the

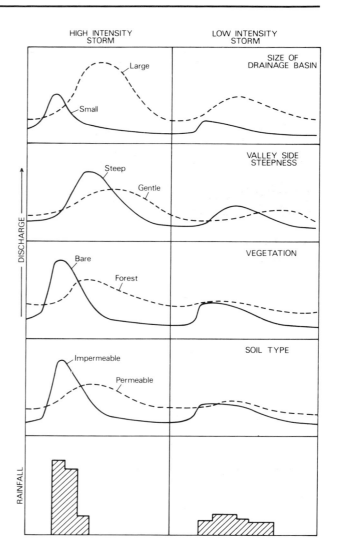

Figure 14.10 Hydrographs for different catchments under two different storm conditions

hydrograph. Figure 14.10 indicates the types of variation that may occur in different conditions.

Base flow

So far we have looked at the shape of the storm hydrograph, but what about the flow that occurs between storms – the base flow? Where is this water coming from and what controls the base flow?

The intriguing thing about the base flow is that it may continue for many months after rainfall; even during periods of prolonged drought it persists with little variation. Thus it is more or less independent of short-term rainfall events, although there may be a slight seasonal fluctuation. As this implies, base flow represents water which is being

221

released from relatively long-term storage, rather than water which is being cycled through the landscape following rainfall. There are two main contributions to base flow. The first is the water held within the soil and released by excessively slow drainage through the small pores in the soil. The second, and often most important, contribution is from the bedrocks; from the groundwater.

Overland flow and man

The water which runs off over the surface, whether as sheetwash, rill and gully flow or streamflow, has many implications for man. It may erode his soils, wash away his fertilizers and seeds, flood his land and his towns. It represents a hazard. Let us look at some of the effects of this hazard.

The terror induced by a flood can only be imagined. Consider the words of an eye-witness in Trinidad, Colorado:

Thursday (17 June 1965) the water was running about normal. . . . I had pulled the blinds down and then I got up to look out the front door, and there it was coming up as fast as everything. By then it was already around the house. It started coming in under the door and I knew it was going to be bad, but by then it was too late to get out.

I tried to call the Sheriff. . . . When I placed the phone back on the stand, I was in water almost up to my waist. . . . Right then with the water pouring in and the noise and everything, I really thought I was a goner.

I put my dog on the bed and I got up there too. I held tight on the head of the bed and it started jumping around, floating here and there. The water just kept coming in and filling things up. I was afraid it would get so high I would drown.

Just about then . . . a part of the wall right in the bedroom broke, and then water was running out as fast as it ran in. I could hear windows breaking and doors were crashing and furniture was swimming around. . . . I began to get awful cold and was shivering wet. It smelled awful, that mud and dirt in the water.

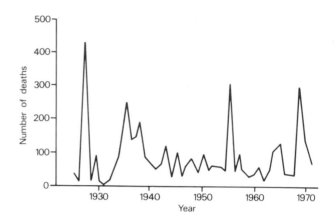

Figure 14.11 Deaths by drowning in the USA, 1925–71 (data from Ward, 1975)

Not everyone survives to tell the tale. In August 1969 Hurricane Camille swept through Virginia, causing floods that killed 152 people. In fact, between 1903 and 1972 an average of 100 people were killed each year by floods in the USA (Figure 14.11). These figures are insignificant compared to the great floods of history, such as those in Hunan Province, China, in 1887, which killed an estimated 1 million people.

The effect on man is not only direct. Floodwater may bury or ruin crops, destroy roads and buildings, sweep away vehicles and cause untold ecological damage. A flood in Rapid City, South Dakota, on 9 June 1972, for example, not only killed 237 people and injured a further 3057, but destroyed 5000 cars, 1335 homes and did a total of $160,000,000 of damage. Often, in the wake of the floods comes famine and disease.

In the face of floods of this nature, man has inevitably learned to take precautionary measures. He may avoid the most flood-prone areas. He may try to build flood-proof houses and factories. He may try to control the stream itself. In these ways he

Table 14.2 *Effects of flood-proofing on flood damage, Lismore, NSW, Australia*

Flood height (m)	Potential damage (without flood-proofing) A$	Actual damage (with flood-proofing) A$	Actual damage as per cent of potential
13.0	25,630,000	5,156,00	20.1
12.0	12,735,000	2,350,000	18.5
11.0	3,614,000	688,000	19.1
10.0	291,000	53,200	18.1

Data from D. I. Smith, 'Actual and potential flood damage: a case study in urban Lismore, NSW, Australia', *Applied Geography*, **1** (1981), pp. 31–9.

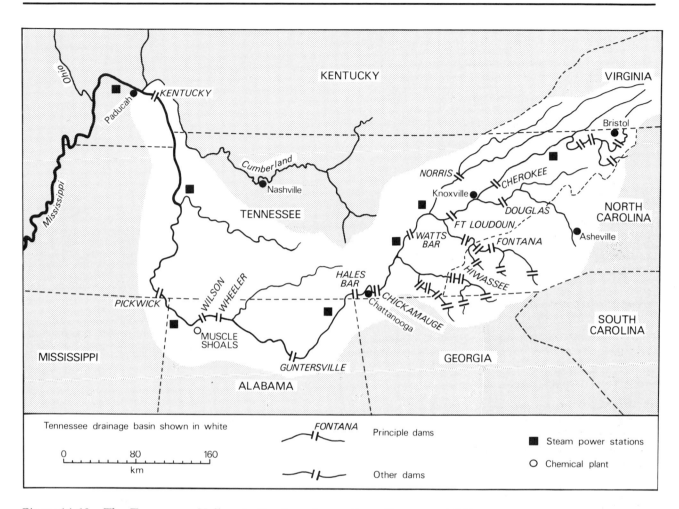

Figure 14.12   The Tennessee Valley Authority scheme (from Paterson, 1975)

can avoid loss of life and greatly reduce flood damage. In the Lismore area of New South Wales, in Australia, for example, it is estimated that flood-proofing has reduced damage to less than 20 per cent of its original level (Table 14.2). In the Fraser River valley in British Columbia the construction of dykes, dams and reservoirs has been proposed to regulate flow and to cope with discharges of as much as 17,000 $m^3 s^{-1}$ – a flow which is expected to occur, on average, only once in 150 years. To date, little progress has been made on this scheme, but the advantages in terms of flood reduction are apparent. As the Tennessee Valley scheme has shown (Figure 14.12) not only can flooding be controlled by such measures, but also the reservoirs may provide hydro-electric power, recreational facilities and a better water supply.

Not all attempts to control flooding are successful, however, and there have been several examples

of man exacerbating the flood problem. The flood at Rapid City, that we have already mentioned, occurred because heavy rainfall resulted in the failure of a reservoir dam. In 1864 the Dale Dyke dam near Sheffield collapsed, causing floods that killed almost 250 people. The consequences of stream control also extend well beyond the effect on flooding. Changing the stream regime may alter the rate of sediment transport with serious effects on deposition and erosion downstream. Ecological conditions, too, may be affected. When the Hoover Dam was built in the Colorado River, for example, changes in the river bed were seen as far as 560 km downstream. Construction of the Aswan Dam and Lake Nasser had even more far-reaching effects. Sediment, which once fertilized the soils in the lower reaches of the Nile, was trapped in the lake, causing siltation behind the dam and reducing soil fertility downstream. To counteract this, farmers have had to buy fertilizers, the cost of which outweighs the value of the electricity produced by the dam. In addition, reduction of the amount of sediment reaching the Nile Delta has led to erosion of the sand bar, which once protected the brackish estuarine water, and the consequent collapse of Egypt's sardine industry. As if this were not enough, the change in the regime of the Nile has resulted in the creation of stagnant water in dykes and pools in which water snails breed. These snails are vital in the life cycle of the Bilharzia fluke which ultimately attacks internal organs of the human body leading to a general debilitation and death. Clearly, man should attempt to manage the hydrological cycle with caution.

Water storage in the soil

If we monitored soil conditions after rainfall, we would find that the water in the soil slowly drained downwards under the effect of gravity for a day or so, after which drainage more or less ceased. At this stage, however, the soil would not be entirely dry. Instead, a considerable amount of water would remain in the soil. Why is this so? How does this water defy the pull of gravity?

The reason is that within the matrix of the soil, in the tiny pore spaces, the water is affected by forces that counteract gravity and operate to retain moisture in the soil. The main force involved is known as the **matric force** and it is due to two processes. Particles attract water molecules to their surface (**adhesion**) while the water molecules are also attracted to each other (**cohesion**). Together these effects retain the water in the soil. We can see these same processes in operation in many other situations as well; they are the reason that water droplets hang on a window-pane during rainfall, for example.

The matric force can be thought of as a negative pressure; one acting against the force of gravity to draw or suck water upwards into the soil. For this reason we often speak of it as a **soil suction** or **soil tension**. Traditionally this has been measured in bars (just as we used to measure atmospheric pressure), but now we use Pascal units (1 Pa = 10^{-4} bars). Close to the surface of the soil particle the matric force is extremely strong, equivalent to about 1×10^9 Pa (almost 10,000 times atmospheric pressure), but away from the particle surface it declines rapidly (Figure 14.13). Within a distance of only

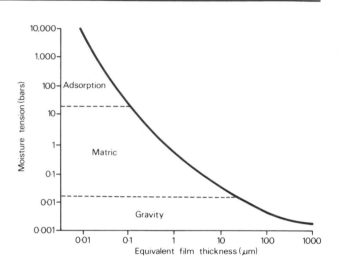

Figure 14.13 The relationship between soil moisture tension and the thickness of the water film. The moisture tensions shown represent the force operating to hold the water to the soil particles at the outer limit of the water film 1 bar = 10^4 Pa (from Briggs, 1977)

0.06 mm it has shrunk to about 5×10^7 Pa and is almost negligible. Thus, it is only the water held very close to the surface of the particles which is affected to any significant extent by the matric force. Water beyond about 0.06 mm from the particle is free to drain under the influence of gravity.

Because of the matric force, the soil remains wet for a long period after rainfall, thin films of water forming around the particles and persisting in the finest pores in the soil. This water is almost immobile, but not quite. It cannot move under the effect of gravity, but it can move through the effect of the matric force itself. Where there are differences in the matric force, the water is drawn from areas where the force is low to areas where it is high. These differences arise because of localized drying of the soil. As the soil dries out, either by evaporation or because plants withdraw moisture from the pores, the water films become thinner. As they become thinner, the matric force operating on the outside of the film increases. Thus water migrates from the wet areas of the soil where the water films are thick to the drier areas, where the films are thin. This movement can be in any direction, upwards, downwards or sideways, and it is entirely independent of gravity.

Movement in this way is known as **capillary**

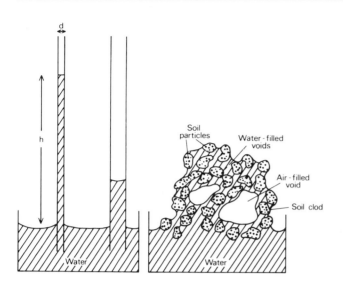

Figure 14.14 Capillarity: the height of capillary rise (h) is related to the diameter of the tube (d) as follows: h = 0.3/d

movement and we can see it in operation when we dip a thin glass tube into water (Figure 14.14). The water is drawn up the tube. It goes on rising until the weight of the water in the tube is just balanced by the force drawing it upwards. It is an extremely important process, for although it operates

relatively slowly in the soil it is one of the main ways by which water moves upwards towards the roots of the plants.

Not all the water retained in the soil is able to move in this way. That which is very close to the particles, closer than about 0.0002 mm, is almost wholly static. It is only removed when the soil is heated and evaporation takes place.

It follows from what we have said that we can define three main types of water in the soil: that which moves freely under the effect of gravity; that which moves more slowly through capillary processes; and that which is totally immobile. These are known as the **gravitational**, **capillary** and **hygroscopic water** respectively. Following rainfall, we find that the gravitational water drains away within about two days. The soil is then said to be at **field capacity**. Afterwards, if there is no more rainfall, the capillary water is slowly redistributed and drawn towards the surface where it is removed by plants and by evaporation. Eventually, almost all the capillary water is lost and the soil is said to be at wilting point. Plants cannot obtain the remaining moisture and they wilt and die. As we can see from Figure 14.15, the tension in the soil at this point is about 15×10^4 Pa. Most of the water which is left is referred to as hygroscopic water.

The soil water table

Let us now look at what happens to the

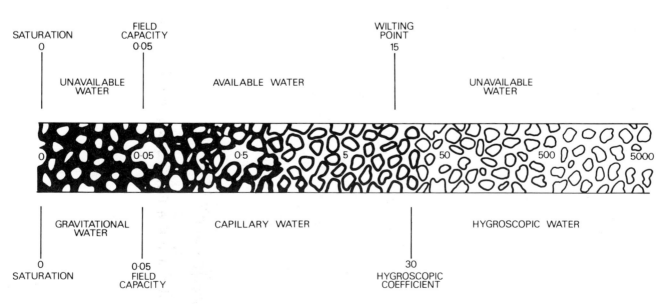

Figure 14.15 Classification of soil moisture according to biological (upper) and physical (lower) criteria. The figures represent soil moisture tensions in 10^4 Pa (from Briggs, 1977)

Figure 14.16 Relationship between the regional water table and surface topography

bottoms to contribute to stream flow. It is this water which is partly responsible for the base flow of streams. This movement is slow, however, and if we mapped the height of the water table in an area we would generally find that it approximately followed the surface topography (Figure 14.16). Thus, the water is trapped here for a considerable time, and soil water storage represents one of the main storage components of the hydrological system.

gravitational water: that which was free to drain downwards through the soil. Where does it go?

We can see the answer in many cases if we dig a hole in the soil. Ultimately, we reach a point where the hole starts to fill with water. We have reached the soil **water table**. Beneath this level the soil is totally saturated, and it is to this level that the gravitational water drains.

The fact that this saturated zone develops deep in the soil indicates that something must be preventing the water from draining downwards, and in many cases the reason is the presence of a dense, impermeable layer in the soil or the underlying bedrock. As we have already seen, the water can only move through the relatively large pores and fissures in the soil, and if a layer develops which has no large pores, then the water is trapped. Drainage ceases and the soil becomes perpetually waterlogged. A layer like this, which prevents water movement, is said to be **impermeable**. It may occur in the soil because of the presence of a very clayey layer, or because of compaction, caused, for example, by machinery.

The water below the water table is not entirely immobile, and over time it tends to flow laterally into lower areas, eventually emerging in the valley

Groundwater

As we have noted, the zone of saturation below the water table develops because downward movement of water is impeded by an impermeable layer. The same situation exists in the bedrocks. Water percolates downwards through the pores and fissures in the rocks until it reaches an impermeable layer. It then becomes trapped and a water table forms in the rocks. The water beneath is known as **groundwater**.

Rocks which store water in this way are termed **aquifers**. Chalk, sandstone and some limestones provide exceptionally good aquifers, and it is from these that man derives much of his groundwater. On the other hand, rocks such as clay, shale and many igneous rocks are impermeable and they are known as **aquicludes**. It is the presence of these rocks within the strata that halt the downward percolation of water and create the aquifer.

Aquifers develop in a variety of situations (Figure 14.17). In some cases alternating beds of permeable and impermeable rocks occur, giving rise to one or more perched aquifers. Water from these escapes as springs along a clearly defined springline (Figure 14.18). In other cases, the aquifer may be warped

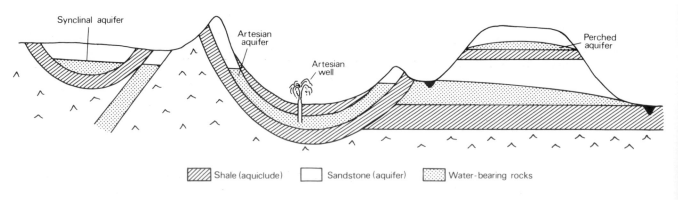

Shale (aquiclude)   Sandstone (aquifer)   Water-bearing rocks

Figure 14.17 Types of aquifer

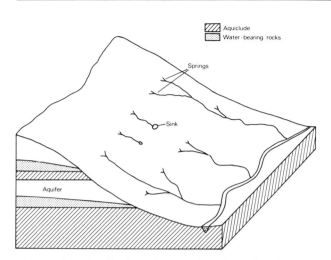

Figure 14.18 Spring lines developed along a perched aquifer. Note that where a stream passes off the impermeable aquiclude back on to the aquifer it may disappear underground at a sink.

into a synclinal basin and sealed by an overlying impermeable layer. Water then collects in the aquifer under considerable pressure, and where the water escapes back to the surface (in a well or along a natural fault) it emerges with considerable force. **Artesian basins** of this type are found adjacent to

mountain ranges, and they may be very important in supplying water in the arid interior of the continent. Probably the most well-known example is the Australian artesian basin.

Groundwater and man

Much of man's incessant need for water is met by use of the natural underground reserves of the aquifers. Over the years man's ability to extract these waters has improved. Simple stone-lined, hand-pumped wells and **qanats** (Figure 14.19) have given way to deep tube-wells from which the water is raised by pumps and hydraulic rams. As a result, the rate at which man abstracts the groundwaters has increased; with it have increased the problems of groundwater exhaustion and falling water tables.

When man draws water from a well, the water table around the well falls, through a process known as **draw-down**. This creates a **cone of depression** around the well (Figure 14.20). Water moves through the aquifer to replace that which man is removing by flow through the rocks. Where the rocks are porous and flow is rapid, replenishment of the well is efficient and the cone of depression is small. Where the aquifer is relatively impermeable and flow is slow, replenishment is much less efficient and the cone of depression is deep. In

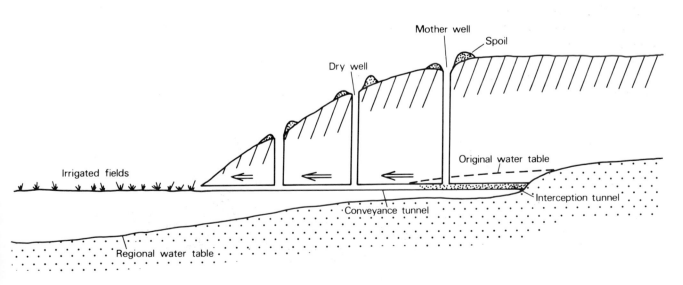

Figure 14.19 Qanat system. The dry shafts are dug to provide access to the conveyance tunnel which is excavated back (at a slope of about 0.5–5.0 m km^{-1}) until it intercepts the water table. As the water table falls due to exploitation, the tunnels are extended

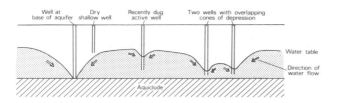

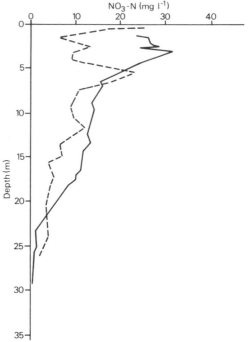

Figure 14.20 Cones of depression developed around wells. Water extraction in the London area has resulted in serious drawdown in this way over the last century. More recently, however, a decline in the rate of water usage has caused the reverse problem – a rise in the water table and a flood risk to subsurface structures such as the Underground

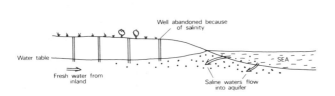

Figure 14.21 Salinization of groundwaters due to falling water table

Figure 14.22 Nitrate–nitrogen profiles in two sandstone aquifers in Britain. The increase in nitrogen contents towards the surface has been interpreted as evidence of increasing pollution by fertilizer applications since the 1950s (from Young and Gray, 1978)

either case, if man draws more water than can be replaced by lateral flow, then the cone of depression becomes deeper and eventually the well runs dry. Its owners then have two options: dig it deeper or drill another well elsewhere. If the well is deepened, then so is the cone of depression, and eventually – if the depth of the well does not make it too costly to raise the water in any case – the well reaches the bottom of the aquifer (Figure 14.20). It then has to be abandoned. If a new well is dug elsewhere, the process repeats itself.

Over time, of course, it is not only the lateral flow of water through the aquifer which is important. This may supply water to the well, but it is only redistributing water within the aquifer. If the water table is not to fall overall, then the aquifer itself must be recharged. As we have seen, this takes place mainly through deep percolation of water from the surface. It is the relative rates of this recharge and man's abstraction of the water which determines whether or not, in the long run, the aquifer is being depleted.

There are many examples of this happening. The apparently vast and limitless bodies of groundwater are often the result of thousands of years of slow percolation with no real extraction. Man extracts the water much more rapidly than it is currently being replenished and the water table falls. In the London area this has been taking place for several centuries and it is estimated that the water table has fallen by up to 45 m in thirty-three years. The same is happening in the Ogallala formation in the High Plains of Texas. The groundwaters here are ancient waters which probably built up over a period of several million years. In 1938, before large-scale pumping started, they were believed to have a volume of about 600,000 km³. Since then these reserves have dropped by about 110,000 km³. Man has been extracting the water far more rapidly than the scanty precipitation can replenish it.

The effect of a falling water table can be disastrous. Where the water table lies close to the surface quite a small fall may lead to the drying up of streams and lakes and problems of drought for the farmers as water contents in the soil decline. Near the coast, water may be drawn into the aquifer from the sea; saline water which contaminates the groundwaters and makes them useless for consumption and even irrigation (Figure 14.21). It is particularly acute in Malta, which obtains almost all its water from aquifers connected to the sea, and under extreme conditions can turn large areas of good agricultural land into virtual desert.

It is not only through water abstraction that man affects the groundwaters. He also influences them by pollution. Agricultural fertilizers in particular may seep into the ground and accumulate in the aquifers, and in time toxic levels of nitrate and phosphate have been known to build up. In parts of Britain it is possible to detect a gradual increase in pollutant levels in recent years as the fertilizers, which have been applied in increasing quantities since the 1950s, make their way into the aquifers (Figure 14.22). Unfortunately, the rate of percolation of this water into the aquifers is so slow that the damage may have already been done by the time the first signs are detected.

Conclusions

Runoff, infiltration and groundwater storage are vital parts of the hydrological system, governing the way in which the rainfall is routed through the landscape and controlling the time it takes for the water to reach the sea.

Waters which are diverted over the surface as overland flow move quickly and may enter the sea within days or even hours of reaching the ground. Water which soaks into the ground and reaches the water table may take months, years or even centuries to complete the cycle. Some indication of the routes which rainwater follows to reach the sea, and what happens to it on the way, is given by the hydrological game in Figure 14.23. Try it!

In addition to the control on the hydrological cycle, runoff and storage also affect many other aspects of the global system. The water which runs across the land surface is a major agent of erosion and transport and is responsible for cutting many of the features of the landscape, as we will see in Chapter 22. The water which soaks into the soil takes part in weathering reactions and it is vital to plants; without it the vegetation would die and the earth would be barren.

Figure 14.23 The hydrological game (see overleaf). (Instructions: players place their counters on the starting point (RAIN CLOUD). Each player throws the die in turn and moves along an arrow into the next box, according to the number thrown. First one to the SEA wins)

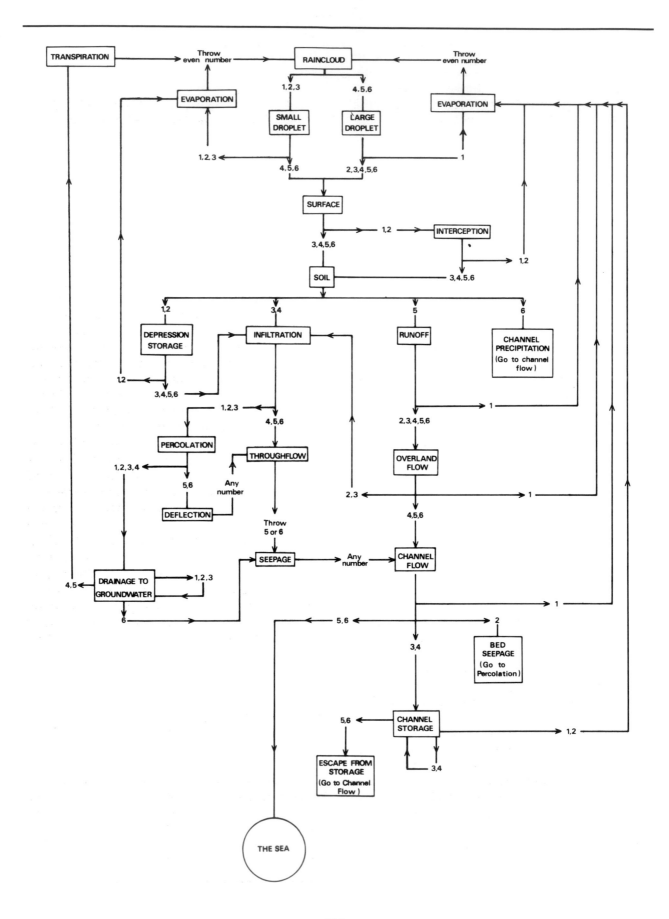

15

Oceans and their circulation

The character of the oceans

In recent years we have discovered a great deal about the oceans, but they still represent the 'great unknown' compared with the other parts of the global or solar systems. Man has set foot on the moon, sent spacecraft to Mars and Venus and photographed Jupiter and Saturn but the pressure of the water has restricted his exploration of the oceans to the uppermost layers. To delve beyond 100 m, man must depend upon instrumental technology. Slowly progress is being made to discover the nature of the deepest oceans, its animal life living in permanent darkness, its low temperatures and peculiar currents. Even on the surface we know less about the controls of water movement than we do about the atmosphere and yet the oceans are the greatest stores of water on earth, contain vast quantities of minerals and nutrients which affect our food supply, and play a fundamental role in moisture and energy cycling.

If so little is known about the oceans, why should they be included in a book about physical geography? Some of the reasons would become clear if we tried to contemplate the world without oceans and imagined what the effect might be. With no oceans, evaporation would become minute on a global scale and the hydrological cycle merely a small exchange between the freshwater store of the atmosphere and evapotranspiration from the ground. It is doubtful if there would be enough water to produce runoff. The earth's climates would all show continental qualities with hot summers and cold winters. In fact the changes would be so great that we cannot really imagine life as we know it without the oceans. Let us have a look, then, at the characteristics of the oceans before considering the actual processes which affect them and the consequences they have on the oceanic circulation.

The shape of the oceans

Geological evidence from sediments suggests that the oceans have existed for at least the past 3000 million years. But we now know from sea-floor spreading (Chapter 18) that the ocean basins we have today are relatively young features of the earth – less than 250 million years old – formed by movement of the crust away from the mid-oceanic ridges. As a result of this, the positions of the continents and therefore the sizes of the oceanic basins have also changed through time. Currently they are interconnected like spokes from the hub of the circum-polar Southern Ocean, northwards to the Pacific, the Atlantic and the Indian Oceans, in total covering some 70 per cent of the earth's surface.

The area of the oceans varies greatly, from the 166 million km² of the Pacific, to the mere 14.3 million km² of the frozen Arctic Ocean (Table 15.1). In terms of depth, the oceans are peculiar. It might be expected that they would gradually deepen towards the furthest point from land, but this is far from true. Some of the deepest parts of the ocean, such as the Marianas Trench which reaches 11,022 m, are quite close to land (Figure 15.1). These

Table 15.1 *Properties of ocean basins*

	Area (sq. km × 10⁶)	Mean depth (m)
Total oceanic area	361	3650
Pacific	165	4270
Atlantic	81	3930
Indian	75	3930
Arctic	14	1250
Minor seas	26	–

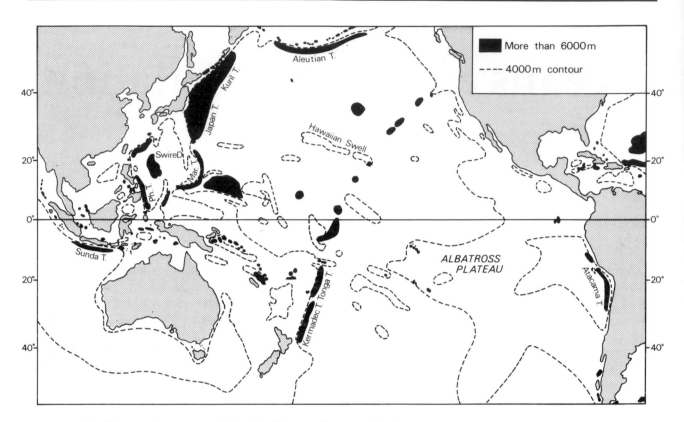

Figure 15.1 The configuration of the Pacific sea-floor and the location of the main trenches (Mar. T. Is the Marianas Trench) (after Monkhouse, 1975)

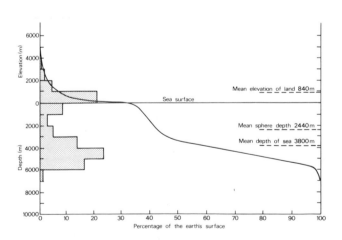

Figure 15.2 Hypsographic curve showing the area of the earth's solid surface above any given level of elevation or depth. At the left is the frequency distributions of elevations and depths for 1000 m intervals (after Lagrula, 1966)

oceanic trenches, as they are known, are most common in the Pacific, running parallel to the island arcs. However, almost all the oceans show relatively shallow seas near the continental coasts, then a sudden deepening down the continental slope before reaching the abyssal plains.

If we plot the frequency with which certain depths would be found in the ocean we obtain the graph shown in Figure 15.2. Nearly a third of the ocean floors are between 4000 and 5000 m below sea-level and 72 per cent between 3000 and 6000 m. The shallow continental shelves, less than 200 m deep, make up about 8 per cent of the oceanic area.

Temperature

Ocean temperatures vary considerably both with depth and laterally. Unfortunately, we lack detailed data on ocean temperatures so it is difficult to get a complete picture of the global pattern. There are few parts of the world where surface temperatures are taken regularly and even less where subsurface temperatures are recorded. Until the development of satellite photography we were dependent upon

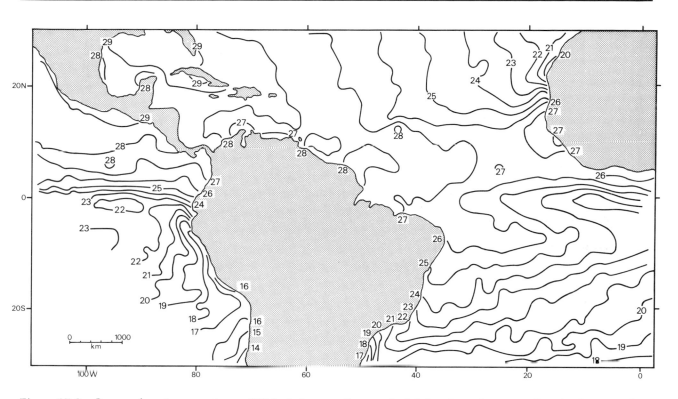

Figure 15.3   Sea surface temperatures (°C) in July over the tropical Atlantic and eastern Pacific Ocean (after Hastenrath and Lamb, 1977)

ships' logs which gave a dense network of readings along the main shipping routes and few elsewhere. Even now, areas with a high frequency of cloud cannot be monitored by satellites so sea surface temperatures are not well documented. Figure 15.3 shows the sea surface temperature for the tropical Atlantic and eastern Pacific based mainly on ship observations over a sixty-year period.

At a global scale we might expect that ocean temperatures would show a general decline away from the equator in response to reduced radiation inputs. To some extent this is true (Table 15.2). The Arctic Ocean is frozen because it does not extend into parts of the world receiving high inputs of radiation, and it is also more or less cut off from the Atlantic. The Red Sea, conversely, is almost entirely enclosed by hot desert lands with high energy receipts; it would be surprising if the waters were not warm. But several other factors affect sea temperatures, so many of the other oceans show a much more complex pattern.

One of the factors which modifies the effect of radiation is advection. This occurs mainly at depth

Table 15.2   *Mean surface temperatures of the oceans by latitude (°C)*

Latitude	Atlantic Ocean	Indian Ocean	Pacific Ocean
(°N)			
70–60	5.6	–	–
60–50	8.7	–	5.7
50–40	13.2	–	10.0
40–30	20.4	–	18.6
30–20	24.2	26.1	23.4
20–10	25.8	27.2	26.4
10–0	26.7	27.9	27.2
0–10	25.2	27.4	26.0
10–20	23.2	25.9	25.1
20–30	21.2	22.5	21.5
30–40	16.9	17.0	17.0
40–50	8.7	8.7	11.2
50–60	1.8	1.6	5.0
60–70	−1.3	−1.5	−1.3
(°S)			

From   H. U. Sverdrup, *et al.*, *The Oceans: their physics, chemistry and general biology*, Englewood Cliff, NJ: Prentice-Hall, 1970.

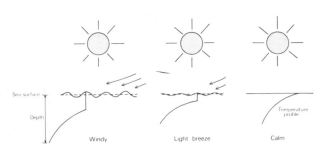

Figure 15.4 The effects of wind speed on surface water temperature profiles on sunny days

Table 15.3 *Concentrations of the major components of sea-water*

	Parts per thousand (‰)	Per cent
Chloride	18.980	55.04
Sodium	10.556	30.61
Sulphate	2.649	7.68
Magnesium	1.272	3.69
Calcium	0.400	1.16
Potassium	0.380	1.10
Bicarbonate	0.140	0.41
Bromide	0.065	0.19
Boric acid	0.026	0.07
Strontium	0.013	0.04
Fluoride	0.001	0.00
Total	34.482	99.99

From D. E. Ingmanson and W. J. Wallace, *Oceanology: an introduction*, Wadsworth, 1973.

and it involves the movement of water from one place to another without any mixing with the atmosphere. At the surface, several other processes influence temperatures. Loss of heat by evaporation is the most important process, though sensible heat convection occurs between the ocean and atmosphere, transferring heat from the air to the water when the sea is colder than the air, or from the oceans to the atmosphere when the air is colder. Condensation of water vapour liberates latent heat into the oceans. In addition, surface winds have an important effect. They can produce a mixing of the upper layers to give a marked discordance in the temperature profile (Figure 15.4); winds blowing off-shore move the waters seaward allowing cool water from the ocean bottom to rise to the surface. This reduces surface temperatures and weakens the temperature gradient. It may also have important effects on the regional climate, as we saw in Chapter 8.

The effects of some of these processes can be seen in Figure 15.3. The influence of upwelling cold water is seen off Chile and Peru, the Canary Islands and southern Africa, for example, and the anomalously cold surface temperatures extend eastwards in both the Atlantic and Pacific Oceans. Apart from modifying the regional climate in these areas they also bring nutrient-rich waters to the surface, stimulating organic activity. Many of the world's major fishing zones are associated with these upwelling waters.

Looking at the vertical pattern, we find a general decline in sea temperatures with depth (Figure 15.5). This occurs for two main reasons. Radiation is absorbed almost entirely in the upper few metres of the ocean, so little heating occurs at depth. In addition, cold water tends to subside, gathering in the lower part of the ocean and reducing temperatures there. Thus, at about 1000 m, sea temperatures are only about 5°C and they continue to fall below this depth to only 1 or 2°C. This temperature structure makes the oceans very stable. Unlike the atmosphere, it experiences very little vertical movement and much of the mixing takes place by small eddy currents and molecular diffusion.

Nevertheless, there are variations to the temperature profile, as can be seen if we look at a north–south cross-section of the Atlantic Ocean. Temperatures at the surface are higher in the tropics and the warm water extends to greater depths between 20 and 30°N of the equator. This is largely due to the intense solar warming in this area.

Salinity

As anyone who has swum in the sea knows, the sea water is salty. It contains, in fact, a variety of dissolved minerals (Table 15.3), but by far the most abundant is common salt (sodium chloride or NaCl).

On average, the sea water contains about 35 g of

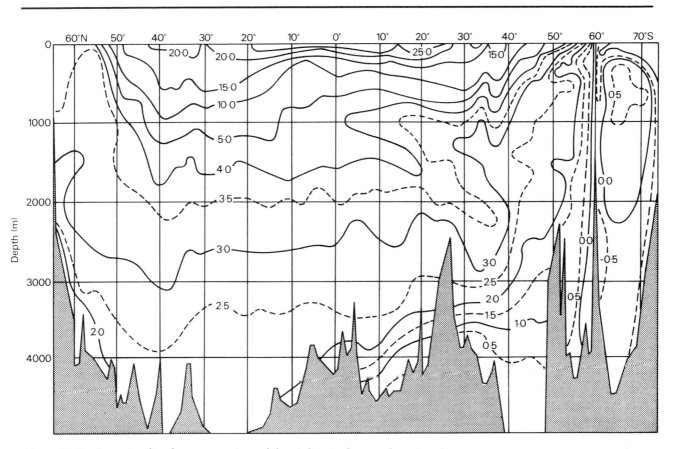

Figure 15.5 Longitudinal cross-section of the Atlantic Ocean showing the temperature structure in relation to depth (°C) (after Ingmanson and Wallace, 1973)

sodium chloride in each litre, but this concentration varies considerably from one part of the oceans to another. Much of the salt has been derived from the land, for over the eons of geological time streams have carried sodium and other solutes into the sea, where they are trapped. It has been calculated that if all the minerals dissolved in the sea were precipitated out and returned to the land, they would produce a layer over 150 m thick. Locally, however, this slow build-up of salt is being counteracted by inputs of freshwater, by precipitation, by ice melt or by runoff, which dilute the concentration. Thus salinity tends to be relatively low near the coast and higher in the mid-ocean areas. It also tends to increase with depth since salty water is generally more dense and sinks.

At a global scale, the highest concentrations of salt are found about 20°N and S of the equator. This narrow belt of high salinity occurs because of the imbalance between the input of freshwater from precipitation and runoff and the output of fresh-

water by evaporation. In these areas of the subtropical anticyclones, precipitation is low and evaporation is high. There is a net loss of freshwater, so salt concentrations increase. In high latitudes, the melting of ice causes a more complicated pattern and a marked seasonal variation in salinity. Where seas are enclosed we find extreme levels. In the Red Sea and Persian Gulf, both precipitation and runoff are low and evaporation high; no mixing occurs, so salinities reach 40‰ (parts per thousand). Indeed, the Dead Sea is so saline and the waters so dense that it is possible to float upon it without sinking (Plate 15.1). In the Baltic Sea, in contrast, precipitation and runoff are appreciable and evaporation is relatively low so we find salinities of as little as 5‰ (e.g. in the Gulf of Bothnia).

As with temperature, the vertical variations of salinity are influenced by the origin of the waters (Figure 15.6). The main feature is a zone of minimum salinity between 700 and 800 m. The

be traced in the Atlantic as far as 20°N of the equator and to the equator in the Indian and Pacific Oceans.

Below this layer, salinity increases to a maximum between 1500 and 4000 m. Deeper again, we find water of slightly lower salinity; cold water which originated in the Antarctic area. Superimposed on this general pattern, however, are smaller-scale variations. The Mediterranean Sea has above average salinity levels. Across the Straits of Gibraltar (Figure 15.7) there is a surface flow from the Atlantic bringing in cool relatively fresh water, while at depth warm saline waters flow in the opposite direction. This lower flow involves water volumes over 100 times those of the Mississippi, and it is not surprising that its effect can be detected well out into the Atlantic, at depths of 1000–2500 m.

Ocean tides and currents

For any visitor to the sea shore it is the waves and the tides which are most apparent. Compared with other water bodies, the sea appears to be much more dynamic and mobile, and the sea shore is a zone of great activity. Waves are continually breaking against the cliffs and, for many seas, the point at which the waves break on the shore varies during the course of the day, depending upon the state of the tide. However, in many seas the tidal range is small. It is less than 1 m in parts of the Mediter-

Plate 15.1 Afloat on the Dead Sea maintained by the high buoyancy of the saline water (photo: G. Rowley)

source of this less saline water is the cold, dense and relatively fresh surface waters of the Antarctic south of 45°S which slowly sink and extend northwards. This tongue of cool, low salinity water can

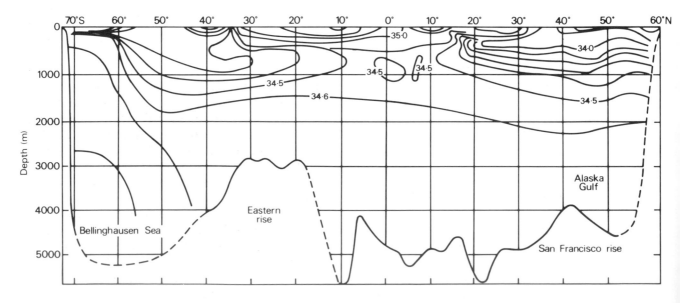

Figure 15.6 Longitudinal cross-section of salinity for the central part of the Pacific Ocean (parts per thousand (after Gerard, 1966)

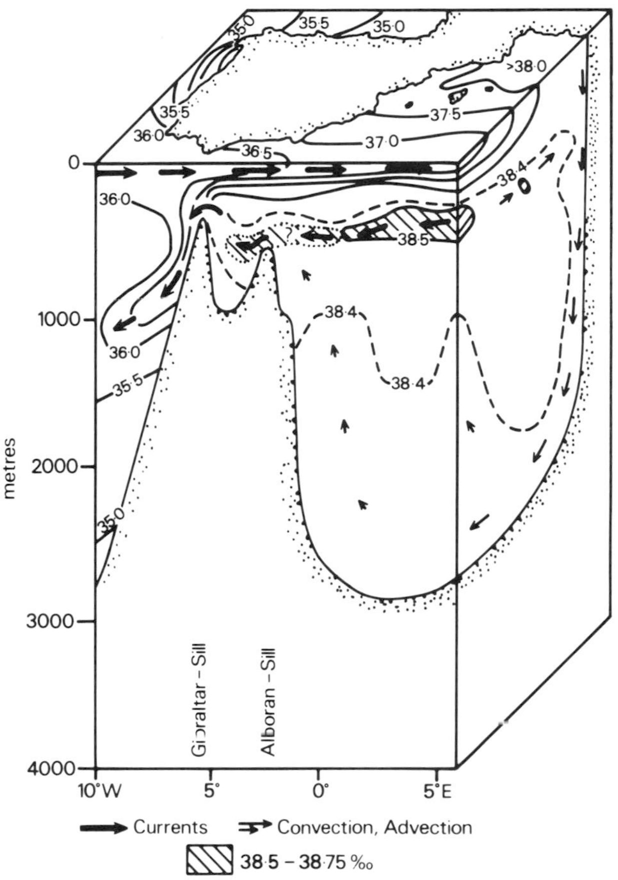

Figure 15.7 Schematic block diagram of vertical circulation and distribution of salinity in the Straits of Gibraltar during winter (after Gerard, 1966)

ranean. Elsewhere the range can be enormous; it is 12.2 m at Avonmouth on the Severn Estuary and up to 15.4 m in the Bay of Fundy, Nova Scotia. This tidal activity is clearly a widespread and powerful process; so powerful in fact that it has been suggested that much of the world's energy could be obtained from it. But how does it happen?

Tides

The tide refers to the rhythmic diurnal or semi-diurnal rise and fall of the sea surface. Unlike the weather, we can predict the tides – and how high they will be at any given time. The reason for this predictability is that the tides are caused by the physical forces of our solar system, primarily those exerted by the sun–earth–moon system.

The details of these forces are beyond the scope of this book, but in a very simple way we can visualize the waters of the oceans responding to the gravi-

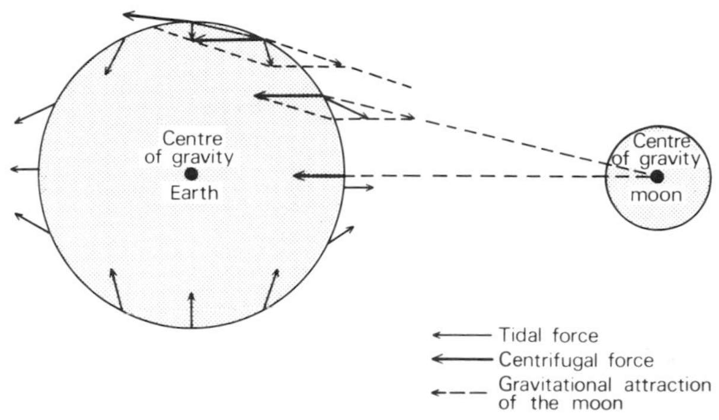

Figure 15.8 The tide-generating forces between the earth and moon (after Weyl, 1970)

tational pull of the moon. This counteracts the earth's own gravity to some extent and, in essence, creates two points where the net gravitational effect of the earth is diminished, one directly beneath the moon and the other on the opposite side of the earth (Figure 15.8). The differences are small but they are sufficient to cause the water to move down the 'gravitational gradient' and it is this movement that we see in the tides.

If the earth rotated more slowly, and if the moon remained above the same point on the earth, the oceans would adjust to this gravitational pull and they would have a permanent slight bulge. There would be a mound about 358 mm high in the middle of the gravitational basins, and a trough 179 mm deep along the intervening zone. This pattern is known as the equilibrium tide. But the speed of the rotation of the earth and the movement of the moon mean that the focus of attraction sweeps around the globe at 3000 km per hour. The equilibrium tide never gets a chance to establish itself, and the oceans constantly pour one way and then the other following the moon's effect. It takes 24 hours and 50 minutes for the moon to return to the same point above the earth, and in this time two tidal rises occur; one when the moon is overhead and the second when the moon is on the opposite side of the earth. It is this that gives the almost twice-daily tides.

If we plot tide levels for several weeks, we would notice that the tidal range is not constant (Figure 15.9). Over a period of about seven days the range builds up to a maximum value, called the **spring tide**. It then declines over the next seven days to give a low value, called the **neap tide**. This modification of tidal pattern is produced by the interaction between the lunar and solar forces. When the sun

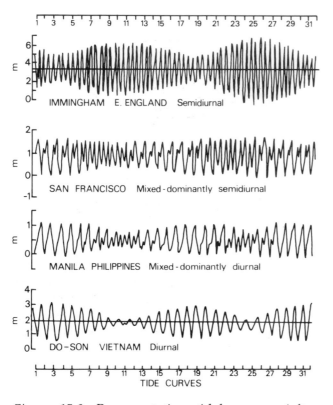

Figure 15.9 Representative tidal curves (after King, 1962)

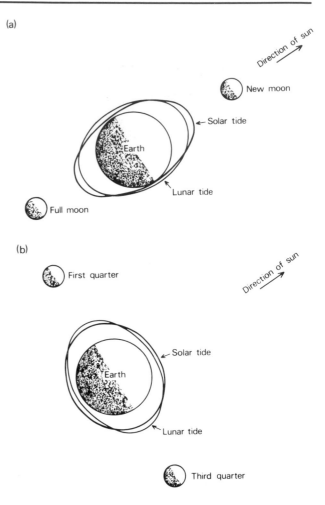

Figure 15.10 The effects of the gravitational pull by the sun and the moon on tides. When both bodies are pulling in the same direction, spring tides result; when at right angles, neap tides are experienced (after Harvey, 1976)

and moon are pulling together (Figure 15.10) we get a greater attractive force operating on the oceans and a larger tidal range. When they are at right angles to each other (at first and third quarters of the moon) the effects are counterbalanced to a certain extent and so the tidal range is less. In even more detail, the tidal patterns over the globe vary slightly depending upon the interactions between the solar and lunar forces and the coastal and sea-bed configuration.

Surface currents

As well as the flux of tides around the earth, we find that the oceans are also affected by net flows of water in one direction – ocean currents. In the surface layers, the currents are developed mainly by wind stresses, the frictional interaction between air movement and the sea surface. These surface movements may also disturb the waters at depth, but more generally deep currents develop due to differences in temperature and salinity.

When a wind blows across the ocean surface it causes a movement of the upper layers of the water, to a depth of about 100 m. Because of the deflection caused by the earth's rotation, this moves at right angles to the direction of the wind, to the right in the northern hemisphere and to the left in the southern hemisphere. The general pattern of oceanic circulation in the north, therefore, produces a piling up of water at about 30°N and a depression at about 60°N (Figure 15.11). A similar pattern develops in the southern hemisphere and, between the two, at the equator, a more complex depression occurs.

Water movement around these **gyres**, as they are known, is similar to air movement around high or low pressure cells. Due to the effect of the Coriolis Force, the water flows at right angles to the surface gradient, giving a more or less circular

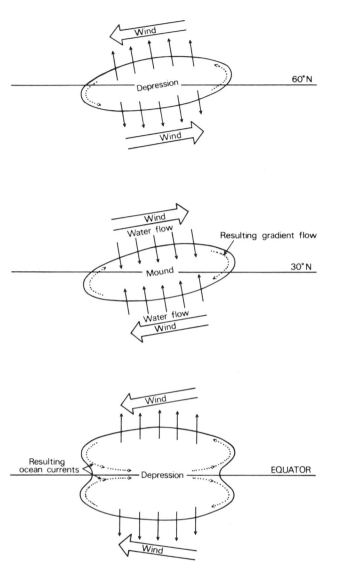

Figure 15.11 Creation of oceanic low and high pressures by the prevailing winds (after Pirie, 1973)

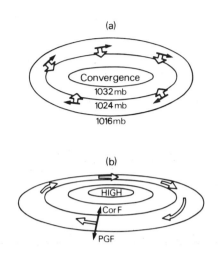

Figure 15.12 Water movements associated with anticyclonic winds in the northern hemisphere: (a) atmospheric pressure and winds and the associated frictional transports to the right; (b) resultant topography of the sea surface and associated gradient currents below the depth of frictional influences (after Harvey, 1976)

The Pacific Ocean

The Pacific Ocean, the largest and most regular of the world's oceans, displays the pattern most clearly. In the north a clear sub-tropical gyre can be seen, consisting of the westward flowing North Equatorial current and the eastward flowing Kuroshio and North Pacific currents. The flow is completed by the California current. This gyre therefore carries warm water northwards along the Japanese coast and then generally eastwards to California. Cooler water flows southwards down the California coast.

A similar sub-tropical gyre is also seen in the South Pacific. This comprises the warm South Equatorial and East Australian currents and the cool Humboldt current which flows northwards along the Peruvian coast.

The equatorial gyre is also visible, although it tends to be displaced slightly north of the geographic equator due to the similar displacement of the atmospheric circulation here. The Equatorial Counter current provides the central line of this double cell, which is completed by the North and South Equatorial currents.

The sub-polar gyres are less clearly developed in the Pacific. In the north the Bering Straits restrict the circulation, and no warm water passes through the straits to complete the gyre. All that can be

(geostrophic) flow around the gyre (Figure 15.12). In the northern hemisphere this is clockwise around the high at 30°N and anticlockwise around the depression. In the southern hemisphere the flows are reversed, while at the equator the two gyres merge to give a flow shaped rather like a figure of 8.

Because the earth's surface is not covered entirely with sea, and because of the irregular configuration of the continents, we do not find quite such a simple pattern in reality, but the basic elements are nevertheless visible in the global ocean currents (Figure 15.13).

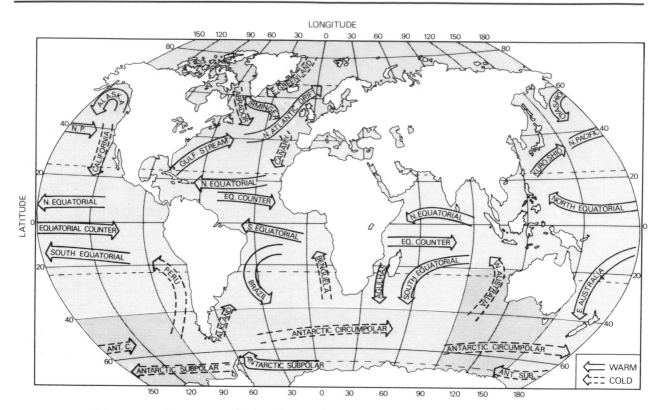

Figure 15.13   Main oceanic currents during the northern hemisphere winter

detected is the southward flowing Oyashio current and the almost separate cell produced by the Alaska current. In the South Pacific the Antarctic Circumpolar and Antarctic Subpolar currents give a more complete gyre at 60°S.

The Atlantic Ocean
In general, the same patterns of circulation can be seen in the Atlantic. The northern sub-tropical gyre is possibly one of the best known, at least to people in the eastern USA and in Britain, for the warm waters of the Gulf Stream greatly modify the climates of these regions. In the North Atlantic it changes its name to the North Atlantic current or Drift, but the waters are the same, and their effect is felt as far north as 69°N, where they keep the Russian port of Murmansk ice-free throughout the year. The gyre is completed by the southward flowing Canary current and the westerly North Equatorial current.

The sub-polar gyre in the North Atlantic also has considerable significance. Cold surface waters flow out of the Arctic Basin as the Labrador current and the East Greenland current. Together they bring cold water into the Newfoundland area, as well as

melting icebergs. Where it meets the warm North Atlantic current near the Grand Banks region, we have a situation of warm and cold currents in close proximity. The prevailing airflow here is from the south so warm air blows across the cold currents. Cooling is rapid, condensation occurs and so the dense fogs for which the Grand Banks fishing area is notorious are produced.

The Indian Ocean
The Indian Ocean produces an interesting anomaly to this general pattern of circulation. This occurs because of the marked reversal of atmospheric circulation associated with the seasonal progress of the trade winds (the monsoon). This reversal of wind directions causes a related change in the oceanic circulation (Figure 15.14). In winter the north-east monsoon blowing across the Indian Ocean generates a current moving in the same direction towards East Africa, termed the North-East Monsoon current. This eventually merges into the Equatorial Counter current just south of the equator. In summer, a clockwise pattern of circulation develops. A weaker reverse flow exists off Sumatra, complicated by the configuration of the

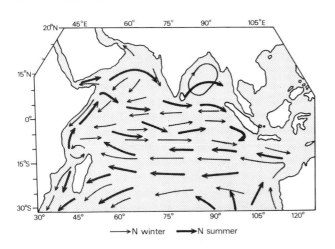

Figure 15.14   The seasonal variation of the surface currents of the Indian Ocean (after Weyl, 1970)

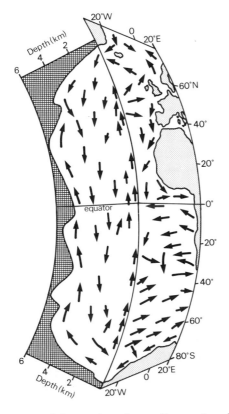

Figure 15.15   Schematic, three-dimensional representation of surface and sub-surface currents in the Atlantic Ocean (after Harvey, 1976)

land. The flow as a whole, however, is generally anticlockwise.

In summer, the pattern reverses and a clockwise circulation develops. Off Africa we find the strong Somali current appearing in April and reaching its peak in July and August. This turns eastwards into the South-West Monsoon current (remember that currents, like winds, are named according to where they have come from, not where they are going to) and then merges into the South Equatorial current (Figure 15.14). By November, this circulation is weakening and the flows start to reverse once more.

Deep ocean currents

Although the surface flows are the most obvious aspects of the oceanic circulation and, certainly in the days of sailing ships, were perhaps the most important, beneath the surface there are other flows. These are usually much slower and we know much less about them. They include both horizontal flows in the ocean depths and also vertical movements.

Subsidence and upwelling

We have already mentioned that the temperature structure of the oceans means that they are much more stable than the atmosphere. Convection currents of the form we find in the air do not occur. Nevertheless, considerable vertical motion must take place if the circulation of the oceans is to be maintained. There are several situations in which these occur. Subsidence takes place because of differences in density of the waters, produced either by differences in salinity or temperature (or, indeed, a combination of the two). Ascent of water (**upwelling**) occurs where surface winds blow coastal waters seawards.

The main areas of subsidence occur in the high latitudes where cold, polar waters are created by the input of ice from the ice caps and by the radiational cooling of the surface. The cool waters sink and, as we have noted, gather in the lower levels of the oceans where they help maintain low temperatures. The effect of these general zones of subsidence can be seen particularly in the Atlantic (Figure 15.15), but they also occur, less markedly, in the Pacific.

More localized subsidence occurs in a number of areas. Cool sediment-laden water from the continents often subsides to some extent as it enters the sea, while the saline waters from the Mediterranean

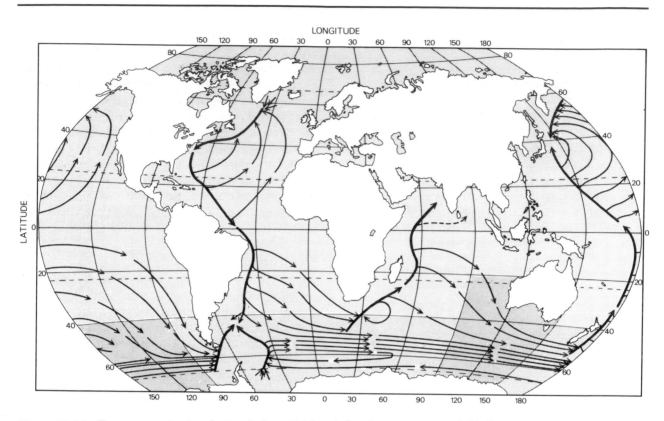

Figure 15.16   Deep oceanic circulation below 2000 m (after Ingmanson and Wallace, 1973)

subside as they enter the Atlantic. In addition, in zones of convergence, where surface currents meet, subsidence tends to occur. The cool Labrador and Greenland currents, for example, sink beneath the warm North Atlantic current in the Newfoundland area.

Upwelling occurs mainly along the deep coastal areas where offshore winds blow water seawards. The water which is removed in this way is replaced by cool bottom waters which rise from the ocean depths. As we have seen (Chapter 8), these greatly modify the local climate and they also influence the ecosystem.

Horizontal flows at depth

We know very little about flows deep in the ocean, although we can, to some extent, deduce what happens from the patterns we find at the surface. Clearly to complete the circulation, these deep flows must feed the patterns of surface movement. In addition, by studying temperature and salinity profiles of the oceans we can get a broad picture of sub-surface movements.

Without doubt the most important flow is that formed from the Antarctic and Arctic zones of subsidence. In the Atlantic, the cold Antarctic waters seem to flow at the very bottom of the ocean, north to 20–40°N. The Arctic waters flow southwards at a slightly higher level (3000–4000 m) to almost 60°S of the equator. In the Pacific the flows are slower and less apparent, in part because the Bering Straits prevent deep, cool bottom water spreading from the Arctic Ocean. The circulation, as far as we know it, is illustrated in Figure 15.16.

Nearer the surface, in what is often called the central zone, more complex patterns of flow occur, with water from a variety of sources mixing and converging. In the Atlantic water from the Mediterranean cuts across the general southerly drift of the sub-surface circulation and extends as much as 2500 km into the ocean. In the Pacific, water from the North Atlantic and Indian Oceans merges with that from the Antarctic to give sluggish, ill-defined flows.

Conclusions

Considering their size, we know remarkably little about the oceans, especially the magnitude of exchanges. However, we have been able to outline the general aspects of their circulation. This pattern is a vital part of the hydrological system, transferring waters through the globe, supplying waters to areas of deficit, removing them from areas of excess. The oceanic circulation involves, also, a transfer of energy which is important to the global energy system.

But possibly the most significant aspect of the oceans is their size. They contain 97.6 per cent of the world's water; they are the major store within the hydrological cycle. Compared with the oceans, the atmosphere, the cryosphere and the ground-waters contain negligible quantities of stored water. Through this role as the overriding store of water, the oceans exert a dominating control over the hydrological cycle. They influence the amount of water available for circulation, they provide the main surface for evaporation, they influence climate and thereby affect rates of precipitation. They may be the last unknown, but they are a part of the world we need to know much more about.

16

Landscape form and process

Introduction

The landscape is not static. Over the years and eons it changes and develops. New landscapes are formed; old ones are destroyed or modified. Sometimes these changes are dramatic. New islands, such as Surtsey, may rise from the ocean; volcano action, earthquakes and land tremors may rip apart and recreate whole regions almost overnight. Sometimes the consequences of these events are immediate and terrible for man; more commonly, the changes are slow and almost imperceptible. It is the process of gradual wear and movement, operating over thousands or even millions of years, that carves and moulds the landscape.

The landscape that we see, therefore, is merely a fragment in time; rather like a still from a film. It represents just one moment in an endless sequence of development. To understand the landscape we need to consider the processes which have been, and still are, acting upon it. It is not enough merely to describe what we see, to classify the landforms; we must try to grasp the 'logic' behind the landscape. The questions we should ask, therefore, are not: 'what is it?', but 'why and how?'. Why is the landscape that shape; how did it form; how is it changing?

These questions raise a myriad of more detailed questions. We may try to explain a landform as the work of ice or frost, of the action of rain or rivers, of wind or gravity, but such explanations are too general. In reality, the landform does not develop through the operation of such general forces; it is a result of countless minute events, many working in opposition to each other. It is the result of the formation of individual ice crystals, of the impact of individual drops of rain, of individual gusts of wind. If we are to understand our landscape fully, it is these details that we must ultimately comprehend.

To tackle the details, however, it is useful to have a more general framework within which to operate.

We need a model of the landscape to guide us. In this context, we might start by noting that the landscape is carved out of, and built from, rock materials, and that all landscape-building processes involve in some way the reorganization, alteration or movement of these materials across the earth's surface. As a basis for our model, therefore, we can define what is referred to as the **cycle of rock materials**; it is this cycle which underlies the development of the landscape.

The rock material cycle

The cycle of rock materials is illustrated in general terms in Figure 16.1. As this indicates, the main input to the cycle comes from magma – molten rock from deep in the earth's interior. This is intruded into the earth's crust or escapes on to the surface, where it cools and **lithifies** (i.e. forms rock). The rocks which are formed in this way are often subjected to powerful earth forces which lift and fold them, while, once exposed at the earth's surface, they are also subjected to attack by wind, water and ice. Thus the rocks are gradually **weathered** and the debris is carried through the landscape by streams and glaciers, by wind and gravity. Ultimately, the **material** is laid to rest when, through the effects of burial and compression, by heat and chemical reactions, it may once more be turned into rock. Then the whole cycle can start anew.

The model we have outlined in Figure 16.1 is, of course, extremely simple. It is important to appreciate that the rock materials do not necessarily follow very precisely the idealized sequence we have described; they may miss out some steps and repeat others. They may, for example, be subjected to many phases of earth movement before being exposed at the surface and weathered. Similarly, they may be repeatedly eroded and redeposited before coming to rest for long enough to be lithified again into rock. Thus the progress of rock materials

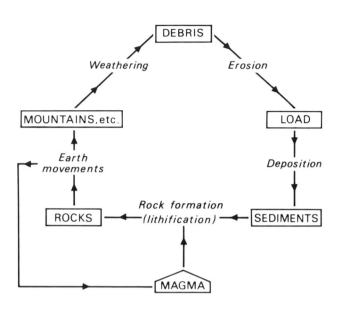

Figure 16.1 A general model of the cycle of rock materials

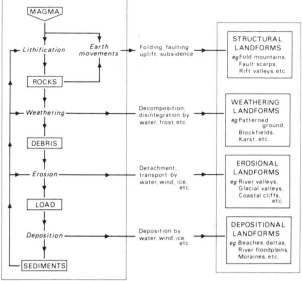

Figure 16.2 Landforms and the cycle of rock materials

through this cycle is not continuous, nor is it rapid. It may take many millions of years for the cycle to be completed. In the process, however, the forces acting upon the materials shape them into the landforms which make up our landscape: the continents and oceans, the mountains and valleys, the hillslopes and plains (Figure 16.2).

Earth movements play a fundamental role. They are responsible not only for the creation of relatively local or regional features, such as fold mountains like Snowdon in Wales, or fault-lines like the Great Glen fault in Scotland. They are also responsible for the broad configuration of the oceans and continents, and for the worldwide distribution of mountains, marine trenches and volcanic activity. All these features are related to the slow but massive displacement of the earth's surface known as plate tectonics. Vast plate-like areas, on which the continents stand, are gradually migrating across the surface, powered by the deep circulation of heat and molten rock in the earth's interior. Where these plates are separating, new ocean floor is being created as magma wells upward and escapes through fissures in the crust. Where the plates are converging, the continental rocks are being subjected to huge compressional forces which fold and lift them into mountain chains, and release volcanic activity from deep in the mantle. Also in these zones of convergence, the crust is being forced

downwards, to be consumed within the earth's interior. Thus, these massive earth movements not only produce **structural landforms**, but they also create and consume crustal materials, and thereby play a vital role in the natural cycling of rock materials.

By comparison, the effects of weathering and erosion might seem mild. Nevertheless, together they provide the main means by which the earth's surface is stripped and lowered – a process known as **landscape denudation** – and by which the materials are redistributed across the earth's surface. In general, weathering provides the preparation for this movement by weakening the rocks and providing debris for transport. Water plays a major part in this process, carrying substances to and from the weathering rocks, acting as an important solvent, and also setting up significant physical stresses within the materials. As the rocks decompose and disintegrate, **landforms of weathering** are produced, none perhaps more spectacular than those of the karstic landscapes of limestone areas such as Yugoslavia and western Ireland.

The weathered materials, however, are subjected to other forces. Water, wind, ice and gravity detach and transport the rock fragments. Material prised from cliffs and hillsides tumbles down the hillside; rain splashes the small particles on the ground surface; running water picks up the debris and carries it down valley; winds transport the fine material. As it is transported the material acts as an agent of erosion in its own right, abrading the surface and causing new debris to be detached and removed. Thus, **erosional landforms** are developed. Streams and glaciers cut valleys; the wind strips the surface and carves the resistant rock masses; on the coast, the storm waves etch the cliffs.

But ultimately the debris transported by these processes comes to rest, and as it does so new landforms are built. Fans of debris accumulate at the foot of the slope; beaches are formed at the margins of the sea; glaciers retreat and leave behind them masses of material in an irregular and chaotic landscape. Sometimes these **depositional landforms** are short lived. The next flood may remove the material, or a new ice advance may override its former deposits. On occasions, however, the deposits remain for long periods. They may be buried; the earth's surface may rise or the sea-level may fall, and the marine sediments may be left high and dry. Then, new changes occur. The deposits are dried, compressed, altered by pressure, heat and chemical reactions, cemented by compounds washed through them by percolating water. They are turned into sedimentary rocks; as we have seen, the whole cycle may start again.

The debris cascade

We can illustrate the cycle of rock materials more anecdotally by following an individual quartz pebble on its journey through the landscape. The pebble we are concerned with started, perhaps, as a piece of vein quartz in the volcanic rocks of the English Lake District. Long ago, as the climate deteriorated during the build-up to one of the Quaternary ice ages, it was prised from its cliff by frost and ice crystals. Released, it bounced down the slope and came to rest at the slope foot. There it lay for many centuries, being attacked by the sands carried in the fierce winds and by the lichen which grew on the exposed ground. But, as the climate worsened further, ice sheets began to develop in the mountains, and a glacier moved down the valley catching up the pebble amid the debris that trundled beneath its belly. So, chipped and scratched by the constant scraping against the bedrocks, the pebble was carried out on to the

lowlands where it was dumped unceremoniously in a pile of moraine.

Yet even there the journey was not over, for meltwaters from the now decaying ice eroded the moraine and washed the pebble downstream. Over the years, hopping from one gravel bar to the next, becoming rounded by wear as it did so, the pebble made its way into the estuary of the Eden. There, it was caught by the tides, for a period became entangled with seaweed and was washed far along the coast. But, eventually, the waves shuffled it on to the beach and then, during a severe spring gale, it was finally flung high on to the storm ridge, where it now lies, smooth and spherical.

The journey is, of course, a hypothetical one. In reality we cannot know the detailed history of an individual pebble in this way. But it is probably a not exceptional example of the sort of history that many pebbles in our landscape have had. And it illustrates what we refer to as the **debris cascade** – the movement of rock materials through the landscape following uplift. It is an apt term, for it indicates the generally downward course of this movement, from source to final resting place, which is often in the sea.

The journey we have described also illustrates several important aspects of this movement. As the pebble – indeed as any material – passed through this cascade, it was affected by many different processes: by the action of wind and water and ice; weathering, transport and deposition. These processes modified the pebble. It was fractured and chipped by the ice, abraded by the water; it ended up probably smaller and rounder than it started. But, in addition, the pebble itself took part in the formation of many different landforms. Now it is part of a scree slope, now a glacial moraine. Next it becomes part of a series of gravel bars. Finally it ends up as a component of a beach. Even when it was mobile, it was affecting the landscape, wearing the bedrock beneath the glacier, battering against other pebbles as it moved down the stream. Thus, the movement of this material through the debris cascade contributes to and affects the landscape through which it passes. The processes operating in the debris cascade act to shape the landscape.

The landscape as a process–response system

This relationship between the processes operating

within the landscape, the materials on which they are acting, and the shape or character of the landforms which develop is central to our understanding of geomorphology, and it will be a theme we will follow in the rest of this section. In the jargon associated with systems theory, we are considering the landscape as a **process–response system**: one in which a **morphological system** (the landforms) and a **cascading system** (the debris cascade) interact. More simply, we are stating that the morphology of the landscape is a function of the processes acting upon it.

It is important, however, to consider some of the implications of this concept – implications which are not always remembered. First, this relationship is not a one-way relationship. Not only do the processes affect the landscape, but so does the landscape influence the processes which operate on it. Thus, glaciers and streams mould the valleys, but the shape of the valleys also influences the way the glaciers and streams behave. Similarly, tides and waves act to build beaches, but the shape of these beaches modifies the flow of the tidal currents and the pattern of the waves. In other words, feedback processes occur. Landscape form and process are intimately and reciprocally related.

Second, though, we must remember that few landforms are simple. Many are both complex and composite. That is to say, landforms are often produced not by a single process acting in isolation, but by a whole range of processes acting together. Waves and wind act in unison to mould a sandy shore. Water, ice and many hillslope processes work together to carve the valleys. Thus, many landforms are **polygenetic**: they owe their origin to several different processes. Moreover, these processes often change over time, so that individual landforms go through several cycles of development. We saw this in the case of the pebble we discussed earlier, and the same is true of larger landforms and whole landscapes. Periods of weathering are followed by phases of glacial action; these in turn are succeeded by fluvial action or the effects of wind. Landforms, therefore, are often **polycyclic**. What we see now is simply one stage in their history, possibly not even an important stage. The landscape as a whole is, in part, a legacy of the past. If we are to understand the landscape we thus need to understand not only the processes operating at present, but also the factors which have operated in the past to make the landforms the way they are now.

It is here that geomorphology and geology come together. What we need to be able to do is to use our geomorphological knowledge of present-day processes as a basis for interpreting more ancient landscapes. We use modern landforms as analogues of the past. But we also need to have an understanding of past environmental conditions and events, to act as a framework within which to place our interpretations. Thus, as background to our study of the landscape, we need a broad understanding of geological history.

The geological column

The general history of the earth is described by what is known as the geological column. A simplified version of the geological column is shown in Table 16.1. As this indicates, the column is divided into four major **eras**: the Precambrian, Palaeozoic, Mesozoic and Caenozoic. These eras are then subdivided into **periods** – phases during which the earth is believed to have experienced reasonably stable conditions, or changed in a reasonably consistent manner. These are further divided into **epochs**, each of which represents a specific chapter of earth history.

This history began some 4600 million years ago, but the early events are obscure, for no rocks from that time have been found. The oldest known rocks, in fact, are about 3800 million years old, from Minnesota in the USA. Life, however, did not appear until considerably later: the earliest evidence comes from rocks in Australia, some 3000 million years old.

None of this remote history appears to be recorded in Britain. We find a few rocks of late **Precambrian** age, but the rest all date from the **Palaeozoic** or later, less than 590 million years ago. Even so, the geological history of Britain is complex and we cannot go into the details of it here. What is apparent, however, is that the geological development of the country illustrates in broad terms the cycle of rock materials that we defined in Figure 16.1. Periods of uplift and mountain building, together with associated volcanic activity, have been followed by phases of denudation and deposition. Close to the mountains, the debris often built up as coarse-grained rocks laid down under terrestrial conditions (e.g. in lakes or river valleys). Further away, fine materials often collected in shallow seas or in deep marine basins (**geosynclines**).

Table 16.1 *A general outline of the geological column in Britain*

Era	Period	Age (m yrs)	Epoch	Rocks and events
CAENOZOIC	Quaternary		Holocene	Peat; alluvium. Rising sea-level; ameliorating climate. Extensive human effects
		0.01	Pleistocene	Till; outwash sands; river terraces; raised beaches. Fluctuating sea-level. Glacial–interglacial climate
		3.0	Pliocene	Denudation; local 'Crag' deposits. Falling sea-level; deteriorating climate
	Tertiary	7.0	Miocene	Denudation. Falling sea-level; deteriorating climate
		26.0	Oligocene	Sands and clays; folding (Alpine orogeny)
		38.0	Eocene	Sands and clays; basalts in Skye, Antrim
		54.0	Palaeocene	Sands and clays; basalts in Skye, Antrim
		65.0		
MESOZOIC	Cretaceous	135		Clay, sand, chalk
	Jurassic	200		Clay, sandstone, limestone
	Triassic	240		Sandstone, marl, conglomerate
PALAEOZOIC	Permian	280		Sandstone, marl, conglomerate, limestone; folding (Hercynian orogeny); volcanics
	Carboniferous	370		Limestone, clay, gritstone, coal; volcanics
	Devonian	415		Sandstone, gritstone, limestone, siltstone, shale; volcanics; folding (Caledonian orogeny)
	Silurian	445		Greywacke, shale, limestone, siltstone; folding
	Ordovician	515		Greywacke, shale; volcanics; folding
	Cambrian	590		Gritstone, shale, quartzite
PRECAMBRIAN				Sandstone (Torridonian) Gneiss, schist (Lewisian and Moinian)

Thus, the **Cambrian, Ordovician** and **Silurian** periods were dominated by the gradual accumulation of materials in a broad geosyncline that ran from south-west to north-east across what is now Britain. The middle of these periods – the Ordovician – was a time of great instability, with widespread volcanic activity, but it was at the end of the Silurian that the real upheavals took place. By then, the geosyncline had become shallower and filled, and the earth was being subjected to massive compressional forces which lifted up huge mountain chains in the north and west. This whole phase of mountain building is known as the **Caledonian** orogeny, and its imprint is still visible in the structure of the Scottish Highlands and Cambrian mountains.

Following this uplift, a long period of denudation took place. During the **Devonian**, this was associated in many parts of Britain with the accumulation of sandstones, laid down under desert conditions close to the mountains, and thus the rocks of this period are often referred to as the Old Red Sandstone. But during the subsequent **Carboniferous** period, large areas of limestone and clay were formed in shallow seas, while gritstones accumulated in large coastal deltas. Finally, as the land surface was reduced further, coal shales developed in vast deltaic swamps.

The whole cycle was more-or-less repeated at the end of the Carboniferous. Renewed mountain building (the **Hercynian** orogeny) took place, thrusting up extensive land masses in the south and causing wide-scale volcanic activity in what is now Devon and Cornwall. In the following **Permian** and **Triassic** periods these land masses were again eroded under desert conditions, giving rise to the New Red Sandstones – though in reality the rocks are neither universally red nor all sandstones. Instead they include dune sands, salt beds formed in desert lakes and lagoons, and clays and limestones laid down in shallow seas.

Deposition continued in the **Jurassic**, with a broad sequence of clays, sandstones and limestones forming in marine basins fed by sediments from extensive land masses to the west. As in the Carboniferous, coastal deltaic swamps developed as the land surface was eroded, though they were nowhere near so widespread on this occasion and coal is rare. By **Cretaceous** times, however, the relief of the neighbouring land masses was apparently much reduced, and shallow marine conditions dominated in which alternating successions of clay and sands accumulated. Finally, deep deposits of chalk were formed in the warm, clear seas which now received only minimal amounts of debris from the land.

The Cretaceous marks the end of the **Mesozoic** Era, some 65 million years ago. In the **Tertiary** period, which represents the opening of the current **Caenozoic** Era, marine sedimentation continued for a while in the Hampshire and London basins, while volcanic rocks were extruded in Skye and Northern Ireland. In mid-Tertiary times, however, folding associated with the **Alpine** orogeny – which further south in Europe, as well as in the Andes, Rockies and Himalayas was throwing up huge mountain chains – raised even these areas above sea-level. Thus, by the beginning of the **Miocene** Epoch, the broad structure of the British land mass had been established, and the history of the area since then has essentially been one of subaerial denudation. Falling sea-level and a gradually deteriorating climate characterized much of the Tertiary, and accelerated the rates of denudation. Thus, though remnants of the old Tertiary land surface are still visible in parts of Britain, denudation since then has generally eradicated much of the original landscape, at no time more so than in the **Pleistocene**, when extensive ice sheets covered large areas of the country. During the subsequent **Holocene** Epoch, geomorphic processes have perhaps been less active – the climate is temporarily more benign – but even so old landforms have been further modified and new ones created. It is also interesting to note that in some cases Tertiary and Quaternary denudation has exhumed even older landforms: the coast of west Wales, for example, appears in part to be an exhumed land surface, originally formed in Permian times.

The lesson, therefore, is clear. If we look at the present landscape we see a system which is dynamic, and which is changing, but which is also in large part ancient and inherited. To understand it, we need to comprehend the events and processes which have taken place in the past. Probably our greatest tool for understanding the past, however, is a knowledge of present-day processes. It is this approach which is enshrined in the so-called '**law of uniformitarianism**': that the present is the key to the past. More fundamentally, however, we need to appreciate the intimate relationship between process and form in the landscape: the way in which processes determine form and, in turn, the way in which landform influences process. It is this relationship which we will examine in subsequent chapters. Initially (Chapters 19–21) we will consider the main stages in the cycle of rock materials. Then we will look more explicitly at the landscape as a process–response system and discuss the ways in which the processes involved in the debris cascade (weathering, erosion and deposition) affect different morphological systems (hillslopes, drainage basins and streams, glaciers, aeolian systems and lakes and seas).

17

The formation of rocks

The role of rock formation

The wide variety of landforms and landscapes we see at the earth's surface are due to many different factors. One of the most important is the variation in rock type which occurs, often over quite short distances. This affects the resistance of the surface to processes of weathering and erosion, and thus influences the rates at which these processes operate. In a general fashion, therefore, the landscape reflects the character of the bedrocks in which it is carved. Rocks which are resistant to attack, for example, tend to survive as upstanding areas of countryside, while less resistant materials are eroded to form vales or low-lying plains. The residual materials produced by weathering and erosion – the debris which is carried by wind, water and ice across the earth's surface and which is ultimately deposited and converted to rock – also owe their characteristics to the rocks from which they were derived. And the nature of these materials, in turn, affects the changes which take place during transport, and thus the character of the depositional landforms. Hence, rock type exerts a fundamental control over both the geomorphic processes operating at the earth's surface, and the landscapes which are formed.

The properties of rocks

The term rock means many things to many people. In geomorphology it is given a relatively wide definition, including the hard rocks (such as sandstone or limestone) and many relatively soft materials, such as clays and sands. A distinction is made, however, between rock and sediment.

Sediment, often the raw material of rocks, is the loose, uncompacted material laid down by agents of deposition such as water, wind or ice. **Rock** is different in that its materials have undergone alteration by pressure, heat or chemical processes and it

has therefore adopted a different structure.

Three main types of rock are commonly recognized:

1 Igneous: formed by solidification of molten magma.
2 Sedimentary: formed by alteration and compression of old rock debris or sediments at the earth's surface.
3 Metamorphic: formed by alteration of existing rocks by intense heat or pressure.

We will consider the processes by which these are formed later. For the study of rocks in the landscape it is useful to examine three main properties: composition, texture and structure.

Composition of rocks

Most rocks are composed of **minerals**. These are amalgamations of inorganic compounds arranged according to a fairly specific and definite pattern. Over 2000 minerals have been recognized in the earth's rocks, but almost all the commonly occurring ones are related to ten major mineral groups (Table 17.1).

Table 17.1 *Major rock-forming minerals*

Minerals	Average abundance in selected rocks (per cent)			
	Granite	Basalt	Sandstone	Limestone
Quartz	31.3	0.0	69.8	3.7
Feldspars	52.3	46.2	8.4	2.2
Micas	11.5	0.0	1.2	0.0
Pyroxenes	t	36.9	0.0	0.0
Chlorites	0.0	0.0	1.1	0.0
Amphiboles	2.4	0.0	0.0	0.0
Olivine	0.0	7.6	0.0	0.0
Carbonates	0.0	0.0	10.6	92.8
Clay minerals	0.0	0.0	6.9	1.0
Iron ores	0.5	2.8	0.3	0.3

Note: t = trace.

Data from A. Holmes, *Principles of Physical Geography*, Nelson, 1965.

The minerals which make up the rocks are composed of **elements**. There are a large number of elements which occur naturally on the earth, but only eight are abundant. These eight – oxygen, silicon, aluminium, iron, calcium, sodium, potassium and magnesium – account for about 99 per cent of the total weight of the earth's rocks.

Thus, eight major elements, arranged to give ten main mineral groups, constitute the majority of the earth's crust. Nevertheless, there are a large number of rocks, for it is clear that the relative proportions of these and the less abundant minerals may vary considerably, and it is these variations which greatly influence the geomorphological significance of the rocks. Minerals such as quartz are highly resistant to weathering and tend to abound in the residual materials of weathered and eroded rocks. Others, such as the olivines and pyroxenes, are far less resistant to attack and are more readily destroyed during weathering (see Table 19.1, p. 293).

Table 17.2 *Size classification of sedimentary particles*

Size (mm)	Size (μm)	Wentworth Grade
		Cobbles
60	60,000	
		Coarse gravel
20	20,000	
		Medium gravel
6	6,000	
		Fine gravel
2	2,000	
		Coarse sand
0.6	600	
		Medium sand
0.2	200	
		Fine sand
0.06	60	
		Coarse silt
0.02	20	
		Medium silt
0.006	6	
		Fine silt
0.002	2	
		Clay

Texture

Not only does the composition of the rock-forming minerals vary; so does the size of the particles which they form. Rocks may be composed of grains varying in size from microscopic particles less than 0.0001 mm in diameter to large crystals several centimetres in size (Table 17.2). Grain-size, or **texture**, is therefore an important feature in classifying rocks. It is important, too, because it affects the behaviour of rocks and is related to many other characteristics such as their porosity, susceptibility to weathering and engineering properties.

Rock structure

The physical arrangement of the grains is also important. In most cases rocks are not composed of a random agglomeration of grains, but are structured; the grains are arranged in layers or groups, between which occur lines of weakness. These lines of weakness may represent **bedding planes**, formed during deposition of the sedimentary materials of which the rock is composed, or **joints** and **cleavage** lines formed during the drying, cooling or compaction of the rock. Along these planes of weakness rocks split or weather more readily.

Formation of igneous rocks

We have already seen that rocks can be divided into three main groups, and certainly the most abundant within the earth's crust is the igneous group. **Igneous rocks** are produced by cooling of magma, molten rock material that originates deep in the earth and rises upwards into the crust. The nature of igneous rocks varies, however, depending upon both the initial composition of the magma and the conditions during cooling. In this latter context, an important distinction can be made between **extrusive igneous rocks**, which are formed from magma escaping on to the earth's surface, and **intrusive igneous rocks** which are formed by cooling of magma in the crust (Figure 17.1). As we will see, these vary considerably in their character and composition.

The origins of magma

Most of the igneous rocks are derived from what is known as **silicate magma**. This forms at depths of about 15 to 25 km in the earth, where temperatures are in the range of 500 to 1200°C, while pressures may be more than 10,000 times those experienced at the earth's surface. Under these conditions, the peridotite which comprises the mantle undergoes partial melting. In the process, a magma is formed of rather different composition to the initial materials: richer in silica but depleted in iron and magnesium. As this magma rises it melts and dissolves some of the rocks which it encounters, so that its composition changes further.

Cooling processes

The magma which enters the crust, or escapes on to the surface, therefore varies in composition depending upon the conditions it experienced as it rose from the mantle. Further changes also take place during subsequent cooling. Some of the constituents then escape in gaseous form, while those which are left solidify to form minerals. What happens at this stage, and the character of the rocks which develop, depends particularly upon the rate of cooling, and this in turn is dependent upon the environment in which cooling takes place. Typically, extrusive magmas cool rapidly, for the temperatures at the surface are lower and dispersal of heat through the surrounding rocks and into the

atmosphere or oceans is relatively efficient. The presence of water will considerably aid cooling, and magmas which are released into the ocean, for example, solidify very rapidly, producing highly characteristic rocks. Conversely, magmas which are trapped within the crust tend to cool more slowly; the temperatures and pressures in the surrounding rock are higher and the opportunity for dispersal of heat is much less.

Reaction series and mineral formation

These differences in cooling environment and rate have fundamental effects on the nature of the rocks which are formed. These effects occur because of the process of **magma fractionation**. The constituents of the magma tend to crystallize at different temperatures, so that, as the magma cools, a sequence of different minerals is formed. Depending upon the rate of cooling, and the initial temperature of the magma, a range of minerals forms. Not all these minerals are stable, however, and some may melt again or be altered before the rock completely solidifies.

The sequence in which minerals form during cooling of the magma was studied by N. L. Bowen in the 1920s. He found that there existed two, separate sequences which have been termed the **Bowen Reaction Series**. One of these sequences tends to occur in discrete steps; as the temperature alters so abrupt changes in the nature of the minerals being formed occur, one mineral disappearing and another appearing. This sequence is called the **discontinuous series**. The other sequence is more transitional; a fall in temperature is accompanied by a progressive change in mineral composition, with one mineral grading into another. This is referred to as the **continuous series**.

Both the continuous and discontinuous reaction series may operate simultaneously, and they tend to converge, at relatively low temperatures, with the formation of a mineral known as **orthoclase**, one of the feldspar type minerals which is rich in sodium. At lower temperatures still, **muscovite** (a form of mica) and **quartz** are produced.

At higher temperatures, the two series produce very different minerals. The discontinuous series commences with the formation of **olivine**, which is replaced during cooling by minerals of the **pyroxene** group. This in turn is replaced by **amphiboles** and then **biotite** (another form of mica). All these minerals are dark in colour and

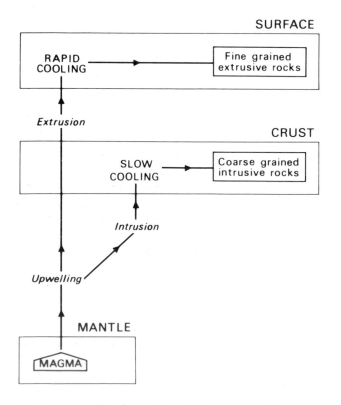

Figure 17.1   The formation of igneous rocks

contain relatively high proportions of magnesium and iron. They are referred to as **mafic** minerals.

The continuous series, in contrast, produces light coloured minerals dominated by **feldspar**. Initially, feldspars rich in calcium (**calcic feldspar**) are formed, but these grade into sodium-rich forms (**plagioclase**). These minerals, together with the low temperature products (orthoclase, muscovite and quartz), are known as **felsic** minerals, since they are dominated by feldspar and silica.

One question that this raises is how do the minerals formed early in the continuous reaction series survive during cooling? After all, we would expect from what we have just said that all the olivine, for example, would have been altered before the magma had solidified. This is a difficult question to answer, for we do not know exactly what happens inside the magma. It appears, however, that during cooling some of the minerals become segregated due to their different densities. Heavy minerals such as olivine and pyroxene may sink, forming a layer at the base of the magma which protects the crystals from alteration; lighter minerals such as the feldspars may float and similarly avoid alteration. In the case of magmas which cool slowly at depth, a further process may operate. Slow cooling allows the gradual growth of very large crystals, or the development of clusters of crystals, and the centres of these may again be protected from alteration. Thus, factors such as the rate of cooling and the relative significance of the two reaction series determine the character of the rocks that form.

In addition to these variations in composition, the rate of cooling affects the texture of igneous rocks. Those which have formed by slow cooling tend to be dominated by coarse-grained minerals, for ample time exists for growth of the crystals. Rapid cooling, on the other hand, generally results in finer-grained and glassy rocks.

This, of course, is a major simplification of a very complex process. As a generalization, however, it is a useful basis for understanding igneous rocks.

The nature of igneous rocks
The almost infinite range of rock types produced by igneous activity makes any rigorous classification highly complex. It is, however, possible to devise a very crude classification (Table 17.3) which identifies the main forms of igneous rocks.

At one extreme in this classification we may see the granitic rocks. These are coarse-grained, acidic rocks containing relatively large quantities of quartz and orthoclase. **Granite** is one of the igneous rocks formed at some depth in the earth, where slow cooling allows the growth of large crystals and a predominance of low-temperature minerals.

A gradation can be seen from granite through to more basic coarse-grained igneous rocks. The proportions of quartz and orthoclase decline and the quantities of other minerals, such as plagioclase and hornblende (one of the amphibole minerals) increase as we progress along this series. Thus granite grades to **granodiorite**, and thence to **diorite** and ultimately **gabbro**. This, unlike granite, is a relatively dark-coloured rock for it contains little or no quartz and orthoclase (both felsic minerals) but is dominated instead by calcic feldspar and the darker, mafic minerals (Plate 17.1).

The other main difference in igneous rocks relates to grain size, and the two end members of the series which we have just outlined grade to fine-grained equivalents. These represent the rapid cooling at the surface of the same magma types. Granite, for example, grades to **rhyolite**, an extrusive igneous rock of similar mineral composition, but much finer texture. Gabbro grades to **basalt**, a very dark, fine-grained material produced by lava flows from volcanic activity (see Chapter 20).

Economic significance of igneous rocks
It is interesting to note that igneous activity produces many of the world's most important minerals. These are generally associated not with the main body of the igneous rock, however, but with the more volatile substances related to them. These **volatiles**, moving through the surrounding rocks as gases or liquids, may become trapped in cracks and fissures within the rock where they solidify to give **veins**. Lead, copper, zinc and many other metalliferous minerals are produced in this way. Other ores of economic value are produced by the segregation of crystals during cooling; heavier minerals may sink to the bottom of the cooling magma to create a deposit of relatively pure material: magnetite (an iron ore) and nickel may both be formed in this way.

One of the main features of man's exploitation of his planet over the years has been his attempt to locate and mine some of these products of igneous activity. It is clear that a knowledge of igneous processes plays a part today, in finding suitable locations.

Table 17.3 *Classification of igneous rocks*

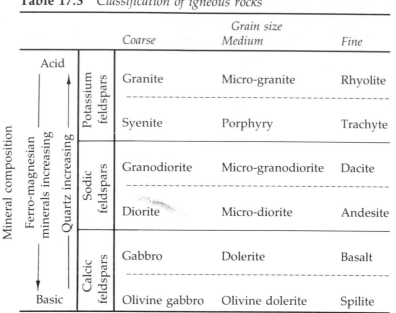

Mineral composition			Coarse	Grain size Medium	Fine
Acid	Quartz increasing	Potassium feldspars	Granite	Micro-granite	Rhyolite
			Syenite	Porphyry	Trachyte
Ferro-magnesian minerals increasing		Sodic feldspars	Granodiorite	Micro-granodiorite	Dacite
			Diorite	Micro-diorite	Andesite
		Calcic feldspars	Gabbro	Dolerite	Basalt
Basic			Olivine gabbro	Olivine dolerite	Spilite

Formation of sedimentary rocks

Sedimentary rocks are produced in many different ways in many different circumstances. Unlike igneous rocks, however, they are all formed at the earth's surface, and they are all composed of old rock debris that has been eroded, transported and redeposited. The range of rocks produced in this way is quite awesome, but we can get some idea of the factors controlling their character from the simple model in Figure 17.2. As this indicates, the nature of sedimentary rocks is a function mainly of the duration and distance of transport and their environment of deposition.

Composition of sedimentary rocks

A large proportion of sedimentary rocks are produced by the compaction and alteration of **detrital material**. This is the residual debris laid down at the end of the erosion process.

Quartz is one of the most resistant detrital minerals and it is certainly the most common. It is composed of silica and oxygen (SiO_2), and is commonly in the form of an almost **amorphous** material, lacking in any cleavage and thus in major lines of weakness. The grains may survive many cycles of erosion, sometimes being converted to sedimentary rocks, then eroded and redeposited

several times. In addition to quartz, mica (especially muscovite), calcite and feldspar are relatively common detrital minerals, although none are so resistant to weathering. Many sedimentary rocks also contain small quantities of **ancillary minerals**. Some of these are of high density and are referred to as **heavy minerals** (e.g. garnet, tourmaline and rutile), and several of these are quite resistant to weathering. The occurrence of these ancillary minerals is often revealing, since they may tell us something about the provenance of the sediments.

It should also be noted that almost all the constituents of these detrital sedimentary rocks are derived, ultimately, from igneous rocks. Only small quantities of minerals are generally formed in sedimentary environments. One of the main exceptions to this rule is **calcite**. This is commonly produced by living creatures, which assimilate calcium from the sea into their bodies to produce shells and skeletal material which sink and collect on the sea floor when the creatures die. Calcite and other minerals may also be produced by precipitation from water containing dissolved substances. This produces a **hydrogenic sediment**. Salt (sodium chloride) is another common example which is formed by evaporation of saline waters.

A third type of sedimentary material may be mentioned. This comprises the decomposition

Plate 17.1 Igneous rocks: (a) grandodiorite; (b) granite; (c) gabbro; (d) diorite; (e) rhyolite; (f) andesite; (g) micro-granite; (h) dolerite; (i) spilite; (j and k) basalt (photo: John Owen)

products of weathering; the minerals which are formed by alteration of the less stable original minerals. Clays are an example of this process, for many **clay minerals** are derived from the breakdown of detrital minerals. Many of the silicate minerals are altered in this way during weathering to give a range of clay minerals. As we will see when we consider weathering processes in more detail, the nature of the clay depends to a great extent upon the type and intensity of weathering.

Lithification

In their original form, detrital materials do not

deserve the name of 'rock'. Rather they are the unconsolidated sediments which must be consolidated and altered to form a true rock. This process of conversion from raw sediment to rock is known as **lithification**, and it comes about in a variety of ways. **Drying** and **desiccation** is one way; another is the effect of **compaction**, usually by the accumulation of overlying materials. Water is expelled, drying the sediment, and forcing the grains closer together. This produces a more permanent bonding between the particles.

Lithification is produced by chemical as well as mechanical processes. Changes in the surrounding environment may result in chemical reactions, to produce compounds which bind the particles together. **Oxidation** of the material, due to the drying of the sediment, may lead to the creation of iron and other oxides which coat the grains and form a cement between them. **Cementation** also occurs because of the **precipitation** of other substances from water percolating through the material. These substances may be washed in from outside, or from overlying sediments, or they may be dissolved from the particles themselves and redeposited around them as a cement. In addition to iron oxides, calcium, silica and aluminium compounds may all act as cements; two of the most common are calcium carbonate ($CaCO_3$) and silica (SiO_2).

Lithification of sedimentary rocks tends to alter the original composition of the materials to some extent. It also imposes on the materials a new structure; for example, **joints** may be produced through shrinkage of the sediments during drying (Plate 17.2). Not all the structures which are found in sedimentary rocks are produced during lithification; many of them are the results of the original depositional processes. Most sedimentary rocks, however, retain evidence of their depositional history in the form of bedding planes, laminations and various sedimentary structures (e.g. ripple marks, cross-bedding, mud cracks and the impressions left by organisms which once lived in or on the accumulating sediment).

The nature of sedimentary rocks

To get an idea of the main types of sedimentary rocks which we find in the earth, let us return to Figure 17.2. As this shows, we can start by assuming a relatively simple situation in which a large mountain range is being attacked by mechanical and chemical weathering. The debris produced by

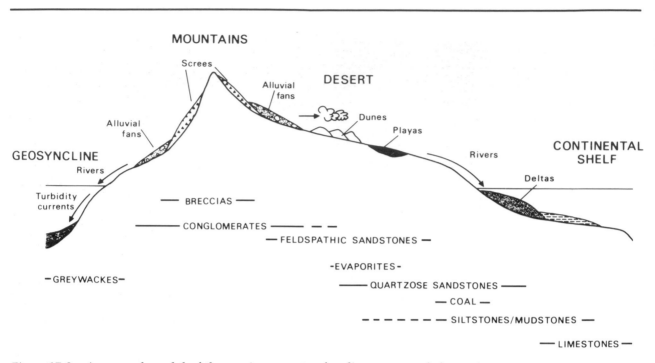

Figure 17.2   A general model of the environments of sedimentary rock formation

Plate 17.2   Jointing in Jurassic mudstones, Watchet, Somerset (photo: Dave Briggs)

these processes is being swept across the land by the action of streams, wind and ice and ultimately much of it is carried into the sea. As we will see in Chapter 20, during transport the debris is worn and sorted accorded to size, so that the material becomes finer as it is carried further from the mountains. Deposition of the debris en route, and its eventual compression and compaction, produces a range of sedimentary rocks.

Close to the mountains, on the valley sides and floors, the coarsest debris accumulates. It consists of angular rock fragments, transported only short distances and hardly worn at all. We can see similar material today accumulating as scree on steep valley slopes (see Chapter 21). When it is lithified it forms a rock known as **breccia**. Material which is transported slightly further and worn more intensively accumulates in the form of rounded boulders and pebbles. As finer material filters into the spaces between these large particles, and as compounds such as iron oxides, silica or calcium carbonate are precipitated around the particles, the debris is cemented into a rock known as **conglomerate** (Plate 17.3); the 'Bunter' conglomerates of the Midlands of England are excellent examples.

Plate 17.3   Sedimentary rocks: (a) conglomerate; (b and c) sandstones (note the different characteristics of the two samples); (d) siltstone; (e) greywacke; (f) limestone; (g) chalk; (h) flint (photo: John Owen)

Further away from the mountains, the material is more worn and finer grained for it has been transported greater distances. Nevertheless, many of the relatively fragile materials may have survived, so in addition to resistant grains of quartz there may remain considerable quantities of minerals such as feldspar. Overall the material is of the size of sand (0.06–2.00 mm in diameter), and when it is lithified it produces a special form of sandstone known as **feldspathic sandstone** (or arkose). Rocks of this sort are often found fringing the old mountain ranges (Figure 17.2). Beyond them, or in areas where the sediments have been more intensively worn during or prior to transport, occur sandstones composed almost wholly of more resistant materials; in particular quartz. As Figure 17.2 indicates, the sandstones are often found in the old coastal areas, for it is here, in the estuaries and in the shallow sea beyond, that the sands carried by the rivers are deposited and then moved around by the tides. But, in continental areas, vast areas of sand may accumulate in deserts, and these too may ultimately be lithified into rock (Plate 17.3). The Permo-Triassic 'Bunter' sandstones were formed in this way.

The finest material – the silt (0.002–0.06 mm diameter) and the clay (<0.002 mm diameter), which are produced by weathering or by prolonged attrition during transport, are generally carried the greatest distance. Much of it reaches the sea and is washed into the deeper waters where it very slowly accumulates on the ocean floor. The silt forms **siltstone** when compressed and dried (Plate 17.3); the clay is changed first to **shale** (when it is partially compacted) and then to **mudstone**.

In addition, where deep ocean basins (geosynclines) occur close to mountain areas, a relatively wide range of materials may be transported down the slope as turbidity currents and accumulate in the basin (see Chapter 25). The poorly sorted sandstones which are formed are known as **greywacke** (or, more generally, turbidites) and are common sedimentary rocks from Lower Palaeozoic strata (e.g. in Wales and the Lake District).

All these rocks are known as **clastic** rocks (from the Greek word meaning broken), for they are produced from fragments of old rocks. However, as we have mentioned, some of the material eroded from the pre-existing rocks may be dissolved and carried away as solutes. This material is ultimately precipitated by chemical or biological processes to produce **non-clastic** rocks.

One of the most common examples of chemical precipitation is the formation of evaporites. Salts dissolved in lake waters are precipitated as the water is evaporated and they accumulate on the lake bed. They may be dominated by a variety of minerals such as sodium (in which case **halite** is formed), or calcium (which produces **gypsum** or **anhydrite**). In addition, similar (though more complex) processes may result in the creation of surface crusts to the land. The **silcretes** (formed of silica), **calcretes** (of calcium carbonate) and **ferricretes** (of iron compounds) which occur in many sub-tropical areas are examples of these (Plate 17.4).

Biological precipitation of calcium carbonate is also a widespread process, and this gives rise to **limestone** rocks. These vary considerably, according to their exact nature of formation. **Chalk**, like that which comprises the North and South Downs in England, is produced by the accumulation of microscopic skeletal remains of marine organisms, particularly **foraminifera**. **Coral** or **reef limestones**, such as the Carboniferous limestone of the Pennines, is built by the accumulation of carbonates secreted by corals and algae growing in clear, shallow, littoral waters (see Chapter 25). Other limestones, however, may be formed by the deposition of materials eroded from older rocks; such **detrital limestones** are really clastic rocks.

Within limestones, it is common to find bands or nodules of **chert** and **flint**. These are composed of silica, and are produced by precipitation both biologically and chemically. Flint, it seems, is precipitated by the concentration of silica derived from

fossil organisms in the chalk by percolating waters. However, not all flint is formed after deposition of the chalk; some, it is argued, may form during deposition by the precipitation of silica around the remains of marine organisms. Chert, too, is produced in this way; deposits of siliceous **radiolarian** remains (minute simple organisms rich in silica) are cemented by precipitated silica to produce a compact splintery rock.

Biological accumulation also produces rocks such as **coal** and **lignite**. Most coal deposits originated as deltaic swamps in which the vegetation died and accumulated. Because of the anaerobic (oxygen deficient) conditions within the swamp, the organic material does not break down very rapidly, but builds up to form peat. A rise in the sea level (or subsidence of the land) ultimately floods the area, killing off the vegetation and allowing muds and silts to be deposited above the peat. Continued burial results in compaction and desiccation of the peat which is gradually changed into lignite. This is brownish in colour and woody in appearance and contains 30–40 per cent of water by volume. Further compaction and heating, however, convert the lignite to **bituminous** coal, which in turn is altered to **anthracite**, or hard coal.

Economic significance of sedimentary rocks

Apart from the obvious use of many sedimentary rocks for building (clays and shales for bricks, sandstones and limestones as stone) many of the rarer products of sedimentary processes have economic value. Flint has been used by prehistoric men to make tools (e.g. arrowheads and hand axes). Many mineral deposits (e.g. phosphates, gypsum, potassium) are used as fertilizers in agriculture; rock salt (halite) is used both for consumption and in industry; sulphur and sodium are employed extensively in the chemical industries. In addition, several of the sedimentary products are used as a source of energy. Coal is the most widespread example, but oil and gas, both of which are associated with the accumulation of organic materials in deep-sea areas, are clearly vital to modern societies. The situations in which oil and gas are formed are shown diagrammatically in Figure 17.3.

Formation of metamorphic rocks

Both igneous activity and lithification create new rocks. Metamorphism, on the other hand, is a

Plate 17.4 Calcrete, near Cape Town, South Africa (photo: Dave Briggs)

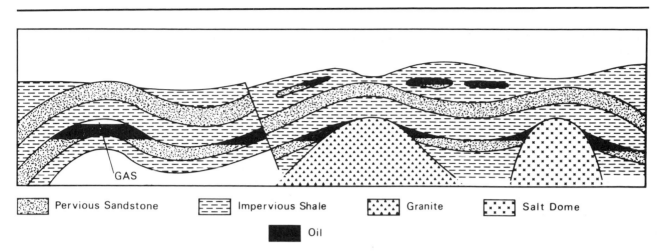

| Pervious Sandstone | Impervious Shale | Granite | Salt Dome |

Oil

Figure 17.3   A diagrammatic representation of geological situations in which oil and gas are formed

process of alteration of existing rocks within the crust rather than a creative process. It is convenient to discuss it here since it shares many characteristics of these processes. As before, the character of the rock which develops depends to a great extent upon the nature of the original materials (in this case the pre-existing rock) and the conditions in which metamorphism occurs.

Processes of metamorphism
Metamorphic alteration of rocks is brought about by excessive heat and pressure, and by chemical changes resulting from the action of hot gases or liquids passing through the rock. As such, metamorphism tends to be associated with igneous activity, since the intrusion of magma into the crust clearly results in considerable changes in the surrounding environment. The magma exerts a pressure upon the adjacent rock, it heats it and volatile substances escaping from the magma permeate the surrounding material.

It is apparent that in many instances all three processes may operate together, and they may affect rocks over a large area. Such **regional metamorphism** tends to show a distinct pattern, however, for the effects are most intense close to the source of the heat and pressure, and die out further away. Close to the source, in fact, the effect of the hot gases and liquids is at its strongest, and local, **contact metamorphism** occurs. This may produce what is known as an **aureole** around the intrusive magma. With increasing distance, this effect becomes less important and heat and pressure cause the major changes. Heating of the original **country rock** may lead to localized melting and

recrystallization of the minerals, while pressure alters the texture and structure of the rocks. Even if temperatures are not sufficient to cause melting, changes in the composition of the rocks may occur due to chemical reactions brought about by the heat.

In some cases, of course, the three processes may operate independently. Localized heating of the rocks may result only in **thermal metamorphism**. Pressures, exerted perhaps by intense mountain-building processes (see Chapter 18), may lead to structural change with little or no alteration in composition. This is known as **dynamic metamorphism**. Alternatively, active gases and liquids permeating through the rocks may result in chemical changes such as the replacement of elements within the minerals through the process of **metasomatic metamorphism**.

The nature of metamorphic rocks
These processes of metamorphism can be considered as a sequence, grading from relatively weak processes which leave intact many of the original features of the country rock to intense processes which obliterate almost all the initial characteristics. It is here that the tripartite classification of rock formation into igneous, sedimentary and metamorphic runs into difficulties, for it is clear that metamorphism grades imperceptibly into the other two processes. Very weak metamorphism is akin to the process of lithification; it may involve no more than pressure upon the surrounding sedimentary rocks. Intense metamorphism, with remelting and recrystallization of the original materials, is similar to certain stages in the formation of igneous rocks.

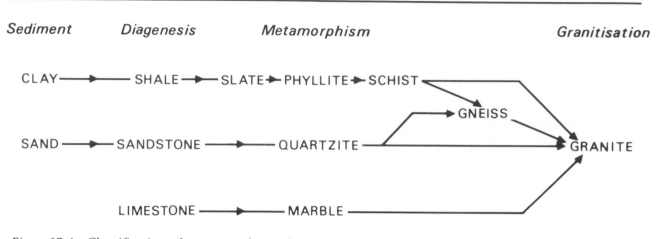

Figure 17.4   Classification of metamorphic rocks

In the case of more extreme metamorphic change, the nature of the original rock is relatively unimportant. Impregnation of these rocks by igneous materials and associated metasomatism may result in rocks totally unlike the original forms. **Granitization** of country rocks produces a rock similar to granite (which is an igneous rock) from a wide range of original materials. Clays, sands and even basalts may all be altered to a similar end-product.

More commonly, metamorphism is less intense and some characteristics of the original rocks are preserved. Commonly the structure is more or less altered, with the imposition of cleavage and various **flow structures** such as **foliation** and **banding** (Plate 17.5). Foliation refers to a layering of the rock (like

Plate 17.5   Metamorphic rocks: (a) gneiss, (b) schist; (c) phyllite; (d) slate; (e) quartzite; (f) marble (photo: John Owen)

the sheets of paper in a book) brought about by reorientation of the particles by pressure and shearing. Banding represents a layering of the rock due to the segregation of minerals into discrete zones. The difference is of some importance since foliation tends to produce weaknesses in the rock which may be exploited by weathering; banding may leave a quite strong, massive rock.

In addition to these structural changes, there may, of course, be changes in composition. Less stable minerals may be altered and new minerals formed. Such alterations, therefore, do not necessarily affect the whole rock and many of the original constituents may be preserved.

Metamorphic rocks vary in character, therefore, in response to differences in the original materials and in the intensity and process of metamorphism. A broad classification is possible by relating the rocks to these two factors (Figure 17.4).

One of the most commonly found series of metamorphic rocks is that resulting from the alteration of clays and shales. Under conditions of relatively weak metamorphism these change to **slate**, a compact, fine-grained rock with a distinct cleavage. This grades into **phyllite** as metamor-

phism becomes more intense, and thence into **schist** (Plate 17.5). This latter is frequently a beautiful rock, with plate-like grains of mica set in thin foliations; within these may occur scattered larger crystals of minerals such as garnet.

The metamorphic sequence exhibited by sandstone produces **quartzite** at all but the most intense levels of alteration. Quartzite is a hard, compact rock produced by the recrystallization of silica. The Cambrian Eriboll quartzites of north-west Scotland are fine examples.

Limestone subjected to relatively low levels of metamorphism is altered to **marble**, a rock of immense variety. Many so-called marbles are, in fact, merely very fine-grained, highly polished limestones. However, when subjected to heat, the calcium carbonate in the original limestone recrystallizes to give crystals of approximately equal size. Pressure may distort these to give a beautiful range of internal patterns to the rock (Plate 17.5).

Another common metamorphic rock is **gneiss**. This, again, is a term which is used rather loosely. Many gneisses are intermediate between granite and other rocks, and as this implies it is a product of

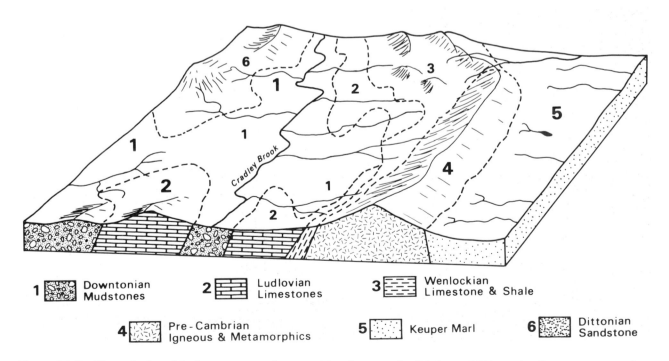

1 Downtonian Mudstones   2 Ludlovian Limestones   3 Wenlockian Limestone & Shale

4 Pre-Cambrian Igneous & Metamorphics   5 Keuper Marl   6 Dittonian Sandstone

Figure 17.5 The relationship between geology and landscape: the Malvern Hills and adjacent areas (after Palmer, 1976)

intense metamorphism, often of pre-existing igneous rocks. It is formed of alternating bands of minerals such as quartz and feldspar with micas and amphiboles. It often grades into schist.

Economic significance of metamorphic rocks

As we have seen, several of the metamorphic rocks are considered to be beautiful, and thus one of their main commercial uses is for ornaments and craftwork. Metamorphism also produces a variety of ore deposits such as copper, zinc and lead. These are often developed in the zone of contact metamorphism where alteration of the country rock by gases and liquids derived from the magma are most intense. Many precious stones are similarly produced by contact metamorphism; diamonds, sapphires, rubies and garnets are all formed in this way. In addition, slate is an important building material, and the finely laminated yet strong slates of Wales

have for long been quarried as roofing materials.

Rock type and landscape form

As we have noted, the nature of the rocks exerts a major influence upon the character of the landscape, and even a cursory analysis shows a general relationship between geology and topography in many areas (Figure 17.5). In some cases this relationship is so faithful that specific types of landscape are seen to be associated with specific rocks. In humid temperate regions, at least, claylands are characteristically gentle and undulating, while areas of Carboniferous limestone are often typified by karstic scenery. Granite areas, such as the moors of Devon and Cornwall, generally comprise rounded hills capped by craggy tors; chalk landscapes are also highly distinctive (Plate 17.6).

On a smaller scale, variations in rock type control the details of the landscape. Local differences in slope form, for example, are often related to the outcrop of rocks of different lithology, while valley side-slope angles frequently vary from one rock type to another (Figure 17.6). Similarly, rates of denudation differ. Rates of surface lowering of limestone in Derbyshire have been estimated at 0.075–0.083 mm y^{-1}; the gritstones of the Pennines, however, are being lowered at probably little more than 0.01 mm y^{-1}. These varying rates of denudation and different rock types are associated also with different processes of erosion. Limestone and chalk are susceptible to chemical weathering, and much of the erosion involves loss of material in solution. Gritstones and sandstones, on the other hand, are attacked mainly by mechanical weathering and debris is removed by streams mainly as solids. Thus, rock type is one of the major exogenous controls on the processes operating within the debris cascade.

Conclusions

Although the processes of rock formation are complex, and a detailed understanding lies beyond the scope of this book, a general picture can be gained by considering the processes and relationships which occur within three broad zones of the earth: the interior (mainly the mantle), the crust and the surface.

Formation of igneous rocks takes place in all three

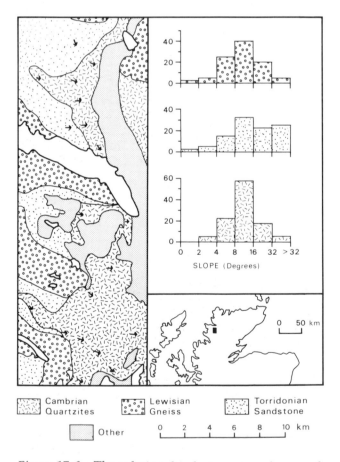

Figure 17.6 The relationship between geology and landscape: the Loch Assynt area (after Gardiner, 1983)

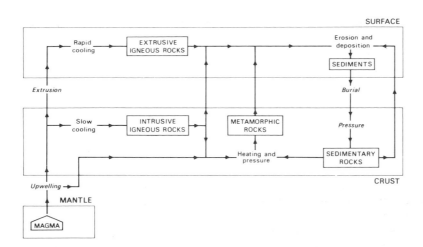

Figure 17.7   A general model of rock formation

Plate 17.6   The relationship between landscape and geology: landscapes in four different rock types: (a) a landscape in Carboniferous limestone, the Burren, County Clare, Ireland (photo: Peter Smithson); (b) a landscape in Chalk, showing a dry valley, Uffington, Berkshire (photo: Rick Cryer); (c) the Cheshire Plain: a lowland vale floored with glacial drifts and underlain by Triassic mudstones (photo: Dave Briggs); (d) the granite landscape of Dartmoor, Devon (photo: Dave Briggs)

zones. Magma from the mantle moves upwards into the denser crustal rocks and may cool there to form intrusive igneous materials. Further upwelling may bring the magma to the surface where rapid cooling results in the formation of extrusive igneous rocks. Factors such as the amount of water and the depth of cooling play an important part in both cases.

Sedimentary rock formation results from a much more varied set of processes. The initial materials collect at the earth's surface, either as clastic sediments or as biogenic accumulations or as evaporites. Burial, compaction and desiccation of these materials results in the imposition of new structures and some mineral alteration. In general these processes of lithification take place in the crust.

Metamorphic rocks generally result from the interaction of igneous activity with pre-existing crustal rocks. The gases, heat and pressure produced in the crust by upwelling magma alter the surrounding rocks and produce both chemical and structural changes.

Looking at the processes of rock formation as a whole, we can see how these various activities interact (Figure 17.7). It is also apparent that the exhumation of rocks within the crust rejuvenates the processes, breaking down the fresh rocks and supplying new materials as sediments for ultimate lithification. In all cases, the rocks which are formed influence the nature of the landscapes which develop at the surface.

Earth-building

Progress towards a flat earth

For over four and a half thousand million years the surface of the earth has been exposed to the action of the atmosphere. The rocks have been attacked by wind, water and ice. Erosion has been wearing away the moutains and hills; deposition has been filling up the valleys and seas. By now we might have expected that the whole surface would have been levelled out.

Despite these processes, however, the earth is far from flat and featureless. The reason is that even as the surface is being worn and reduced so it is also being uplifted and rebuilt. We speak of **earth-building processes**: the processes of uplift and subsidence, vertical and lateral shifts of the crust, folding and faulting. These are the processes by which the earth rejuvenates itself, and by which the landscape is kept alive. They are vital processes, for weathering and erosion can only continue if energy is available to drive them. Earth-building provides that energy. It uses energy derived from deep within the earth to raise the earth's crust in one area, to create huge basins elsewhere. The difference in level provides **potential energy** to the higher surfaces which can be used in the processes of erosion. Through the action of gravity, either directly or via the effects of water and ice, the rocks may be carried downwards. The debris cascade is initiated.

Structure of the earth

Before we can tackle the question of how these processes of earth-building occur, it is useful to discuss the more fundamental issue of the structure of the earth. Many of the earth-building processes originate deep within the earth and owe their effects to the inner character of our planet.

The earth is an oblate spheroid: that is to say, it is a sphere, slightly flattened at the poles. It is com-

posed of a number of different layers, the outer ones of which have been observed either directly or by deep borings (e.g. the Moho Project), or can be deduced from study of the materials reaching the surface, while the inner layers are deduced from **seismic** evidence, such as the behaviour of sound and shock waves, and from astronomical measurements.

A highly simplified diagram of the earth's structure is given in Figure 18.1. This shows that the innermost layer of the earth – the **inner core** – is believed to be solid, and to be separated from the liquid outer core by a transitional zone, referred to as the **F layer**. The density of the inner core is assumed to be in the order of 14 g cm^{-3}; the **outer core** is believed to have a density of about 10 g cm^{-3}. These densities imply that the core is composed of iron and small proportions of nickel.

Around the core is the **mantle**. There appears to be an abrupt change at the boundary of these two layers, and the mantle is assumed to consist of solid rock dominated by the mineral olivine. The density of this is about 5 g cm^{-3}, decreasing to about 3–4 g cm^{-3} in the upper mantle (Figure 18.1). This change in density seems to be related to a change in

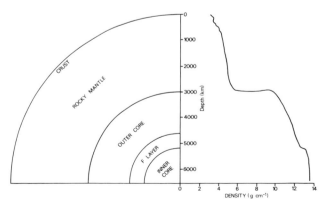

Figure 18.1 Internal structure of the earth, showing changes in density with depth

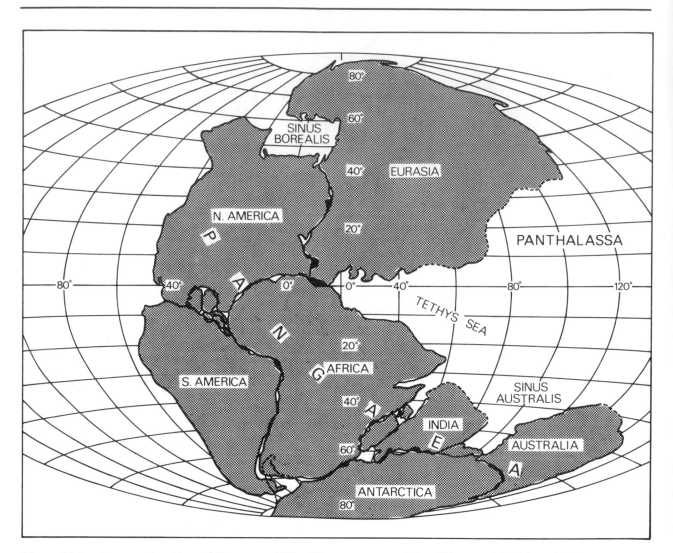

Figure 18.2   A reconstruction of Pangaea, 200 million years ago (after Dietz and Holden, 1970)

composition, and **peridotite**, a rock comprising the minerals olivine and pyroxene, probably dominates in the upper layer.

At the top of the mantle occurs another marked discontinuity. This is referred to as the **Mohorovicic boundary** (or Moho) after the Yugoslavian seismologist who discovered it early this century. This separates the mantle from the crust, although the zone of separation is irregular. Beneath the continents, the crust is relatively deep and the 'roots' of the land-masses sink into the underlying mantle. Beneath the oceans the crust is much thinner. Until recently it was thought to be only 5–6 km deep, but now it is believed to average a thickness of about 8–9 km. The continental crust is relatively light, with a

density of about 2.85 g cm^{-3}, while the oceanic crust has a density of 3 g cm^{-3} on average. Thus the continents can be seen to float upon the denser mantle rather like ice cubes in water, with the bulk of their volume below the surface.

On the basis of recent evidence, however, it is believed that the upper layers of the earth are rather more complex than this simple model suggests. The crust and part of the mantle beneath are thought to form large plate-like areas of rigid **lithosphere** which float upon the soft and plastic **asthenosphere** below. Thus, the crust and uppermost part of the mantle act as a single unit. As we will see, these lithospheric plates play a fundamental role in earth-building processes.

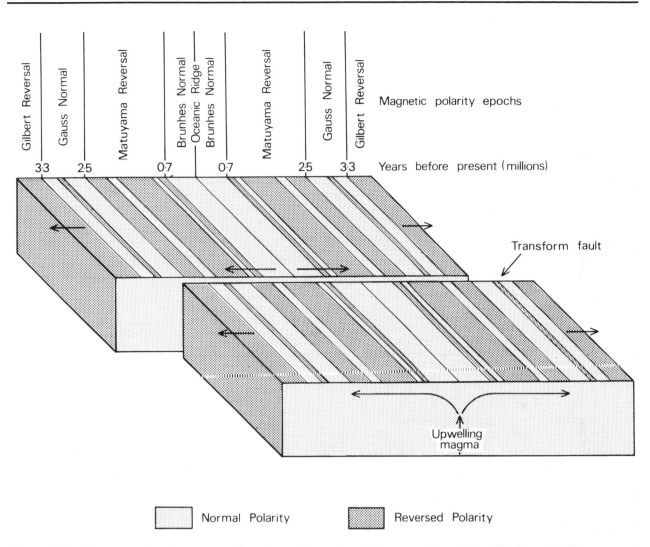

Normal Polarity   Reversed Polarity

Figure 18.3   Patterns of palaeomagnetism from the rocks of the ocean floor (after Hurley, 1968)

Plate tectonics

The discovery of plate tectonics

It was realized as long ago as the sixteenth century that the coastlines of the continents showed a remarkable conformity, and Francis Bacon was probably the first to indicate that they could be fitted together to form a single super-continent. It was during the twentieth century, however, that these ideas developed into a clear concept of what became known as **continental drift**. Then, two geologists, F. B. Taylor in the USA and Alfred Wegener in Germany, came independently to the same general idea: that the continents were not fixed but were adrift.

It is to Wegener that the concept of continental drift is normally attributed. Like Bacon, he realized that the continents could be fitted together in the manner of a jig-saw puzzle (Figure 18.2). He conjectured that from the Carboniferous Age, some 250 million years ago, until some time in the Pleistocene, about 1 million years ago, a single continent, which he called **Pangaea**, slowly broke up and drifted apart. Thus were born the continents of the world.

A wide range of evidence seemed to support this interpretation. Rare, identical fossils, such as the Cambrian coral-like organism *Archaeocyatha*, are found in rocks on different continents, now separated by thousands of kilometres of ocean.

Glaciations which are known to have occurred during Carboniferous times can be seen to have affected areas which become contiguous if the continents are fitted back into the Pangaean land-mass. Similarly, rocks, mountain ranges and fold belts are all consistent within this reconstructed super-continent.

For a time, therefore, the concept of continental drift was widely accepted, but gradually doubts re-emerged. The main problem was that no mechanism was known which could have caused Pangaea to break up and which could have moved the continents such vast distances, in different directions. Over time, therefore, geologists abandoned the notion; or rather, because such is the way of scientific progress, the new generation of geologists simply refused to accept the ideas. However, the question of what hypothesis was to replace these ideas remained.

An answer began to emerge during the 1950s and 1960s, derived from studies of the ocean floors. As volcanic rocks cool, they adopt a magnetic orientation which reflects that of the earth's magnetic field. At times, however, the magnetic field of the earth reverses (the positive and negative poles swop) so that rocks formed at different times show opposite magnetic orientations. Taking a transect across the floor of the ocean, geologists found that not only did the palaeomagnetism of the rocks vary, as if the ocean floor had been formed at different times, but the patterns either side of the **mid-oceanic ridge** were almost identical (Figure 18.3). Moreover, it became clear that rocks relating to each magnetic phase could be traced linearly along the ocean floor like a band parallel to the mid-oceanic ridge. In addition, the width of each band was found to be proportional to the duration of the palaeomagnetic phases established from studies of rocks of known age on land. Comparisons of the sequences on land and on the ocean floor thus enabled the rocks either side of the mid-oceanic ridges to be dated. And, startlingly, it was found that they became progressively older further away from the mid-oceanic ridge. The implication was clear: as the crust at the mid-oceanic ridges split, and the ocean floor moved apart, new rocks were being created in the gap.

So, the idea of **sea-floor spreading** developed. Further evidence supported the existence of the process. Drilling in the ocean floor, for example, showed that the sediments above the volcanic rocks increased in thickness away from the mid-oceanic ridges, indicating that longer periods of time had been available for deposition. But what was the mechanism of spreading? One suggestion was that the earth as a whole was expanding, and that, as it did so, the rigid crust fractured and was pulled apart. Other geologists, however, argued that the size of the earth was relatively constant, and that the crust was in fact being carried across the ocean floor. In this case, somewhere, the crust must be destroyed at approximately the same rate at which it is being formed. One possibility was that this destruction took place at the deep marine trenches, such as those which fringe the western Pacific. But, on the other hand, the fact that the continents, many of them formed of rocks millions of years old, had clearly retained their general shape (as indicated by the way they could be fitted together) showed that the continents themselves were not being destroyed.

Tentatively, explanations began to emerge. What if the crust consists of huge, dense plates on which the lighter continents are perched? Could it not be these plates that are moving across the surface, carrying the continents with them, then sinking into the mantle along the line of the ocean trenches to leave the land-masses bobbing at the surface? It might seem a wild idea, but it was also an attractive one, for it seemed to offer explanations of many phenomena. It explained the regular pattern of palaeomagnetic bands on the ocean floor, and the distribution of sediments. Simply, as the plates moved apart, new magma welled up along the mid-oceanic ridges and cooled, adopting the magnetic orientation of the earth at that time and creating a new band of ocean floor. On either side, the crust moved away, gathering sediment as it did so. It also seemed to explain some of the more major features of the earth's surface structure. Where two plates converged and collided, immense compressive forces must be generated which would buckle the continental rocks and produce fold mountains. Here, too, we might expect volcanic activity. Such indeed is the case, for the marine trenches of the Pacific are associated with some of the world's most active volcanoes.

An outline of plate tectonics

Thus was born a new concept: **plate tectonics**. A highly simplified picture of this concept is given in Figure 18.4, showing the way in which the lithospheric plates ride above the semi-plastic asthenosphere. At the mid-oceanic ridges the earth is split-

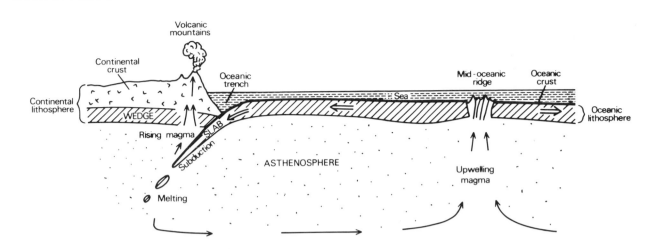

Figure 18.4 A generalized model of plate tectonics

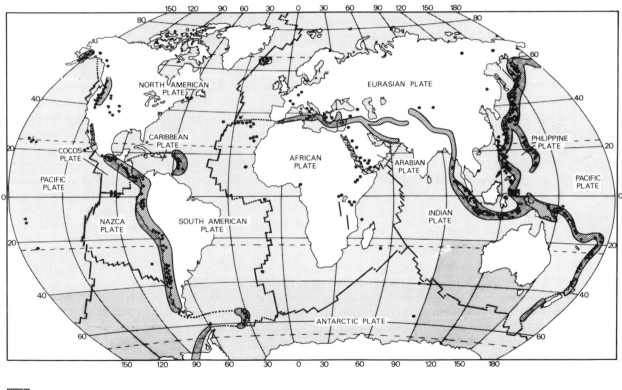

▨ Subduction zones   ── Rifts (and oceanic ridges)   ········ Transform faults   • Active and recently active volcanoes

Figure 18.5 The distribution of active and recently active volcanoes in relation to the world's plates (after R. and B. Decker 1982)

ting, allowing magma to rise through the asthenosphere. On either side the lithospheric plates are moving apart, across the asthenosphere. Where two plates converge, one, referred to as the **slab**, sinks – a process known as **plate subduction**. In the process, the slab experiences massive increases in pressure and temperature until, at a depth of 400–600 km, it melts and is absorbed into the mantle. Melting and friction release pockets of basaltic and andesitic magma which rise up through the overriding plate (the **wedge**) and create volcanoes. The stresses caused by subduction also trigger off earthquakes: shallow ones where the crust is stretched as it dips beneath the surface, and deeper ones where friction and compression start to destroy the slab.

Where two plates collide in this manner, it may be expected that the denser plate will subside and the lighter, more buoyant plate will remain at the surface. Where two plates carrying only thin rocks of the ocean floor meet, therefore, either plate may be subducted. On the other hand, where an oceanic plate meets a plate carrying continental crust, the tendency will be for the oceanic plate to subside. This is because, as we have seen, the continental crust is deeper and lighter than the oceanic crust, and is thus more buoyant. It is largely for this reason that the continents have survived subduction.

In some cases, however, two continents must collide. What happens then? The answer is that neither subsides. In fact, subduction tends to cease and the continents become fixed together. If this happens, though, rates of subduction across the earth as a whole no longer match rates of formation of new crust. Consequently, adjustments in subduction rates must occur elsewhere to maintain a balance within the system.

One other situation must also be mentioned, for in some circumstances adjacent plates slide past each other. Here, interactions between the two plates are less severe, but the tearing action at the plate boundaries causes the development of transform faults, at right angles to the mid-oceanic ridges but parallel to the direction of movement (Figure 18.5).

Processes and mechanisms

If this general picture of plate tectonics is accurate, it should be possible to identify the earth's plate structure from the evidence of the earth's surface features and associated seismic activity. There is some disagreement about this evidence, but the broad layout indicated in Figure 18.5 is widely accepted. This, of course, is not a static picture: it merely represents the current state of play in plate tectonic activity. Over the millennia, the plates have been moving and, in the forseeable future, they will continue to do so. Rates of plate movement can be calculated from the age–distance relationship of the rocks either side of the mid-oceanic ridges. These indicate rates of movement up to 8–9 cm y^{-1}, highest in the Pacific and least in the Atlantic and Indian Oceans (Figure 18.6).

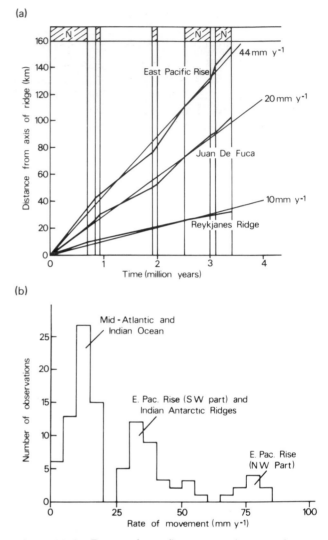

Figure 18.6   Rates of sea-floor spreading as shown by distance–age relationships for (a) transects across three mid-oceanic ridges; (b) measurements at each 5° length of ridge from the Atlantic, Pacific, Indian and Antarctic Oceans (after Oxburgh, 1974)

This movement has certainly been going on for many millions of years, for the rocks of the ocean floor, although nowhere near as ancient as those of the continents, date back at least to Jurassic times. As Wegener postulated, therefore, Pangaea probably started to break up some 200–250 million years ago, since when the continents have been transported to their present positions (Figure 18.7). The questions nevertheless remain: what is the mechanism for plate movement? What caused the break-up of Pangaea? And why did Pangaea not break up before?

The answers appear to lie in the existence of convection currents within the mantle. If they do exist, these currents are certainly not regular nor permanent. They are affected, for example, by surface conditions so that the positions of the plates modify the distribution of the currents – a process of feedback. As we have seen, when two plates become locked, subduction ceases and the convection pattern must be modified. None the less, it appears that activity at the mid-oceanic ridges is associated with decay of radioactive materials within the earth which releases energy and heats the surrounding materials. These then rise towards the surface. In the zones of subduction the converse is happening, for older, cool and dense surface material is sinking. Beneath the crust a reverse flow of material must be taking place to maintain the system (Figure 18.4).

Plate tectonics and mountain building

One of the most appealing aspects of the concept of plate tectonics is that it appears to explain so many of the major characteristics of the earth. We have already noted the fact that the distribution of both volcanic activity and mountain building can be related to the pattern of plate movement. More detailed studies of individual areas, however, show the ways in which these processes operate. The Himalayas, for example, appear to have been developed initially by the collision of the Australian and Indian plates which thus became locked and now, as a single plate, are being forced against the Eurasian plate to the north. Similarly, the eastern Alps result from the collision of the Eurasian and African continents. In both cases, the continental boundaries are not wholly conformable, so that while the land-masses are in contact and undergoing folding and uplift in some places, gaps exist between them elsewhere. The Mediterranean and Black Sea are possibly gaps of this type.

In the case of the Andes and Rockies rather different processes must have been at work, for in neither case do the converging plates both carry continents. The Andes, for example, lie at the junction of the oceanic Nazca plate and the continental edge of the Atlantic plate. Their history seems to have been complex. It seems that initially subduction of the Nazca plate during Triassic and Jurassic times led to the creation of an arc of volcanoes off the South American coast. As subduction con-

tinued, during the Cretaceous period, volcanic activity became more intense and huge bodies of andesite were intruded into the sedimentary rocks, raising the surface into what is now the Western Cordillera. At the same time, the forces associated with this activity rippled eastwards, folding the rocks and thrusting up the fold mountains of the Eastern Cordillera. Since then, erosion of the two ridges has led to deep infilling of the intervening area to produce the altiplano (Figure 18.8).

Plate tectonic activity can thus be related to the formation of most of the younger mountain ranges of the world – those which have formed since the break-up of Pangaea. But what of the older periods of mountain building? Were they, perhaps, formed during earlier phases when the continents were adrift prior to their accretion into the Pangaean land-mass? Have there been several phases of continental accretion and break-up? We do not know for certain, but there is reason to speculate that the processes of plate tectonics are not confined solely to the last 250 million years. What we have is possibly a process which has been operative for much of the earth's history. It has been suggested, for example, that the Appalachians were formed by subduction and continental collision during the Palaeozoic era, possibly as Pangaea came together.

One thing is certain: plate movement has been taking place for a long time and, at one place or another, continental collision and mountain building have been occurring at intervals throughout this period. Such phases of mountain building are nevertheless often brief, a mere 50–100 million years, and they do not occur everywhere at the same time. Thus, the traditional concept of phases of mountain building (orogenies) as synchronous, worldwide events is no longer tenable.

Plate tectonics and earthquakes

As an understanding of plate tectonics has developed, it has become clear that many of the world's main earthquake zones are associated with plate boundaries. Shallow earthquakes (<70 km depth) occur along the marine trenches where plate subduction is taking place, along the mid-oceanic ridges and in the slip-fault zones where two plates are sliding past each other. Intermediate (70–300 km) and deep (>300 km) earthquakes, on the other hand, are related almost wholly to areas of subduction (Figure 18.9). Indeed, analyses of the focal

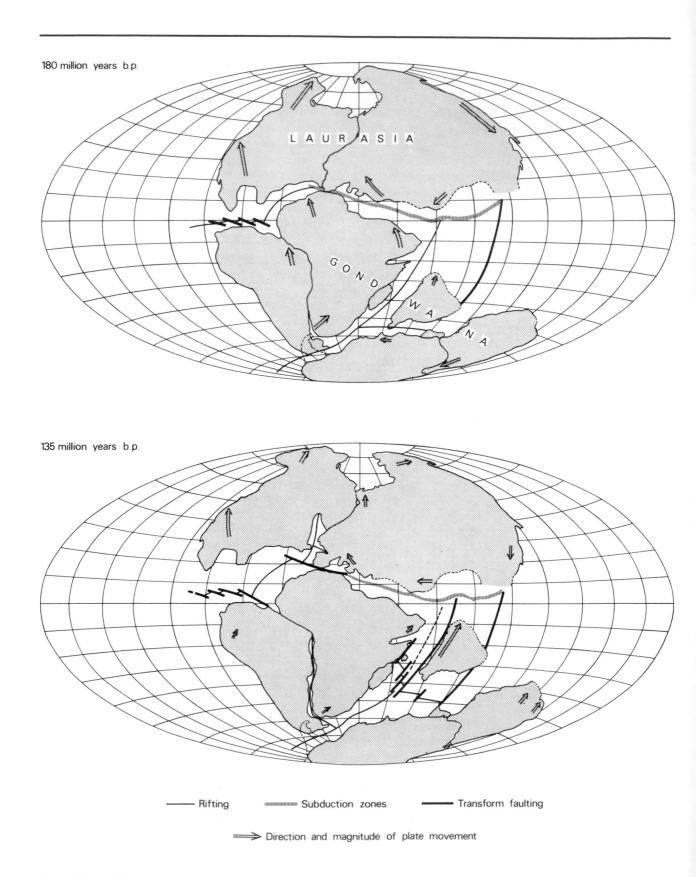

Figure 18.7 The break-up of Pangaea (after Dietz and Holden, 1970)

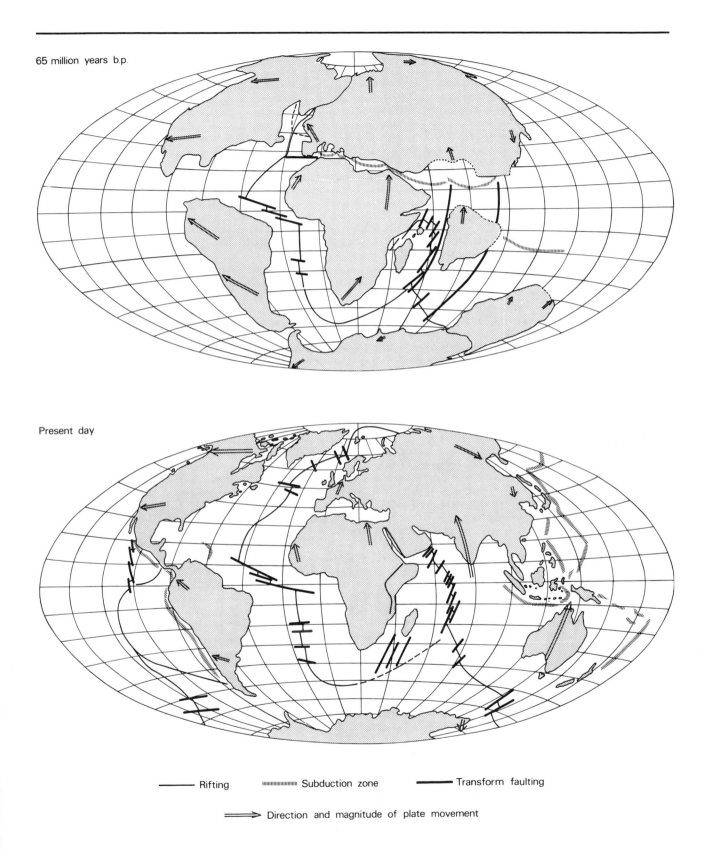

65 million years b.p.

Present day

——— Rifting ᴜᴜᴜᴜᴜᴜᴜᴜ Subduction zone ━━━ Transform faulting

⟹ Direction and magnitude of plate movement

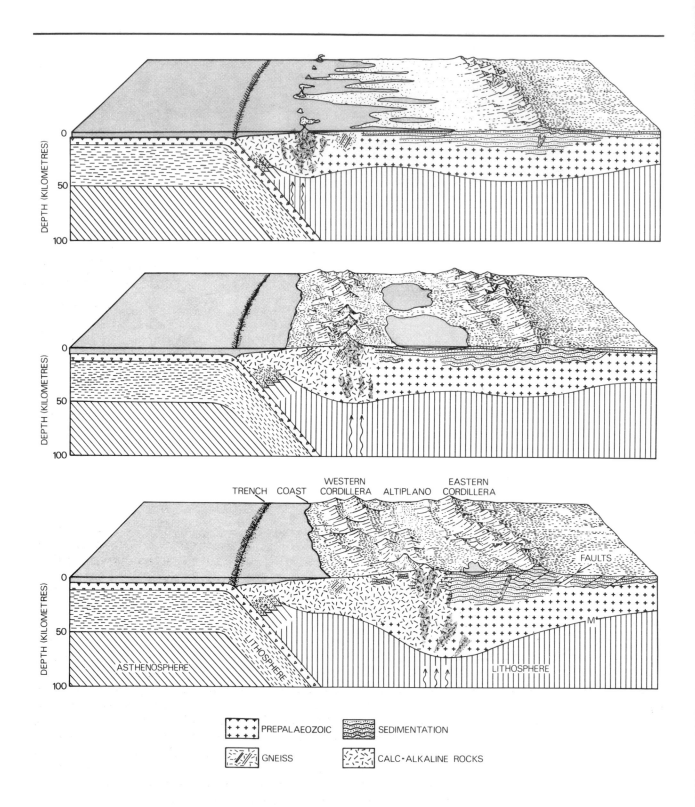

Figure 18.8 The development of the Andes (from James, 1973)

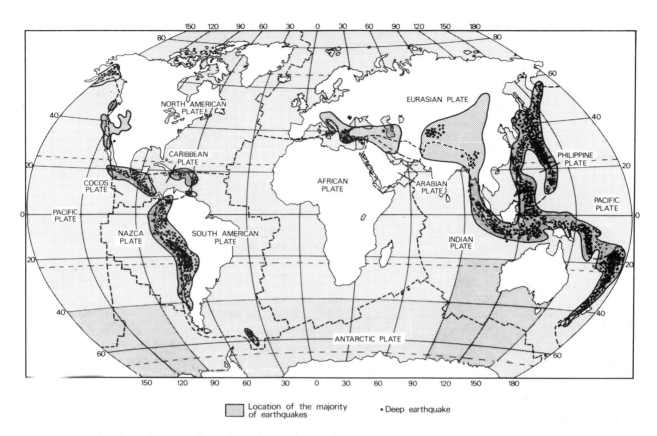

Figure 18.9 The distribution of earthquakes (after Toksoz, 1975)

points of earthquakes in the subduction zones have enabled the actual shape of the subsiding slab of oceanic plate to be identified and have shown the various ways in which subduction occurs (Figure 18.10).

An earthquake is a sudden and massive fracturing and movement of the rock along a distinct **failure plane**. This failure plane tends to develop where the stresses within the rock are greatest relative to the strength of the rock. Rocks derive their strength from the interlocking and cementation of the grains or crystals of which they are built. The stresses, on the other hand, are set up by tension or compression due to differential movements of the lithosphere. Where plates are descending into the mantle along the lines of subduction, for example, the stresses within the slab build up with depth until they are able to overcome the inherent strength of the rock. The rock then fails and an earthquake occurs. This generally happens at depths of less than 700 km, for by then the plate material has heated to around 2000°C and has been more or less assimilated into the mantle: it is prob-

ably too plastic to undergo brittle failure. Consequently, earthquakes do not seem to originate beyond 700 km. In fact, most earthquakes are shallow, occurring at a depth of 5–15 km.

Earthquakes are surprisingly common. Fortunately only a few make news through the destruction they cause and most occur in remote areas where they do not affect man. The energy released by earthquakes may be vast, however, and the largest events may produce as much as 10^{19} Joules. The magnitude of earthquakes is normally measured, in practice, on the **Richter scale** (named after the seismologist who devised it). This rates earthquakes on a scale from 0 to 9. The largest earthquake recorded, in Chile in 1960, reached a magnitude of 8.9, and three events of 8.6 have been measured since 1900. Most major earthquakes, however, have magnitudes of about 6.5.

The energy released by earthquakes is expended in the form of shock waves. These travel through the rocks, from the centre of the earthquake (the **focus**), either through the earth's interior or along its surface. In many cases there are initial shock

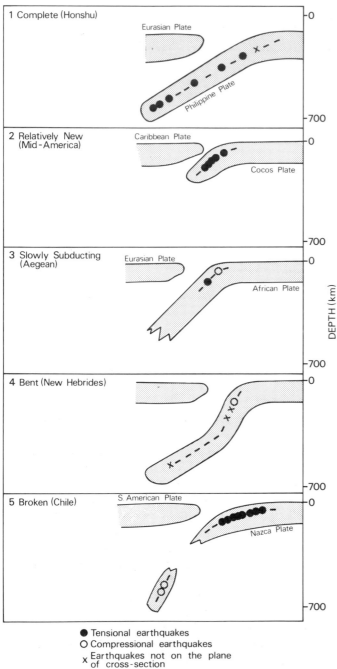

DEPTH (km)

● Tensional earthquakes
○ Compressional earthquakes
x Earthquakes not on the plane of cross-section

Figure 18.10 Earthquake depths and postulated patterns of subduction at five convergent plate boundaries (after Toksoz, 1975)

waves which precede the main shock. These **fore-shocks** relate to the shattering of obstructions or bonds along the failure plane. The **principal shock**, which follows, is the most severe, but it may last for

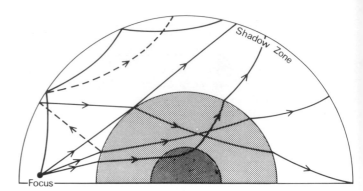

Figure 18.11 Patterns of P (solid lines) and S (dashed lines) waves emanating from an earthquake (from Bullen, 1955)

no more than a few seconds or, at the most, minutes. Afterwards, further shocks occur due to minor movements, triggered off by the main displacement.

The shock waves which are produced by the earthquake travel through the earth in three main ways. Deep waves travel rapidly through the inner earth in the form of preliminary, **push-pull** (*P*) **waves** and secondary, **shake** (*S*) **waves**. The P waves involve the vibration of particles to and fro in the same direction as wave propagation. The S waves involve transverse motion of the particles. Whereas P waves can pass through solids, liquid and air, the S waves can only pass through solids. They travel more slowly than P waves, at about 60 per cent of the speed.

A third type of waves is also generated by the earthquake. These have a longer wavelength and are referred to as **L (long) waves**. They are the slowest of all the waves and travel at the surface. Both P and S waves are reflected by the discontinuities within the earth, and as a result there develops a shadow zone which does not feel the effect of these waves following an earthquake. L waves, however, may travel right round the earth.

As we have seen, the point of maximum displacement – the focus – from which these shock waves originate is often deep in the earth's crust. Vertically above this, at the surface, is the location known as the **epicentre**. Often, severe earthquakes occur so deep in the earth that no surface displacement is seen. Nevertheless, as the shock waves erupt at the surface they cause immense disturb-

Table 18.1 *Major earthquakes during the twentieth century*

Date	Location	Richter magnitude	Death toll (where known)
1906	San Francisco	8.3	315 (+ 352 unaccounted for)
1906	Andes of Colombia, Equador	8.6	
1906	Valparaiso, Chile	8.4	
1911	Tien Shan, Sinkiang, China	8.4	
1920	Kansu, China	8.5	
1923	Tokyo, Japan	8.3	140,000
1933	Japanese Trench	8.5	
1949	Garm, Tadzhikistan, USSR	7.9	12,000
1950	North Assam, India	8.6	
1960	Chile	8.3–8.9	
1964	Alaska	8.6	115
1975	Haicheng, China	7.3	300
1976	Antigua, Guatemala	7.5	
1976	Tangshen, China	8.5	750,000
1978	Oaxaca, Mexico	7.8	0
1980	Southern Italy	6.8	3,000

ancc of the land; huge ripples may spread out as the waves run across the surface; the land may bulge and crack, then close up again. It is these land waves, occurring close to the epicentre, that occasionally cause considerable damage; they may trigger off rockfalls, landslides and avalanches which take as great a human toll as the original earthquake. In addition, the secondary effects of collapsing buildings, ruptured dams and gas or water pipes, and consequent fires and floods may be devastating. The San Francisco earthquake of 1906 was one of the more appalling examples, but during the present century there have been a number of devastating earthquakes (Table 18.1).

Another consequence of earthquakes is the generation of huge sea waves or **tsunami**. These are produced by displacements in the ocean floor, either as a direct effect of the displacement caused by the earthquake, or, indirectly, by submarine slumping of sediments around the coast and, especially, on the edges of the submarine trenches. In 1958 a tsunami some 30 m in height approached the Alaskan shore at a speed of over 200 km per hour and devastated large areas. Whole series of waves may be set up by these events, and they may have wavelengths of several hundred kilometres. Their effect is most intense where they enter shallow water, and thus they are most damaging where the coastline slopes gently, allowing the energy which is being transmitted to create a steep-fronted wave.

Volcanoes and igneous activity

Volcanoes

As we have seen, the earth's subduction zones and mid-oceanic ridges are areas of intense volcanic activity. Between them, they account for about 95 per cent of all the active volcanoes, the remaining 5 per cent occurring in areas associated with what are called **mantle plumes**, in the interior of the plates.

The types of volcanic activity occurring in these three areas vary markedly. In the case of the mantle plumes, it seems that basaltic magma rises to the surface from permanent **hot spots** beneath the crust, possibly associated with stable convection cells in the mantle. In oceanic areas, where the crust is thin, the magma breaks through to the sea floor where it is rapidly cooled and produces **pillow** or **columnar** lavas like those in Plate 18.1. Typically, the lavas build up to form a gently sloping **shield volcano** (Plate 18.2) and, in some instances, this may rise high enough to reach above sea level. A volcanic island is thus created. Over time, as the plate drifts across the hot spot, a succession of such volcanoes is formed, producing an **island chain**. The best known is the Hawaiian chain, the oldest member of which is Midway (about 18 million years old) in the north-west, while the youngest is Hawaii in the south-east, which is still active.

Volcanic activity is less common where mantle plumes lie beneath continents, for the magmas

Plate 18.1   Types of volcanic tephra and lava: (a) columnar basalt, Aldeyjarfoss, Iceland (photo: A. R. J. Briggs); (b) edge of 1981 lava flow, Krafla, Iceland (photo: A. R. J. Briggs); (c) ropy lava, Reykjahlid, Iceland (photo: A. R. J. Briggs); (d) volcanic bomb, Fljotsdalur, Iceland (photo: D. J. Briggs); (e) volcanic ash exposed in a road cutting, Thorolfsfell, Iceland (photo: P. A. Smithson)

there are more likely to be trapped in the crust. In these cases, intrusive igneous features are formed, as we will discuss later. None the less, where the magmas do break through to the surface, vast areas of basalt may be produced, forming extensive plateaux. Possibly the most spectacular example is the Deccan plateau of India, which covers 700,000 km², while similar **flood basalts** form an area of

about 300,000 km² in the Columbian plateau of the north-west USA.

Extensive lava flows are also associated with the mid-oceanic ridges. Here, magma rises into the long fissures created by separation of the plates which make up the ocean floor. Commonly the site of eruption is only temporary, and over time it migrates along the fissure so that a series of

Plate 18.2 Types of volcano: (a) a shield volcano, El Misti, Peru (photo: Jill Ulmanis); (b) a cinder cone, near Krafla, Iceland (photo: D. J. Briggs)

volcanoes is formed, each one often being marked by the development of a **cinder-cone** built up from the ash ejected during the eruption (Plate 18.2). Nowhere are the effects better seen than in Iceland. This country lies astride the mid-Atlantic ridge and is built almost entirely of basaltic lavas of Tertiary and Quaternary Age. (Incidentally, the Tertiary basalts of Skye and Northern Ireland are part of the same sequence.) It is still highly active and, in 1783, for example, a series of eruptions occurred along a 32 km fissure at Laki, creating a line of over a hundred small cinder-cones. More recently, in 1964, eruptions occurred in the Vestmannaeyjar off the south coast. Ash and lava piled up until they rose above the surface of the sea and created an entirely new island – Surtsey. A few years later an eruption occurred on the nearby island of Heimaey. Vast amounts of ash and lava were ejected, burying large parts of the island's only town and almost blocking the entrance to the harbour.

If we examine a map of the world's volcanoes we find that about 80 per cent lie along the zones of subduction – notably in a great arc of islands around the western side of the Pacific and along the line of the Andes and Rockies (Figure 18.5). Volcanoes along the island arc of the Pacific seem to coincide with areas where the subducted plate is 100–200 km deep. Whether activity is a result of melting of the plate during subduction, or whether magma from the mantle escapes along fissures created as the plate subsides into the mantle is not clear. It is apparent, however, that island arcs such as this are common where two oceanic plates collide. In these cases the volcanoes which are formed are typically **strato-volcanoes**, consisting of alternating layers of ash and lava. They are composed of a range of rock types, including basalt, andesite and rhyolite.

On occasions, these volcanoes are truly explosive. In 1883, for example, Krakatoa, a volcanic island in Indonesia, exploded, ejecting 75 km³ of material into the air and leaving a huge **caldera**, some 7 km × 6 km in diameter. The event is vividly described by A. Holmes in the *Principles of Physical Geology*, p. 336, 1964:

The climax was reached during the last week of August. On the 26th formidable detonations were heard every ten minutes. Dense volcanic clouds reached a height of 17 miles, and ashes, converted into stifling mud by incessant rain, fell over Batavia (now Djakarta) which was plunged into thick darkness, relieved only by vivid flashes of lightning. On the morning of the 27th came four stupendous explosions, the greatest of which was heard 3000 miles away in Australia, and a vast cloud of incandescent pumice and ashes rose 50 miles into the air.

Giant tsunami were generated by the explosion and 36,000 people died in Java and Sumatra.

Several other calderas of this nature are found close to the zones of plate subduction, all of them testifying to massive volcanic explosions. Crater Lake, in Oregon, USA, is a spectacular example, some 10 km × 10 km in diameter. Even larger is Lake Toba in Sumatra (about 1000 km²).

As all these examples show, volcanoes vary

Figure 18.12   Types of volcanic eruption (based on Holmes, 1965)

A   Fissure or Icelandic
Very fluid basaltic magma gives almost horizontal flows

B   Hawaiian
Molten lava erupts from a 'lake' in the central crater. Occasional bursts of incandescent spray; little tephra. Produces shield volcanoes

C   Strombolian
Sporadic eruptions of gas; lava forms bombs; tephra mainly bombs and lava

D   Vulcanian
Viscous lava which rapidly solidifies; leads to violent eruption of gases; abundant dust and bombs

E   Vesuvian and Plinean
Very violent expulsion of magma charged with gas, following prolonged quiessence. Abundant dust and bombs. May reach great height in Plinean eruptions

F   Pelean
Very highly viscous lava results in blockage of main vent. Lava charged with gas escapes from weak points as nuées ardentes

Table 18.2 *Types of tephra*

State	Name	Description
Gas	Fume	
Liquid	Aa	Rough, blocky lava
	Pahoehoe	Smooth, ropy lava
Solid	Dust	<0.06 mm
	Ash	0.06–2.0 mm
	Cinders	2.0–64 mm
	Blocks	>64 mm, solid
	Bombs	>64 mm, plastic
Flows	Pyroclastic	Hot, fluidized flows
	Mudflows	Flows fluidized by rainfall, ice melt, snow or water released from crater lakes

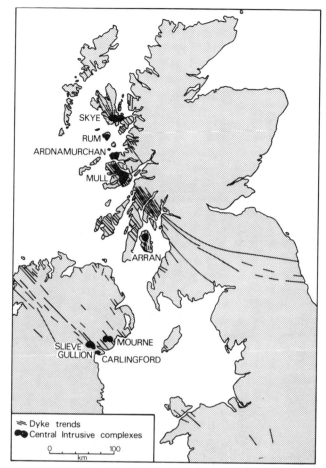

Figure 18.13   Dyke swarms emanating from Tertiary igneous intrusions in northern Britain (from Sparks, 1971, and Richey, 1964)

considerably both in the nature of their eruptions and in the character of the **ejecta** they produce. A general classification of types of volcanic eruption is given in Figure 18.12, while types of ejecta are shown in Table 18.2 and Plate 18.1. Depending upon these two factors, the morphology of the volcano also varies. Some examples are illustrated in Plate 18.2.

Intrusive igneous activity

We noted earlier that magma rising from the mantle does not always reach the surface of the earth but may be confined within the crustal rocks. Within the crust, the magma cools relatively slowly to give intrusive igneous rocks (see Chapter 17).

The intrusive magmas may produce many different features. Thin vertical veins within the crust may be exploited by the magma, forced apart by the pressures it generates, and ultimately filled with the solidified rock. These **dykes**, as they are known, often disseminate out from the centres of intrusion (Figure 18.13). Because the dykes have a large surface area relative to their volume, the magma in them cools fairly quickly and fine-grained rocks predominate. Sometimes the dykes feed into surface vents such as volcanoes and they thus act as conduits by which magma reaches the surface.

Where magma seeping along dykes is able to exploit a horizontal plane of weakness within the rock, a feature known as a **sill** may develop. Occasionally, these sills remain concordant with the strata for considerable distances, but, where vertical discontinuities or weaknesses occur, they

may break through the strata and exploit different levels in the rock (Figure 18.14).

More viscous magma may spread less readily through the rocks and may, instead, create dome-shaped masses of more limited extent. **Laccoliths** of this form occur in the Henry Mountains of southern Utah. Larger-scale masses, formed by the upwelling of magma are known as **batholiths** (Figure 18.14). These are typically formed of granitic rocks and they incorporate and displace large quantities of the surrounding country rocks. The series of moorlands in south-western England which include Dartmoor and Bodmin Moor are examples of batholiths (Figure 18.15).

All these subsurface igneous forms may influence the surface topography when they are exhumed by erosion. Dykes may be either more or less resistant than the surrounding rocks, and may thus form thin walls or trenches across the

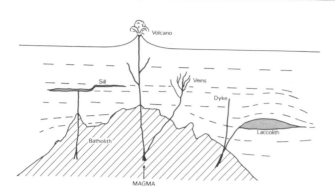

Figure 18.14   Types of igneous intrusion

landscape (Plate 18.3). Sills, such as the Whin Sill in northern England, similarly form dominant structural landforms (Plate 18.4). Exhumed laccoliths and batholiths may also impose their character upon the landscape.

Folding and faulting

The upheavals created by plate movement, subduction, earthquakes and volcanic activity set up huge stresses in the earth's crust. These often deform the rocks, either by **folding** where the rocks are able to deform plastically, or by **faulting** where brittle failure occurs.

Folds

At its simplest, a fold represents a single 'crumple', rather like that which can be formed by pushing gently on a tablecloth. Depending upon the degree and direction of pressure, this fold may be symmetrical or asymmetrical (Figure 18.16). Where the fold produces an arch-like form the feature is termed an **anticline**; where the rocks are warped downwards it is referred to as a **syncline**. A series of folds often develops to give a number of parallel anticlines and synclines. Minor folding may also be superimposed upon larger features to give **anticlinoriums** and **synclinoriums** (Figure 18.16).

Where the lateral pressures are greater, the folds become more complex. Anticlines may be **overturned** as one limb of the fold passes the vertical; they may become **recumbent**, when the rocks are forced over upon themselves. Associated with folding, fracturing of the rocks may occur. The rocks may shear along a line of weakness, generally running through the axis of the anticline, to produce **overthrusting** (Figure 18.16). All these features vary greatly in size, and folds may be as small as a few centimetres or as large as many kilometres in width and height.

In landscapes which have not yet been drastically modified by erosion, these processes may be reflected by the topography; anticlines form hills and

Plate 18.3 (above)   A volcanic dyke forming a rock wall, County Antrim, Northern Ireland (photo: D. J. Briggs)

Plate 18.4 (below)   The Whin Sill. Note the columnar structure of the basaltic rocks (photo: J. Rose)

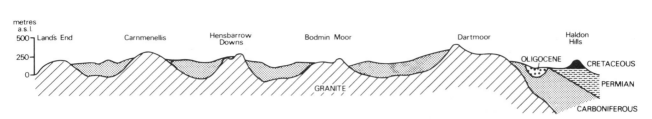

Figure 18.15 The granite intrusions of Devon and Cornwall (after Eastwood, 1964)

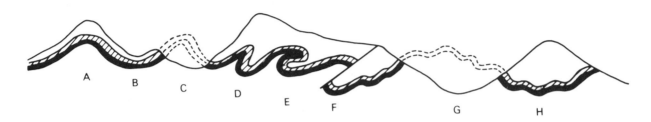

Figure 18.16 Types of fold: (A) symmetrical anticline; (B) symmetrical syncline; (C) asymmetrical anticline; (D) inclined anticline; (E) recumbent fold; (F) overthrust fold; (G) anticlinorium; (H) synclinorium

synclines valleys (Plate 18.5). In time, however, erosion of the rocks usually destroys the original surface expression of the folds, by preferentially removing weaker rocks. Indeed, in some cases inversion of the landscape may occur as the anticlines are worn away and the more compressed and resistant strata in the syncline are preserved as hills (such as Snowdon) (Figure 18.16). In areas of complex folding, such as the European Alps and the fold mountains of South Africa, it may be very difficult to identify the original pattern of folding, due to the equal complexities of erosion. In addition, because folding may entirely reverse the sequence of rocks exposed at the surface, it can lead to problems in reconstructing the chronology of the area.

Although folding is often closely associated with major orogenic activity, there are many examples of folding produced by smaller scale events. The intrusion of magma into the crust may cause localized folding, while, at an even smaller scale, salt domes, created by the rise of relatively low density salt into denser overlying rocks, may produce similar features. Localized folding also arises from the subsidence of dense material into less dense substrata (**load structures**), and from the growth of ice wedges and lenses beneath the ground surface.

Faults

When the relatively rigid rocks of the crust are subjected to tensional stress, they tend to fail along distinct failure planes. One of the consequences of this failure is the creation of faults.

Plate 18.5 An anticline: Whaleback Rock, Bude Harbour, Cornwall (photo: Peter Smithson)

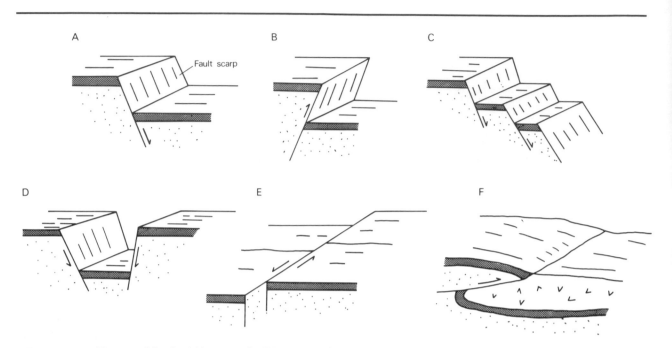

Figure 18.17   Types of fault: (A) normal; (B) reversed; (C) stepped; (D) rift; (E) transcurrent; (F) thrust

Various types of faults occur, depending upon the nature of the stresses which caused them. The simplest form, the **normal fault**, is created by tensional forces operating in opposite directions (Figure 18.17a). Characteristically, the rocks break along a steeply inclined fault plane and one mass of rock slips vertically relative to the other. Such faults

Plate 18.6   Aerial view of the San Andreas Fault (photo: © Dr Georg Gerster/John Hilleson Agency)

give rise to simple **fault scarps**. These may show a considerable range in both height and length.

More complex faults arise where failure occurs not along a single plane, but along a number of parallel planes (Figure 18.17c); this produces a step-like scarp. In addition, tensional stresses may result in the subsidence of a block of land, as the rocks on either side are pulled apart, creating a steep-sided trench in the landscape, referred to as a **graben**. On a large scale these features are known as **rift valleys** (Figure 18.17d).

Where the stresses are exerted parallel to each other, **transcurrent faults** may develop (Figure 18.17e). A well-known example is the San Andreas fault in California (Plate 18.6); another is the Great Glen fault in Scotland. As we have seen, transcurrent faults are also common in the mid-oceanic ridges, where differential horizontal movement of the oceanic crust is occurring.

Faults are not only produced by tensional stresses, however, for compression may also result in brittle failure. The overthrusting of rocks involved in intense folding is an expression of this process, but overthrusting need not always be associated with folding. Characteristically, **overthrust faults** have a gently dipping fault plane. Where the failure occurs along a steeper plane, and the adjacent rocks do not move over each other, compressional forces may result in a **reverse fault** (Figure 18.17b). Here, the tendency is for one side of the fault to be

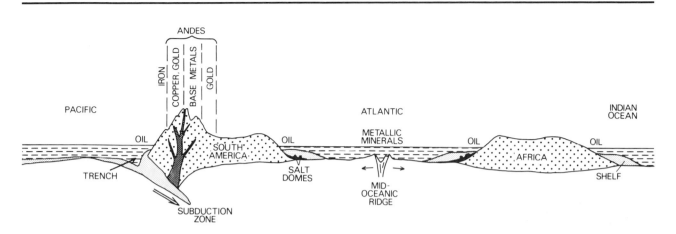

Figure 18.18   The relationship of mineral resources to plate tectonics (from Rona, 1973)

upthrust relative to the other. As with normal faults, these may occur together, and blocks of land may be upthrust to give a feature known as a **horst**. On a large scale, the Black Forest area of Germany represents a horst.

Earth-building processes and man

Plate tectonics and mineral resources

The processes of earth-building have considerable significance for man. In particular, it is now apparent that processes of plate tectonics account for the global distribution of many mineral resources. Many of the world's major deposits of metallic minerals, for example, are associated with former zones of subduction. In this situation, the reactions which occur as the subducted plate descends into the mantle release mineral-rich solutions which rise into the crust and precipitate in the form of mineral deposits. Metallic sulphides (i.e. metals which have combined with sulphur) are especially significant in these areas, and a large proportion of the world's iron, copper and other base metals are located in zones of plate convergence. Gold is often associated with these reserves, while the deep marine trenches created by subduction favour the accumulation of organic debris which is subsequently buried by sediments and converted to oil (Figure 18.18).

Plate convergence, as we have seen, often occurs either at continental margins or in association with island arcs. In both cases, the mineral resources formed in these areas tend to be relatively accessible and have been known and exploited for many

centuries. The concept of plate tectonics, however, has drawn attention to a second area of rich mineral resources: the mid-oceanic ridges. Here, too, volcanic activity results in the release of hydrothermal solutions which have deposited large concentrations of minerals, including iron, copper, lead, uranium and mercury. Because of the problems of exploration on the sea floor, the full worth of these resources is not yet known, but an indication of their potential is given by the mineral resources found in the Troodos Massif of Cyprus. This is believed to be an old mid-oceanic ridge which has subsequently been uplifted. It contains enormously rich deposits of copper (from which the island gets its name) as well as iron, zinc, gold and silver. It is not certain that these formed while the area was part of the mid-oceanic ridge, but if so they illustrate the mineral wealth which may be awaiting discovery beneath the sea floor.

Earth-building as hazard

Not all the effects of earth-building are beneficial for man. As we have seen, earth-building processes involve immense upheavals of the earth, and in some cases they may be serious hazards. In 1980 the inhabitants of the area around Mount St Helens in Washington State, USA, saw some of these dangers most forcibly. Few of them are likely to forget the experience of the mountain erupting in a deluge of rock and cinders and sulphurous smoke. The mountain, which had last erupted in 1857, began to stir on Thursday, 20 March. The first sign was an earthquake on the north side of the mountain, but soon earthquakes became frequent. On 27 March a small eruption occurred, producing a new

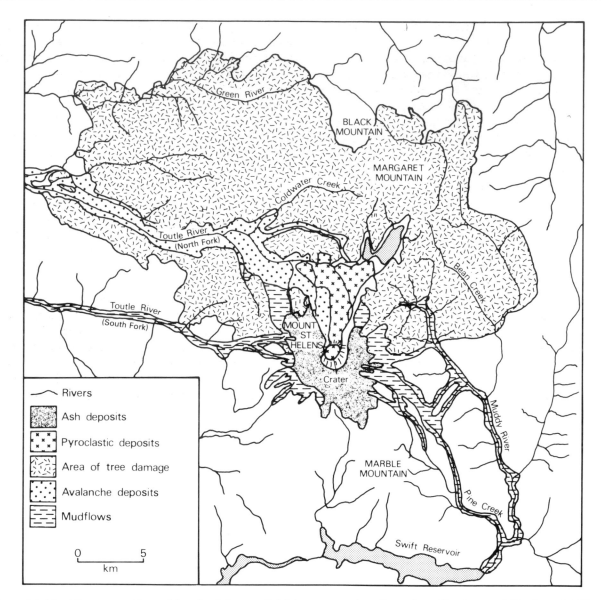

Figure 18.19   The area affected by the Mount St Helens eruption (based on Decker and Decker, 1982)

crater 65 m wide on the summit. Small eruptions followed, and five days later a second fresh crater appeared beside the first. Slowly, the north side of the mountain began to bulge; by mid April it had risen 100 m. On 12 June, and again on 22 July, vast explosions occurred as the mountain, split by a magnitude 5 earthquake, was suddenly relieved of the confining pressures. Over 400 m of the summit was removed and the cloud of dust and debris flattened forests over 32 km away, scattering ash as far afield as Portland and laying waste in total to

560 km² of the surrounding countryside (Figure 18.19).

Afterwards, the full impact of the event became clear. The mountain had changed shape. An estimated 2.7 km³ of material had been blown off; 600,000 tonnes of ash had fallen on Yakima to the north-west; one town had been abandoned and in all sixty-three people were dead. All around, dams had burst, rivers had changed course, roads had been blocked and crops smothered. Both the direct and indirect effects of such an eruption are clearly catastrophic, yet this, by geological standards, was a relatively minor event!

The same is true of earthquakes. As all who live in the San Francisco area know, this great city is built upon an active fault line. The fault moved in 1906 and San Francisco was all but destroyed. It is expected to move again in the near future. In an attempt to minimize the effects, water is being pumped into the fault on the hypothesis that this will lubricate the slip line and encourage it to move in a series of small, gradual slips rather than a single, catastrophic one.

Given the threat they impose, it would clearly be useful to predict events such as earthquakes and volcanic eruptions. To some extent, we can. In many cases they take place with a rough sort of regularity. Earthquakes, for example, require the build-up of critical stresses in the rock and these can be computed from a knowledge of the rock structure and the forces operating in the crust. That is why we know that another earthquake is likely in California. Moreover, we also know now that the spatial distribution of earthquake and volcanic activity is related to the pattern of plate tectonics, so we can define areas where such hazards are most likely to occur. Nevertheless, we are still a long way from predicting such events with any precision. The processes involved are not understood clearly enough, and the data needed for prediction are not normally available.

There are, however, signs that prediction is possible. The Chinese, for example, managed to predict the earthquake that wiped out Haicheng in 1975. They did so not only by monitoring the build up of pressures in the rocks, but by analysing changes in the magnetic and gravitational properties of the area, variations in the water levels in wells, and even unusual behaviour in animals. As a result, 100,000 people were moved out of the city before the earthquake and disaster was averted.

Whether similar success will be achieved elsewhere is doubtful. Earthquakes and volcanoes vary considerably in character, and methods that work in one case do not necessarily work everywhere. But the necessity to understand these tremendous processes of the earth is clear. Hopefully, the knowledge will be achieved in time to cope with the next San Francisco earthquake.

19
Weathering

The basis of weathering

The role of weathering in the environment
Weathering is a process by which rocks adjust to their environment. It is an equilibrium process; it originates from the fact that rocks, formed in one environment, are introduced to different environments.

The implications of weathering are manifold. The products of weathering become available for erosion; they are thence transported through the landscape, deposited and ultimately lithified to form new rocks. Weathered residues may lie on the surface of the ground and contribute to the soil. Substances extracted from the rocks by weathering may be dissolved in the waters which percolate into the ground and then be carried to streams or taken up by plants. It is a process, therefore, which is fundamental to many other aspects of the environment.

The causes of weathering
The disequilibrium which initiates weathering processes arises for a number of reasons. One of the most common we have discussed in the previous chapter. Earth movements, such as folding and faulting, generally associated with the collision of crustal plates, force the rocks into mountain chains. Rocks which may have been buried deep within the crust or sediments laid down deep beneath the sea may be thrust to the surface; the change in the conditions – the exposure of these rocks to the water, gases, heat and dissolved substances at the surface – results in changes within the rocks. Constituents which are unstable in these conditions are altered to more stable forms, or are destroyed.

Another common cause of disequilibrium is volcanic activity. The magmas which well up from the mantle are similarly exposed to a strange, subaerial environment, and become subject to weathering. Erosion, too, may uncover rocks which once lay deep in the crust and again a state of

disequilibrium may be induced, while a fall in sea level may expose marine sediments to subaerial conditions.

It is not only changes in the relative position of the rocks which stimulates weathering, however; the surrounding environment may also change. Fluctuations in climate, such as increased rainfall or reduced temperature, may lead to alterations in the nature of weathering. Even changes in the vegetation may have an effect, for many of the substances which take part in weathering are derived, directly or indirectly, from the vegetation. Man also changes the weathering environment; pollutants entering the atmosphere may be particularly effective as agents of weathering.

As a rule, this disequilibrium is most acute at the start of weathering and diminishes thereafter. As a result, the motive power behind the process tends to decline over time, and as the rocks approach their equilibrium forms, the rate of change declines (Figure 19.1). This is only part of the story,

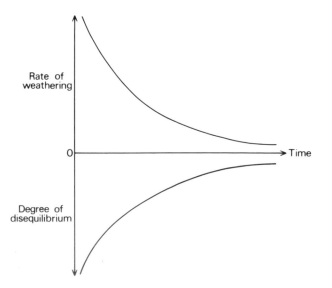

Figure 19.1 The general relationship between rate of weathering and disequilibrium

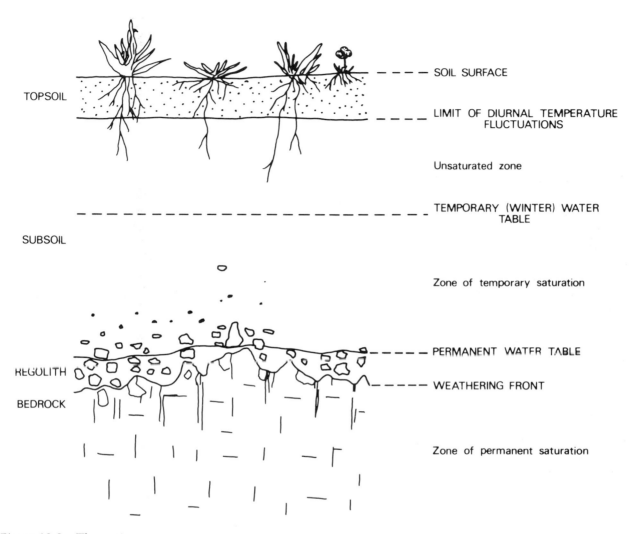

Figure 19.2 The active zone

however, for changes in the initial character of the rocks, brought about by weathering, may provide more suitable conditions for the operation of other processes.

The location of weathering

Weathering, in general, is confined to a relatively narrow zone of the earth's crust. The forces which are involved in weathering are largely related to the atmosphere, and thus it is mainly rocks which are in close contact with the atmosphere that are subject to weathering. The weathering zone, therefore, extends from the soil surface to the maximum depth of penetration of mobile water, gases and energy derived from the atmosphere. There are exceptions

to this, as we will see, but in most cases this zone is between a few centimetres and several metres in thickness. Within it, weathering tends to be most intense only in the upper layers; but it may also operate irregularly throughout the active zone depending upon local conditions.

This active zone comprises several rather different environments (Figure 19.2). It includes the soil – which is largely a product of the weathering processes acting upon the original materials – a zone, in some cases, of unsaturated material; an upper zone of temporary saturation, where the water is derived mainly from downward movement of rain – and a lower zone of temporary saturation affected by the upward movement of ground-waters. The base of the active zone is generally marked by the water table, beneath which the ground remains permanently saturated. Beneath the water table weathering may operate, but it tends to do so at a much reduced rate, largely because of the slow turnover of the groundwater (Chapter 14) and the consequent stability of conditions; after a while the groundwaters and the rocks come into a state of almost static equilibrium.

Weathering agents

The rocks and soils within the active zone are acted upon by **weathering agents**. These may be considered as the inputs to the weathering system.

The main weathering agents are water, heat and dissolved substances (reactants). Most of these are initially derived from the atmosphere, but there is clearly a tendency for all three to work their way downwards through the active zone (Figure 19.2). In the process changes occur, and materials washed from one zone are transported to lower layers. This may have considerable importance, particularly because substances dissolved in percolating waters often act as significant agents of weathering. It is a situation which can be seen microcosmically, and somewhat artificially, on many buildings. Where, say, a sandstone has been used as a window-ledge in a building otherwise composed of limestone, there is a tendency for weathering to be increased below the ledge, largely due to the action of substances dissolved from the sandstone.

Heat

One of the main agents of weathering is heat. It operates in two main ways: directly, through fluctuations in temperature which exert stresses upon the rocks; and indirectly through its influence on chemical reactions, most of which are accelerated by an increase in temperature.

The action of heat is most intense close to the surface, for few soils or rocks are good conductors of heat, and extreme temperatures are rarely felt below a few centimetres depth. In fact, it is this lack of heat conductance which encourages weathering processes due to temperature fluctuations. The inability of rocks to transmit heat downwards means that the surfaces of the rocks heat up considerably during hot periods, and cool rapidly during cold periods. Thus, the skin of exposed boulders and rocks in desert areas may be subjected to considerable stresses induced by these temperature changes. Even a shallow covering of soil or weathered material tends to reduce the effect by insulating the rocks from the heat of the sun.

In polar and sub-polar environments heat plays a rather different role, for extremely low temperatures may allow the development of a layer of permanently frozen soil (permafrost) near the surface of the ground. In this zone stresses may be set up by the freezing of water, which expands by about 9 per cent of its volume as it changes to ice. In confined areas this expansion may set up considerable stresses. Above the zone of permafrost occurs the active layer, in which seasonal or shorter-term melting may occur. Here, the fluctuations in temperature around freezing point, and the consequent alternations of ice formation and melting, may cause even more intense weathering.

Water and dissolved substances

As this implies, it is often the interaction between heat and water which is responsible for weathering, and instrumental in this is the ability of water to freeze and expand at temperatures which are not uncommon in nature. Most of the water entering the active weathering zone of the earth is derived from rainfall, however, and it is the chemical as well as physical character of this water which is important.

Pure water is composed of hydrogen and oxygen in the ratio of two hydrogen ions to one oxygen ion (H_2O). (An ion is an atom or molecule bearing a net electrical charge. This may be positive, e.g. H^+ – hydrogen or NH_4^+ – ammonium, or negative, e.g. HCO_3^- – hydrogen carbonate, or O^- – oxygen.) Under natural conditions, the combination of these elements is not always stable, and the water molecules tend to separate (dissociate) to produce individual hydrogen (H^+) and hydroxyl (OH^-) ions. The concentration of these dissociated or **free hydrogen ions** determines whether the water is acid or alkaline. Water which contains an excess of free hydrogen ions is **acid**, and it is generally able to weather rock minerals more readily. Acidity, incidentally, is measured on what is known as the pH scale. This ranges from 0 to 14; liquids with a pH below 7 are acid, those with a pH above 7 are alkaline. We will encounter this scale again when we discuss soils in Chapter 27.

Water is rendered acid either by the introduction of extra hydrogen ions, or by the dissociation of the water, which increases the concentration of free hydrogen ions. These processes are brought about in a variety of ways. Hydrogen ions, for example, may be released by the vegetation; consequently rainwater dripping through the vegetation and percolating through the soil tends to be made slightly acid. Similarly, carbon dioxide dissolved from the atmosphere tends to react to produce free hydrogen ions in the rainwater:

$$H_2O + CO_2 \rightarrow H^+ + HCO_3^-$$

water   carbon   free   hydrogen
        dioxide hydrogen carbonate
                 ion   ion

This converts the rainwater to a weak solution of **carbonic acid**, which is able to attack many rock minerals.

Many other substances may also become dissolved in the rainwater, increasing its weathering ability. Pollutants such as sulphur dioxide may convert the water to a weak sulphuric acid, while organic compounds washed from the vegetation produce weak organic acids. As the water percolates through the soil and the surface rocks, it picks up various other organic and inorganic substances which may similarly make it more active as an agent of weathering.

Against the inclusion of these **reactants** within the water must be set the tendency for the water to absorb, and eventually become saturated with, the products of weathering. Water percolating through limestone, for example, tends to dissolve the calcium carbonate. Ultimately, it will contain so much calcium that it will be unable to absorb any more. It thus loses its ability to weather the limestone with which it subsequently comes into contact. Indeed, the water may end up precipitating the dissolved substances, a phenomenon which contributes to the formation of the beautiful features such as **stalagmites** and **stalactites** which adorn many cave systems (Plate 19.1). The details, both of limestone weathering and of deposition, are more complex than this simple picture indicates, but it does demonstrate the principle that the percolating waters may ultimately reach chemical equilibrium with the rocks they are encountering.

The ability of water to carry out weathering is therefore dependent to a great extent upon its chemistry. As we have seen, however, the chemistry of the water changes as it carries out weathering, and this introduces an important principle of weathering. This is that active weathering requires the constant regeneration of the weathering agent. Thus, water must have reasonably free entry to and egress from the materials which are being weathered. It is essential that the weathering agent does not remain around so long that it comes into equilibrium with the rocks it is attacking; if it does so the rate of weathering declines.

Another important factor relating to the role of water in weathering is clearly the quantity of water passing through the weathering zone. This is controlled partly by climatic conditions – particularly the balance between rainfall and evapotranspiration – and partly by the nature of the surface rocks and soils. On the whole, rocks and soils which allow ready and rapid percolation of water are more susceptible to weathering than those which are essentially impermeable. Nevertheless, the quantities of water percolating through the active zone tend to fluctuate over time, and this undoubtedly alters the rate of weathering. The relationship is not simple, however, for as we shall see later it is often the alternation of wet and dry conditions which encourages change in the rock and soil materials.

The weathering materials

All this goes to demonstrate that the processes of

Plate 19.1 Stalagmites and stalactites in Easter Grotto, in the Ease Gill Cave System of the north Pennines (photo: A. Waltham)

weathering are influenced by the nature of the **weathering materials** as well as the agents of weathering. There are three main properties of the weathering materials which may be mentioned, although these often interact closely and other factors may be locally important.

Structure

As our discussion of the percolation of water implies, one of the constraints upon the rate and nature of weathering is the permeability of the weathering material. This is affected most markedly by the structure. Joints, bedding planes and faults in rocks, for example, often control the pathways taken by percolating water, and the quantities of water which can move through the rock. Similarly, as we noted in Chapter 14, it is the structure which largely controls water movement through the soil.

The importance of structure is twofold, however. It not only influences the rate or quantities of water passing through the weathering material, but also determines the volume of that material which is available to be weathered. The surface area exposed to weathering in a well-jointed rock is far greater than that exposed in a massive rock. As a result, the same volume of water can cause far more weathering in the former than the latter. By the same token, the joints and fissures allow the removal of the weathering products, and prevent the water coming into equilibrium with the rock.

Texture

The texture, or particle size, of the weathering materials is often related to the structure, and has a similar effect. Small particles present a larger surface area in relation to their volume to the agents of weathering than do large particles. All else being equal, therefore, finer-grained materials weather more rapidly than coarser-grained ones. In reality, all else is rarely equal, and rock or soil texture is often related to both structure and composition. Indeed, fine-grained materials are frequently composed of relatively stable materials which resist weathering, and are often tightly bound together so that there are few pore spaces between the particles through which water can flow. The effect of texture, therefore, cannot always be distinguished from related effects of other factors.

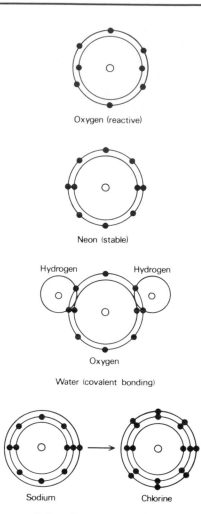

Figure 19.3 Structure of the atom. Atoms consist of a nucleus surrounded by a series of 'shells' containing electrons. The first shell contains up to two electrons; outer shells contain up to eight electrons. The atom is at its most stable when the outermost shell is full. Thus, the oxygen atom, with only six electrons in the outer shell, is reactive, while neon is stable. Bonding between atoms occurs either by sharing electrons (covalent bonding as in water) or by transferring electrons (ionic bonding, as in sodium chloride)

Rock composition

If we were to be precise, it is not the rocks or soils which weather, but the mineral grains of which they are composed. The chemical composition of these minerals, therefore, and the proportions of the different minerals in the weathering material, influence the overall nature of weathering processes.

To some extent, the principles of mineral composition are similar to those we have mentioned previously. The presence of structural weaknesses within the minerals, for example, is likely to encourage weathering either by allowing breakage to occur more easily, or by allowing the penetration of water and reactants into the mineral grain. Beyond this, however, it is the detailed chemistry of the mineral, and the arrangement of the **molecules** and the structure of the **atoms** of which it is composed which are critical to weathering.

Considered very simply, the constituent elements of a mineral can be seen to occur as atoms. Each atom is composed of a nucleus, consisting of protons and neutrons. Around the nucleus there circulates one or more electrons, arranged in a number of shells (Figure 19.3). It is the character of these outer shells which determines how active the element is, for it is through the interaction of the **electron shells** that elements react chemically. Thus, minerals composed of the more reactive elements are relatively unstable, but superimposed upon this is the effect of the structural arrangement of the atoms within the mineral grain. In general, these two factors determine the susceptibility of different minerals to weathering.

Several attempts have been made to define sequences of mineral stability. In detail, specialized series have to be defined because there are many different forms of weathering, and each mineral reacts rather differently according to the weathering environment. Nevertheless, a general picture of mineral stability can be attained (Table 19.1) and, as this implies, variations in the composition of the weathering material play a major role in weathering processes. It is notable that this sequence is approximately the reverse of the sequence of crystallization seen in igneous rocks (Chapter 17).

Interactions between weathering agents and materials

The processes of weathering depend upon the interaction between the agents of weathering and the materials with which they come into contact. Different processes of weathering therefore occur, depending upon these two factors. It is useful, however, to define two main types of weathering: mechanical and chemical. As we will see, these often interact closely, but they provide a basis for discussing the more detailed processes involved in weathering.

Mechanical weathering

The process common to all forms of **mechanical weathering** is the fracturing of the weathering material by stresses acting upon or within the rock. As we noted in the previous chapter, when discussing faulting, rocks are able to withstand such stresses to a certain extent, due to the bonds within and between their constituent particles. When the stresses exceed the strength of the rock, however, failure occurs.

Unlike the situation in faulting, failure in weathering reactions tends to occur somewhat irregularly within the rock mass, rather than along discrete and continuous failure planes. This is because most of the stresses exerted by weathering processes are more randomly and locally distributed within the weathering material. In addition, these stresses are relatively small, and it is only at particularly weak points within the rock that failure occurs.

Table 19.1 *Mineral weathering series: the relative resistance of common minerals to weathering*

	Heavy minerals	Light minerals
Most resistant ↑	Rutile	
	Zircon	Quartz
	Tourmaline	↑
	Garnet	
	Staurolite	
	Kyanite	Micas
	Amphiboles	↑
	Sphene	
	Pyroxenes	Feldspars
Least resistant	Olivine	

As can be imagined, the main weaknesses in most rocks are associated with joints and bedding planes, and it is along these that failure most often occurs. Nevertheless, stresses caused by mechanical weathering may also prise individual minerals from the rock mass, and, particularly where previous chemical weathering has weakened them, the mineral grains themselves may be shattered.

Freeze–thaw

The stresses applied to rocks during mechanical weathering derive from a number of sources. As we have seen, water is often a main agent of weathering, especially when it freezes. **Freeze–thaw** activity is thus a major process of mechanical weathering. Freezing is a desiccation process; it dries the rock locally, creating a moisture tension gradient from the frozen zone (high tension) to the unfrozen zone (low moisture tension). Water moves along this gradient, towards the area of high tension and is thus apparently attracted towards the growing body of ice. In this way, ice crystal growth is encouraged and, under suitable conditions of ample moisture supply, large crystals or lenses of ice may develop. The increase in volume of the water as it freezes may set up stresses of as much as 7×10^7 Pa on the surrounding materials.

In practice, these pressures are rarely reached, for freezing does not normally occur in a wholly confined space. The access the water uses to enter the crevice in the rock, for example, provides an escape for the growing ice body and relieves some of the pressure which would otherwise develop. Nevertheless, ice growth is often at least partially confined, for freezing generally takes place from the surface of the crevice inwards. Thus, the surface ice seals the crevice and subsequent freezing of the remaining water may build up considerable pressures on the surrounding materials.

If freezing is a one-off event, the amount of disintegration is likely to be small, for most rocks are able to yield to such pressures to some extent without rupturing. Repeated freezing and thawing, however, increase the degree of rock disintegration. Each time the ice thaws, more water is able to enter the crevice, refilling it, and ensuring that during the next freezing cycle stresses will be maximized. Because of this, it is generally the number of freeze–thaw cycles rather than the intensity of freezing which controls the degree of weathering. In addition, of course, disintegration is encouraged by an abundant supply of water, so that freeze–thaw processes tend to be most active in moist sites, such as those fed by lateral movements of groundwater, and in porous rocks which are able to absorb large quantities of water. Data on rates of rock disintegration from experimental simulations of freeze–thaw processes under a range of conditions are given in Table 19.2.

Thermal weathering

Similar stresses may be set up by the expansion of rocks upon heating. This is a process which is believed to operate mainly in desert environments where great extremes in air temperature occur, and where there is little or no vegetation to protect the surface. Under these conditions, diurnal surface temperature ranges of 100°C are not uncommon. It has been argued that the effects may be dramatic. During the daytime, the outer skin of exposed rocks may be heated intensely, while the interior of the rocks remains cool. The differential expansion which results may cause flakes of rock to crack away from the surface, producing the effect of **onion-weathering** or **exfoliation**. Repeated heating and cooling exacerbates this process. Differential heating and expansion may also occur on a smaller scale, for differences in the thermal properties of dark and light materials mean that mafic minerals may be heated preferentially with the result that stresses are set up between adjacent grains. In this case, individual grains may be prised from the rock.

Table 19.2 *Rates of weathering by freeze–thaw processes under two artificial climatic regimes*

Rock type	Percentage of material removed	
	Icelandic regime	Siberian regime
Slate	1.16	0.16
Quartzite	0.02–0.19	0.007
Mica schist	0.25	0.04
Gneiss	0.65	0.01
Porphyritic granite	0.29	0.19
Granite	0.15	0.07

Icelandic regime: 36 cycles from −7°C to +6°C
Siberian regime: 9 cycles from −30°C to +15°C

Data   from S. Wiman, 'A preliminary study of experimental frost weathering', *Geografiska Annaler*, A45 (1963), pp. 113–21.

Plate 19.2 Desert boulders showing the effects of exfoliation, Encanto Valley, Chile. Note the rounded shape of the boulders and the 'flake' of rock spalling from the side of the large boulder in the centre of the picture (photo: H. R. Singleton)

Plate 19.3 Algal-bored limestone in the inter-tidal zone near Fanore, Co. Clare, Ireland. Note also the effect of jointing within the rock (photo: J. D. Hansom)

Such, at least, is the theory. Unfortunately, confirmation of the effect of these processes has not been attained. In possibly the most thorough experimental investigation of the phenomena, Griggs (1936) subjected granite to 89,400 heating and cooling cycles (equivalent to about 245 years of thermal weathering) – with no detectable effect! What is clear, however, is that desert areas are often characterized by an abundance of shattered blocks, and by rounded boulders from which skins of material appear to be flaking (Plate 19.2). The circumstantial evidence, at least, for the existence of such processes remains strong.

Biological weathering

Plant roots have also been held responsible for mechanical weathering. Roots may develop down joints and along bedding planes, and, as the lifting of pavements demonstrates, quite large stresses may be exerted on the surrounding rock. However, the ability of roots to act against confining pressures is severely limited, and the process is often less effective than imagined. In many cases, the roots are simply exploiting cracks opened up by other weathering processes. In addition, in marine areas, blue-green algae may bore into the rocks, creating an irregular fretted surface (Plate 19.3). Indeed, in the inter-tidal zone this may be the dominant process of limestone weathering.

Salt weathering

As we mentioned earlier, the distinction between mechanical and chemical weathering is not always clear-cut. This is seen in particular in the case of salt weathering, a process of considerable importance in many desert areas. In part, this is a purely mechanical process in that, where saline waters are evaporating, crystals of salt may grow in crevices within the rock. As with the development of ice crystals, these tend to grow from the surface downwards so that crystallization occurs in partially confined conditions. Again, the stresses exerted on the surrounding materials are maximized by repeated saturation and drying of the rock, for this allows

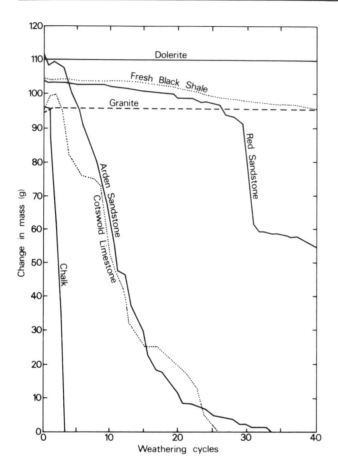

Figure 19.4 Breakdown of rocks by salt weathering. Each of the rock samples was subject to up to forty salt-weathering cycles in a solution of sodium sulphate (from Goudie *et al.*, 1970)

are treated with different types of salt solution: sodium sulphate, for example, is far more effective than sodium chloride. In part, at least, this indicates that salt weathering is not such a simple process as is often assumed, and involves chemical reactions as well as mechanical stresses. Sulphate solutions seem to exacerbate weathering because the sulphur present initiates complex chemical reactions including hydration which weaken the rock and make it more susceptible to the effects of crystal growth. Similarly, failure of buildings in desert areas is often the result of chemical reactions due to the inclusion of salt in the mortar.

Wetting-and-drying

Similar complexity occurs in the case of processes of **wetting-and-drying**. This involves a number of different effects, one of which is due to the **slaking** action of water. Simply, as the soil or a soft, porous rock is wetted, the water traps air bubbles within the pore spaces. When wetting is rapid the air cannot diffuse outwards but ultimately explodes, causing minute ruptures in the material. Chemical reactions such as hydration and hydrolysis also occur, however, so that the process is not simply one of mechanical weathering.

Pressure-release

Finally, we might mention the process of **pressure-release** (or **dilatation** as it is sometimes called). In many ways this has been an under-rated process, though it tends to occur on a larger scale than the other processes we have discussed. It results from the unloading of rocks during exhumation. As the overlying materials are removed, the consolidating pressures are released and the rock tends to rebound. The stresses created open up joints and bedding planes. Only rarely does the process result in direct disintegration, but it seems to be a widespread means by which rocks are weakened and made more susceptible to other weathering processes. As we will see later, for example, it is believed to be a process involved in the formation of tors.

Chemical weathering

Chemical reactions within the soil or rock normally involve water. Three general types of reaction can be identified: those in which ions from the rock mineral are removed and enter the water; those in which ions from the water enter and combine with the mineral; and those in which an exchange of ions takes place.

continual increments of salt to be added to the rock, permitting the growth of large crystals which fill the available space. The effectiveness of the process is also dependent upon the degree of supersaturation of the water, for this controls the ability of the crystals to continue growing against the confining pressure of the surrounding materials.

Experimental studies have shown that the rate of rock disintegration by salt weathering varies markedly between rocks. As with freeze–thaw effects, salt weathering is most effective in porous rocks such as chalk and open-fabric limestones or sandstones. More massive rocks such as diorite and granite are almost wholly resistant to the process (Figure 19.4). It has also been found, however, that rates of disintegration vary when rock specimens

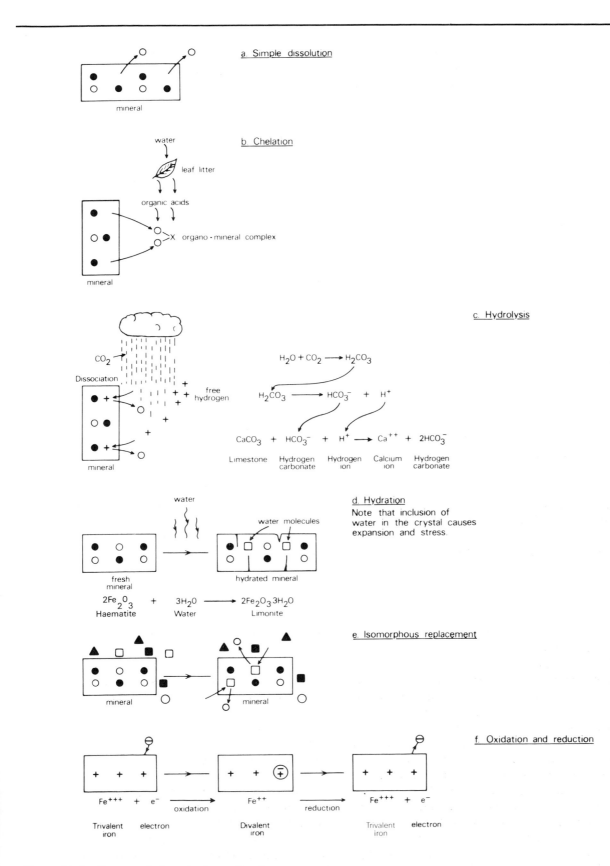

Figure 19.5 Forms of chemical weathering

Solutional processes

The first type of process is often referred to rather loosely as **solution**. In reality it involves a number of slightly different reactions. Some substances are soluble in pure water (salt is an example) and dissolve by the migration of ions from the mineral into a dispersed form in the water (simple dissolution). Other substances enter into solution largely through the action of reactants dissolved within the water. In one way or another they are combined with ions or compounds in the water, and essentially can be considered to leave the mineral to take part in this combination (Figure 19.5). One example of this is the process known as **chelation**. In this, ions leave the mineral to combine with organic compounds within the water.

Hydrolysis and hydration

Hydrolysis is possibly one of the most important chemical reactions involved in weathering. It occurs when free hydrogen or hydroxyl ions in the water enter into the mineral structure and create a new compound. The process operates to some extent in pure water, for this dissociates to yield H^+ and OH^- ions, but the presence of carbon dioxide greatly accelerates the reaction by forming a weak carbonic acid:

$$CO_2 \;+\; H_2O \;\rightarrow\; H_2CO_3$$

| Carbon dioxide | Water | Carbonic acid |

This immediately dissociates to form hydrogen carbonate (once referred to as bicarbonate) and hydrogen ions:

$$H_2CO_3 \;\rightarrow\; HCO_3^- \;+\; H^+$$

| Carbonic acid | Hydrogen carbonate | Hydrogen ion |

These ions, in turn, take part in mineral weathering. The weathering of limestone by this process (sometimes referred to as carbonation) is illustrated in Figure 19.5. A similar process also occurs in the weathering of orthoclase feldspar to kaolinite:

$$2KAlSi_3O_8 \;+\; 2H_2O \;+\; CO_2 \;\rightarrow\; Al_2Si_2O_5(OH)_4 \;+$$

| Orthoclase feldspar | Water | Carbon dioxide | Kaolinite |

$$K_2CO_3 \;+\; 4SiO_2$$

| Potassium carbonate | Silica |

The potassium carbonate produced in the process is removed in drainage waters.

The carbon dioxide necessary to initiate this process comes either from the free atmosphere or, in soils, from respiration by plants and soil organisms. As a result, hydrolysis is favoured by biological processes, and is most active in or just beneath the root zone, where CO_2 concentrations are high.

Hydration is a related process in which not just the hydrogen or hydroxyl ion, but the whole water molecule combines with the mineral. It is exemplified by the conversion of haematite to limonite (Figure 19.5), and, as this shows, involves the development of considerable physical stresses due to expansion of the mineral.

Isomorphous replacement

Ion exchange processes, in which ions are lost from the surface of the mineral and are replaced by others from the water, are common in many chemical reactions, but they are readily reversible and cannot be considered as true weathering reactions. On the other hand, **isomorphous replacement** of ions within minerals may occur, and this process is less readily reversed and fundamentally alters the composition of the mineral. It is a basic process in clay formation. In general terms it operates through the exchange of ions held within the structure of silicate minerals by similarly sized ions from the water. The control upon the process is the size relationship of the ions involved. Ionic size is normally measured in nanometres (1 nm = 0.000001 mm).

Silica, for example, with a diameter of 0.39 nm, may be replaced by aluminium, which is slightly larger, with a diameter of 0.57 nm. This, in turn, may be substituted by magnesium (0.78 nm). In the process the chemical composition of the mineral changes, but, in addition, the inclusion of ions of somewhat different size may result in stresses being set up which rupture the mineral. Physical breakdown as well as chemical alteration therefore occurs.

Oxidation and reduction

We might mention here two final processes of widespread significance. **Oxidation** is a process by which compounds lose an electron. In simple terms, **electrons** (negatively-charged particles) are present within the air in the soil and weathered mantle. Under conditions of high oxygen availability, these electrons become attached to oxygen within the air, and there is a tendency for a progressive release of electrons from compounds within the minerals. Iron, for example, may release electrons in this way; in the process it is said to be oxidized.

The reverse process is known as **reduction**. Here, electrons, derived in many cases from the respiration of plants, become attached to compounds such as iron in the soil. The compound is thus reduced (Figure 19.5).

We will examine the full implications of oxidation and reduction, together with many other weathering processes, later (Chapter 28); the point to note here is that oxidation and reduction are closely related (they are often occurring simultaneously within the weathering material), and they represent processes by which the composition of the minerals is altered. On the whole, the com-

pounds which are produced by oxidation are relatively stable; those produced by reduction are less stable and may be further altered by, for example, solutional processes.

Products of weathering

Weathering residues

The processes of weathering result in the loss of certain compounds from the weathering material, the alteration of others to new forms and the leaving of certain, more stable compounds, in their original state. The **residues** of weathering consist of the altered and unaltered materials left by the process.

These residues take a variety of forms. Among

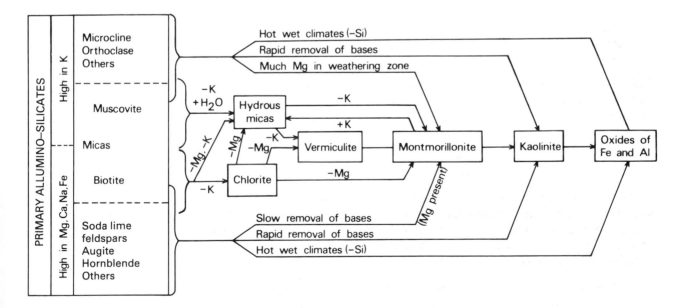

Figure 19.6 Processes of clay mineral formation. Under relatively mild or brief weathering conditions, the primary alumino-silicate minerals tend to break down to hydrous micas or chlorite. Under more intense weathering conditions, vermiculite and montmorillonite tend to form, while kaolinite and oxides of iron and aluminium are generally produced by intense or prolonged weathering. In each case, clay formation involves the loss of soluble elements such as potassium, sodium, calcium and magnesium (from Brady, 1973)

the most common unaltered residues are the more resistant minerals, particularly quartz. It is partly because of the stability of quartz in the face of many weathering processes that it dominates many detrital sediments. It derives this stability from its atomic and mineral structure.

The other more stable minerals may also escape weathering under all but the most intense conditions. They tend, however, to be less abundant within the environment anyway, and although enrichment of weathered residues with minerals such as garnet, tourmaline, rutile and zircon may occur, these are rarely more than accessory minerals within the residues.

Many of the residues of weathering are alteration products of the processes which were active. Even quartz, under extreme weathering conditions, may be dissolved and reprecipitated in the form of non-crystalline opal or chalcedony. The less resistant silicate minerals tend to break down more readily, especially under the effect of chemical weathering, to produce **clay minerals** (Figure 19.6). The nature of the clay mineral which forms depends upon the intensity and duration of weathering, as well as the nature of the original mineral.

The nature of the residual products also depends upon the nature of the weathering process and for this reason broad generalizations are difficult. Some indication of the relationships between weathering processes, weathering material and residual products is given in Figure 19.7. In general, it is clear that mechanical weathering tends to produce chemically unaltered residues, but results in changes in the size and shape of the materials; chemical weathering influences the composition of the materials. In reality, of course, a combination of both chemical and mechanical weathering occurs, and the residues are characteristically a mixture of shattered and rotted material.

The products of weathering which are dissolved within the water, or released as gases to the atmosphere, are lost more or less irretrievably from the materials. They thus represent outputs from the weathering system. They consist, in most cases, of the more soluble and least stable components of the original material. Most of these products are carried away as dissolved materials within the water, and thereby enter streams, lakes and groundwaters. Here they may act as a pollutant as far as man is concerned; high levels of lead and cadmium, for example, may be washed into streams from the weathering of ore deposits.

More importantly, in most cases, these weathering products become inputs to other parts of the environment. They may be taken up by plants as nutrients; they may become concentrated and eventually accumulate as precipitates in lakes or in the sea or in cave systems beneath the ground.

Landforms of weathering

Weathering also alters the morphology of the active zone within which it operates, and thereby produces weathering landforms. In truth, many of these landforms are only expressed at the surface through the subsequent effects of erosion. Ice or water, for example, may strip away the weathered residues to leave the fresh, unweathered topography. In Scotland, it has been suggested that the barren upland landscapes represent such a topography, created by the removal of once deep weathered residues by ice sheets during the Quaternary.

Similar interactions of weathering and exhumation seem to be involved in the formation of **tors**, without doubt some of the most enigmatic and controversial landforms found in Britain. They consist of irregular piles or masses of rock, often perched on hilltops or on plateau edges (Plate 19.4), though sometimes also found on valley side-slopes. They are particularly common in granite areas, but may similarly occur in sandstones such as the Millstone Grit of Yorkshire.

The debate about their formation has been long and, on occasions, vitriolic. At one stage two opposing schools of thought existed, one arguing that tors were entirely a product of chemical weathering and exhumation in a tropical environment, the other claiming that they result from frost-shattering in polar conditions. Unfortunately, neither explanation seemed universally applicable, for tors (or tor-like features) are to be found well outside areas affected by periglacial conditions, yet they also occur in regions which have, at least recently, been subjected to intense periglacial activity.

In reality, it seems likely that what we call tors comprise a wide range of features, produced by a variety of different processes. In almost all cases, however, weathering processes seem to have played a vital part. Thus, some tors apparently reflect the effects of deep chemical weathering which are concentrated in areas of close jointing,

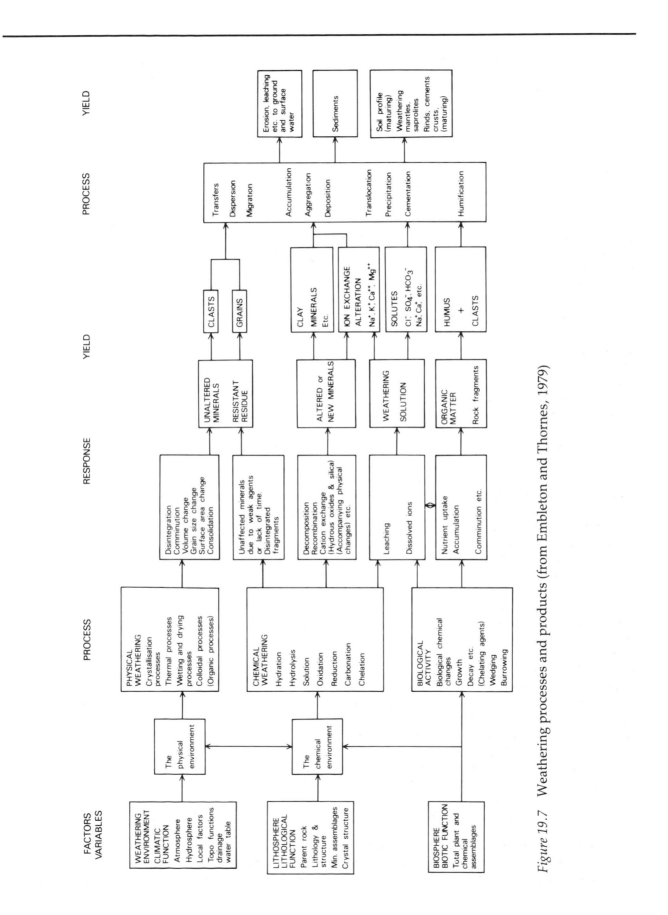

Figure 19.7 Weathering processes and products (from Embleton and Thornes, 1979)

followed by removal of the weathered debris, either by fluvial action or by solifluction and other periglacial processes (Figure 19.8). Other tors possibly result from disintegration by freeze–thaw and associated mass movement of the debris downslope. It is also suggested that tors may result from pressure-release following fluvial incision (Figure 19.9). There is no reason why all these processes might not have been operative, not only individually but, in some instances, in combination. In other words, tors may testify to the existence of **equifinality** in the landscape: the production of similar forms by different processes.

Limestone landforms

Among the most intriguing and, in many cases, the most beautiful landforms of weathering are those which develop in areas of limestone solution. Preferential weathering in relatively massive limestones often occurs along joints and bedding planes and may produce a varied range of features, including the **clints** and **grikes** which characterize many **limestone pavements** (Figure 19.10). On the surfaces between the joints, various **weathering pits** and grooves and runnels (referred to as **rillenkarren** and **rinnenkarren** according to their size) also develop. Beneath the surface, solution results in the formation of networks of caves. All these features are characteristic of what are referred to as **karst** landscapes, the name deriving from the Karst area of Yugoslavia where these features are particularly finely developed. In Britain the Pennines, and in Ireland the Burren, display similar scenery.

Periglacial landforms

Weathering landforms are also characteristic of many polar and sub-polar areas. Here, mechanical weathering is often dominant, with active freeze–thaw and frost-shattering occurring. Associated with these are processes which act to sort the weathered material according to size, producing a range of **patterned ground** features, including stone polygons, stone stripes and stone garlands. We will discuss these in detail when we examine glacial and periglacial processes in Chapter 23.

Weathering and climate

Between the polar extremes and the tropics, the deserts and humid oceanic regions of the world, there are great variations in climatic conditions.

Plate 19.4   A tor formed in Millstone Grit: Mother Cap Tor, Derbyshire. Cross-bedding is distinctly visible in the rock and is being picked out by weathering (photo: D. J. Briggs)

1 Original form. Variations in joint frequency due to inherent variations in granite

2 Deep weathering under a warm temperate climate; most intense in closely jointed granite.

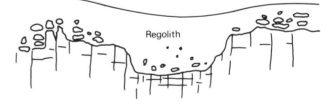

3 Removal of regolith by mass movement and river action to leave exposed tors.

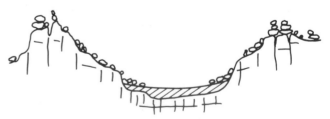

Figure 19.8   The development of tors according to the theory of Linton

1 Development of unloading dome due to erosion of overlying material.

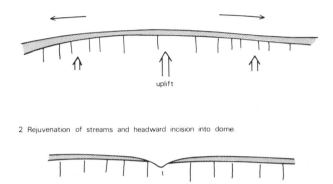

2 Rejuvenation of streams and headward incision into dome.

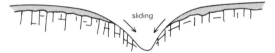

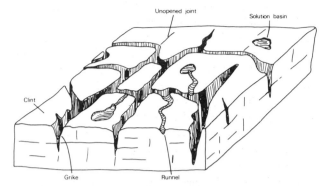

Figure 19.10   Features of a limestone pavement

3 Valley development and opening up of joints due to lateral release of pressure and some downslope sliding of granite.

4 Chemical attack opens joints further and creates deep weathered mantle.

5 Solifluction and frost action strips the weathered mantle and exposes the tors.

Figure 19.9   A theory of tor formation involving the combined action of unloading, chemical weathering and solifluction

Since water and heat play such vital roles in weathering, we might expect that these variations are reflected in the pattern of weathering processes and associated landforms across the globe.

To some extent this is the case. At a broad scale we can indeed define gradients in the rates or effectiveness of individual weathering processes, related to changes in factors such as the amount of annual rainfall or temperature conditions. In the USA, for example, the number of frost cycles per year shows a marked regional trend, and this almost certainly influences the intensity of freeze–thaw processes (Figure 19.11). Similarly, it has been argued that the significance of chemical weathering varies geographically in relation to rainfall and mean annual temperature (Figure 19.12). On the basis of such relationships, attempts have thus been made to identify weathering regions. The most noted is the scheme devised by Peltier (1950), which is illustrated in Figure 19.13. As might be anticipated, this shows that mechanical weathering is most effective in cold moist regions, while chemical weathering is at its most intense in humid tropical regions. All forms of weathering tend to decline with increasing aridity.

Peltier's classification of weathering regions has often been quoted and it is certainly valid in broad terms. It has to be admitted, however, that it is, at best, a highly intuitive and generalized description of the distribution of weathering processes. At a smaller scale, it has limited significance, for other factors – including topography, geology, vegetation cover, soil conditions and microclimate – all affect rates of weathering. A transect across a limestone coast, for example, will reveal marked variations in the intensity and character of weathering within distances of a few metres, biological weathering being active in the lower intertidal zone, wetting and drying and salt-weathering being dominant near the tidal limit (Figure 19.14). Similarly, considerable differences in weathering

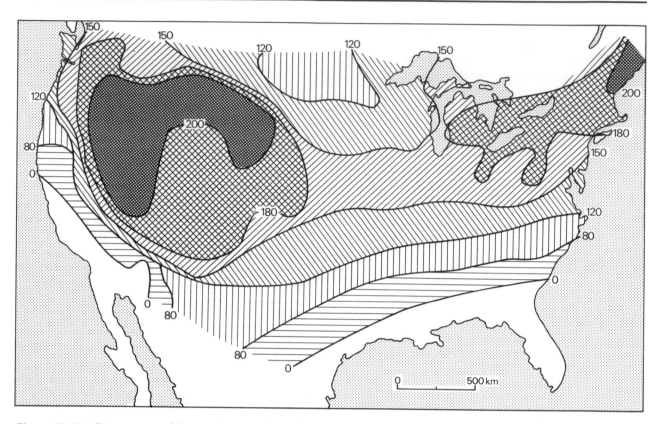

Figure 19.11 Frequency of freeze-thaw in the USA. Figures show the number of freeze-thaw cycles per year (from Embleton and Thornes, 1979, after Visher, 1945)

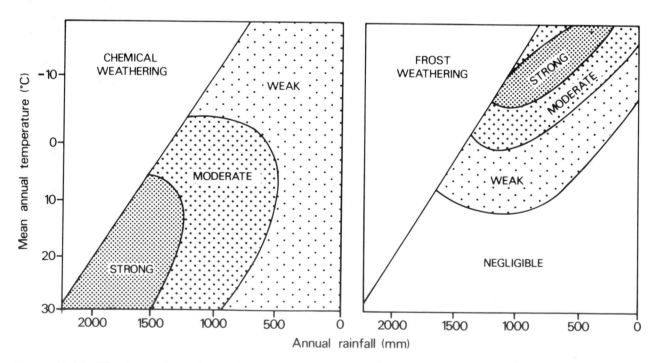

Figure 19.12 The intensity of weathering processes in relation to annual rainfall and mean annual temperature (after Peltier, 1950)

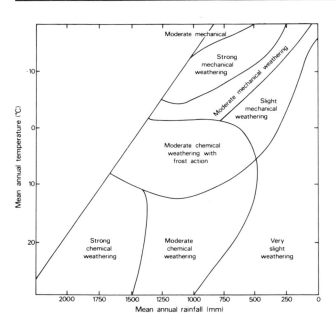

Figure 19.13 Weathering regions of the world in relation to mean annual temperature and rainfall (after Peltier, 1950)

processes may exist between slopes of different aspect. In the mid-altitude zone of the Tatra Mountains of Poland, for example, frost action is far more active on east-facing slopes since these experience more frequent fluctuations across freezing point. Finally, we also have to remember that the climate has itself changed fundamentally during the Quaternary, and many of the effects of weathering that we see in the landscape around us are legacies from previous episodes, when weathering processes may have been different. Thus, relationships

between present-day climatic conditions and landforms of weathering are often relatively weak.

Weathering and man

The products and processes of weathering have considerable implications for man. Several of the materials produced by weathering have economic importance, among them **bauxite**. This is a product of hydrolysis of plagioclase feldspars, and is the main source of the world's aluminium. **Kaolinite**, produced by the rotting of granite, apparently by hot briny waters rising through the rock (**hydrothermal weathering**), is used in the paper and ceramics industries (it is sometimes referred to as **china clay**).

Weathering is also a fundamental process in soil formation, and thus much of man's use of the land for food production is influenced by the products of weathering. In addition, the weathered mantle of the earth is important in relation to engineering, for it is necessary to construct buildings on a firm foundation. The effect of weathering, however, is often to weaken the surface materials, and thus, particularly where the residues are deep, the creation of stable foundations may present a problem.

Roadcuttings and embankments must also be planned with due regard to the effects of weathering upon the stability of the materials. There are a number of occasions when failure of slopes has occurred because slopes cut through weathered rocks were too steep. A similar problem may arise if weathering continues after construction; this may, in extreme cases, undermine foundations

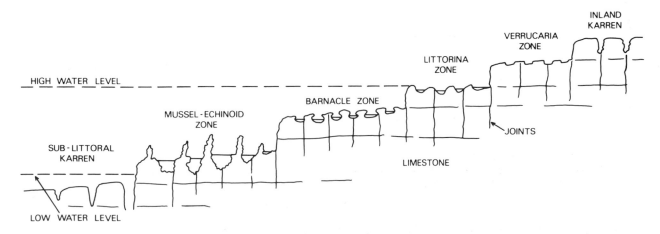

Figure 19.14 Weathering zones across a limestone shoreline, Co. Clare, Ireland (from Lundberg, 1977)

sufficiently to cause collapse, or de-stabilize slopes. The disaster at Aberfan, in South Wales, was an example of this. Mine spoil, which had weathered since being tipped, was ultimately affected by heavy rains and sludged down the hillside, enveloping the school and killing many children. There are also instances of dams collapsing because of weathering along weaknesses such as joints and faults; the failure of the Dale Dyke Dam near Sheffield in 1864 was apparently because of weathering and seepage along lines of weakness in the underlying clay. The ensuing disaster killed 250 people.

Fortunately, weathering processes are generally so slow that it is rare for constructions created by man to suffer from their subsequent effects. It has been shown, for example, that weathering of clays in roadside embankments is unlikely to reduce the strength of the material to cause collapse within the normal life of the feature. To put it another way, man's constructions generally fall down or become obsolete before they are destroyed by weathering! More direct effects of weathering can be seen in many buildings, however, on a smaller scale; many ancient buildings are worn by the effects of solution and hydrolysis.

In many cases, too, man is responsible for accentuating the weathering process. Salt, used to keep highways free of ice, splashes against buildings and weathers the stonework. Atmospheric pollutants encourage chemical weathering by rainfall (Plate 19.5). As we have seen, sulphur dioxide, which is

Plate 19.5 The effects of atmospheric pollution on St Paul's Cathedral. When built in 1720, the lead plugs were flush with the surface of the stones. Since then, weathering has lowered the stones by almost 2 cm. (The lightning conductor which runs along the centre of the wall is more recent) (photo: R. U. Cooke)

emitted by power stations for example, may produce a dilute solution of sulphuric acid in the atmosphere. This can result in markedly increased rates of weathering. In these ways, man is encouraging the destruction of his architectural heritage and, to a small degree at least, affecting rates of weathering in the landscape as a whole.

20

Erosion and deposition

Principles of erosion

The nature of erosion

The evidence of erosion is all around us. The hills and valleys, the polished or carved surface of exposed rocks, the shape of the coastline – all are products of erosion by wind, water and ice. Not only the landforms testify to these processes; we can watch the processes at work. If we look into a stream after a period of heavy rain, we see the murkiness of fine particles of soil and sediment being washed down-valley by the swollen waters. On a windy day we may be able to feel the dust and grit in the air. We can hear the crunch and rattle of pebbles on the beach as the ebb and flow of the sea sorts and moves the shingle. All around us, the weathered residues of the earth's surface are in motion.

These processes of erosion represent an important part in the sequence of landscape development. They redistribute the products of weathering; they wear down the features of earth-building; they act to re-establish an equilibrium between the shape of the landscape and the forces acting upon it. They also have immense significance for man. The main impact of erosion is borne by the surface layers of the earth – by the soil. The removal and destruction of this soil, from which man derives most of his food, may be catastrophic; it can all but destroy societies, as it did in the Tigris-Euphrates area; it can cause untold suffering (see, for example, Steinbeck's description of conditions in the 'Dust Bowl' in his novel *Grapes of Wrath*).

The energy of erosion

The disequilibrium between the earth's landscape and the forces acting at the surface arises largely from the potential energy imparted to the materials of the surface during mountain building. Uplift of the rocks provides them with the potential for downward movement under the influence of gravity. Water on these uplifted surfaces also has a potential energy; it tends to flow downhill and so its potential energy is converted to kinetic energy. It is this kinetic energy – the energy of movement – which provides the water with the ability to transport rock particles, and to erode.

In addition to the potential energy derived from the position of the surface materials, there is another source of kinetic energy which acts upon these materials. This is derived from atmospheric processes. Rainwater has a kinetic energy; due to the lifting of water into the atmosphere through the processes of evaporation and convection, raindrops develop a potential energy which is expended in motion through the atmosphere. The impact of raindrops upon the ground is a vital process in erosion, as we saw in Chapter 12; the energy of impact is derived from the kinetic energy of rainfall. Similarly, the air itself has a kinetic energy, which is expressed by the action of winds. These winds are able to pick up and transport rock particles; they act as agents of erosion.

The sequence of erosion

The energy available for erosion operates in a variety of ways. Much of the energy is used in carrying the rock particles through the landscape in the process of transport. This, however, is only one step in the overall cycle of erosion. Prior to transport, the particles have to be picked up; they have to be entrained. Before they can be entrained, in many cases, they have to be detached from the rock or soil masses. Thus erosion can be seen as three processes: **detachment**, **entrainment** and **transport** (Figure 20.1). As we will discover, these three processes are often closely related, and a single particle moving through the landscape may undergo repeated detachment, entrainment and transport. For these reasons, perhaps, it is not entirely valid to visualize them as a true sequence of events. Nevertheless, for the sake of simplicity, it is useful to consider each step in the process separately and in sequence.

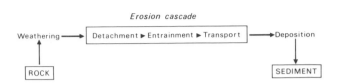

Figure 20.1   General model of the erosion cascade

Detachment

The forces of attachment

The ability of the wind, of water or of ice to pick up and transport rock particles is controlled to a great extent by the size of those particles relative to the energy available. In general, smaller particles are eroded more readily than larger ones (although we will qualify this statement later), and it is only when the rocks are broken down into relatively small fragments that erosion proceeds. For this reason, the detachment of materials represents a vital step in the process of erosion.

Detachment involves the breaking of bonds which hold the particles together. The exact nature of these bonds depends upon the type of material we are dealing with. In many sedimentary rocks, for example, the bonds arise from the cementing effect of compounds such as iron oxides, silica or calcium carbonate which coat the grains and fill the interstices between them. In many igneous rocks these bonds are derived from the intricate intergrowth of crystals, and from the fusion of the crystals during cooling. In both cases, the bonds are relatively strong. By contrast, materials which have been weathered (such as the soil) are weak. Chemical reactions and the mechanical effects of frost action, wetting-and-drying and heating and cooling have broken or weakened the bonds between the particles. Often all that holds the particles together is cohesion, provided by thin films of water, the weak cementing effects of organic compounds, calcium carbonate and iron oxides, and the minute electro-chemical forces which occur between clay particles. Many of these forces are impermanent, and are considerably reduced by wetting of the material, which dissolves some of the cements and ultimately starts to force the particles apart. Consequently, weathered materials are generally more susceptible than fresh rock to detachment, and it is partly for this reason that the soil is prone to erosion.

The forces of detachment

From what we have said, it is clear that detachment of particles from rocks and soil masses is often accomplished by weathering. In addition, however, the agents of erosion exert their own forces of detachment upon the surface rocks and soil. The exact nature of these forces depends upon the character of the erosive agent, but two main processes can be defined:

1   **Quarrying:** the detachment of particles by the action of the erosive agent itself.
2   **Abrasion:** the removal of particles by material carried by the agent of erosion (i.e. by the sediment load).

Quarrying

Although it might logically be considered as the major cause of detachment, quarrying is often relatively ineffectual. Even in glaciers, the ability of the ice itself to tear particles from the rock surface is limited, although it does operate in the process known as **plucking**: ice freezes on to the surface, particularly in cracks and crevices, and plucks the weakened fragments from the surface of the rock.

Flowing water is also able to carry out quarrying to a limited extent. Where the flow is confined, and flow velocities are high, a process known as **cavitation** may occur. The impact of the water on the rock causes the water molecules to break; minute airless bubbles are formed and as these implode they exert considerable forces upon the adjacent surface. Evidence for the operation of these processes is limited.

Possibly one of the most widespread and important processes of detachment is that caused by **raindrop impact**. The force of a raindrop falling on to a soil or weathered rock surface is often sufficient to break the bonds which link the particles. The importance of this process is derived from two factors: first, the surprising strength of the force and, second, its frequency and extent of operation. The strength of the force (i.e. the kinetic energy) is related to the velocity of the raindrop when it reaches the ground (its *terminal velocity*) and the *mass* of the raindrop. This relationship is expressed by the equation $E = \frac{1}{2}V^2 \cdot M$, where E is the kinetic energy, V is the terminal velocity and M is the mass. The terminal velocity of large drops often approaches 8 or 9 metres per second, and in heavy storms droplets of as much as 5 mm in diameter, with a mass of about 0.05 g may be generated. In a heavy storm such as those occurring in the Mid West of the USA this may represent a total energy input of 2500 J m^{-2} during a period of about thirty minutes.

The frequency and extent of raindrop impact is related to the character of the storm. Intense storms often only affect small areas, but the predominance of large droplets and the large number of droplets makes them particularly effective. Gentler rain which may affect much wider areas has little ability to detach particles since the droplets tend to be small and to have a lower terminal velocity (2–3 m s^{-1}).

Abrasion

Probably the most effective force for detachment is abrasion. This results from the impact and friction exerted by the material being transported by the agent of erosion. Abrasion occurs at the base and edge of glaciers; for example, rock fragments held in the ice scrape the surrounding rock. Similarly, boulders or sand grains bouncing along the stream detach fragments from the bed and banks. Wind also causes abrasion, for sand transported by the wind tends to detach particles as it bounces on the soil surface.

The strength of this force relates to the velocity of movement of the particles, their size (mass) and their concentration at the rock surface. The force seems to be particularly active in glaciers, where the particles may be very firmly trapped by the ice. However, its effect may be seen in a variety of other environments; witness, for example, the shattering caused by pebbles thrown by the sea against a cliff or beach – there is a record of a 60 kg boulder being thrown 30 m on to a beach in Oregon – or consider the force which large boulders being dragged along the bed of a stream must impose.

Entrainment

Entrainment is the action of picking up of a particle by the agent of erosion. In many cases, entrainment and detachment are so closely related that they become one. The force needed to detach a particle from a rock or soil mass is considerably more than that necessary to cause entrainment; thus if conditions are sufficient to allow detachment, entrainment automatically follows. Not all erosion, however, involves the removal of small fragments or particles from larger rocks. Often it is uncon-

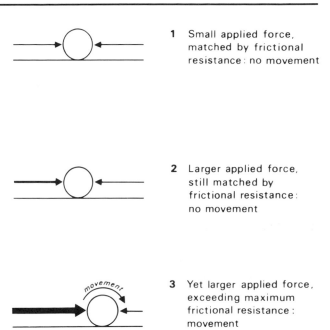

1 Small applied force, matched by frictional resistance: no movement

2 Larger applied force, still matched by frictional resistance: no movement

3 Yet larger applied force, exceeding maximum frictional resistance: movement

Figure 20.2 Initiation of particle movement

solidated sediments or soils which are being eroded. In these cases entrainment may act independently; loose particles are lifted from the surface by the agents of erosion.

Forces resisting entrainment

There are several forces which provide particles with a resistance to entrainment. The most important in many cases is a frictional force, which arises from the interaction between the particle and its surroundings. Frictional resistance only operates when the particle is being placed under stress, and the magnitude of the force increases as the applied stress increases. Thus, a particle at rest on a horizontal surface has no frictional strength. If stress is applied (for example, by a force trying to push the particle), the frictional resistance starts to operate. Initially, this frictional resistance will prevent the particle from moving, but as the applied stress increases there will come a point at which the frictional resistance reaches a maximum value. Additional forces applied to the particle will cause movement (Figure 20.2).

The maximum frictional strength of particles can be expressed as an angle, since it is clear that by tilting the surface on which the particle rests a point will be found at which spontaneous sliding occurs. This angle is referred to as the **angle of sliding**

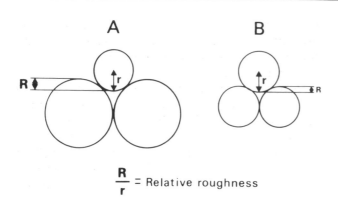

Figure 20.3 The effect of surface roughness on frictional resistance. For particle A, the relative roughness is 0.7 and the frictional resistance is high; for particle B the relative roughness is 0.3 and the frictional resistance is low.

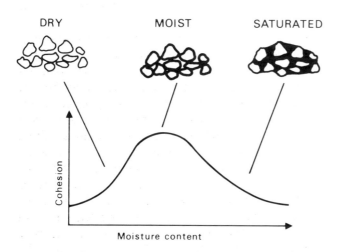

Figure 20.4 The effect of moisture content on cohesive strength

friction. It depends upon two factors: the weight of the particle – and thus the downward force it exerts on the surface – and the roughness of the contact between the particle and the surface. Thus, heavier particles have a greater frictional resistance to movement while rougher grain-surface contacts also increase the frictional resistance. This latter factor becomes particularly important in many erosional processes because grains do not normally lie on a perfectly flat surface, but tend to be lodged within depressions. In this case, the roughness of the particle–surface contact is expressed as the ratio of the radius of the depression in which the grain lies to the particle radius (Figure 20.3). We will see the significance of this, in particular, when we consider erosion in streams.

In addition to the effects of friction, particles derive a degree of resistance to entrainment from cohesion. This arises from the electro-chemical forces which operate between the grain and its surrounding particles or the surface on which it lies. Water may play a part in this cohesive strength, since thin veneers of water around and between the grains may help to hold particles together by surface tension effects (Figure 20.4).

Forces of entrainment

Forces causing movement vary according to the nature of the eroding agent. In general, two main forces can be seen to operate: the gravitational force which is a function of slope angle and particle weight, and the fluid drag exerted by a medium (such as water, air or ice) moving past the particle.

The effect of gravity

The gravitational force is relatively easily explained. All bodies lying on the earth are subject to a gravitational 'pull' which acts to hold the particle on the surface. The magnitude of this gravitational force is related to the weight of the body.

When a particle rests on a horizontal surface this gravitational force acts downwards into the earth; it can be seen, in fact, that it is involved in giving the particle a frictional resistance to movement, as we have just explained. If, however, the surface is tilted, part of the gravitational force operates downslope. It can be demonstrated that the magnitude of the force acting in a downslope direction is proportional to the sine of the slope angle:

$$F = w \sin \phi$$

where F is the downslope force, w is the weight of the particle and ϕ is the slope angle. There comes a point, therefore, where the downslope force derived from this gravitational effect is sufficient to overcome the frictional resistance of the particle: at this point the particle will slide. This critical angle is referred to as the **critical angle of sliding** (ϕ_{crit}). It is clear that the critical angle of sliding equals the angle of sliding friction (μ_{crit}).

$$\phi_{crit} = \mu_{crit}$$

In other words, particles under the effect of gravitational forces will move when the downslope component of that force is just sufficient to overcome frictional resistance.

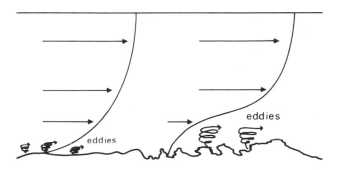

Figure 20.5 Velocity profiles and eddying above a smooth and rough bed. Although the rougher surface reduces velocity immediately above the bed, it may encourage entrainment because of the turbulence which lifts particles into the flow, and because of the steep velocity gradient above the eddy layer

Fluid forces

However, particles do not only move through the effect of gravity. Flowing water, air or ice play a major part in entrainment. These exert both a horizontal **drag** on the particle and a vertical **lift**. The horizontal force is derived from the 'push' of the agent against the particle. If this push is sufficient to overcome the frictional and cohesive resistance, then the particle rolls or slides forward. The strength of this horizontal force is related to flow velocity and thus faster flowing air or water is able to move larger (or more resistant) particles. In addition, however, the strength of the fluid forces depends upon the density of the medium itself. Water, for example, is some 9000 times as dense as air and can therefore exert forces 9000 times as great.

The vertical lift is derived from **turbulence** and from the **buoyancy** given to the particles. Turbulence tends to occur due to irregularities in the surface which cause **eddies** within the flow (Figure 20.5). These tend to carry the grain upwards. Buoyancy occurs because the density of the fluid is, to some extent, able to reduce the weight of the grain. Thus buoyancy acts to diminish the resistance of the particle by apparently giving it a degree of lift. If it were possible slowly to increase the density of the fluid, there would, of course, come a point at which the particle would float. Once the particle is lifted from the surface, its frictional and cohesive resistance derived from contact with the bed declines to zero; only the gravitational resistance remains.

In the case of transport by fluids, there exists a critical condition, therefore, at which entrainment commences. In reality, this depends upon many factors, but a general and dominant relationship exists between entrainment and flow velocity. This allows us to define a critical entrainment velocity; that is, the velocity at which entrainment commences.

The **critical entrainment velocity** varies according to the resistance of the particles. Thus, critical entrainment velocity can be related to particle size. For entrainment by flowing water the relationship is similar to that shown in Figure 20.6. This assumes that the particles have a constant density of 2.65 g cm^{-3} and that the density of the water is 1 g cm^{-3}.

Of course this relationship is highly generalized. In practice it does not hold true exactly. The critical entrainment velocity is affected by a variety of other factors, such as grain shape and slope angle (both of which influence frictional resistance). In addition, most sediments consist of a mixture of materials of different size; this too influences the point at which entrainment occurs. The diagram does, however, illustrate one point of particular interest. We can see that, although there is a tendency for smaller particles to require lower flow velocities to cause entrainment (as we might predict from what we have said), this relationship is not true for very small particles. Clearly other factors must be important here.

The explanation relates to the nature of resistances to movement in fine-grained material. Fine silt and clay particles tend to have a strong cohesive resistance, which overshadows the effect of frictional resistance. It is this cohesion which makes them more resistant to entrainment. In addition, very small particles do not protrude so far into the flow, and thus are protected by each other from the forces of entrainment; they present an essentially smooth surface to the flow.

These relationships between particle size and the flow velocity are valid, in broad terms, for both water and wind (Figure 20.6). Several factors complicate the picture in reality, however. As we have noted, turbulence is important in controlling entrainment and in very turbulent conditions entrainment may occur at much lower velocities,

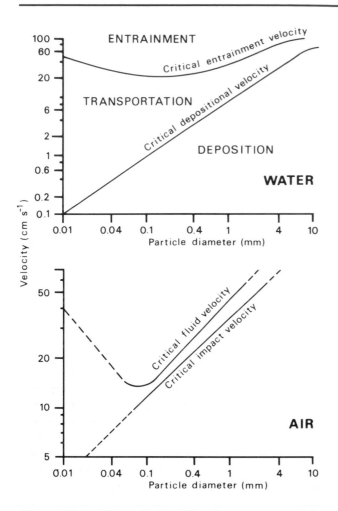

Figure 20.6 The relationships between particle diameter and critical entrainment velocities for water and wind. Note that these relationships assume spherical particles and uniform sediment size (adapted from Hjulstrom, 1935, and Bagnold, 1954)

simply because the turbulence tends to lift the particle upwards; it reduces, in effect, the weight of the grain and thus its resistance to entrainment. Moreover, the impact of particles already in motion may encourage entrainment. In the case of wind transport this may be very important (Figure 20.6). On a stream bed boulders bouncing along the surface tend to push others into movement. Raindrops falling through a thin wash of water may have the same effect. What this means in practice is that once particles are brought into motion, there is a tendency for further entrainment to occur relatively easily.

Entrainment and transport

We noted earlier that detachment is often a limiting factor for erosion in consolidated materials. So too does entrainment often limit the erosion of unconsolidated particles. The reason is simply that once a particle is in motion it requires a lower force (i.e. a lower velocity) to keep it in motion than it does to initially entrain it. The truth of this can readily be demonstrated by trying to push a cricket roller. It needs a lot of effort to set it in motion but, once it is moving, everyone else can go and get their tea! This phenomenon arises because a body in motion has a lower resistance to motion than does a static one.

Transport

Processes of transport

It follows from what we have just said that once a particle is entrained, it tends to be transported, at least for some distance. Transport can occur in four main ways:

1 in suspension;
2 by saltation;
3 by traction;
4 in solution.

Suspension is best exemplified by materials in a stream. The suspended particles are carried along by the water and do not touch the stream bed.

Movement by **saltation** involves the action of bouncing, by which particles are lifted off the surface, but tend to fall back again (Figure 20.7). Either immediately or soon afterwards they are picked up again and carried forward. It is clear that this process may be important in causing entrainment, for we have seen that the impact of bouncing particles tends to throw material into motion.

Movement by **traction** refers to the rolling, sliding and shuffling action of particles almost permanently in contact with the surface. **Solutional** transport occurs mainly in aqueous environments, where substances dissolved in the water are carried along as individual ions.

Factors affecting transport

The weight of the particles, or, more generally, their size, plays an important part in determining which of these processes operates. Larger particles tend to move by traction, and only the finest

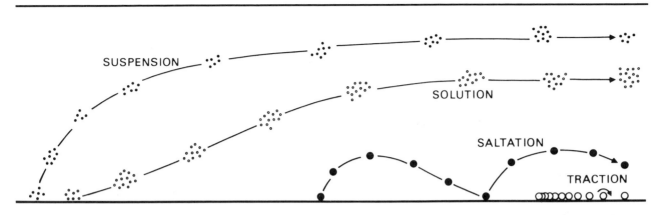

Figure 20.7 Processes of transport. The diagram shows characteristic paths taken by particles moving by traction, saltation, suspension and solution, as might be revealed by time-lapse photography. Note the different distances the material will have moved in the same time

particles are normally light enough to be carried in suspension. This has some significance, for it is clear that material moving in suspension or by solution tends to move faster than either saltation or, slowest of all, traction material. The nature of transport thus determines how fast, and how far, the particles are carried. The different processes lead to a sorting of the material by size; soluble, fine or light particles are generally carried further and faster than large or heavy particles.

In addition, however, the nature of the agent of erosion and the regional climate are significant. On the whole transport by solution is most important in temperate and tropical areas, while solids are dominant in colder regions. In the case of transport by winds, there is a tendency for relatively large quantities of material to be moved by saltation. In the case of transport by water, rather less moves by saltation, and suspension movement is more important, largely because turbulence may be greater and because the water itself gives the particles a degree of buoyancy. Much movement also occurs in solution (Table 20.1). Transport by ice, in glaciers, is less easy to define, and much more difficult to monitor. Probably most of the material transported by glaciers is dragged along

Table 20.1 *Sediment loads, discharge and area of major rivers*

River	Climate	Discharge ($m^3 s^{-1}$)	Area ($km^2 \times 10^3$)	Total load ($t\, y^{-1} \times 10^6$)	Dissolved (per cent)	Suspended (per cent)
Amazon	Tropical rainy	175,000	6,300	498	37	63
Congo	Tropical rainy	39,200	770	79	30	70
Colorado	Cool dry desert	640	635	55	3	97
Nile	Hot dry desert	2,830	3,000	111	13	87
Brahmaputra	Mountain – seasonal tropical	19,300	580	795	9	91
Danube	Moist cool temperate	6,430	805	68	47	53
Mississippi	Moist warm temperate	2,830	3,000	307	30	70
St Lawrence	Moist cool boreal	10,700	1,025	5	91	9
Volga	Moist cold boreal	8,400	1,350	26	75	25
Yukon	Moist cold boreal	6,200	770	79	30	70

Data from A. D. Knighton, *Fluvial forms and processes*, Edward Arnold, 1984.

the bed of the ice sheet as traction load. In the case of movement under the effect of gravity, most transport occurs in contact with the surface. On steep slopes, however, material may bounce in a form of saltation.

Material moving by traction, saltation and suspension influences the processes of transport by exerting a frictional drag upon the transporting agent. Particles carried by the wind, for example, cause friction with the air. This has the effect of damping down turbulence and reducing to some extent the flow velocity. Thus, as the concentration of sediment being carried increases, so the ability of the agent to transport material declines. Clearly the two tend to approach a condition of dynamic equilibrium in which, as additional material is brought into motion, equivalent quantities of sediment are dropped from the load. The maximum quantity (or size) of material which can be transported is referred to as the **competence** of the transporting agent. In the case of transport by solution, competence relates to the amount of the material that can be dissolved before saturation of the water occurs.

Deposition

We only have to watch sand being blown across a dune, or washed across the beach by a shallow stream to appreciate that transport of material is rarely continuous. Instead, individual particles are entrained, carried a short distance and deposited. They rest there for a short while and may then be swept on again. Why is it that the material acts in this way?

The balance of forces

The main reason for this behaviour is that transport depends upon an appropriate balance of forces within the transporting medium. A reduction in the competence of the stream or wind, or an increase in the resistance of the particles, may upset this balance and cause deposition.

Reductions in competence come about in a variety of ways. Velocity or turbulence may be diminished locally by the sheltering effect of large boulders, walls, hedgerows or other obstructions. Additionally, competence changes quite markedly over time.

Wind velocity, for example, varies considerably over a period of seconds, and as a result its ability to transport material is constantly changing. In the longer term, the wind may die down completely as the storm passes, or stream discharge may fall as the flood wanes, and again competence declines. Wherever the ability of the transporting medium to carry material falls in this way, deposition occurs (Figure 20.6).

Increases in the resistance of the material to transport may also occur. Particles may become lodged between larger particles on the surface, for example, or trapped by vegetation. By chance, they may also adopt a more stable orientation, governed by their shape. Rod-shaped particles tend to be transported mainly by rolling about their long axis; if they fall into a position with their long axis parallel to the flow they are less easily moved and become stabilized.

Long-term changes in transporting conditions

The processes of deposition we have mentioned so far are short term in their influence. The next gust of wind or the next storm may again dislodge the particle and transport will start once more. In the longer term, deposition may occur because of much more general changes in the erosional environment. If the climate becomes drier rivers may lose their power to carry material (although this is not always the case since this change may be compensated by other changes in conditions). More impressively, an amelioration of the climate may lead to widespread melting of glacial ice, with mass deposition of the debris it carries. This is what occurred at the end of the Quaternary Period, and it is responsible for the widespread glacial deposits found in the northern hemisphere.

Precipitation and flocculation

We must also mention precipitation and flocculation as processes of deposition. Material carried in solution is deposited mainly through precipitation, a process brought about in most cases by marginal changes in the character of the water. Thus, changes in the chemistry, temperature or gas content of the water may lead to precipitation of dissolved substances. These processes are responsible for many of the beautiful depositional features seen in caves. It is not, as is often supposed, evaporation of the water which accounts for the growth of stalagmites and stalactites; rather it is a process of degassing, as carbon dioxide escapes from the water and alters its chemical equilibrium (Plate 19.1, p. 291).

314

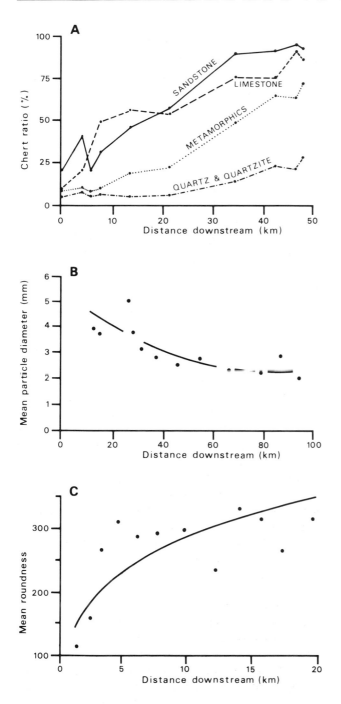

Figure 20.8 Changes in the character of stream sediments in relation to distance of transport: (A) lithology of terrace gravels in Rapid Creek, South Dakota (after Plumley, 1948); (B) particle size of terrace gravels in the upper Thames (D. J. Briggs, unpublished data); (C) roundness of sandstone pebbles in the River Noe, Derbyshire (after Knighton, 1981)

Flocculation is also a chemical process. Many of the fine muds and clays deposited in estuarine or near-shore areas are produced by this process. Salt within the water leads to the agglomeration of minute clay particles into fluffy masses which are heavy enough to sink and collect on the bed. Were it not for flocculation much of this material would be so small that it would remain in suspension almost indefinitely.

Sediment sorting during erosion and deposition

The processes of erosion and deposition tend to lead to a marked change in the character of the material being carried through the landscape. Part of this change is due to the wear and breakage of the particles during transport. The continual collision of particles against each other and with the rock surface leads to their slow attrition. During transport, therefore, the particles tend to become smaller in size and more rounded as sharp corners and protrusions are knocked off (Figure 20.8c).

In addition – and often more important – there is a change in character due to preferential transport and deposition of certain particles. Both shape and size may be important in this **sorting** process. Smaller particles, as we have seen, tend to be transported faster and further; they therefore accumulate in more distant areas; larger particles travel slowly and are deposited more readily. Consequently they accumulate close to their source.

Certain shapes are also more susceptible to transport than others. Spherical or near-spherical particles may roll more readily over the surface, while flat, disc-shaped particles tend to be more buoyant. These may therefore be carried further than very irregular, angular particles.

Changes in composition may similarly occur as a result of transport. Due to the constant physical wear of the particles, only the most resistant survive long distant transport, and softer rocks and minerals are often destroyed close to their source.

All these processes operate together, of course, and it is often difficult to separate their effects. By studying changes in the nature of the materials as they are transported, however, it is possible to distinguish some of the effects of transportation agents upon the sedimentary materials (Figure 20.8).

Landforms of erosion and deposition

Erosional landforms

The landforms left behind by the agents of erosion are the erosional landforms. We will look in detail at many of these landforms in later chapters, but a few generalizations are in order here.

First, we might note that erosional landforms are typically composed of relatively fresh rock. The weathered residues are removed, and only the more resistant materials remain. Thus, the glacial landscapes of northern Canada and the fluvially dissected topography of the Appalachian granites have a common feature; the upstanding blocks represent the resistant cores from which more weathered material has been stripped.

By the same token, many erosional landforms are guided by the structure of the original rocks. The wind-carved landforms of arid and semi-arid areas often show this effect clearly, with the ancient bedding of the rocks picked out by the abrasive effect of the sand. Coastal erosion tends to attack the joints and fissures in the rocks preferentially,

Plate 20.1 A sea-stack at the Cliffs of Moher, County Clare, Ireland. The stack is about 25 m high (photo: J. D. Hansom)

cutting deep inlets along these lines of weakness and sometimes carving a magnificent scenery of stacks and caves and arches (Plate 20.1). Similarly, the karstic scenery of limestone country often shows the effect of structural controls on erosion.

Depositional landforms

Dynamic features

Many of the landforms associated with sediment deposition are relatively transient in character but often very active. They represent equilibrium forms in which material is temporarily stored as part of a natural cycle of erosion and deposition. Thus, during the transport of materials down a stream channel, there is a tendency for temporary storage of sediment in a variety of features. With an increase in stream discharge, due to heavy rainfall or melting of snow for example, the materials may be reworked more generally and carried further down the stream. Similarly, material is laid down within the floodplain of the river, subsequently to be reactivated as the channel shifts laterally and erodes the alluvium. In a stream, therefore, the progress of material down-valley tends to occur not as a continual movement from one end of the valley to the other, but in the form of discrete steps. This is particularly true of material carried by traction or saltation.

The same principle is true of most other environments. Material blown by the wind is alternately deposited and transported, even within a single gale. Between gales, the sediments may lie at rest for some time. Erosion and deposition by ice, by the sea and by gravity operate in a similar cyclical fashion. The landforms which are produced by these cyclical events may persist for some time, since they are often being constantly renewed and regenerated, but they are active. Sometimes, as in the case of dunes, ripples and bars in streams, the features migrate while retaining their form. They are, in the context of the system, an expression of the dynamic equilibrium in the landscape.

Relict landforms

These active landforms may be contrasted with relatively inactive forms which were created, normally under different conditions, during the past. Our landscape is cluttered with such features: river terraces, glacial landforms, the extensive wind-blown loessial plains of China and much of the Mid West of the USA, the vegetated sand-dunes of

Plate 20.2 Depositional landforms: (a) river terraces beside Glen Roy, Scotland (photo: J. Rose); (b) a raised **beach** on the Isle of Jura, Scotland (photo: J. D. Hansom)

South Africa, the fossil raised beaches which surround much of the coastline of Britain (Plate 20.2). All these are relict depositional landforms which have been preserved since they were deposited.

The fact that such features have been preserved indicates, in most cases, that the environmental conditions must have changed since they were formed. Otherwise, the fluctuations in the competence of the transporting agent would have led to reworking of the deposits. Thus, one of the main prerequisites for the long-term survival of depositional landscapes is that they form under conditions of waning erosive power. This may be due to climatic changes, such as those which caused the retreat of the ice sheets that left behind the extensive glacial landforms; it may be due to morphological changes, such as a diversion in the course of the river due to erosion elsewhere in the valley; or it may even be due to the actions of man. In all cases, however, it is the change in the depositional and erosional environment which allows these features to be preserved.

Erosion and man

Needless to say, man is greatly affected by erosion. **Soil erosion**, in particular, represents a major threat, and it is a process in which man himself often plays a major part. Through clearance of the vegetation for agriculture, and through repeated

Plate 20.3 Gullying, Lesotho. Note the overland flow spilling over the gully headwall and the springs issuing from the base of the headwall (photo: D. J. Briggs)

317

Plate 20.4 Undermining of a coastal cliff at Skipsea, Holderness. The till cliffs have retreated about 100 m in the last 50 years (photo: J. D. Hansom)

Plate 20.5 Blow-out in a coastal dune, Banna Strand, County Kerry, Ireland. The blow-out was probably initiated by a combination of over-grazing by cattle and trampling by holiday-makers (photo: D. J. Briggs)

tillage of the soil, he has exposed the soil to the forces of wind and water and, in many cases, reduced the resistance of the soil to these forces. The implications are manifold. Erosion may take his livelihood from him; the eroded particles may damage crops already in the soil; the sediment accumulates in roads and ditches and streams and causes a problem of pollution. Once initiated, erosion often spreads, feeding on itself, for the sand grains carried by the wind encourage the entrainment of more material in neighbouring areas, while deposition of the sands may bury the vegetation.

In desert margins, in particular, erosion represents a serious threat. Yet even in relatively humid environments the problem can be severe. The 'Dust Bowl' in the United States was an extreme example of the damage caused by wind erosion; gullying caused by water erosion is widespread in many semi-arid areas (Plate 20.3). In almost all these instances the problem has been provoked by man's attempt to over-utilize his environment; to farm too intensively.

However, erosion is not only a hazard in relation to agriculture. Coastal areas, for example, are prone to rapid erosion (Plate 20.4). Often, in attempting to control such processes in one area, the problem is aggravated elsewhere, for the materials being washed down the coast are a part of a delicate, interrelated system. Coastal dunes are also subject to severe wind erosion, often induced by use of these areas for recreation (Plate 20.5). Trampling of the vegetation may leave the loose sand open to the winds which blow in from the sea, and rapid deflation may occur.

Looked at in the long time-scale, erosion has played a major part in man's development. There are many areas where excessive soil erosion seems to have been at least partly responsible for the movement or decay of civilization. Today, in a world where the production of food is of overwhelming importance, soil erosion still presents a constant threat to many societies. We must be able to understand and control the process to safeguard our future.

21

Hillslopes

Introduction

Hillslopes are a vital part of our landscape. The whole land surface, in fact, can be thought of as comprising a mosaic of slope units, ranging from vertical mountain and marine cliffs to gentle or almost horizontal plains. Each of these units acts as an important system, through which energy and matter are moved (Figure 21.1), often under the direct influence of gravity. The outputs from these hillslope systems in turn become inputs to other systems – to streams, glaciers or the sea – and thus processes acting on slopes exert fundamental controls on other parts of the landscape. Moreover, the form of the hillslope is itself dependent on the processes acting upon it. Hillslopes can therefore be thought of as process–response systems, the morphology of which responds to, and also affects, the processes operating within the system.

Hillslopes are also important in practical terms. They exert a direct effect upon human activity, governing the type of agriculture which can be carried out and the suitability of the land for construction. The use of tractors and combine harvesters, for example, is limited by slope angle, so in Britain cultivation of land steeper than about 11° for cereal crops is normally not possible. Interestingly, in other countries cultivation is often carried out on much steeper slopes: up to 25° in Spain and 18.5° in Malaysia. These differences reflect partly the stability of the soils in the different areas, and partly the differences in farming methods and population pressure.

As this implies, slope conditions also affect land use indirectly, for hazards such as soil erosion and hillslope failure are important constraints. In general, as slope angles become steeper the amount of runoff and the intensity of erosion increase at an increasing rate (Figure 21.2). This, too, exerts a major control on the suitability of land for cultivation for where excessively steep slopes are ploughed gullying often occurs (see Plate 20.3, p.

317). Similarly, the likelihood of slope failures such as landslides and rockfalls depends on slope conditions. This is significant in relation not only to man's use of natural slopes, but also to the ways in which he constructs artificial slopes such as spoil tips, road cuttings and embankments. If the slope is too steep, failure may occur, often with disastrous results.

The hillslope system

Inputs and outputs

The hillslope system illustrated in Figure 21.1 receives inputs of energy and materials from a variety of sources. Inputs include solar radiation, precipitation and dissolved substances and solids washed from the atmosphere, as well as debris derived from weathering of the parent materials.

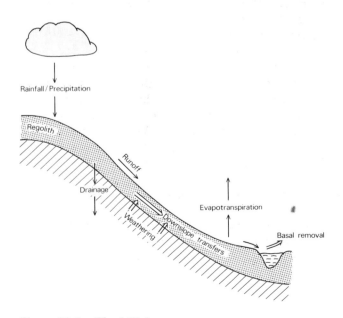

Figure 21.1 The hillslope system

319

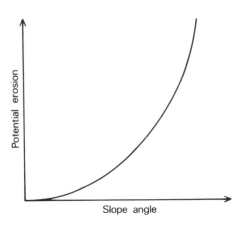

Figure 21.2   The relationship between slope angle and potential rate of erosion

Outputs occur by evapotranspiration, by percolation of water and dissolved substances into the bedrock and, most important of all, by removal from the slope foot in streams, by glaciers or by the sea.

The magnitude of these inputs and outputs depends upon a number of factors, including the geology, climate and relationship of the slope to the wider landscape (Figure 21.3). Inputs of rock materials, for example, are controlled to a great extent by rates of weathering and these, as we have seen, depend in turn upon the character of the bedrock and the local climate. Weathering on resistant, acid crystalline rocks such as quartzites is slow and as a consequence inputs of debris to the slope system are limited; the quantity of debris moving down the slope and being released at the slope foot is therefore limited also. Conversely, in many unconsolidated or soft rocks, weathering is rapid, inputs are high, and the slopes are much more dynamic. Slopes in arid areas may also be relatively less active than those in polar or humid temperate regions due to the lower rates of weathering and the restricted inputs of debris to the system.

Outputs of debris from hillslope systems are controlled primarily by slope foot conditions. The presence of an active stream, for example, encourages removal of debris from the base of the slope and helps to keep the hillslope active. Where removal of debris from the slope foot is limited, perhaps because the stream is too small or too far distant to carry away the material, the debris accumulates at the base of the slope. This clearly alters the slope form, as we will see later in the chapter, but it also affects the overall sediment budget of the slope system and may ultimately affect the processes of weathering and debris transport operating on the slope. As the material builds up, for example, it acts to blanket the bedrock and protect it from weathering. Thus, negative feedback occurs and inputs of debris to the slope are reduced.

As this implies, the balance between inputs to and outputs from the hillslope system exert a major control over slope form and process. In time, a balance between inputs and outputs tends to develop, determined by the more limiting of the two processes. Where inputs are the controlling factor – as on hard rocks which weather very slowly – the slope is said to be **weathering limited**. Conversely, where the potential for weathering is high

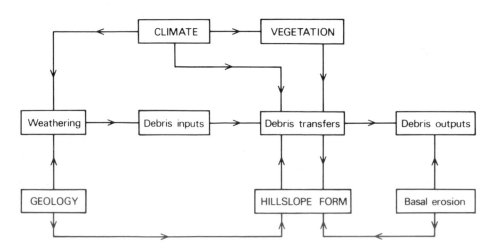

Figure 21.3   The hillslope as a process–response system

but outputs from the hillslope are restricted (e.g. due to the lack of an active slope foot stream) the system is said to be **transport limited**.

Transfer processes and hillslope stability

The processes by which materials are moved through the hillslope system are many and varied. Running water plays a significant role, though we will discuss this more fully in Chapter 22 when we examine stream systems. More important in many cases, however, are the group of processes known as **mass movement**. These involve the transport of debris for the most part not as individual particles but **en masse**, under the direct influence of gravity.

The operation of mass movement processes depends upon the development of instability in the hillslope. Under these conditions, failure of the slope materials occurs, either as a sudden, spontaneous event or as a slow, imperceptible and more or less continuous process. The type of failure which occurs, and the speed at which it operates, is a function among other things of the degree of instability which is generated. Rapid, large-scale failures tend to occur when the stresses exerted on the slope materials greatly exceed their strength – a condition often produced by the operation of short-term **trigger mechanisms**. Slow failure normally occurs when the stresses only just exceed the strength but are maintained over a long time period.

What, then, are the sources of the stresses and strength acting within hillslope materials? As we have noted, a major source of stress is the gravitational force. We also saw in the last chapter that the magnitude of this force is related to the slope angle and the weight of the material:

$$F \propto w \cdot \sin \phi$$

where F is the downslope gravitational force, w is the weight of the material acting at any point on the slope and ϕ is the slope angle.

A little thought will show that the magnitude of this force is not constant over the whole slope surface, but increases downslope and down into the soil. This is because the weight of the overlying material increases with its thickness, making the value of w in the equation progressively larger. It is also clear that the gravitational force can be increased by loading of the hillslope – for example, by the construction of a building or by the passage of a vehicle.

The strength of the hillslope materials, on the other hand, varies according to the character of the rock or soil. In the case of unconsolidated materials such as gravels or sand, the main source of strength is derived from frictional resistance. This depends upon the size, shape and packing of the particles. Soils, fine-grained sediments such as clays, and solid rocks, however, obtain their strength largely from cohesion. As we have seen in the previous chapter, this arises in part from surface tension effects associated with the presence of thin water films between the particles. More important, though, are the effects of interparticle fusion, chemical cements and electro-chemical bonds.

In fresh, massive rocks, the cohesive strength may be extremely high and if this was the only factor involved such rocks would be able to support vertical cliffs several thousand metres in height. In practice, however, most rocks contain joints, bedding planes and faults across which the cohesive strength is relatively low. These therefore act as lines of weakness along which failure tends to occur, and which limit the stresses the hillslope can stand. In addition, the strength of the material is often diminished by weathering, while saturation of the rock leads to the development of positive **pore water pressures** as the water films grow and push the particles apart. Water may also dissolve some of the cements binding the materials, so that after prolonged rain their strength may fall markedly.

The stability of the hillslope depends upon the relationship between the stresses and strength of the material. These vary with depth, due to differences in the weight of the overlying material and the degree of compression (Figure 21.4). They also vary according to the nature of the material and the slope form. More important, however, both stress and strength tend to fluctuate over time, and it is these fluctuations which often trigger off slope failures.

Many factors may act as trigger mechanisms. One of the most common, as we have indicated, is prolonged or heavy rainfall, for this results in saturation of the hillslope materials and a consequent loss in strength. In the longer term, weathering also reduces the strength of the materials. Changes in the magnitude of hillslope stresses similarly occur. Earthquakes or volcanic activity, for example, may cause tremors which give rise to short-lived but large increases in stress. A nearby landslide or rockfall may have the same effect, as may explosions – whether caused by activities such as quarry-

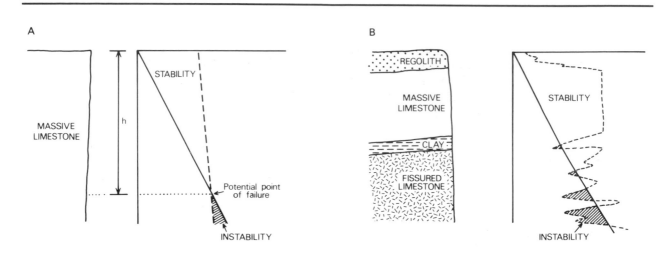

Figure 21.4 Generalized stress-strength relationships in vertical slopes. A The simple situation of a massive limestone slope. If the rock were completely intact the stable height of the cliff (h) would be up to 8000 m. B A composite sequence. Note that failure could occur at several points in the cliff associated with the clay band and fissures in the limestone

blasting or dropping bombs. Even the passage of a heavy vehicle may generate minor shock waves which trigger off slope failure. In addition, as we have noted, loading of the slope by buildings may increase the stress, while oversteepening of the hillslope – by natural erosion or by excavation (e.g. for a road-cutting) – increases the gravitational force operating on the material and reduces slope stability.

In reality, these various effects do not act wholly independently, and slope failures are often a result not of a single trigger mechanism but of the cumulative influence of several different processes. Thus, the development of instability may be associated with a specific combination of events: for example, a major earthquake following a period of prolonged rainfall (A in Figure 21.5); or oversteepening of slope materials whose strength has already been reduced by intense weathering (B in Figure 21.5).

Hillslope processes

As we have noted, numerous processes are involved in the transfer of debris through hillslope systems. These processes vary considerably in relation to the character of the slope materials and it is therefore useful to consider separately slopes formed in non-cohesive materials (e.g. gravels, sands), semi-cohesive materials (e.g. soils and clay), and hard rocks. In the first, strength is derived largely from frictional resistance and is therefore low; as a result, failure often occurs at relatively low slope angles. In the latter two, cohesion provides an additional strength and thus even steep slopes may be relatively stable.

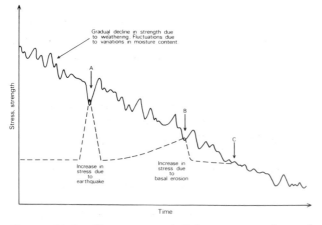

Figure 21.5 Changes in hillslope strength and shear stress over time, and the initiation of landslides. A, B and C = potential landslide episodes

Slope processes in non-cohesive materials

Slopes formed from non-cohesive materials are characteristic of many landforms constructed of coarse-grained sediments, such as alluvial fans, screes, sand dunes and glacial outwash features.

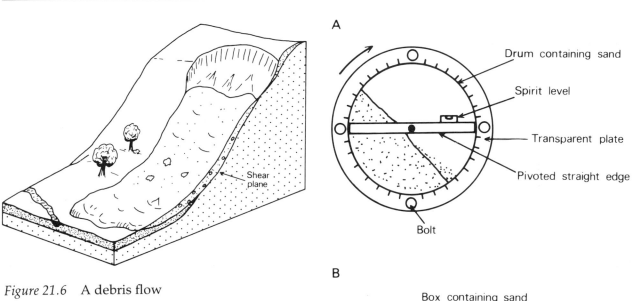

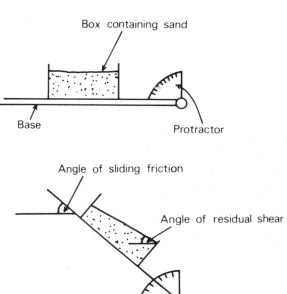

Figure 21.6   A debris flow

Figure 21.7   Measurement of the angle of residual shear and angle of sliding friction in unconsolidated materials. A   A rotating drum. B   A sand box

Many man-made slopes are similarly composed of non-cohesive materials. Mine spoil, for example, is often in this form.

On slopes of this nature movement occurs largely through the sliding or rolling of individual particles, as localized instabilities arise, or through a process of **shallow sliding**. In the latter, planes of weakness develop near the surface which reduce the resistance (or shear strength) of the material and create instability. Failure of the material along this plane may be triggered by minor events such as rainfall or vibration, and the surface material rapidly slides *en-masse* downslope. The **shear plane**, along which the material moves, tends to be relatively planar and shallow (Figure 21.6).

Processes acting upon unconsolidated materials can be simply studied by the methods shown in Figure 21.7a. Studies of this type indicate that the material tends to adopt a relatively constant slope angle after failure. This angle is dependent upon the nature of the material and is referred to as the **angle of repose**, or the **angle of residual shear**. As the slope angle is steepened – for example, by rotating the drum in Figure 21.7a or through undercutting of natural slopes – it will be seen that the material remains stable for a period. Then individual grains start to move as local instabilities are created (e.g. due to the random arrangement of the particles). Finally, again at a constant angle, mass sliding occurs. This starts near the base of the slope, where the gravitational force is at the critical

level. The angle at which movement is initiated is referred to as the **angle of initial yield**.

The reason for these different angles is related to the character of the strength of the material. When the material is static (as it is before sliding), its frictional resistance is relatively high. When the particles are moving, however, their frictional resistance is slightly reduced. This results in a lower angle of stability; that is to say, the material only restabilizes when the stresses, which are a function of slope angle, are relatively low.

Slope processes in clays and soils

Slopes formed in clays, or covered with a mantle of soil, display rather different processes of mass movement. In these cases, the materials have a degree of cohesion which makes them potentially more stable than non-cohesive material (though markedly less so than many solid rocks). This cohesion is derived mainly from the electro-chemical bonds which operate between the fine particles and the surface tension effects of water films contained in the pore spaces.

Both these sources of cohesion are dependent upon moisture content, and, consequently, the strength of these materials tends to be controlled to a large degree by moisture conditions. This has considerable significance, for it provides a means by which stability may be reduced in the short term, without any change in slope form (e.g. basal under-cutting). Repeated failure may therefore be possible as moisture contents increase and the strength of the material falls. This gives rise to a range of slope processes, some of which only operate slowly, others which occur rapidly.

Rotational slips and mudflows

Among the forms of rapid movement are rotational slips and mudflows. **Rotational slips** occur along clearly-defined planes of weakness which develop concave to the surface (Figure 21.8). The generation of these failure planes is due to the distribution of stresses within the material. Rotational slips are common forms of failure, especially in clays and shales (Plate 21.1), although they are rarely perfect in form. Clearly structural weaknesses within the material result in the deflection of the failure along these planes (Figure 21.8).

In some cases failure may not occur until the

Plate 21.1 A rotational slip, Mam Tor, Derbyshire. Until 1977 there was a road across the main unit of the slip, but repeated damage by minor slips during 1977–9 led to the road being closed (photo: D. J. Briggs)

material is so saturated that it acts almost like a fluid. In this condition a variety of processes, generally referred to as **mudflows**, may occur. The saturated material flows like a thick slurry downslope until, as the water is lost through seepage, the flow solidifies and comes to a halt (Figure 21.9). Mudflows may operate on very low slope angles, for the high moisture content of the material reduces the frictional resistance and the cohesion almost

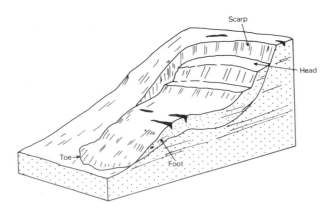

Figure 21.8 A rotational slump

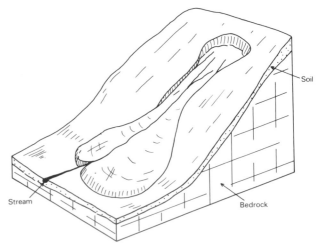

Figure 21.9 A mudflow

to zero. Thus, only small gravitational stresses are necessary to initiate movement. In addition, rates of movement may vary considerably, depending upon slope angle and moisture content. In some cases almost instantaneous failure may occur as the material is rapidly saturated; in other cases slower, long-term movement may take place, often mainly during the wetter periods of the year. In Chile, in 1949, over 1.5 million m³ of material was swept down the southern slopes of Carro Oadillal into the sea within a day. On the other hand, many coastal mudflows operate much more slowly, with flow rates of no more than 10–20 m per year (Table 21.1).

Cohesive materials, like unconsolidated sediments, may be subject to shallow sliding when surface materials are weakened by weathering or saturation. Indeed, it can be seen that shallow slides represent an intermediate form between the deeper rotational slip and the superficial mudslide.

Soil creep and solifluction

Many of the slope processes acting upon cohesive materials operate more slowly. One of the most widespread and intriguing processes is **soil creep**. Material in the weathered surface layers moves gradually downslope in a series of jerky steps. The process may be powered by a variety of mechanisms. The effect of gravity, temperature fluctuations and variations in moisture content within the soil may all act to cause displacement of particles. We can distinguish between various types of soil creep. Often the process acts discontinuously and seasonally and is therefore known as discontinuous

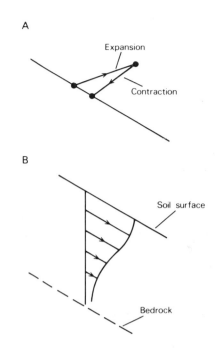

Figure 21.10 Discontinuous creep. A The theoretical pathway of a particle during a single expansion–contraction cycle. B Theoretical velocity profiles

creep. The two most important mechanisms in this case are the heave and settlement of particles due to freeze–thaw activity and the shrinkage and expansion of material due to wetting and drying (Figure 21.10).

Continuous soil creep is largely independent of

Table 21.1 *Rates of mass movement*

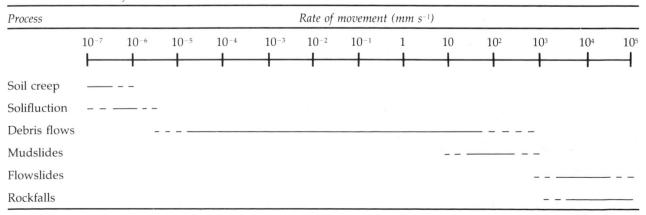

Process	Rate of movement (mm s⁻¹)
	10^{-7} 10^{-6} 10^{-5} 10^{-4} 10^{-3} 10^{-2} 10^{-1} 1 10 10^2 10^3 10^4 10^5
Soil creep	
Solifluction	
Debris flows	
Mudslides	
Flowslides	
Rockfalls	

moisture or temperature fluctuations, and is due, instead, to the gravitational force. It occurs when the stress due to gravitation effects is between the **yield strength** and **residual strength** of the material, in which case a form of viscous failure occurs. The process operates, therefore, where stress and strength are almost in balance. Rates of movement are in the order of a few centimetres per year, and, as with discontinuous creep, there is a tendency for movement to be most rapid at the surface, where the strength of the material is generally lowest (Figure 21.11).

Slow movement caused by freeze–thaw processes is often referred to as **solifluction**, and it is a widespread and important phenomenon in polar and sub-polar areas (Plate 21.2). The process is often aided by wetting and drying, and by the occurrence below the surface of an almost permanently frozen zone above which, during the wetter months, thawing may produce relatively rapid, mudflow-type movements. In other words, solifluction, like most other processes, tends to occur in conjunction with a variety of related activi-

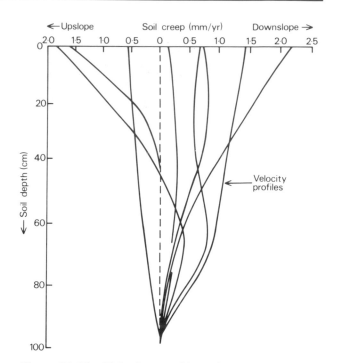

Figure 21.11   Velocity profiles of soil creep in the Mendip Hills, Somerset (from Finlayson and Statham 1980)

ties. Measurements of solifluction, however, show it to be a slow process. In Alaska, rates of 2–6 cm y^{-1} have been determined; elsewhere rates range from less than 1 cm to over 30 cm y^{-1} (Table 21.1). It should also be noted that solifluction may occur on relatively low slope angles. It was also a process of much greater significance during the Pleistocene when sub-polar conditions were considerably more extensive.

Slope processes in hard-rock slopes

Failure of hard-rock slopes is often spectacular and rapid. The reason is in part that such rocks are able to form relatively stable slopes of considerable steepness and height. The spectacular amphitheatre in the Drakensberg Mountains of South Africa, for example, is formed of basalt cliffs over 500 m in height. When instabilities in these materials do develop, therefore, the potential for failure is considerable.

As we have seen, hard rocks derive their strength almost entirely from the strong inter-granular bonds, but weaknesses occur along bedding planes and joints. Rock structure consequently exerts a fundamental control over processes of mass movement in these rocks. In relatively massive rocks, small fragments may be prised loose by gravi-

Plate 21.2   Solifluction lobes, S. Georgia, Antarctica. The turf banks at the front of each lobe are clearly visible, showing where the vegetation is rolled beneath the debris as it flows downslope (photo: J. D. Hansom)

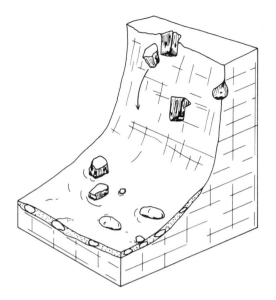

Figure 21.12 Granular disintegration

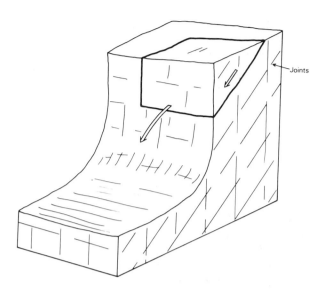

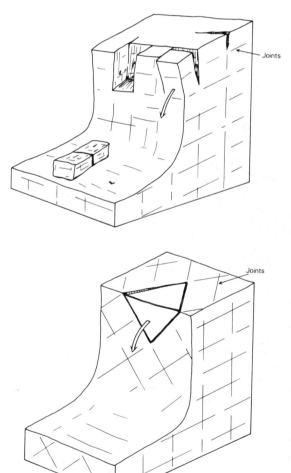

tational stresses, aided perhaps by freeze–thaw processes, to give **rockfalls** or **granular disintegration** (Figure 21.12). The material then builds up as a scree slope at the foot of the slope.

Larger scale failure may occur where movement occurs along well-defined joints or bedding planes. **Toppling failure** takes place where joints are vertical, **slab failure** and **wedge failure** where the weaknesses are inclined (Figure 21.13). In addition to the original joints and bedding planes in the rock, **tension cracks** may develop due to the pressure release during erosion. In the case of relatively massive rocks, failure often involves the extension of irregular, disconnected weaknesses (Figure 21.14). Massive rockslides may then occur (Plate 21.3).

Not all movements in solid rocks are rapid, and a variety of slower, more continuous processes may operate. Among these are **cambering** and associated processes. These occur most commonly where solid (competent) rocks overlie clays or other incompetent materials. Under the weight of the overlying materials the clays may be deformed and squeezed out into the valley floor. At the same time, over-steepening of the slope allows slabs of rock to slide slowly downslope. The joints in the rock are gradually opened up to create **gulls**, and the rocks appear to dip into the valley (Figure 21.15). During the Pleistocene, these processes were active in many parts of Britain, for seasonal meltwaters and

Figure 21.13 Types of large-scale failure in hard rocks: A slab failure; B toppling failure; C wedge failure

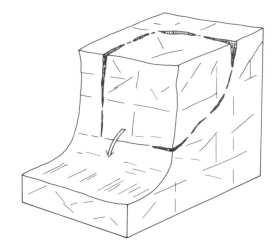

Figure 21.14 Large-scale rockfall, controlled by jointing

Plate 21.3 Rockfall in basalt at Dettifoss, Iceland. The rock has collapsed on to a sand drift (photo: D. J. Briggs)

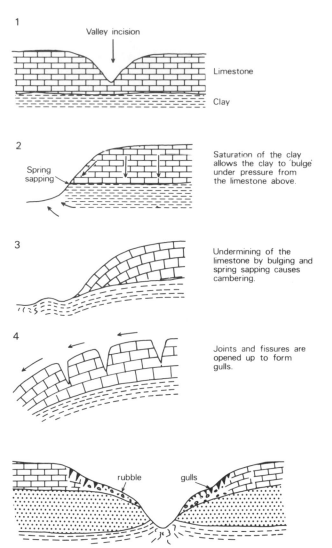

Figure 21.15 Gulling and cambering

glacial outwash swelled the streams and encouraged fluvial incision. In the Cotswolds, for example, this produced deep, steep-sided valleys and exposed the incompetent clays in the valley floor, creating ideal conditions for cambering and valley-bulging.

The role of water on hillslopes

As we noted earlier, water plays a significant role in the development of many slopes, especially in humid temperate regions. **Rainsplash** is a particularly widespread and effective process. The impact of raindrops on the soil detaches individual particles and throws them into the air. On a horizontal surface, the effect is to redistribute the material without any net transport, but on a slope there is a tendency for more material to be splashed downslope than upslope (Figure 21.16). As a result, rainsplash leads to a net downslope movement of the material. Indeed, on slopes of about 25° or more, almost all the splash occurs in a downslope direction.

In addition, considerable transport of material takes place by surface runoff. On relatively smooth surfaces, runoff occurs in the form of **sheetwash**. The erosive power of this is normally limited because the sheet of water is shallow and non-turbulent and cannot readily entrain material. Rain

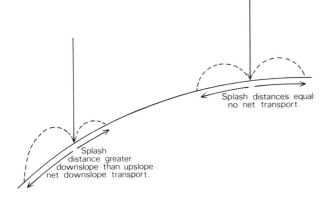

Figure 21.16   Particle movement by rainsplash

falling through the flow, however, may detach soil particles and encourage some transport.

Sheetwash is rarely stable over long distances or for long periods of time because the turbulence caused by rainfall, or by local topographic irregularities, stimulates scour and leads to the development of small depressions. These rapidly extend into rills which, by a process of positive feedback, further concentrate flow and grow to form gullies. Over time, coalescence of gullies may result in large, permanent channels along which large quantities of sediment may be transported. Much of this material is disgorged at the slope foot, where the gullies open out on to the valley floor, and here the debris may build up to form an alluvial fan.

Slope development

If we look at the landscape around us, we can detect many different types of slope. Their angles vary; so does their curvature. Some are straight (or **rectilinear**), some are concave, some convex. Many are composite in form.

At first, there may seem to be little logic to the distribution of slope forms. Yet, if hillslopes truly represent process–response systems, there should exist a close relationship between the processes acting on them and the character of the slope form which develops. Different processes should give rise to different slope forms. Over time, slopes should develop towards predictable, equilibrium forms. The question therefore arises: how do slopes evolve, and what governs their evolution? The search for an answer is one which has occupied

geomorphologists for many years. Three general processes of evolution have been postulated: slope decline, parallel retreat, and slope replacement.

Slope decline

The concept of **slope decline** was proposed by W. M. Davis at the beginning of this century as part of a general model of landscape development. He argued that landscapes undergo a cyclical pattern of evolution, commencing with tectonic uplift and progressing through a phase of fluvial incision into a final phase of **peneplanation** (Figure 21.17). During fluvial incision, relatively steep slopes are formed, but as the river reaches what he called a graded profile downcutting ceases and hillslope processes lead to a gradual waning of the valley sides. Davis was never very explicit about the exact nature of these processes, but he showed that the consequence is the development of concavo–convex slopes of progressively lower angle (Figure 21.17).

Although Davis's concept of landscape development is now generally dismissed, such slope forms are, in fact, common in many humid temperate regions. It is also apparent that different hillslope processes tend to operate on different parts of the slope, and together these may result in slope decline (Figure 21.18). Thus, soil creep is active on the convex crest area. Here there is no input of material from upslope, while rates of transport increase as the slope gets steeper downslope. Thus, soil creep is able to transport the debris as it weathers, little material builds up and, over time, this zone undergoes gradual lowering. By contrast, the steeper, rectilinear mid-slope receives inputs of material from upslope as well as from weathering of the underlying bedrock. Larger quantities of material must be transported through this segment, therefore, and rapid mass movements often occur due to periodic oversteepening of the slope and the development of thick, unstable mantles of debris. Over time, this segment is also lowered while

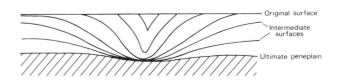

Figure 21.17   The sequence of slope decline according to the concept of W. M. Davis

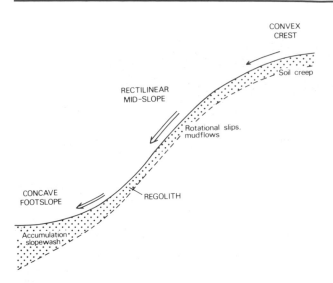

Figure 21.18 Processes operating on concavo-convex slopes

maintaining its essentially straight form.

At the slope foot, a basal concavity evolves due to the slower rate of decline. This segment receives all the debris removed from the whole of the slope length, which accumulates on the footslope and protects the underlying bedrock from weathering. Transport of material through this zone is primarily by slopewash, and removal of debris is controlled by basal erosion. Where basal erosion is active, the concavity does not develop, but where erosion is limited inputs to the footslope exceed outputs and the concavity is marked (Figure 21.18).

Parallel retreat

Concavo–convex slopes are, of course, by no means ubiquitous. In many semi-arid areas, for example, more complex profiles are common, comprising an upper convexity, a steep cliff or **free-face**, a rectilinear **debris slope** and a gently concave **pediment** (Figure 21.19).

The evolution of these slopes has been considered to take place by a process of parallel retreat. This idea was most strongly advocated by the South African geomorphologist, L. C. King. According to King, each of these slope segments is characterized by different processes. Thus, rainsplash and soil creep are the main processes on the crest, while rockfalls and landsliding are active on the free-face. On the debris slope, sliding and avalanching of the unconsolidated materials occur, so that the slope angle is determined by the angle of residual shear of the debris. The pediment is maintained largely debris-free by the process of sheetwash.

Slope development in this situation, King argued, is controlled by loss of material from the free-face. As this takes place, the debris slope also moves back (though by a process which is not always clear), with the result that the whole slope undergoes parallel retreat (Figure 21.19). As a consequence, the pediment is gradually extended while the hilltops diminish. Late in the cycle of erosion the hills are left as isolated, steep-sided relicts, called inselbergs in Africa, or buttes and mesas (according to size) in North America (Plate 21.4).

King's model of slope evolution appears to be broadly valid in the semi-arid and geologically stable environment of southern Africa. Its relevance elsewhere, however, remains controversial, although King himself has claimed that parallel retreat and 'pedimentation' is an almost ubiquitous mechanism of landscape development. Whether or

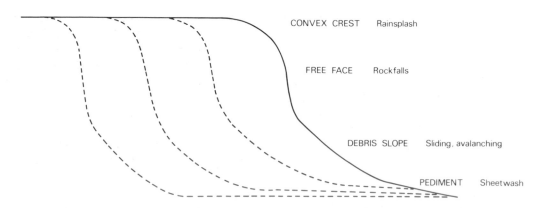

Figure 21.19 Parallel retreat according to the concept of L. C. King

Plate 21.4 A granite inselberg: Paarl Rock, Cape Province, South Africa. The rock is about 30 m high (photo: D. J. Briggs)

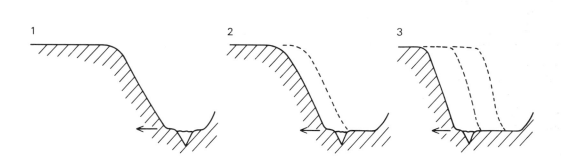

Figure 21.20 Parallel retreat due to basal undercutting of a valley side-slope by a stream

not this is so, parallel retreat can certainly be envisaged in other situations, notably where cliffs or rectilinear hillslopes are being affected by lateral basal erosion. This might take place, for example, where a stream is undercutting a hillside, or where the sea is attacking a coastal cliff. So long as the rate of undercutting is balanced by the rate of debris transport on the slope, and greatly exceeds the rate of weathering and slope lowering, a form of parallel retreat may occur (Figure 21.20).

Slope replacement

One of the weaknesses in King's model of slope development is the process by which the debris

slope retreats in pace with the free-face. In the absence of active basal erosion it is difficult to see why this should happen, and modern studies of cliff-scree slopes indicate that in many cases it does not. Instead, a process of slope replacement occurs. Simply, as the cliff retreats, the scree (equivalent to King's debris slope) extends until it totally covers the surface.

The mechanism by which this happens is shown in Figure 21.21. Initially, material falling from the cliff face accumulates as an ill-sorted pile of debris at the base. As this pile grows, debris falling on to it is able to bounce and roll away from the cliff due to the high impact velocity of the particles. This results in

331

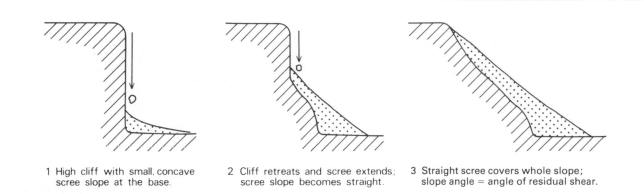

1 High cliff with small, concave scree slope at the base.

2 Cliff retreats and scree extends; scree slope becomes straight.

3 Straight scree covers whole slope; slope angle = angle of residual shear.

Figure 21.21 Slope replacement: the example of scree slope development

the development of a concave, low-angle scree slope. In time, as the cliff continues to weather and retreat, the fall height – and thus the impact velocity – of the material declines. More debris therefore gets trapped on the scree slope, steepening the angle and eliminating the concavity. Ultimately, as the cliff height declines to zero, debris is released on to the scree with a very low impact velocity; movement then takes place by sliding and avalanching, and the scree tends towards the angle of residual shear.

Environmental influences on hillslope development

As mentioned earlier, much attention has been given to finding a universally applicable model of slope development. None of the models so far proposed, however, seems to have general validity, and different slopes appear to evolve along various pathways to different equilibrium forms. Why is this so?

One reason is that processes of hillslope development vary with climate. In humid temperate regions, for example, chemical weathering tends to produce a deep soil mantle which supports a dense soil and vegetation cover. Moreover, rainfall intensities are low. As a consequence, infiltration capacities are high relative to rainfall inputs, little overland flow is generated and the surface is protected from rainsplash. On the other hand, retention of water in the soil favours processes such as soil creep and shallow sliding. In contrast, in arid regions there is little vegetation cover, soils are often thin and processes such as rainsplash and sheetwash predominate. In periglacial environments, seasonal

saturation of the slope materials encourages processes such as mudflows and solifluction.

In different climatic regions, therefore, different processes tend to dominate, and these give rise to different slope forms. Nevertheless, climate is not the only control on slope development. Geology also exerts an important influence. As we noted earlier, this governs the susceptibility of the material to weathering and thus influences the rate of debris input. Rock structure also controls the strength of the material and the range of angles over

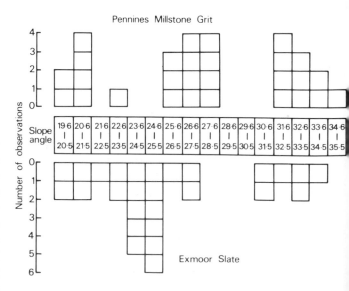

Figure 21.22 Valley side-slope angles on different lithologies (from Finlayson and Stalham, 1980, after Carson and Petley, 1970)

which it will form stable slopes. The significance of geology can readily be seen, therefore, by comparing the distribution of slope angles in areas of different lithology (Figure 21.22).

It should also be remembered that slopes take a long time to develop, and that environmental conditions themselves do not always remain stable. The effects of climatic variation, tectonic activity or changes in sea level may all be reflected in the present form of hillslopes. Indeed, because of repeated environmental changes some slopes may never achieve complete equilibrium with the prevailing conditions.

In general, therefore, we need to consider slope development not in terms of universal processes, acting towards a single form, but as the adjustments of hillslope systems to spatially and temporally varying inputs and outputs. In other words, as in all process–response systems, the form which hillslopes progress towards is a function of the processes acting upon them. These, in turn, are dependent upon the interaction between the hillslope materials and the surrounding environment.

Slopes and man

It is apparent that slope failure may have terrible consequences for man. On a small scale, collapse of embankments and localized landslips may cause inconvenience and endanger life (Plate 21.5); but on a larger scale the effects may be catastrophic. In 1970, for example, in Peru a spate of small rockslides was triggered off by an earthquake and these caused local damage. More tragically, a major failure was initiated on the near vertical face of Mount Huascaran. The material fell a distance of 650 m and crashed into loose glacial debris at an altitude of about 6000 m above sea level. Much of this debris was picked up as the material then tumbled down the mountain side, taking only two minutes to travel 14.5 km – a speed of 400 km hr^{-1}. Before the material slowed down, it had destroyed all in its path, including the lives of 21,000 people. Even after it had slowed to a mere 25 km hr^{-1}, it continued for a further 50 km.

A similarly horrific story can be told of the Langarone landslide in northern Italy. The rocks of the Langarone valley dip inwards, towards the valley with the result that they have very low shear strengths. Despite this, a huge reservoir – Lake Vaiont – had been constructed. Under the effect of

Plate 21.5 A rockfall blocking the George to Knysna railway line in Cape Province, South Africa. The fall occurred about 30 minutes before the photo was taken, narrowly missing a train which was due to pass this section (photo: D. J. Briggs)

the positive pore water pressures of this lake, and lubricated by prolonged and intensive rainfall, part of the valley side began to slide in the spring of 1963. Although attempts were made to monitor the slide in order to assess the problem, the scale of the event was initially underestimated, for the slide was so big that it was taking with it the supposedly fixed markers against which its progress was being measured! The danger of the situation was realized too late and attempts were made to avert disaster by draining the reservoir. On 9 October 1963 an area of the valley side covering 1.8 × 1.6 km collapsed into Lake Vaiont, creating a flood wave which over-topped the dam and swept down-valley. 2600 people were drowned.

As this example indicates, hillslopes can be hazards, and man is often responsible for increasing the hazard by mismanagement of the land. The truth of this statement is also illustrated by the landslide at Turtle Mountain in the Canadian Rockies, in 1903. The mountain is composed of Devonian and Carboniferous limestones which have been thrust folded so that they dip at 50–65°. The main joint system is inclined at 25–40° into the valley, and the valley sides have been over-steepened by glacial erosion. The whole area is potentially unstable, therefore; but despite this, a coal-mine was opened in the hillside in 1901. By October of the following year, repeated small col-lapses were occurring and gangways and chambers were having to be constantly repropped or even abandoned. Nevertheless, mining continued and chambers up to 120 m in height were excavated. Then, on 29 April 1903, the roofs began to cave in. The miners escaped to the surface just in time, while behind them 9 million tonnes of rock fell almost 800 m down the mountain side. The rock hit the ground and rebounded over a sandstone ridge into the Old Man River. There, it ploughed on, climbing almost 120 m up the opposite valley side and covering an area of 2.5 km². Seventy people were killed.

Fortunately, not all of man's effects are so catastrophic. In some cases, as at Folkestone Warren in Kent, where repeated failures of the coastal cliffs have occurred, we are learning how to control landslides by improving drainage and by constructing concrete supports. None the less, the lesson is clear: we need to understand how hillslopes function if we are to cope with the hazards they present.

Streams

Streams and the landscape

The Rio Grande

In all its career the Rio Grande knows several typical kinds of landscape. . . . It springs from tremendous mountains. . . . It often lies hidden and inaccessible in canyons. . . . From such forbidding obscurities it emerges again and again into pastoral valleys of bounty and grace. . . . In such fertile passages all is green, and the shade of cottonwoods and willows is blue and cool, and there is reward for life in water and field. But . . . the desert closes against the river, and the gritty wastelands crumble into its very banks, and nothing lives but creatures of the dry and hot; and nothing grows but desert plants of thirsty pod, or wooden stem, or spiny defense. But at last the river comes to the coastal plain. . . .

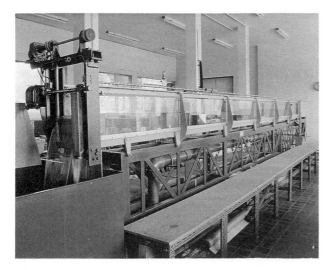

Plate 22.1 A laboratory flume. The flume is used to simulate flow and sediment movement in a stream channel. Sediment is placed on the bed and the flow depth and velocity of the water adjusted by controlling the rate of input and the height of the gate at the foot of the flume section (photo: P. Morley)

After turbulence in mountains, bafflement in canyons, and exhaustion in deserts, the river finds peaceful delivery into the sea, winding its last miles slowly through marshy bends, having come nearly one thousand nine hundred miles from mountains nearly three miles high.

That, in the words of a traveller, is the Rio Grande in Texas. And as these words demonstrate so vividly, this mighty river – the second largest in the United States – is an active and intimate part of the landscape. It acts as a major artery for water moving from land to sea, and as a vital component of the debris cascade. It is not difficult to accept, when we see the Rio Grande, that rivers are important agents in carving and moulding the landscape.

Yet we do not need to go to the great rivers of the world to see these processes in action, for, within broad limits, the action of running water in the landscape is similar whatever its scale. A trickle of water across a pile of sand, a gully on a hillside, a mountain stream or a huge river such as the Mississippi or Rhine are all governed by the same principles, and they all interact with the landscape in a similar fashion. Indeed, much of our knowledge about fluvial processes has been derived not from studies of large rivers, but from small streams like Rosebarn Brook in Devon; even from artificial streams created in laboratory **flumes** (Plate 22.1). Streams at this scale are much easier to study!

The work of streams

Moving water and sediment alter the landscape. The stream erodes its bed and banks, undercutting cliffs, carving deep channels, gorges and canyons. It picks out weaknesses in the rocks and creates potholes, rapids and waterfalls (Plate 22.2). It deposits vast spreads of alluvial silt and clay at its margins to produce extensive, flat, often boggy floodplains (Plate 22.3). It constructs smaller features within its channel – gravel bars, dunes, ripples (see Plate 22.7). Over time, the action of the streams, ever changing, with alternate periods of

Plate 22.2 Aldeyjarfoss, Iceland: a 40 m high waterfall over columnar basalt (photo: A. R. J. Briggs)

Plate 22.3 The floodplain of the River Adur in Sussex. Note the tributary channels, the standing water on the floodplain, and the 'rigg and furrow' system which has been dug (probably in the eighteenth–nineteenth centuries) in an attempt to drain the land (photo: D. J. Briggs)

erosion and deposition, results in the gradual shift of the channel across the surface, the development of meanders, the diversion of streams to leave ox-bow lakes and abandoned meanders. Over even longer periods of time, net aggradation may be superseded by net erosion of the stream bed, and the old floodplains may be converted into extensive river terraces. These and many other features are characteristic of fluvial landscapes – we will examine them in detail later in this chapter. Many of the landforms we see around us, even in the most arid regions of the world, owe their origins, at least in part, to the action of running water.

Stream channel process

The stream channel

If we follow a stream to its mouth we can see the subtle changes in the form and size of the channel. It becomes larger, deeper and wider. Typically, its slope declines, and in many cases the channel becomes increasingly sinuous in its lower reaches as huge and beautiful meanders develop (Plate 22.4). All these changes are intricate adjustments to the processes operating within and upon the channel.

Stream channels are systems. They receive, transport, modify and lose material. Water enters

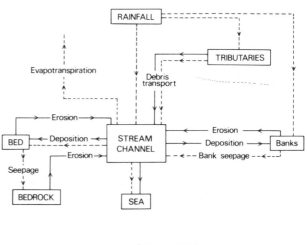

Figure 22.1 The stream system

the channel by direct precipitation, by inflow from tributaries and by seepage from the banks. It brings with it dissolved and solid materials. Additional debris is eroded from the stream banks and bed. These sediments are carried down the channel

Plate 22.4 Meanders in the Spjelfjelldalen, Okstinden, Norway. Note that meandering is not confined to lowland rivers; in this example, large meanders have developed on a glacially eroded bench where the gradient of the river is reduced (photo: J. Rose)

towards the sea; on the way, they are occasionally deposited and stored. Losses of water occur throughout the length of the channel, by seepage into the bedrock and by evaporation to the atmosphere (Figure 22.1). The channel system, like almost all environmental systems, is open to both inputs and outputs.

Stream flow

One of the main processes controlling the nature of the channel system is the flow of water through the channel. This is defined in general by the **stream discharge**. As we saw in Chapter 13, stream discharge is measured in cubic metres per second (cumecs), and corresponds to the velocity of flow multiplied by the cross-sectional area of the channel:

$$Q = v \cdot w \cdot d$$

where Q is the discharge, v is the velocity, w is the average width and d is the average depth of the flow.

The discharge of the stream varies over both time and space. Generally it increases downstream as more water enters the channel from tributaries and from **bank seepage**. This increase in discharge is catered for by corresponding adjustments in velocity, depth and width. It is commonly believed that stream velocity declines downstream, the waters becoming slower and more sluggish, but this is rarely true. It is a misconception that arises because the streams tend to be less turbulent and more muddy in their lower reaches, both conditions disguising the swiftness of flow. In most cases, velocity increases slightly downstream; depth and width increase more markedly (Figure 22.2).

Variations in discharge also occur **at-a-station** (that is, at any single point along the stream) over time. These variations are a response to changes in the inputs of water from rainfall, snowmelt, drainage from the surrounding land and release from reservoirs. We have only to watch a stream during a rainstorm to see the way this increase in discharge occurs. Velocity and depth both increase dramatically. Width changes only slightly, for most streams are steep-sided, so even as they fill up they do not become much wider. However, when the banks are overtopped and the waters spill out on to the floodplain it is clear that the width expands suddenly and often catastrophically (Plate 22.5).

Velocity and turbulence

Looking at a stream in cross-section, we can see a number of other characteristics of the flow. If we measured stream velocity at different depths we

337

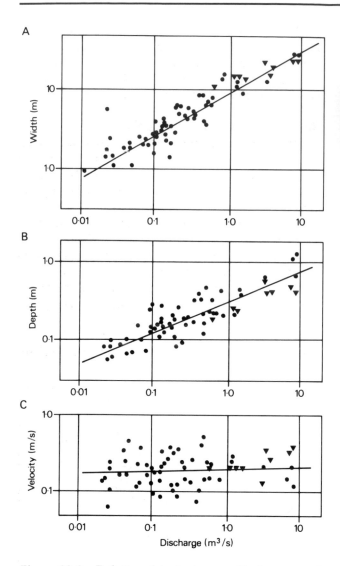

Plate 22.5 Flood in the Nene Valley, Northamptonshire (photo: J. Allan Cash Ltd)

velocity (the **thalweg**) weaves from side to side, wriggling snake-like down the channel as it is deflected from one bank to another (Figure 22.4).

Variations in velocity and the effects of bed and bank roughness result in the water flowing not as horizontal streams (**laminar flow**) but as discrete eddies and vortices, known as **turbulent flow**. In some cases the flow may even spiral, to give what is called **helical flow**. This seems to occur particularly where the channel is sinuous and it plays an important part in sediment transport and deposition and in the formation of point bars.

Overbank flow

Flow is not always contained within the channel and during periods of excessive stream discharge **overbank flow** may occur. Water then inundates the floodplain which borders the stream and complex patterns of flow may develop around obstacles and features on the surface.

Figure 22.2 Relationship between discharge and channel width (A), depth (B) and velocity (C) as discharge increases downstream (from Wolman, 1955). ▲ = main sampling points; ● = single measurements sample

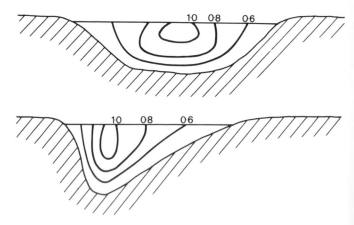

Figure 22.3 Velocity distribution in stream channels (Velocity isolines (isovels) shown in m s^{-1})

would find that velocity was at a maximum close to the surface and at a minimum near the bed; a similar pattern would occur laterally (Figure 22.3). The reason for this is that the material in the channel bed and banks, together with the sediment being dragged or washed along the channel, exerts a frictional drag on the water which reduces its velocity. The coarser the material, the rougher the channel, or the greater the concentration of sediment being transported by the stream, the greater is the reduction in velocity.

Seen in three dimensions, the line of maximum

338

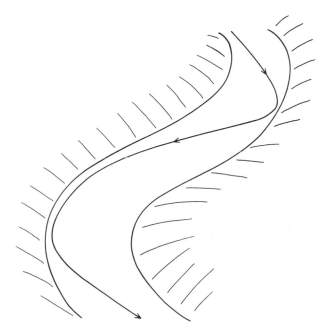

Figure 22.4   The thalweg of a stream channel

The vegetation on the floodplain helps to reduce flow velocities, but even so the force of these flood waters may be surprisingly strong, and walls and buildings may be demolished while the floodplain sediments are eroded and carried downstream.

Sediment in stream channels

At one time or another most streams carry sediments which have either been washed into the channel from the surrounding land, or have been eroded from the channel floor and banks by the stream. The quantity of this material varies considerably from time to time due to changes in discharge. As discharge and velocity increase, the amount of sediment being transported rises correspondingly. A relationship between the discharge and sediment load can be calculated for any single station and a **sediment rating curve** produced (Figure 22.5). Significantly, however, the changes in concentration are not simple, but vary markedly from one storm to another (Figure 22.6).

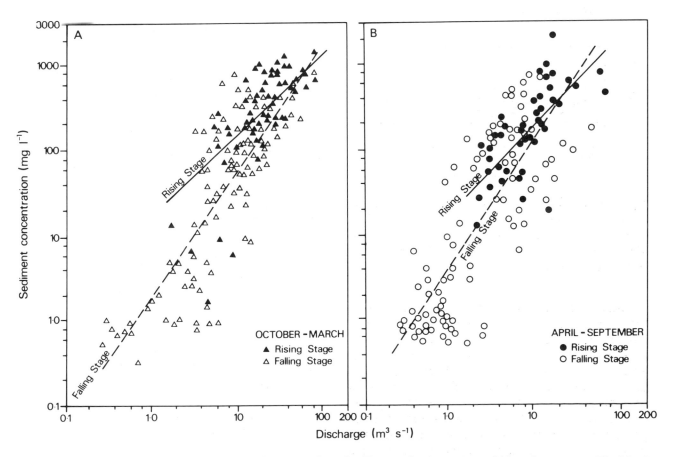

Figure 22.5   Sediment rating curves for the River Creedy, Devon during winter (A) and summer (B). Note that the relationship varies both with season and with stage (from Walling, 1978)

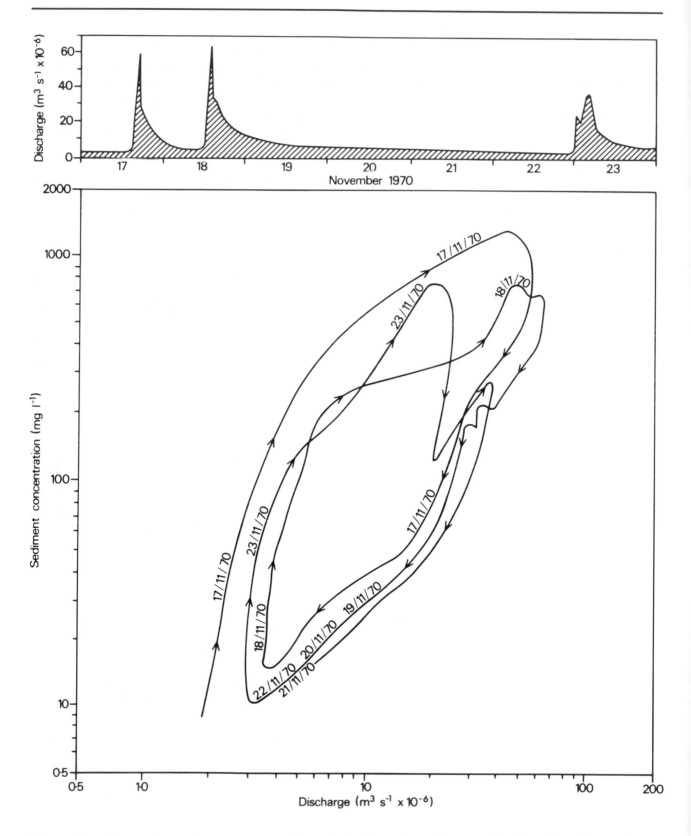

Figure 22.6 Changes in sediment concentration and discharge in a tributary of the River Exe, Devon during three storms in 1970 (from Walling, 1974)

All the sediment rating curve shows is the overall relationship.

During periods of low flow, relatively little sediment movement takes place; as a consequence river channels tend to be stable in these phases. Conversely, as the flow increases, more and more sediment is entrained from the floor and stream banks. Loose sediment on the bed is picked up largely because of the fluid drag exerted by the flowing water. Material from the stream bank is eroded mainly by the action of **bank-caving**. The waters in the channel may undermine the bank and cause collapse (Plate 22.6); but more important in many instances are the processes of freeze–thaw and wetting and drying which prise material from the bank during periods of low flow. Later, when the water level rises, it sweeps this material into motion.

Much of the material which is carried during periods of high discharge is therefore old material reworked from the bed or banks. Most of it moves intermittently, with often long periods of storage in the boundary sediments or in the floodplain alluvium, interrupted by short periods of transport. New material is also moved. Undercutting of bedrock slopes may create cliffs from which boulders are eroded; rocks may tumble and bounce down scree slopes into the channel; slower processes of creep and solifluction carry material down the hillside into the stream. During storms, overland flow may also develop on the valley sides, and sediment may be washed by sheet-flow, and in temporary rills and gullies, into the channel.

Sediments and sorting

The stream carries a wide variety of material: logs and branches and leaves from the surrounding vegetation; plastic bottles, paper and old boots cast aside by man. It carries also pebbles and sand which move along the bed as bedload or traction material, silts and clays in suspension (suspended load) and material in solution (dissolved load). Both the

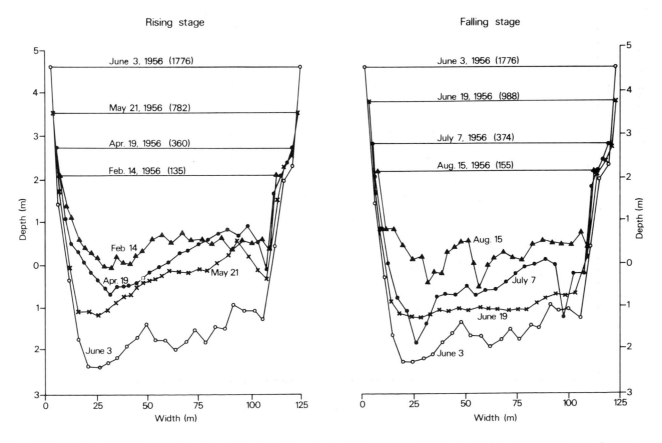

Figure 22.7 Scour and fill in the Colorado River in Arizona, 1956. Figures in brackets represent discharge in m³ s⁻¹ (from Leopold *et al.*, 1964)

absolute quantities and the relative proportions of these components vary from one stream to another (Table 20.1), and, perhaps more significantly, from one time to another. The coarser particles, for example, move only when the discharge is very high, and many of the larger boulders entering the channel may be too big to be moved by all but the greatest floods. They may accumulate as a lag deposit. On the other hand, the finer silts and clays may be in almost perpetual motion; certainly dissolved materials, such as fertilizers, animal wastes and soluble compounds washed or weathered from the soil are constantly being carried downstream.

There are two important implications of this phenomenon. The first is that the material is actively sorted by size during transport; the second is that different components of the stream sediment are in equilibrium with different flow conditions. For this reason, there are great difficulties in defining a single equilibrium condition for the channel as a whole.

The downstream sorting of sediment can be observed along the course of almost any stream. From the headwaters to the mouth there is a decrease in the size of the sediments (see Figure 20.8, p. 315). Because the finer particles travel more frequently and more rapidly they are washed further downstream, while the slowly, infrequently moving coarser materials are left in the upper reaches. It should be noted again that the commonly held idea that material is progressively deposited – coarser particles first and finer particles later, because the velocity declines downstream – is generally false; in most cases velocity increases or at least remains constant downstream.

As we have noted, movement of sediment is not continuous, but involves intermittent transport and deposition. Even during a single flood, erosion and deposition may be occurring simultaneously as the sediment-laden waters drop some of their load and pick up new material. This results in complex processes of **scour and fill** during floods (Figure 22.7).

Sediment deposition in channels

The intermittent movement and deposition of sediments within the stream channel is associated with the development of a variety of **sedimentary structures** and **bedforms**. Many of these are dependent upon the complex interplay between stream velocity and particle size, so that as the discharge increases during a storm the character of the bedforms may change (Figure 22.8); the bedforms, like

Plate 22.6 Bank-caving, Snaizeholme, North Yorkshire. This probably represents one of the most important processes of sediment input into many streams. Note also the box which has been constructed in the stream bed to act as a sediment trap. Material moving along as bedload is collected in this way so that rates of bedload transport can be assessed (photo: D. J. Briggs)

other features of the channel adjust to equilibrium with prevailing flow conditions.

Gravel bars

In streams carrying mainly coarse materials – gravels and coarse sands – **bars** develop. These take a range of forms. **Point bars** are possibly familiar, for we can see them on the inside of meander bends in many streams (Plate 22.7). They develop where the flow is inhibited by the increased frictional resistance and reduced depth of the water and by the tendency of the main thalweg to be thrown towards the outside of the bend. As a result, the ability of the water to carry sediment (its competence) is reduced and the coarser particles accumulate on the inside of the bend. Helical flow may also occur and this seems to encourage bar formation. The feature which is formed is characteristically lobe-shaped and shows an interesting distribution of particle sizes; coarser material is concentrated on the outer margins of the feature where the flow is most active, while finer particles accumulate on the inside of the bar.

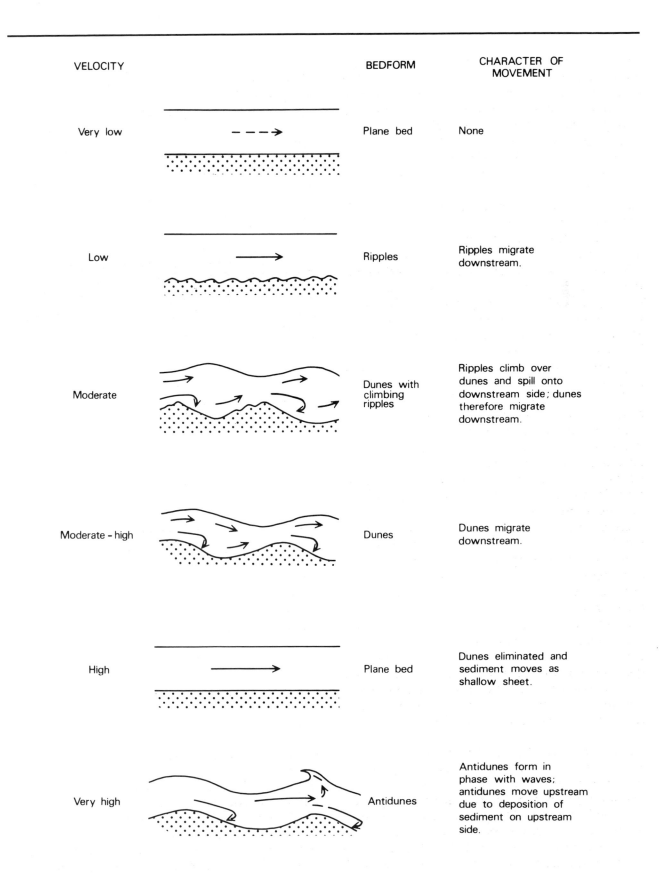

VELOCITY		BEDFORM	CHARACTER OF MOVEMENT
Very low		Plane bed	None
Low		Ripples	Ripples migrate downstream.
Moderate		Dunes with climbing ripples	Ripples climb over dunes and spill onto downstream side; dunes therefore migrate downstream.
Moderate – high		Dunes	Dunes migrate downstream.
High		Plane bed	Dunes eliminated and sediment moves as shallow sheet.
Very high		Antidunes	Antidunes form in phase with waves; antidunes move upstream due to deposition of sediment on upstream side.

Figure 22.8 The relationship between stream velocity, bedform and sediment movement

Plate 22.7 River bars: (left) a point bar on the River Wharfe at Burnsall, North Yorkshire (photo: D. J. Briggs); (right) a longitudinal bar on the River Wharfe near Grassington, North Yorkshire (photo: D. J. Briggs)

Pools and riffles

In straight streams, bars tend to develop in a rather different fashion. Material builds up in regularly spaced **riffles** along the channel. These typically slope alternately towards one bank and then the other, and the thalweg winds between them. Between the riffles occur deeper **pools**; the whole sequence is known as a **pool-and-riffle** (Figure 22.9). The spacing of these features is associated with the size of the channel; the distance between the top of one riffle and the next is, on average, in the order of five to seven times the channel width.

The reason for the development of pool-and-riffle features is not entirely clear. It is apparent that the material on the riffle tends to be coarser than that in the pool. It is also clear that the form of the features tends to remain constant, although material moves into and out of each riffle. This has been demonstrated by placing painted pebbles on the surface of riffles; during periods of high flow they are seen to hop downstream and are commonly found resting on the next riffle. In other words, the features are in a condition of steady-state equilibrium with stream flow.

Dunes and ripples

In channels composed mainly of finer materials –

sand and silt – the main features are **dunes** and **ripples**. These are similar in formation, but whereas the dunes are normally in the order of 10 or more centimetres in height (and some may be several metres) and often a metre or so apart, ripples are much smaller (often no more than a few centimetres in height and spacing).

The development of dunes and ripples is complex, for they involve the interaction of the sediment being carried by the stream, the feature itself and the character of the flow. As the feature develops it tends to disturb the flow and create conditions which favour the growth of further dunes or ripples. We will examine one aspect of this

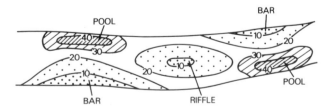

Figure 22.9 Pool-and-riffle sequence showing depth of bed below water surface (all contours labelled in cm)

344

process when we consider wind-formed features in the next chapter. Nevertheless, it is clear that, because of this close relationship, the size and character of the dune tends to be closely associated with flow conditions. Many dunes, for example, are related to water depth. A general relationship is described by the formula:

$$H = 0.086 \, d^{1.19}$$

where H is dune height (cm) and d is water depth (cm).

One implication of this relationship is that it is possible to estimate the depth of water from fossil dunes preserved in fluvial sediments.

Like gravel bars, dunes and ripples are dynamic features. They tend constantly to receive material from upstream and lose material downstream. The exact nature of this gain and loss is interesting, and it results in a migration of the dune or ripple. In general, material on the gently sloping **stoss-side** of the dune (the side facing into the flow) creeps up the slope under the effect of water flow, until it reaches the crest. The particles then avalanche down the steeper **lee-side** and collect at the foot. A degree of sorting according to size occurs here, for the coarser particles tend to roll further. Finer material may also be swept off the dune and hop downstream; as it lands it sets other particles into motion. Over time, the dune appears to travel downstream.

Floodplain development

In northern California, during three days in December 1964, about 280 mm of rain fell in a single storm. The small Coffee Creek stream was swelled by the runoff that the storm generated, and, as the discharge reached a peak with a recurrence interval of over 100 years, the stream burst its banks and flooded out on to the surrounding land (Figure 22.10). The waters scoured out new channels, infilled old ones and caused extensive erosion of the valley floor. As the flood subsided, the eroded sediments were deposited across the floodplain, covering 70 per cent of the surface to an average depth of 0.4 m. The power of the stream to erode and transport debris, and to build its floodplain, was amply demonstrated.

This occurrence was an exception; more commonly, floods are less dramatic, less catastrophic; normally it is only finer material which sweeps across the floodplain in this way. Nevertheless, the principle is clear: during times when the streams over-top their banks, they act to shape their floodplain.

Four stages in floodplain inundation can normally be recognized:

1 Initial spilling of waters on to the floodplain.
2 The development of continuous flow across the floodplain.
3 Waning of the flood and retreat of the water.
4 Drying out of the floodplain surface.

During the first stage erosion is particularly active. As the waters rise and spill over their banks they come into contact with the **levées** which border the channel. These are sandy or gravelly ridges, often in the order of one half to four times the channel width in diameter, which run parallel to the stream. They are intersected at intervals by narrow gaps called **crevasses**, and it is through these that the waters pour on to the floodplain. Erosion in these crevasses is rapid, and large parts of the levée may be destroyed.

Gradually the floodplain becomes drowned as the water washes through the levées. Depressions and abandoned channels become filled and then these are connected up as the water level rises and spills over the intervening ridges and mounds. To some extent the vegetation protects the floodplain itself from erosion, but where flow is rapid and turbulent the plants may be ripped from the soil and the sediment beneath scoured out and washed away. Erosion also occurs where obstacles such as large trees, walls or buildings cause the flow to eddy or restrict its passage downstream. On the whole, however, the second phase of flooding is marked by widespread deposition on the floodplain surface. Erosion continues at the margins of the channel, for there the water remains turbulent, but elsewhere the flow is generally more passive, shallower and constrained by vegetation. Fine materials, carried in suspension across the levée, start to fall to the bottom of the water; clay, silt and sand accumulate on the floodplain. This material is known as **alluvium**.

In time, as the flood wanes, the waters subside and the higher parts of the floodplain re-emerge. In the isolated pools and ponds which are left in the depressions the finest materials continue to settle out. Winds disturb the surface of these waters and the material on the floor is moulded by the turbulence into ripples, while at the margins of the pools small beaches are formed. Finally, evaporation and drainage result in the emptying of the

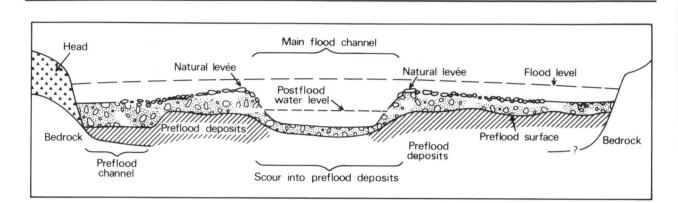

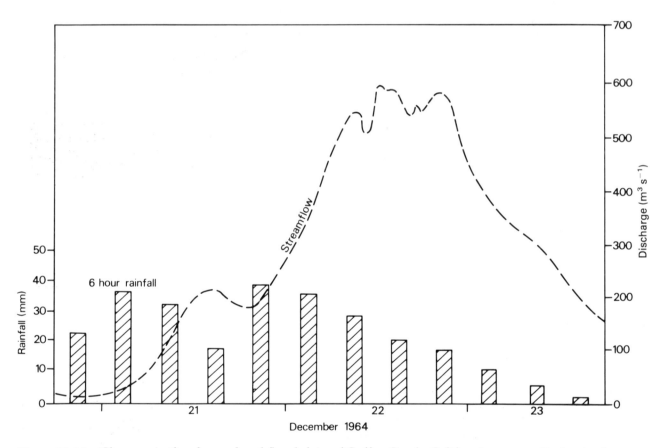

Figure 22.10 Changes in the channel and floodplain of Coffee Creek, California as a result of a single storm in 1964 (from Stewart and Lamarche, 1967)

pools and the floodplain surface dries out. Vegetation encroaches on to the new material and soil development commences. A new floodplain surface is created.

Even between floods, however, the floodplain is being reworked and modified. The stream channels slowly shift laterally, undermining their banks and eroding the floodplain sediments. On the outside of meander bends, where the thalweg impinges against the channel side, erosion is particularly active and the meander becomes accentuated. Eventually, the neck of the meander may be breached, the old channel is abandoned and an **ox-bow lake** forms. At the same time, on the inside of the meander bend, the point bar builds up, while fine-grained sediments accumulate in the **swale** behind the bar. During floods, the point bar is buried by alluvium, so that the floodplain sediments comprise a varied range of materials (Figure 22.11).

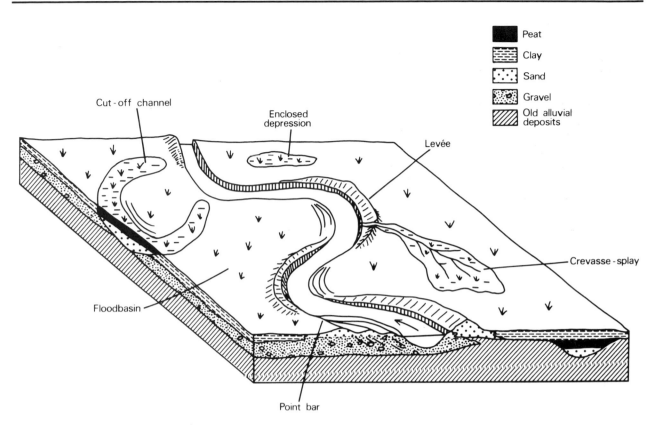

Figure 22.11   The structure of a meandering river floodplain

Stream channel form

Channel shape

Water and sediment moving down the stream channel interact with the bed and bank materials to modify the shape of the channel. Thus, the form of the system adjusts in response to the internal processes. In hydrodynamic terms, a perfect channel is one which is parabolic in shape. Few channels, however, are perfectly parabolic; instead they range from deep, narrow forms to wide shallow ones.

In general wide, shallow channels develop where the materials which make up the bank are loose and non-cohesive. Sands and gravels, for example, are unstable at steep angles. They collapse if the stream cuts deeply into them. Where the banks are composed of more cohesive materials, such as silt and clay, the channel is often deeper and narrower. The ratio of width to depth of stream channels is therefore inversely related to the silt and clay content of the bank material (Figure 22.12). For the same reason, channels cut through bedrock tend to be

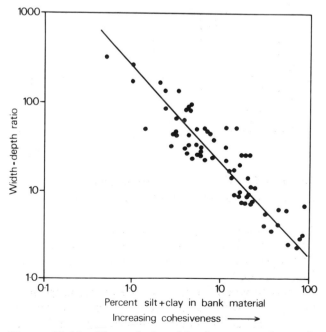

Figure 22.12   The relationship between channel shape and the clay + silt content of the bank materials (after Schumm, 1977)

347

Plate 22.8 A slot gorge in Drakensberg Mountains, Natal, South Africa (photo: D. J. Briggs)

deep and narrow; **slot gorges**, such as that cut by the Colorado are an example (Plate 22.8).

The shape of the channel is not a fixed feature of the stream, but varies according to flow conditions. As we have seen, scour and deposition may occur during a single flood, altering the shape of the channel in a complicated manner. In most cases, the channel returns to a stable form during low-flow conditions, when discharges are too small to undermine the banks, and too little sediment is being transported to significantly modify the shape of the channel.

Meandering channels

Looked at from the air or on a map, stream channels can be seen to be highly **sinuous**; only rarely are they straight for any distance. The degree of sinuosity can be determined by calculating the ratio of the actual channel length to the straight-line distance between two points (Figure 22.13). Channels in which this index of sinuosity exceeds 1.5 are referred to as **meandering** (see Plate 22.4).

The reason for the development of meanders is still something of a mystery. A number of explanations have been proposed over the years, including the occurrence of random obstacles (such as

resistant materials in the floodplain or large boulders), the need for the stream to lose energy as its sediment size declines but its discharge increases, and even the effect of the earth's rotation.

Typically a meandering stream consists of a series of curves of approximately similar form. The character of the meanders can be described in terms of their average **amplitude**, **wavelength** and **radius of curvature** (Figure 22.13). Numerous studies have shown that these parameters are closely related to the size and character of the stream. For example, meander wavelength increases as the radius of curvature or width of the channel increases (Figure 22.14). These relationships seem to reflect a more fundamental association between the size of the meanders and the stream discharge.

The consistency of the pattern shown by meanders indicates that they cannot be regarded as products solely of random conditions, such as changes in the resistance of the bed or bank material. Instead, it seems to be some internal property of the channel system which controls their development. One of the most likely explanations is that they develop in response to an excess of **free energy** in the stream. Downstream, the discharge tends to increase, while the sediment becomes finer. Thus the stream has more energy, but

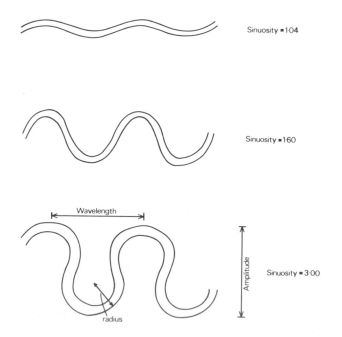

Figure 22.13 Characteristics of meandering stream channels

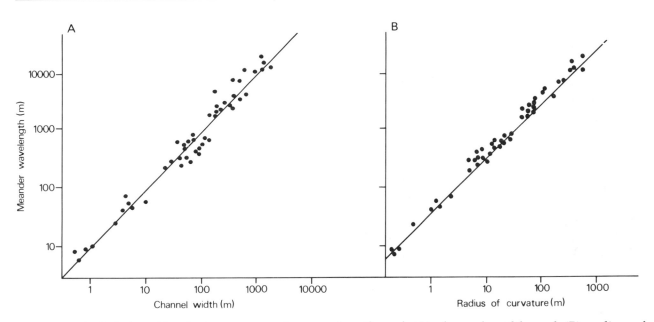

Figure 22.14 Relationships between meander wavelength and (A) channel width and (B) radius of curvature for a range of natural and flume streams (from Leopold *et al.*, 1964)

relatively less work to do. This results in there being more free energy within the system; in a sense the stream is underworked. Meanders are a means of carrying out work and expending this free energy uniformly throughout the stream: a way of maintaining a balance between the capacity to do work and the amount of work done. Interestingly, the shape of curve which best achieves this is what is known as a sine-generated curve. Many meanders approximate to curves of this nature (Figure 22.15).

One reason for favouring this explanation is that it helps to account for the occurrence of meanders under a wide variety of circumstances. Since meanders are found not only in alluvial streams but also in streams developed on glaciers and in the Gulf Stream, it certainly appears that meandering is a fundamental response of flowing water, independent of sediment or bank conditions.

Braided channels

Many channels are not only sinuous but also **braided**. Braiding refers to the separation of the main channel into a number of smaller, interlocking channels. Streams draining ice sheets are frequently braided (Plate 22.9); so too, in parts, is the River Meander in Turkey, from which the term meandering is derived!

Streams of this sort are considered to be highly active and, often, unstable. Active they certainly

are, for they frequently experience great variations in discharge, turning innocuous, partly dry channels into a raging torrent; but they are not necessarily unstable. Individual channels may be abandoned, buried by vast quantities of sediment carried by the streams, or eroded, but the overall character of the stream is retained. The pattern

Plate 22.9 A braided river: the River Lech in Austria (photo: D. J. Briggs)

349

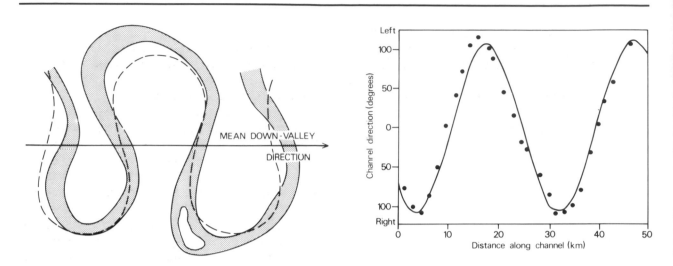

Figure 22.15 Relationship between meander pattern and a sine-generated curve: the Mississippi River near Grenville, Mississippi (from Leopold and Langbein, 1979)

persists. There is, moreover, a marked contrast between the sediments accumulating in braided stream systems and those associated with meandering channels (Figure 22.16).

Conditions favouring development of braided channels include an easily erodible bank (low silt and clay content), a high discharge relative to slope (Figure 22.17), a high sediment load, and, in some

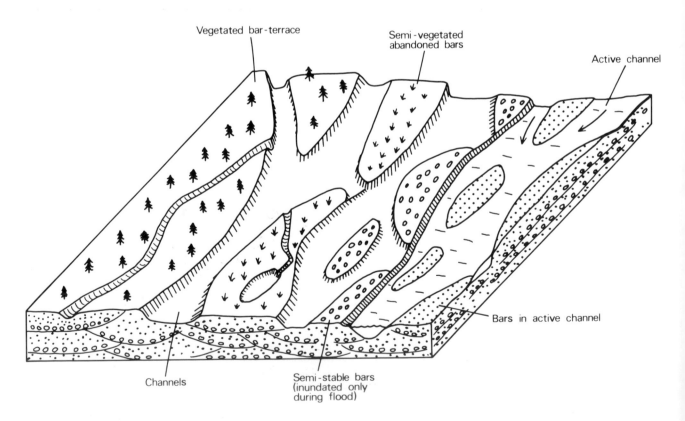

Figure 22.16 The structure of a braided river floodplain

cases, a variable discharge. These are all conditions represented by streams draining from glaciers and ice sheets, for they commonly cross wide, flat expanses of gravel and sand, and are subject to marked changes in discharge as the ice undergoes seasonal or longer periods of melting and advance. It is not surprising, then, that braided streams are common in these environments, and there is reason to believe that they were more widespread during the Quaternary Period when many rivers in what are now cool temperate regions were affected by meltwaters. The legacy of these can clearly be seen, for example, in the Prosna district of north Poland (Figure 22.18).

Two other observations on braiding may be noted. Braided channels are common on beaches, where shallow flows cross the relatively coarse and loose sand and in arid areas, where great seasonal variations in discharge occur. Experiments in artificial flumes have also shown that braiding can easily be generated by introducing poorly sorted sediments into an existing channel. Coarse material apparently builds up as bars within the channel, causing the flow to split (Figure 22.19). Sediment

characteristics are therefore important in the development of braided channels, although the exact process of formation remains a mystery.

The long profile of streams

We are all acquainted with the change in slope displayed by most streams. In their headwaters the channels are steep; as they get closer to the sea they become more gently sloping until, as far as the eye can tell, they flow across a flat surface. In other words, the **long profile** of most streams is **concave-upwards** (Figure 22.20).

This phenomenon has for long intrigued geomorphologists, and many explanations have been proposed for it. At one time it was regarded as a product of the different ages of the various parts of the stream. It was believed that the stream gradually wore away the land to give a flat **peneplain**. In the process the gradient of the stream slowly declined, while its headwaters continued to eat back into the uplands. Within the course of a stream it was thought that there was a sequence from the youthful, steep headwaters to the old or even senile and meandering lower reaches (Figure 22.21).

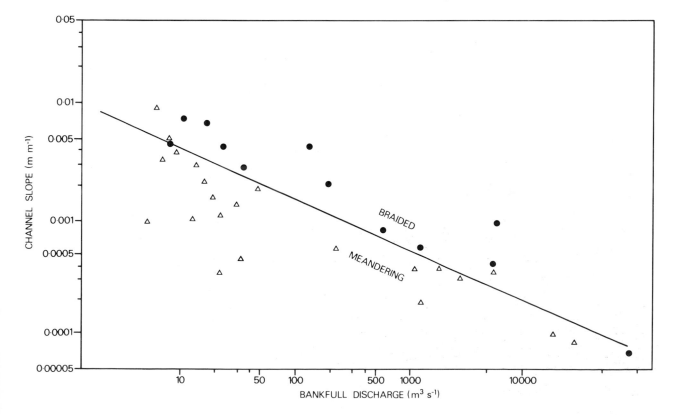

Figure 22.17 The relationship between bankfull discharge and channel slope for meandering and braided streams (after Gregory and Walling, 1973)

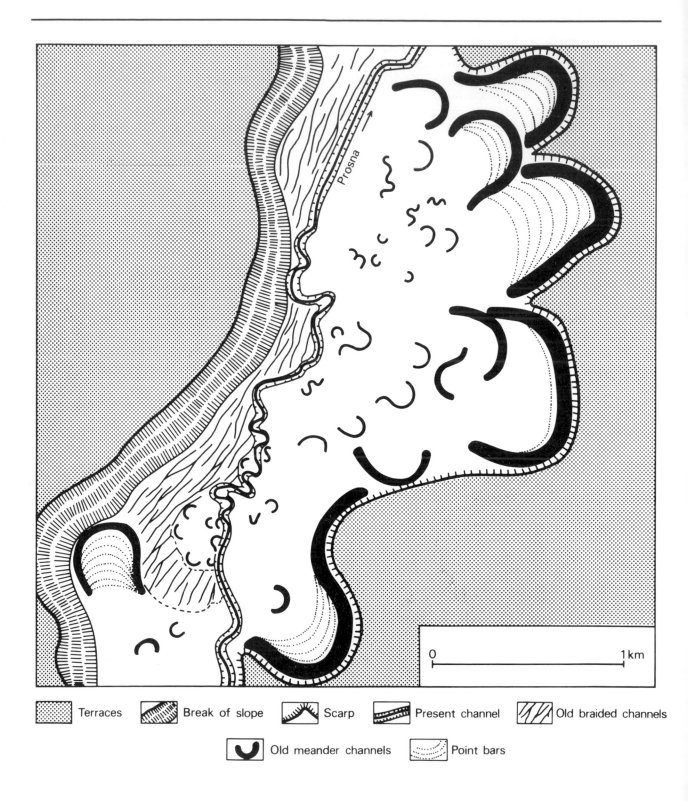

Figure 22.18   Channel patterns in the Prosna valley, Poland (from Kozarski and Rotnicki, 1983)

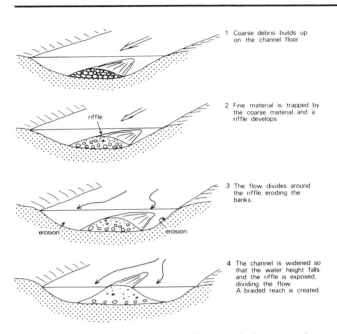

1 Coarse debris builds up on the channel floor.

2 Fine material is trapped by the coarse material, and a riffle develops

3 The flow divides around the riffle, eroding the banks.

4 The channel is widened so that the water height falls and the riffle is exposed, dividing the flow. A braided reach is created.

Figure 22.19 Development of a braided stream by longitudinal bar formation

This idea is no longer accepted. Instead, the long profile of the stream can be seen as a function of the intricate relationships operating within the channel, themselves responses to the distribution of energy throughout the system.

In the upper reaches of the stream, the discharge is low and the sediment coarse. A steep slope is therefore necessary to transport the material. Downslope, the discharge increases while the sediment becomes finer; thus more energy is available, but the losses of energy through friction have not increased proportionately. The sediments can therefore be transported over gentler slopes. Viewed simply, the stream has a greater capacity to carry out work; in order to maintain a balance between this capacity and the amount of work to be done, the stream tends to adopt a lower slope angle.

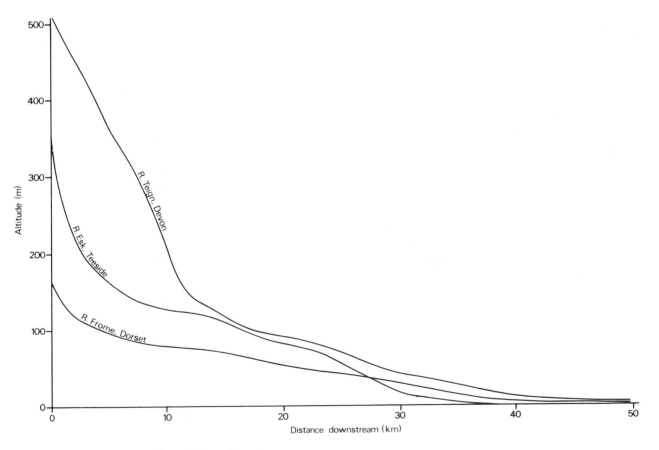

Figure 22.20 Long profiles of three British streams

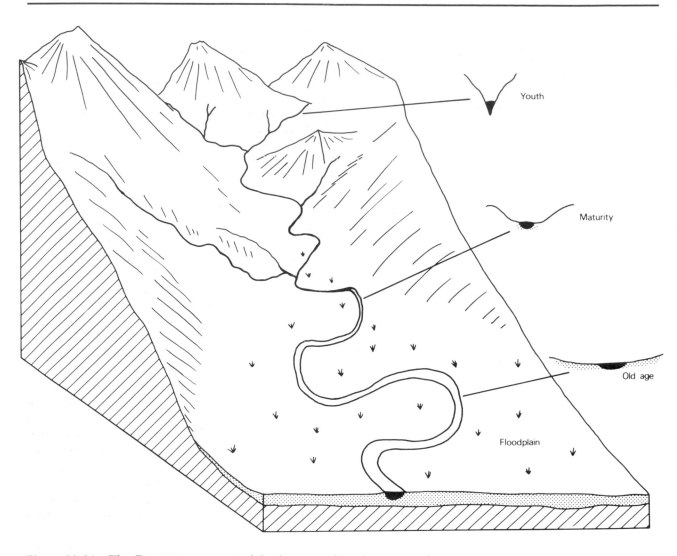

Youth

Maturity

Old age

Floodplain

Figure 22.21   The Davisian concept of the long profile of a stream

In fact, there is no fundamental reason why streams should adopt a concave long profile, and in some conditions, where the relationship between stream energy and sediment load is suitable, straight or even convex profiles may develop. Where discharge declines downstream – due, for example, to excessive evaporation, seepage into the bedrock, or extraction by man – then the degree of concavity diminishes and convex profiles can occur. It has also been shown that if sediment size increases markedly downstream – as happens when coarse material is introduced in the lower reaches from a cliff or scree slope, or from an active tributary – the long profile tends towards a convex form.

As these examples indicate, the long profile of the stream owes much to the relationship between discharge and sediment size, or, more generally, available energy and workrate. Indeed, the whole stream system can be conceived as a set of inter-dependent variables which respond mutually to external conditions such as climate, geology and regional relief (Figure 22.22). The fact that there are so many variables, all interacting, means that it is difficult to predict precisely the ways in which the stream system will respond, and as Leopold and Langbein (1979) have stressed, introduces a degree of **indeterminacy** to stream systems. Stated simply, the large number of variables, the almost infinite number of streams and environmental conditions,

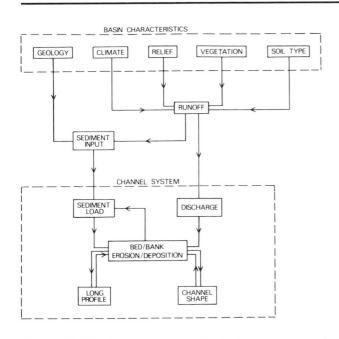

Figure 22.22 Inter-relationships between channel variables and basin characteristics

channel per unit area of drainage basin:

$$D_d = \Sigma L/A$$

where D_d is the drainage density (km km^{-2}), L is stream length (km) and A is the area of the drainage basin (km^2). Drainage density varies in relation to drainage basin characteristics and also, by influencing the time it takes for water to concentrate in the stream channel, affects stream discharge and channel characteristics.

One of the most important controls on drainage density is exerted by the infiltration capacity of the soil. As we saw in Chapter 14, this influences the amount of water available for surface runoff and thus has an effect on stream channel development. Where infiltration capacity is high, little overland flow occurs and channel development is inhibited. Where infiltration capacity is low, most of the water runs off as overland flow; rills, gullies and ultimately major streams develop and a dense network of channels emerges. Drainage density is therefore high.

Many different factors influence infiltration capacity, so these processes vary considerably from one area to another. One of the main influences is that of geology. Rocks such as limestone, gravel, sand or well-jointed igneous rocks tend to have high infiltration capacities, so they are associated with low drainage densities. Impermeable rocks, such as shale, clay or massive igneous rocks give rise to rapid runoff and higher drainage densities (Figure 22.23).

Topography is also important. Runoff tends to be encouraged on steep slopes; it is reduced where the land is flat. Vegetation has an additional effect. Dense vegetation helps to intercept rainfall, to increase the infiltration capacity of the soil and to prevent overland flow. Consequently it inhibits channel development. Where vegetation is sparse drainage densities are much higher. The effect is often seen on agricultural land, and, indeed, reflects a major problem in intensively farmed areas. Where vegetation is removed, dense networks of eroded gullies may be formed.

Geology, relief and vegetation affect what happens to the water when it reaches the ground, but the input of water itself clearly depends upon rainfall. Climatic factors, also, influence drainage basin character. We might guess that drainage density is highest in areas of high rainfall, and to some extent this is true. But it is not invariably so; for three reasons. The first is that it is not simply rainfall, but

and the wide range of possible adjustments that each may make to a particular event mean that it is impossible to forecast in any single situation what the outcome will be.

Streams and drainage basins

Drainage density

So far we have looked mainly at what goes on within the stream channel and its immediate flood-plain. Streams, however, can be viewed in relation to more general systems: as parts of **drainage basins**. The landscape, in fact, is made up of a mosaic of drainage basins, of varying size and shape, each divided by a **watershed** and draining to a separate stream system.

The drainage basin and the streams within it are intimately related, for the basin represents the gathering ground (or **catchment area**) for the water which flows through the stream. Variations in the character of the drainage basin – in its geology, climate or topography, for example – affect the character of the stream. In this context, one of the most important aspects of the drainage basin is the drainage density: the average length of stream

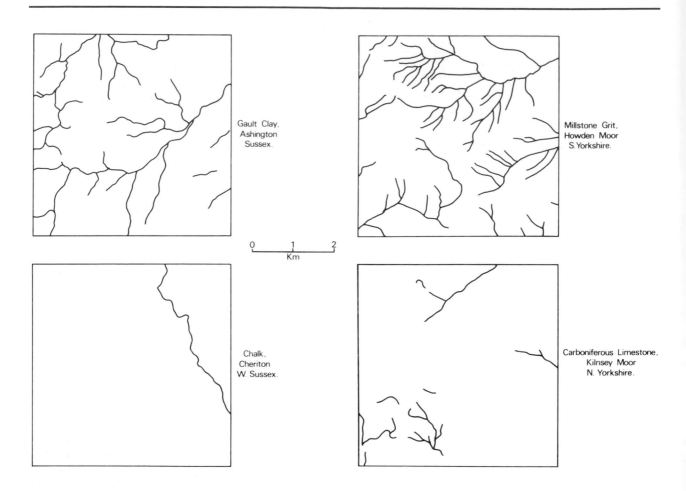

Figure 22.23 Drainage densities on different bedrock lithologies

the balance between rainfall and evapotranspiration which is important. Where evapotranspiration is high less rainfall is available for erosion and channel formation. The second reason is that rainfall and vegetation tend to be closely related, so that areas of high rainfall are often characterized by a dense vegetation cover; as we have seen, the vegetation encourages infiltration and reduces runoff. Finally, it is not so much the total amount of rainfall which is critical, as the intensity of rainfall. Short, intense storms lead to more overland flow than prolonged, light rain, as we discovered in Chapter 12. Thus, the relationships between climate and drainage density are often complex.

Drainage pattern

It is not only drainage density which varies in response to basin characteristics. So, too, does the shape and size of the drainage basin. Drainage basins in areas of high relief, for example, tend to be relatively long and thin; broad, short basins are typical of newly exposed coastal areas. In addition, the **drainage pattern** within the basin varies markedly in relation to factors such as geology, climate and relief. **Trellised** drainage patterns tend to develop where there is strong structural control upon streams, the main channels being aligned parallel to the 'grain' of the country with minor tributaries feeding in at right angles. **Rectangular** patterns evolve where faults or joints guide stream channels, producing a characteristic 'grid-iron' pattern with right-angle bends and a distinct and regular arrangement. **Parallel** patterns are often associated with areas of steep relief or non-cohesive materials (e.g. gullies on spoil tips often adopt this form), while **dendritic** patterns are typical of better adjusted systems on uniformly dipping strata. Finally, **deranged** drainage patterns are found in areas where the original drainage network has been disturbed, for example, by glacial activity. All these patterns are illustrated in Figure 22.24, though it is apparent that relatively rarely do real-world drainage patterns fit exactly the idealized forms!

Development of a trellised drainage pattern
Although we may identify these broad relationships between drainage pattern and basin characteristics, the question remains: how do these systems develop? One answer to this was proposed by W. M. Davis in his theory of the cycle of erosion. He considered a relatively simple situation in which uplift, tilting and then planation of the land (e.g. by the sea) gave rise to a gently sloping surface which cut across rocks of different type (Figure 22.25).

In a very general way, he argued that initially the main stream would flow down the regional slope (the **dipslope**); he called this a **consequent** stream. In time, tributaries would cut laterally along the weaker bands of rock, at right angles to this main channel; these **subsequent** streams would erode rapidly and produce **vales** within the weaker strata; they would also undermine the more resistant rocks to produce a characteristic dip and scarp topography. Further development would lead to the extension of secondary consequents down the dipslope of the surface rocks, while smaller **obsequent** streams would drain the **scarp** slopes. In this way a trellised drainage pattern would emerge.

Development of particularly active subsequent streams by headward extension might result in the **capture** of consequent streams, leaving the old, abandoned valleys through the scarps as **windgaps**. Further down these valleys, the streams which had lost their headwaters would be left as small channels lying within larger valleys. These streams are said to be **underfit**. Given time, lateral migration of the increasingly complex drainage pattern would reduce the whole area to a flat landscape – a **peneplain**.

With a degree of imagination it is possible to see the results of processes very much like this scheme in certain parts of the world. In the Weald of southeast England a broadly trellised drainage pattern exists and characteristic features such as **elbows of capture**, windgaps and underfit streams can be recognized (Figure 22.25). Similar patterns, resulting from the **super-imposition** of drainage from an old sedimentary cover, are seen in the Appalachians. Here, powerful subsequent streams have dissected the mountains into long, parallel ridges, while major consequents cut at right angles through deep watergaps. Many of these valleys are vital routeways into what would otherwise be impenetrable mountain areas.

Although the relationship between trellised drainage patterns and the regional geological structure is often notable, the validity of Davis's theory of drainage development is far from certain. As with his ideas on slope development (see Chapter 21), the processes were never explicitly described. In particular, the mechanisms of river capture at this scale are difficult to understand. It can be imagined, however, that, as the stream extends its headwaters into the divide with an adjacent drainage basin, the area of land available to provide water to the active stream declines. Ultimately, therefore, there must come a point at which headward extension ceases because the divide is too small to generate sufficient runoff for further erosion. How exactly the divide is then breached is not clear, though hillslope processes such as soil creep and rainsplash may help to lower the watershed.

The development of arroyo systems
In recent years, the limitations of the Davisian concept of the cycle of erosion have become apparent. More attention is therefore being given to understanding the **processes** of drainage pattern development. Studying processes on a large scale is difficult, however, so we need, instead, to examine smaller scale stream systems, such as the **arroyos** which characterize many semi-arid areas.

Where vegetation is lacking and where infiltration capacity is low compared to rainfall intensity, water tends to move downslope through the action of sheet-wash. As we saw in Chapter 21, sheet-wash is generally ineffective as an agent of erosion, but the impact of raindrops upon the soil surface may dislodge particles which are then swept downslope by the thin flow of water. Minor unevenness in the surface disrupts the flow, causing concentration at certain points and encouraging turbulence. As a result, the water is able to erode shallow channels called **rills**. These collect water, erode further, become larger downslope as more runoff feeds into them, and ultimately become quite large **gullies** or arroyos.

Over time the gullies extend headward as water scours the steep headwall. Gully capture takes place as stronger gullies erode either headward or laterally and break into smaller, weaker channels. The extra water, diverted into the larger gully, increases the rate of erosion and stimulates its

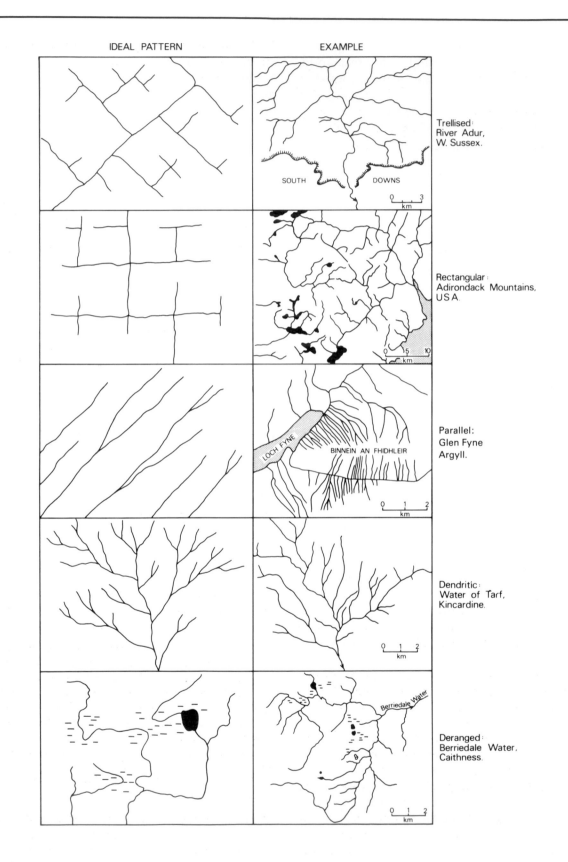

Figure 22.24   Types of drainage pattern: theoretical and actual

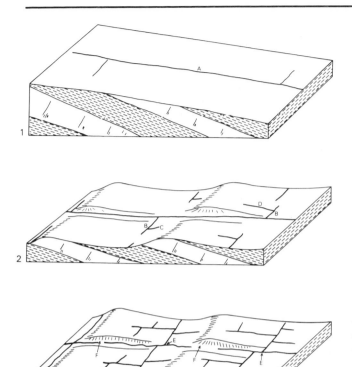

1 Uplift, tilting and planation produce a gently sloping surface down which a consequent stream (A) develops

2 Subsequent streams (B) develop at right angles to this and other consequent streams and erode the bands of weaker rock to produce a vale and cuesta topography. Small obsequent streams (C) cut back into the scarp, while secondary consequents (D) develop on the dip slope

3 Extension of the subsequent streams results in capture of the original consequents and the formation of elbows of capture (E) and wind-gaps (F)

Figure 22.25   Development of a trellized drainage pattern according to the Davisian concept of the cycle of erosion

further extension. Occasionally the gullies are linked or fed by **pipes** in the soil, through which the water flows. As these pipes are scoured they collapse and themselves form gullies. Sometimes the slope of the gully is less than that of the surrounding land, so that the water pours out on to the surface and deposits its load as a fan. Erosion of the gully mouth may also occur in these cases, and the channel may extend downslope until it invades the next gully; in this way **discontinuous gullies** coalesce to form continuous ones.

The developments which take place during evolution of arroyos represent adjustments towards equilibrium. They include both the creation of new gullies, as rills are excavated or headward erosion results in **bifurcation** (splitting) of the main channel, and the loss of gullies either by capture, lateral erosion or a fall in the water table. They include, also, changes in the drainage pattern or geometry. Initially, in the early stages of development, the gullies tend to be separate, and to run parallel to each other downslope. Through the processes of capture and extension, however, this pattern gradually changes. Tributaries grow, and,

root-like, extend across the surface. The drainage pattern becomes **dendritic**.

Fluvial landscapes

Landscapes of fluvial erosion
In 1858 Lieutenant Joseph Christmas Ives entered the Grand Canyon. In his diary he wrote,

The famous 'Big Cañon' was before us; and for a long time we paused in wondering delight, surveying that stupendous formation through which the Colorado and its tributaries break . . . the corresponding depth and gloom of the gaping chasms into which we were plunging imparted an unearthly character to a way that might have resembled the portals of the infernal regions. Harsh screams issuing from aerial recesses in the cañon sides and apparitions of goblin-like figures perched in the rifts and hollows of the impending cliffs, gave an odd reality to this impression.

Without doubt, the landscapes produced by fluvial erosion include some of the most spectacular and awe-inspiring scenery in the world. Without doubt, too, **gorges** and **canyons** such as the Grand Canyon

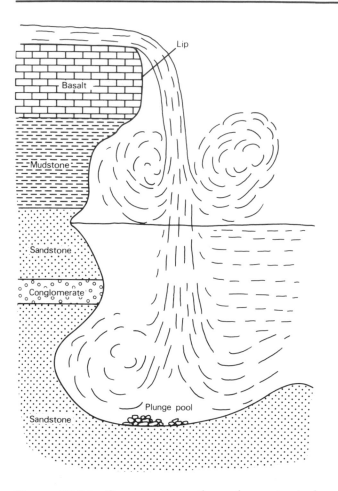

Figure 22.26 Cross-section through a typical waterfall, showing the effects of rock type

them at different rates. The harder rocks themselves produce waterfalls and rapids so that the developing gorge has an irregular long profile. Given time, these irregularities might be removed by the constant action of the water; given time, also, slope processes may act upon the valley sides to reduce them to more gentle forms. But conditions do not always remain constant for long enough to allow this smooth profile to develop. Renewed mountain uplift may stimulate the processes once more; changes in the course of the stream may cause it to abandon its old route, and rapids and waterfalls may be left literally high and dry. What we see in the Grand Canyon is a landscape which is still adjusting to uplift which probably took place some 65 million years ago!

Mountain uplift is not always rapid, and slow earth movements may allow the initial character of the stream system to be preserved. Large, meandering streams once flowing across low-lying landscapes may be slowly rejuvenated so that they cut downwards into the gradually rising land surface. The meanders themselves may be incised; **entrenched meanders** such as those of the Shenandoah and the River Dovey develop.

Throughout geological history the earth's surface has experienced constant uplift, warping and subsidence, while sea-levels have fluctuated in response to changes in the size and shape of the ocean basins and the growth and decay of the polar ice caps. Repeated fluvial incision has therefore occurred. These events have left their mark on almost every landscape; to some degree, we can attribute the overwhelming majority of our landscape to the action of running water. Nevertheless, streams rarely act alone. The valleys they carve are modified by slope processes such as mass movement. Valley side slopes wane and retreat as a result. The ultimate form of the fluvial landscape therefore owes as much – if not more – to the work of slope action as it does to the work of streams.

Depositional landforms

We have already discussed many of the depositional features which occur within the stream channel and the development of the floodplain. In addition, extensive deposition often occurs where rivers disgorge their sediment either into the sea or a lake, or into an open lowland area. In the former case, the deposits build up as a **delta**. This represents, ideally, a triangular shaped feature, which extends from the river mouth. The Nile delta in Egypt is a

of the Colorado, or the Iron Gates on the Danube represent remarkable examples of the power of streams as agents of erosion. But how do landscapes such as this develop?

One reason is mountain building and uplift. As the land surface is raised, the potential energy of the streams is increased; they are rejuvenated and they erode their beds in the attempt to regain equilibrium with their surroundings. Where uplift involves little tilting or warping of the rocks, a more or less flat plateau is formed. Waters flowing across this surface tumble over the margins to create waterfalls and rapids. At the base of the waterfall they erode a deep, smooth **plunge-pool** (Figure 22.26).

Erosion at the plateau edge is active, and the waterfall eats back into the scarp. As it does so it exposes a range of rocks within the valley; some may be soft, others resistant, and the stream attacks

Plate 22.10 An alluvial fan on the flanks of Fjardardalur, Iceland (photo: D. J. Briggs)

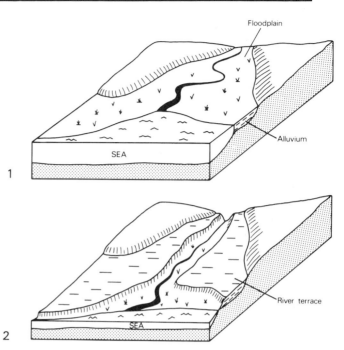

Figure 22.27 Development of thalassostatic river terraces: (1) stream builds up an alluvial floodplain adjusted to sea level; (2) sea-level falls and the stream cuts down through the alluvium to a new level, leaving the old floodplain as a river terrace

good example of this type; the more irregular bird's foot delta at the mouth of the Mississippi is another well-known example. Sediment swept on to the delta by the stream slumps and avalanches down the delta-front to produce steeply-dipping beds of material which slowly encroach into the sea.

Similar features develop subaerially where streams empty on to broad lowland plains. Waters emerging from the mountains cascade down the slope on to the flat pediment area. There they deposit their coarse debris. Over time this builds up to produce an **alluvial fan** (Plate 22.10). The stream, flowing now over this, percolates between the boulders, loses its power to transport its sediment, deposits its load. The fan continues to accumulate, until the sediment supply is cut off or the stream is abandoned, at which stage the fan becomes fossilized.

More familiar to many of us are **river terraces**. Many valleys are flanked by sequences of terraces, which provide flat, fertile land for agriculture and ideal conditions for settlement. As a result they have often acted as sites for the development of towns; Oxford, in the Thames valley, is a fine example.

Terraces develop in a variety of ways. Some are produced by changes in sea-level. A rise in the sea-level causes the stream to deposit its load as it adjusts to the lower slope angle. Aggradation occurs. When the sea-level falls, the stream is

rejuvenated and cuts through the sediment, leaving a bench-like feature (Figure 22.27). This is known as a **thalassostatic terrace** (from the Greek word *thalassa* meaning *sea*).

River terraces are also formed by changes in the supply of sediment and water to the stream. During periods of rapid sediment supply, the streams may become choked with debris, so that, over time, they aggrade their bed. This happened widely in the glacial periods when frost activity encouraged weathering of the rocks and solifluction transported the material down the valley sides. For much of the time the streams were unable to cope with this input of sediment and aggradation occurred. When the climate improved, however, and the waters locked up in the snow and ice cover were released into the streams, discharge increased, and so did the competence of the streams. At the same time, the establishment of vegetation upon the slopes stabilized the valley sides and sediment inputs diminished. The streams eroded their beds and cut through the sediments to produce what are known as **climatic terraces** (Figure 22.28).

wind & water borne deposits which build up the land surface.

361

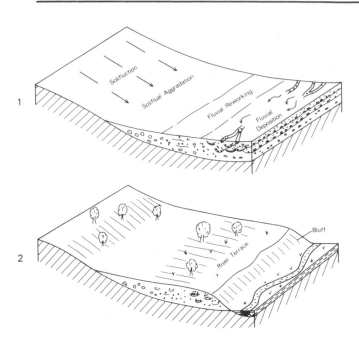

Figure 22.28 Development of climatic river terraces: (1) a braided stream, fed by solifluction and mass movement of hillslope materials in a periglacial environment, builds up a floodplain of coarse debris; (2) amelioration of the climate results in the development of a vegetation cover, stabilization of the hillslopes and the concentration of the streamflow into a single meandering channel. During the relatively short period of climatic change the stream is able to cut downwards, leaving the old floodplain as a river terrace

Streams and man

The relationship between streams and man is intimate. Over the centuries, streams have exerted a fundamental influence upon man's economies and politics. They have helped to fashion the distribution of his settlement; they have acted as barriers to cultural diffusion and as routeways for his trade and movement. They have been both a focus of civilization and a threat to it. In return, man has modified the streams, sometimes deliberately, often accidentally.

We have already discussed the hydrological aspects of this intimate relationship between streams and man (Chapter 14). As we saw then, man derives water from streams and also uses them as a method of removing and diluting his waste products. But the landscapes of fluvial action are also important, as are the geomorphological processes operating within stream channels.

It was to the extensive flat and fertile floodplains of major rivers that prehistoric man was attracted, and it was here that the great civilizations of the world developed: the Babylonians and Sumerians on the Tigris and Euphrates, the Egyptians on the Nile, the Chinese dynasties on the Yellow River. It was not just the presence of water that drew these peoples and permitted them to establish permanent, sophisticated settlements. It was also the agricultural potential of the land; a potential maintained by the silts which the rivers deposited on the floodplain each time they flooded. Even today, a large proportion of the world's population is directly dependent upon this process for their livelihood. Where natural flooding is too irregular or too limited, artificial means of inducing floods are used. Irrigation channels and ditches are cut across the floodplain, and the silt-laden waters are diverted into these, then dammed and allowed to flood the land. In this way, the soil is revitalized by the sediment washed from lands upstream.

Elsewhere, as we have seen, it has been to the drier, higher river terraces that man has been attracted. These, again, provided fertile, agricultural land, especially in the temperate regions where the floodplains were often boggy and inaccessible. They also provided firm foundations of gravel and sand upon which to build, and today these same deposits are being excavated for the construction industry around many major towns.

Today, also, fluvial landscapes attract man for another purpose: for tourism and recreation. Features of fluvial erosion, such as the gorges of the Grand Canyon and the Rhône, waterfalls like Niagara Falls and Victoria Falls, or the rapids of Cascade Canyon are spectacular and beautiful.

But streams are also hazards. The danger of floods we have already noted in Chapter 14. Stream erosion and sudden channel changes may be similarly catastrophic. In 1938, for example, three days of continuous rain resulted in extensive floods in California. The rate of erosion is estimated to have been as much as 287,500 m^3 of sediment per square kilometre, most of it fertile soil from agricultural land. Seven hundred new gullies were formed within an area of about 1600 km^2. The loss to agriculture was devastating.

During events of this type, the stream may

abandon its channel and carve itself a new route. The Yellow River – 'China's Sorrow' – has done so on numerous occasions with terrible cost to human life. It has been estimated that the river has changed course some twenty times during the last 4250 years, and many of these have involved substantial alterations (Figure 22.29). In 1855 a quarter of a million people died as the stream abandoned one course and found a new one. These **avulsions**, as they are known, may occur with surprising rapidity.

The hazards provided by streams are not always entirely natural in origins. Often they derive from the effects of man upon channel conditions and processes. Gullying, for example, induced by over-grazing or vegetation clearance in the steep head-water areas, results in erosion and extension of the drainage network in the upper reaches of the drainage basin; we saw this in the last chapter. But it also delivers extra sediment to the main channels. This may initially encourage erosion of the channel,

but, downstream, the effect is to accelerate fluvial deposition. Over time, changes in land use, involving forest clearance, cultivation and even urbanization, have greatly altered the supply of sediment to stream channels and resulted in periods of aggradation and periods of erosion (Figure 22.30).

As we can imagine, changes in the supply of sediment to channels influence all the characteristics of the stream – its width, depth, velocity, turbulence and profile. Under extreme conditions, as the flume experiments we referred to earlier indicated (Figure 22.19), the input of new sediment may convert a meandering channel to a braided form. But man's effect upon the landscape does not only cause changes in sediment inputs; it also influences the rate and quantity of water supply to the channel. Urbanization, for example, may significantly increase the amount of surface runoff and lead to a marked rise in peak flows. This, too, will cause adjustments in the stream channel.

Changes in channel form produced by human

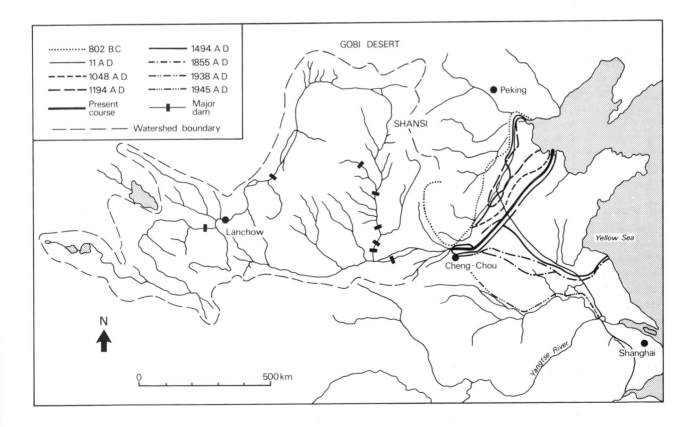

Figure 22.29   Changes in the course of the Yellow River, China since 800 BC (from Tsung-Lien Chan, 1976)

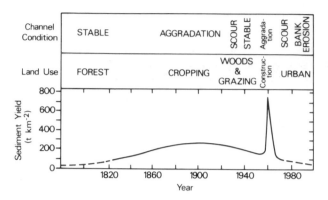

Figure 22.30 Effects of land use change on stream sedimentation in NE USA (from Wolman, 1967)

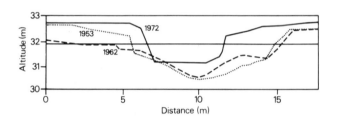

Figure 22.31 Changes in the channel of Watts Branch, Maryland during urbanization (from Leopold, 1973)

impacts have been called **stream metamorphosis**. Changes in the shape of the channel of Watts Branch in Maryland are shown in Figure 22.31. During building operations in the area, the channel became smaller as sediment disturbed during construction was washed into the stream. Subsequently, the capacity of the channel increased as the sediment supply declined, but the rate of surface runoff rose in response to the more extensive area of impermeable urban land. During severe storms, the flood of water from urban land may greatly increase stream erosion, and the channel may adjust accordingly. Reduction of sediment downstream of dams can also increase the energy available for erosion. It has been shown that 9.87 million m³ of bottom sediment were scoured from the channel of the Colorado downstream of the Glen Canyon Dam between 1963 and 1965, before the channel stabilized.

Not all the effects of human interference with stream channels and drainage basins are as dramatic as these examples indicate. Often, the consequences are subtle. Nevertheless, man's use of the drainage basin almost invariably has some impact on the stream system, and frequently these impacts eventually feed back to affect man himself. The need, therefore, is to understand the ways in which stream systems operate and to manage them not in a piecemeal fashion but as integrated parts of whole drainage basins.

23

Glacial and periglacial systems

Distribution and character

Ice covers some 10 per cent of the world's land surface. It provides some of the most hostile and least explored areas of the earth. In Greenland and the Antarctic, for example, the landscape is dominated by almost uninterrupted expanses of ice in the form of huge **ice caps**. Here, temperatures at the ground surface lie below freezing for almost the whole year; polar winds sweep across the surface, driving snow before them until the whole landscape is erased in a white-out; during the coldest months as much as 70 degrees of frost may be experienced. Smaller ice caps occur in Iceland, while in the mountain regions of the Alps, Rockies, Andes and Himalayas spectacular **valley glaciers** carve deep into the rocks.

The effect of ice, however, is not confined to these glacial areas. In the bleak, cold tundra zones which fringe the polar ice caps **ground ice** prevails, underlying a further 10 per cent of the earth's surface. In these **periglacial** areas the subsoil and bedrock are permanently frozen, while at the surface seasonal ice growth and repeated freezing and thawing fracture the rocks and rearrange the material into strange patterns. Here, too, the meltwaters from the ice sheets and glaciers erode and transport the glacial materials; and the barren, unvegetated surface is prey to the scouring wind.

These conditions are not unique to the present. In the past, there have been many periods when the ice caps extended far beyond their present limits, and when periglacial conditions reached across a large area of the middle latitudes (Figure 23.1). We find evidence of past glacial periods in the Permian and Carboniferous tillites (ancient glacial deposits) of South Africa and Brazil, for example. And, more recently, during the Pleistocene Epoch, there were long periods when much of Canada, the northern USA, Britain, Scandinavia and northern Europe lay beneath vast ice sheets. These Pleistocene glacial periods have left a clear imprint on the present

landscape; to understand this legacy of glaciation we need, therefore, to examine the processes and landforms of modern-day glacial and periglacial regions.

The glacial system

Accumulation

Ice sheets and glaciers differ considerably in scale, shape and, as we will see, the detailed processes of movement, erosion and deposition. As a consequence, they are associated with rather different landforms. Nevertheless, at a simple level, both can be thought of as open systems which receive inputs in the form of snow and rock debris and lose outputs through melting and sediment deposition (Figure 23.2).

The main input of snow is normally by direct snowfall on to the ice surface, though small quantities may also be added by freezing of runoff or rainwater, by rime and by blowing of snow from adjacent areas. Winter snowfall may occur over the whole ice surface. Typically, however, summer melting removes the snow cover from the lower and milder areas, so that net accumulation is restricted to higher parts of the glacier or ice sheet. Even in this **accumulation zone** some seasonal melting occurs, but at the end of the summer a residue of **firn** remains which represents the net input to the glacier system. Because of the effects of compaction and partial melting and refreezing, the firn is somewhat denser than the original snow, with a density of about 0.4–0.8 g cm^{-3} compared to 0.1 g cm^{-3} for snow.

Over time, the firn accumulates and undergoes further compaction and recrystallization until it ultimately produces ice. The thickness of firn needed to form ice varies from one area to another depending on the rate of snowfall and melting and the temperature of the snow. On Seward Glacier in Alaska, for example, the firn-ice boundary lies at

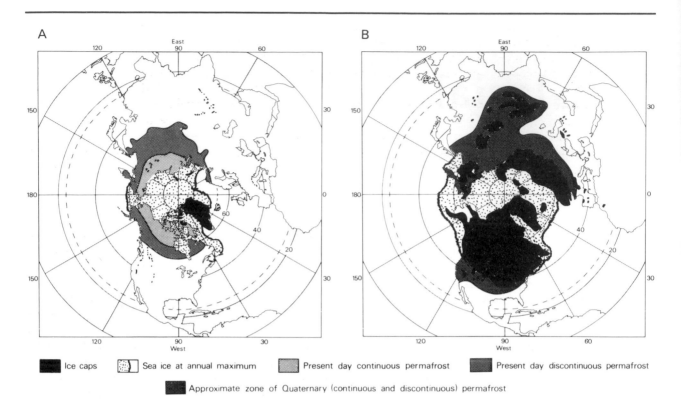

Figure 23.1   The extent of glacial and periglacial conditions at present (A) and during the Quaternary glacial maximum (B) (based on Flint, 1971)

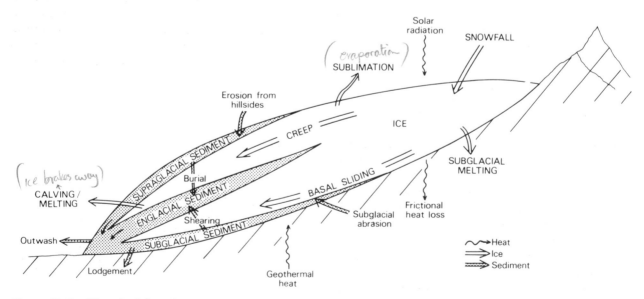

Figure 23.2   The glacial system

only 13 m depth, and ice is formed in about three to five years. In contrast, at the much colder and drier Plateau Station in Antarctica, ice formation takes about 3500 years and the boundary is at 160 m.

Ablation

During the summer, much of the snow which falls on to the ice is lost by **ablation**. This involves three main processes: sublimation, melting and calving.

366

Sublimation is the direct evaporation of the ice to form water vapour. In most cases it is almost negligible, for it requires much more energy to evaporate ice than to melt it.

Melting occurs through a variety of effects. **Surface melting** takes place largely because of the warmth from the sun and atmosphere, though water flowing over the ice surface may also be warm enough to cause some melting. **Subsurface melting** is also caused by flowing water which percolates into the ice and flows along shear planes, crevasses and tunnels. More important in most glaciers and ice sheets is **pressure melting**. This occurs because the weight of the overlying ice reduces the melting point of the ice, at a rate of about 1°C for each increase in pressure of 140×10^5 N m^{-2}. Thus, in places the ice temperature may be above the **pressure melting point**, with the result that melting occurs. Whether or not it does depends on the ice temperature as well as the overburden pressure of the ice. The ice temperature, in turn, is a function of the rate of firn accumulation, for the firn acts to carry cold material into the body of the ice mass. At Byrd Station in Antarctica, for example, pressure melting occurs at a temperature of -1.6°C at the base of the ice, which is 2164 m deep.

Ice temperature and the occurrence of subsurface melting are also influenced by inputs of heat from geothermal sources and by friction. On average, geothermal heat flows into the base of the ice at a rate of about 60 mW m^{-2}, sufficient to melt about 6 mm of ice at pressure melting point each year. Locally, however, geothermal inputs may be much greater. Many Icelandic ice caps, for example, are developed above active volcanoes and, at times, these release vast quantities of heat into the ice, causing massive and often catastrophic melting. The term **jokulhlaup** (= glacier burst) is used to describe the flood of water which these subglacial volcanic events produce. By comparison, release of heat by friction between the bed and ice base is small, though ice flowing at a rate of 20 m y^{-1} produces as much energy as the average geothermal input.

Where glaciers or ice sheets terminate in the sea or in deep lakes, losses also occur through the process of **calving**. Large masses of ice break away as **icebergs** which ultimately melt or evaporate. In the case of ice sheets such as the Antarctic, which occupy permanently cold regions and which are bounded on all sides by the sea, calving is by far the major process of ablation.

The mass balance

Over a period of time, the glacier or ice sheet receives and loses water. The changes in volume of the ice body as a result of these processes are referred to as the **mass balance**. At any point on the glacier, the balance may be negative or positive. Where accumulation exceeds ablation, a **positive net balance** is said to exist; where losses are greater than gains the net balance is negative. Over the ice sheet or glacier as a whole, a negative balance results in shrinkage or retreat while a positive balance causes an expansion or advance of the ice. Within the glacier or ice sheet, however, a consistent pattern tends to exist with a positive balance in the higher, accumulation zone and a negative balance in the lower, ablation zone. These two zones are divided by the so-called **equilibrium line**, where the net balance is zero. Glacier flow maintains the system by transferring the excess ice from the accumulation to the ablation zone (Figure 23.3).

The areas of the accumulation and ablation zones are not fixed, but vary from year to year in response to fluctuations in snowfall and temperature. Thus the position of the equilibrium line also varies around a mean position. Similarly, the mass balance of the glacier may vary from year to year as a result of annual climatic variations. Figure 23.4, for example, shows the mass balance of Storglaciären in Sweden and, as can be seen, there have been some years with a positive balance although the overall tendency is for a negative net balance.

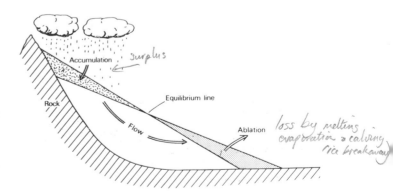

Figure 23.3 The ice budget of a glacial system. In the upper part of the glacier net accumulation takes place; in the lower zone net ablation occurs. The equilibrium line occurs where ablation and accumulation are in balance. Glacial flow maintains the balance either side of the equilibrium line and maintains an equilibrium surface profile

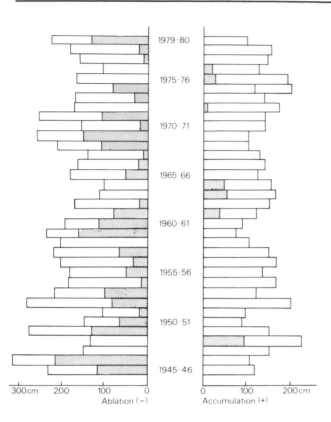

300cm 200 100 0   0 100 200cm
Ablation (−)   Accumulation (+)

Figure *23.4* Mass balance of Storglaciären, Sweden, 1945–80. The net balance (positive or negative) is shown by the shaded area (data from Schytt, 1981)

In the longer term, progressive changes in ice volume may occur due to a bias towards a negative or positive balance. This was seen on a grand scale in the Quaternary, when there were periods of several thousands of years during which positive net balances predominated and ice sheets extended far into what are now temperate regions. Subsequently, the mass balance became negative for long periods and, during the interglacial phases, the ice sheets retreated. The reasons for these large-scale variations in ice-sheet distribution were discussed in Chapter 10. It is worth pointing out, however, that expansion and retreat of ice sheets tend to occur at different rates depending on the source of imbalance in the glacial system. Where imbalances occur due to a change in snowfall amount (i.e. in accumulation rate), a considerable time period may elapse before the effect is felt at the ice margin, for the change has to be transmitted through the length of the glacier or ice sheet. Conversely, when the imbalance is due to a change in

the ablation rate, the response is felt immediately at the ice margin and very little time lag is experienced. In other words, the position of the ice margin is much more responsive to fluctuations in ablation rate than it is to variations in accumulation rate.

Ice movement

The mechanisms of ice movement are complex. It was once thought that the pressure of the overlying ice squeezed the basal ice outwards – rather like toothpaste out of a tube – a process known as **extrusion flow**. This is now known to be false. Instead, glacier flow is seen to involve three main processes: creep, basal sliding and fracture.

Creep takes place due to the pressure of the overlying ice and the gravitational forces related to the ice gradient. Because ice acts rather like a plastic, it does not deform until the combined stresses reach a critical level; then it deforms continuously. Deformation involves the movement of individual ice crystals relative to each other. This process is aided by the partial melting of the ice at depth for the water lubricates the crystal surfaces and enables them to slide past each other. Rates of creep vary through the ice mass, but are normally greatest at the base where pressure melting provides water to lubricate the crystals and where the stresses from the overlying ice are greatest.

Water also plays a major part in **basal sliding**. This process is due in part to the slippage of ice over a thin water layer at the ice–bedrock interface. The water reduces friction between the rock surface and the ice and thus allows the glacier to slide under the influence of the gravitational and overburden pressures. Locally accelerated rates of basal sliding tend to occur wherever basal melting of the ice takes place. One factor which encourages this is the presence of irregularities on the bedrock surface. Increased pressure on the upslope side of small obstacles (less than about 1 m in height) reduces the pressure melting point of the ice and causes melting. The water lubricates the base of the ice and facilitates flow across the obstacle. Larger obstacles inhibit basal sliding, but result in **enhanced basal creep** due to locally increased stresses.

Because of variations in bedrock topography and discontinuities in the rate of ice flow, considerable internal stresses may develop within the glacier which cannot always be accommodated by gradual deformation. Under these conditions, the ice may fracture along clearly defined **shear planes** (Plate

Plate 23.1 Shear planes in glacier ice: Gigjökull, Iceland (photo: D. J. Briggs)

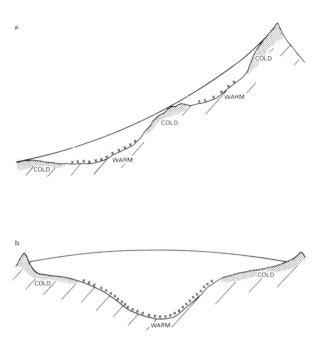

Figure 23.5 Distribution of warm- and cold-based ice in a glacier in response to variations in ice thickness: (a) in longitudinal section down a glacier; (b) in plan section across a glacier or ice-cap covering a pre-existing valley. Crosses show zones of pressure melting; stippled areas show where the glacier bed is subject to permafrost.

23.1). These are typically concave upwards and result in the transport of basal ice towards the surface of the glacier. They are most common where the ice is relatively thin and where creep is limited.

Rates of ice movement

Together, the processes of creep, basal sliding and fracture account for the movement of ice from the accumulation zone to the ice margin. As we have seen, however, rates of movement vary within the ice mass due to differences in the stresses to which the ice is subject and to the availability of water. The role of pressure melting in providing water is fundamental and results in marked differences in flow rate between glaciers which are warm enough at the base to experience enhanced creep and sliding and those which are frozen throughout their depth. The former are called, somewhat euphemistically, **warm-based** glaciers; the latter are referred to as **cold-based** glaciers. In reality, these terms are slightly misleading, for it is not always the whole glacier which is warm- or cold-based. As noted earlier, pressure melting occurs largely as a function of ice thickness. Variations in ice thickness within a single glacier, therefore, result in transitions from warm-based to cold-based ice (Figure 23.5).

The implication of what has been said is that creep and sliding are both most active at the base of the ice where pressure melting and ice stresses are

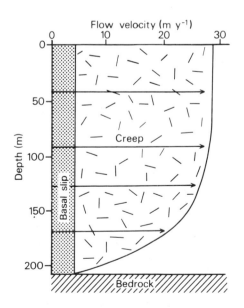

Figure 23.6 The distribution of flow velocity with depth in the Athabasca glacier, Canada (after Sugden and John, 1977, data from Savage and Paterson, 1963)

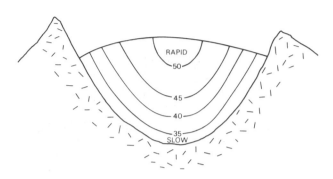

Figure 23.7 Cross-sectional distribution of flow velocity in a glacier; lines show bands of equal flow velocity (approximate velocities are given in m yr^{-1})

at a maximum. It may seem paradoxical, therefore, to learn that rates of ice movement are greatest at the surface of the glacier (Figure 23.6). The reason, however, is quite simple. Although the amount of creep declines towards the ice surface, the total rate of flow increases because the ice is being carried along by the movement of the underlying layers: the small amount of internal deformation is additional to the basal flow. Movement is also restricted at the sides and base of valley glaciers by friction with the rock wall (Figure 23.7).

Similar variations occur along the length of a glacier. In the accumulation zone, flow tends to increase downslope due to the progressive increase in stress. In this area flow is said to be **extending**. In contrast, flow in the ablation zone is said to be **compressive**, for flow rates decline downslope as the ice spreads out on to the lowlands and loses volume by ablation. In addition, local variations in flow rate occur as a result of local differences in ice slope, bedrock topography and ice thickness (Figure 23.8). These variations encourage the development of **crevasses** in the ice. **Transverse crevasses** form where the ice is subject to tension due to obstructions in the floor or changes in bedrock slope. **Longitudinal crevasses** are produced by differential flow rates (e.g. where friction at the valley wall retards flow) or where the ice fans out as lateral constraints are removed (Figure 23.9).

Marked changes in flow rate also occur over time. To a great extent these changes reflect differences in the mass balance of the glacier; increased snowfall or ablation, for example, both lead to imbalances between the amount of ice in the accumulation zone and the amount in the ablation zone. As a result, more ice tends to be transferred down glacier, causing major **surges** at the glacier margin. Rates of flow may be increased 10–100 times during the surge, and the glacier may advance at a phenomenal rate: in 1963–4, the Bruarjökull in Iceland advanced up to 8 km at speeds of as much as 5 m h^{-1}.

Because of these variations, it is difficult to

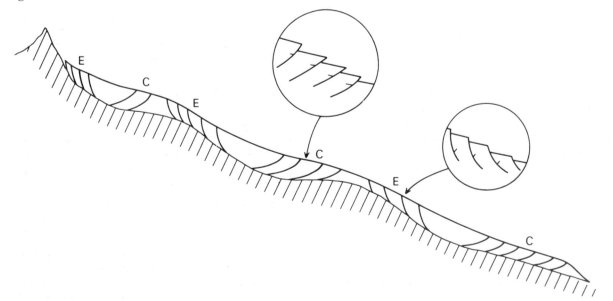

Figure 23.8 The pattern of extending (E) and compressive (C) flow in a glacier and its effect on the development of shear planes. Note that compressive flow results in upward shearing of the ice; extending flow encourages slippage

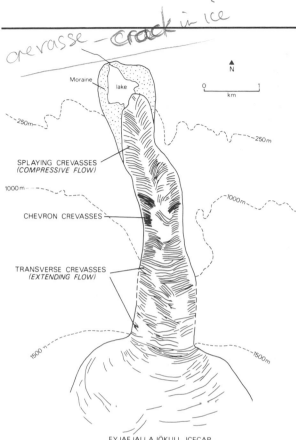

crevasse – crack in ice

Figure 23.9 The pattern of crevasses on a valley glacier: Gigjökull, Iceland (from J. D. Hansom, pers. comm.)

Plate 23.2 A dirt cone on Breiđermerkurjökull, Iceland, showing the ice core and drape of debris. The cone has partially collapsed into a crevasse (photo: D. J. Briggs)

generalize about glacier flow rates. Well-lubricated, warm-based glaciers on the west coast of Greenland, however, have average flows of 4–5 m d⁻¹, whereas glaciers in the Alps commonly flow at rates of about 0.1 m d⁻¹. Modern ice sheets are almost all cold-based and have even lower flow rates, generally in the order of only 0.01 m d⁻¹. The same is not necessarily true of ice sheets during the Pleistocene, however, for the vast thicknesses and the different climatic conditions probably led to much greater internal melting and much larger flow rates.

Glacial erosion

Processes of erosion

One of the surprising things about many glaciers is their dirtiness. Far from being clear and clean, as we might expect, the ice is often cloudy with fine sediment, while marked concentrations of debris are found along shear planes and the surface of the ice is covered with piles of material (Plate 23.2). If we are able to get into the ice close to its base we also find large blocks frozen in the ice mass. All these materials represent the sediment load of the glacier. But how did it get there?

Detachment processes

As with any other agent of transport, we can think of glacial erosion as a three-stage process involving detachment, entrainment and transport of debris. Detachment is a result for the most part of preglacial processes such as chemical weathering, dilatation (pressure release) and frost shattering. These processes loosen the bedrock and provide debris which can be relatively easily picked up by the glacier. In addition, detachment occurs through the process of abrasion. Debris held in the base of the glacier grinds against the bed, prising away small fragments of rock which add to the basal load. The efficiency of this process is determined primarily by the concentration of debris in the base of the glacier,

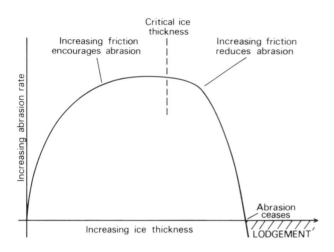

Figure 23.10   The effect of ice thickness on rates of abrasion and lodgement of till (from Sugden and John, 1977)

the velocity of ice movement and the thickness of the ice. Clearly the amount of abrasion tends to increase as basal sediment concentration and ice flow increase, for a greater area of the bedrock

surface will be affected in any period. However, over time the abrasive capacity of the ice declines unless new material is being added to the base of the glacier, for the particles protruding from the ice are gradually worn by attrition. Renewal of the basal load occurs both by entrainment of new debris and by the down-melting of material through the ice, towards the bed. Abrasion also tends to increase with mounting ice thickness due to the increase in applied pressure. Beyond a critical thickness, however, the relationship is reversed: first, because pressure melting may produce a sufficiently deep layer of basal water to buoy up the glacier and reduce the extent of contact with the bed; and second, because very high pressures at the base of the ice may cause particles to lodge in the bed, increasing friction, reducing ice velocity and ultimately causing deposition of the basal load (Figure 23.10).

Detachment may also occur at the glacier bed by processes of dilatation and freeze–thaw. The idea of pressure release operating beneath a huge glacier may seem unlikely, but it is theoretically possible. As the ice cuts into the bedrock it ultimately reaches

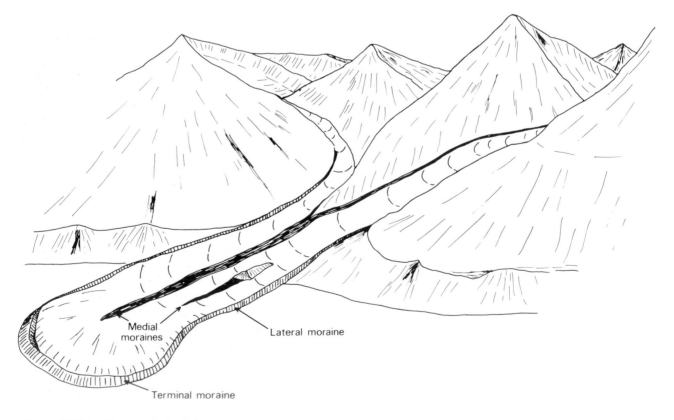

Figure 23.11   Forms of glacial moraine

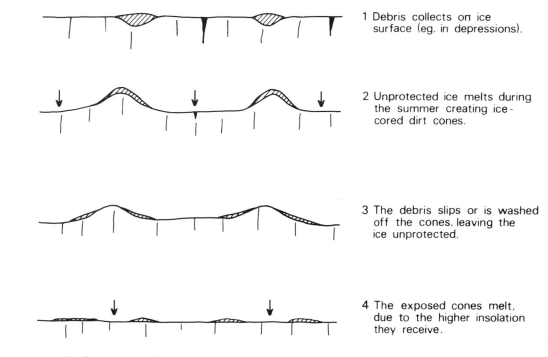

1 Debris collects on ice surface (eg. in depressions).

2 Unprotected ice melts during the summer creating ice-cored dirt cones.

3 The debris slips or is washed off the cones, leaving the ice unprotected.

4 The exposed cones melt, due to the higher insolation they receive.

Figure 23.12 The formation of dirt cones

a position where the weight of material removed is greater than the weight of the overlying ice. From then on, the rocks are subjected to progressive unloading as erosion continues, with the result that dilatation occurs, opening up joints and bedding planes and encouraging detachment. Freeze–thaw operates only beneath warm-based glaciers, where pressure melting and percolation of surface melt-waters provide water which is able to flow into joints in the bedrock and refreeze. Because pressure melting only occurs where the ice is relatively thick, freeze–thaw is probably most active beneath the main body of the glacier and declines in significance towards the ice margin.

Entrainment — pickingup & settingdown of particals.

Entrainment involves a large number of processes. Blocks falling on to the glacier surface from the valley walls or carried on to the ice by streams or wind are automatically entrained by the moving ice. Basal entrainment also occurs as a result of traction and regelation. Ice sliding across a bedrock surface exerts a tractive force on material on that surface and, if this is sufficient to overcome the frictional resistance of the debris, entrainment takes place. This process seems to account for most of the

coarse debris in the base of the ice.

Regelation is probably most effective in warm-based ice, where water produced by pressure melting refreezes to create a **regelation layer**. Fine debris caught up within this layer may be plucked from the surface and carried along by the ice.

At the ice front entrainment also takes place through the ploughing action of the snout and through squeezing of material from beneath the ice. Both processes are most effective when the ice front is advancing over relatively soft or saturated materials.

Transport

Once entrained, glacial debris is transported in three main ways: as surface (**supraglacial**), internal (**englacial**) or basal (**subglacial**) material. Debris falling on to the glacier from the valley walls commonly forms a long ridge of material at the glacier edge, known as **lateral moraine**. Where two or more ice streams merge, these moraines may coalesce in the centre of the glacier to form a **medial moraine**. Similar features develop where ice diverges and then rejoins around a rock outcrop (Figure 23.11).

Debris also reaches the surface through upward

shearing of the ice and by downwasting which releases englacial material. These sediments typically accumulate as a thin layer of sediment which acts to insulate the underlying ice. As a result, beneath the debris, summer melting is inhibited and a small mound or cone of ice is produced, veneered with sediment. This is known as a **dirt**

Plate 23.3 Striations on a glacially eroded rock surface, Llyn Peris, North Wales. The striations were formed during the Last (Devensian) Glaciation (photo: J. D. Hansom)

Plate 23.4 A glacier groove in basalt, near Seydisfjordur, Iceland (photo: J. D. Hansom)

cone (Plate 23.2). Eventually, the sediment washes or slips off the cone and the exposed ice melts (Figure 23.12).

Englacial debris originates from a number of sources. Some reaches the ice interior along shear planes. Other material is washed into the ice down crevasses. Over time, some material works its way downwards through the ice as a result of localized melting and regelation, while debris may also be buried by overriding ice. Large volumes of englacial debris are concentrated by meltwaters in ice tunnels.

The basal debris of a glacier comprises both material frozen into the ice and debris which is being dragged along beneath the ice as a result of tractive forces. Traction of subglacial material is particularly effective where the ice is crossing relatively loose sediments or plastic bedrocks such as clay, and may extend to a considerable depth. The effects are shown by deformation of the underlying strata into complex folds.

Landforms of glacial erosion

The processes of glacial erosion commonly leave a clear imprint on the landscape. At the detailed scale, particles caught in the glacier base cut gouges in the bedrock, referred to as **striations** (Plate 23.3). **Chatter marks** may occur where rocks protruding from the ice have prised thin flakes or chips from the bedrock surface. In addition, the continual wear of the rock produces a smooth, polished surface. Wear is often accentuated along pre-existing joints and concentration of the abrasive action of the debris in this way may create **glacial grooves** (Plate 23.4). In contrast, detachment of blocks by regelation and freeze–thaw processes often creates angular, plucked surfaces. These are particularly common in the lee of obstacles protruding into the base of warm-based ice.

On a larger scale, glacial erosion generates a wide range of landforms. These vary depending on whether or not the flow of ice is confined by the topography. Where it is, as in valley glaciers, **linear erosion** occurs; where it is unconfined, as in ice sheets or **piedmont glaciers** (glaciers fanning out on to lowlands at the end of a valley), **areal scour** tends to take place.

Landforms associated with valley glaciers

Valley glaciers generally originate in distinct, basin-shaped depressions called **corries** or **cirques**. These develop through the accumulation of semi-

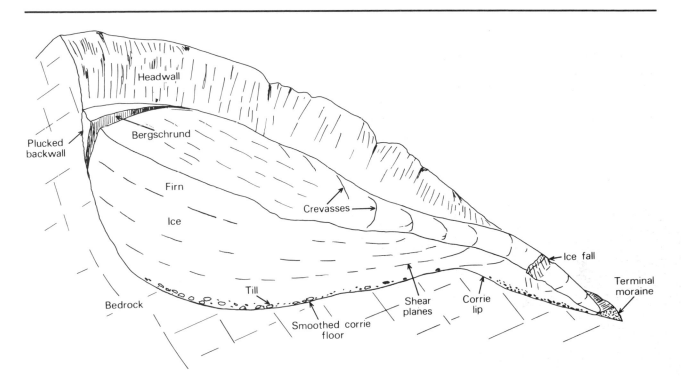

Figure 23.13 Cross-section through a corrie glacier

Plate 23.5 A glacial arête: Striding Edge, Cumbria (photo: P. A. Smithson)

permanent snow patches. Freeze–thaw action beneath, and at the edge of, the snow gradually excavates a shallow depression called a **nivation hollow**. This triggers off a positive feedback cycle, encouraging the survival of the snow through the summer, intensifying and prolonging the effects of freeze–thaw, and leading eventually to the formation of a deep basin. Under conditions of deteriorating climate, snow builds up in the basin, is converted to firn and ultimately to ice. Rotational flow of the ice abrades the corrie floor, accentuating the basin shape, and creating a shallow **corrie-lip**. Meanwhile, frost action and regelation on the backwall lead to the development of steep, angular rock faces (Figure 23.13). Eventually, the ice may spill across the lip and start to flow downslope, at which time it becomes a true valley glacier.

Corries tend to develop just above the snowline, with the result that in any area distinct levels of corrie formation can be seen. These corrie levels are not necessarily horizontal over any great distance, however, for the height of the snowline varies in relation to aspect and distance from the sea (continentality). In the Alps, therefore, corrie levels tend to be lower on north-facing slopes than on south-facing slopes. In the Faroes, corrie levels rise progressively eastwards due to the gradual increase

Plate 23.6 Gigjökull, Iceland: an outlet glacier from an ice cap centred on the crater of a volcano. Note the crevasse patterns and the push moraine on the left of the ice front. See also Figure 23.9 (photo: D. J. Briggs)

in continentality and the reduced snow supply during the Pleistocene. Changes in the level of the snowline also occur over time due to variations in climate, so that several different corrie levels may

exist in a single area, each representing development at a different period. Two levels of corrie formation can be detected on the Dingle Peninsula in south-western Ireland, for example, reflecting a lowering of the snowline as the climate deteriorated during the last glacial period. As the climate improved at the end of the Devensian Glaciation, however, the snowline rose again, the lower corries were abandoned and the higher ones were briefly reactivated.

Numerous features are found in association with corries. Where several corries are cut into a single mountain, steep-sided, jagged peaks called **horns** are created. Many Alpine mountains, such as the Matterhorn and Mont Blanc, provide examples. The narrow ridges between adjacent corries are known as **aretês**: Striding Edge on Helvelyn is a fine example (Plate 23.5).

Not all valley glaciers, it should be noted, form in corries. Some, such as Gigjökull in Iceland (Plate 23.6), are **outlet glaciers**, formed by ice spilling over the edge of a plateau from an ice cap. None the less, however it has originated, as the glacier moves downslope it abrades the bed and gradually over-deepens the valley, eventually producing a deep **glacier trough**. These are often broadly U-shaped in cross-section and are thus referred to as **U-shaped valleys** (Plate 23.7). Often, the valleys are relatively

Plate 23.7 A 'U'-shaped valley: High Cup Nick, Cumbria. The valley has apparently been over-deepened by a glacier flowing *up*-valley from the vale, to create an intrusive trough (photo: D. J. Briggs)

Plate 23.8 A glacial stairway: north of Dundas, Southern Ontario, Canada. Ice-flow was towards the camera (photo: A. Straw)

Plate 23.9 Geirangerfjord, Norway (photo: J. Allan Cash Ltd)

Plate 23.10 A roche-mountonée, Fimmvörđuháls, Iceland. Note the steep, plucked downstream end and smooth, gently sloping upstream end of the feature. Ice-flow was from right to left (photo: D. J. Briggs)

straight in plan, for ice attacks and **truncates** any spurs which extend into the flow. Variations in rock hardness result in different degrees of basal scour, so that the long profile may be stepped in the manner of a **glacial stairway** like that in the Yosemite Valley in California (Plate 23.8).

The degree of valley erosion is a function, among other things, of ice depth with the result that large valleys tend to be deepened more than small ones.

Tributary valleys which carry only small glaciers may therefore be left as **hanging valleys** perched on the hillside. Locally increased abrasion, or excavation of softer rocks in the valley floor, similarly produces **rock basins**. These vary from a few metres to many kilometres in length and, after the ice has melted, they often form lakes. Many of the lakes in the English Lake District are believed to have formed in this way.

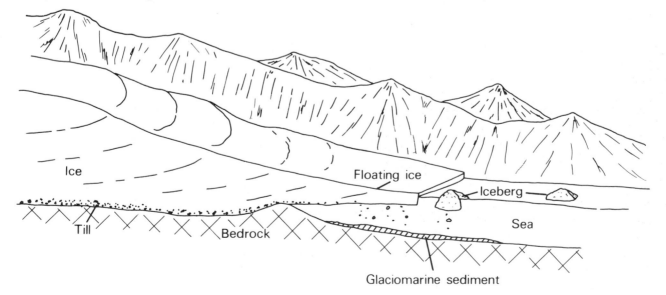

Figure 23.14 Formation of a fjord by floating ice

Plate 23.11   Areal scour seen from the summit of Suilven, NW Scotland (photo: J. D. Hansom)

Plate 23.12   Lodgement till exposed in a stream bank in Derbyshire. Note the great range in sediment size and the lack of bedding (photo: J. D. Hansom)

Similar variations in the intensity of basal erosion occur where valley glaciers enter the sea. The sea acts to buoy up the ice, gradually reducing its erosive power so that the valley becomes progressively shallower. Ultimately, where the water is deep enough to allow the ice to float, erosion ceases and a small lip is formed (Figure 23.14). Subsequently, when the ice melts and sea level rises the valley is partially flooded to create a **fjord**. Spectacular examples are seen on the Norwegian and Scottish coasts (Plate 23.9). These form excellent harbours because the shallow lip and long, deep inlet provide sheltered waters which do not readily freeze over in the winter.

Within the glaciated valleys, various smaller erosional forms are produced. Hillocks are smoothed and streamlined parallel to the ice flow to form what are called **whalebacks**; resistant rock outcrops are shaped by abrasion into **rock drumlins**; smaller mounds are smoothed on the upslope (stoss) side and plucked on the downslope (lee) side to form **roches moutonées** (Plate 23.10).

Landforms associated with ice sheets
To a great extent, the landforms created by ice sheets are similar to those associated with valley glaciers: striations, grooves, roches moutonées, rock drumlins, whalebacks and rock basins are all formed. Even glacial troughs may be produced where erosion is concentrated along pre-existing valleys or areas of weaker strata. Nevertheless, the broader character of the landscape formed beneath ice sheets is highly distinctive. Areal scour operates where the ice is thick enough to cause pressure melting, producing extensive, smooth, abraded surfaces (Plate 23.11). Where the ice is thinner, erosion is limited and is affected to a much greater extent by pre-existing relief. Pressure melting occurs only in valley bottoms, where the ice is deep, and these are over-deepened to form troughs. Elsewhere, basal ice temperatures are low, movement is slight and erosion may be negligible. As a result, interfluves may be almost unmarked and relatively delicate pre-glacial features such as tors may remain intact (Figure 23.15).

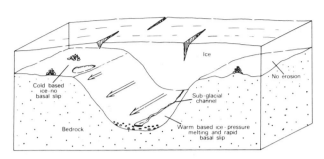

Figure 23.15 The effect of ice depth and basal thermal conditions on sub-glacial erosion

Depositional processes and landforms

The stripped and scoured relief of glaciated uplands testifies to the massive erosive power of glacial ice. But this is only half the story of glacial impacts on the landscape, for all the debris that has been scraped, chipped and plucked off the surface is available for deposition.

The deposits laid down by ice can broadly be classified into three, depending upon the environment and agent of deposition. **Tills** represent the materials deposited directly from the ice. **Glacio-fluvial** deposits are laid down by meltwater flowing beneath or from the ice. **Glacio-lacustrine** deposits accumulate in lakes ponded up at the edge of the glaciers. Each of these deposits produces its own, distinctive landforms.

Tills

Tills are formed both beneath the active ice and as a result of down-wasting. It is, of course, difficult to investigate processes operating beneath glaciers, but it appears that several processes are important. Pressure melting releases fine material which accumulates on the bed. In addition, friction between the debris in the glacier base and the bedrock results in particles being plastered on to the underlying surface. Shearing of ice may also occur so that blocks of ice and contained debris become

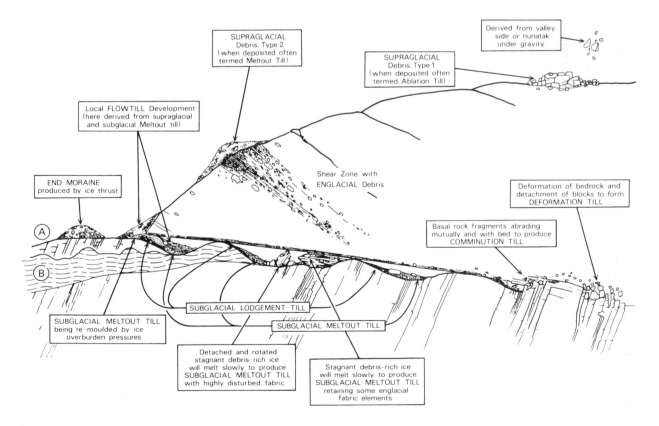

Figure 23.16 Types of till in a warm-based glacier. Two cases of terminal moraine formation are illustrated: at A the glacier creates a ridge of till as it slides over bedrock; at B thick, saturated till is squeezed out from beneath the glacier snout (from Derbyshire *et al.*, 1979, after McGowan and Derbyshire, 1977).

Plate 23.13 Annual moraine in front of Solheim-jökull, Iceland. The moraine is ice-cored and covered with ablation debris left as the ice melted back during the summer. Overall, however, the glacier is advancing; hence, the steep ice-front (photo: D. J. Briggs)

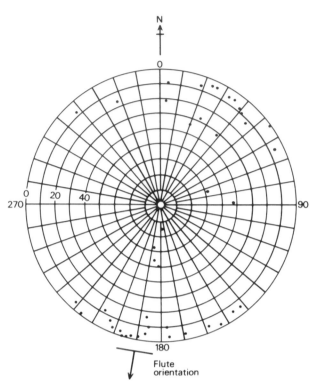

Figure 23.17 Till fabric from fluted moraine, Norway. Each dot shows the angle of dip (on the radius) and long axis orientation (on the circumference) of a rock clast. The preferred orientation of the clasts is clearly almost parallel to the orientation of the flute (from Briggs, 1976)

detached from the glacier and lodge on the bed, while fine, saturated material may be squeezed into cavities beneath the ice (Figure 23.16). The deposits that accumulate in these ways are known as **lodgement till** and they are typically compact, unbedded and ill-sorted, ranging in size from fine clays to large boulders (Plate 23.12). Shearing and plastic flow of the debris due to the pressure of the ice may create shear planes and tend to orientate the larger clasts parallel to the direction of flow. Indeed, the preferred orientation of the clasts in lodgement till (the **till fabric** as it is known) can be used to deduce the direction of ice movement (Figure 23.17).

Lodgement till may be shaped into a variety of landforms. Where the ice is relatively thick and actively flowing, extensive, more or less flat **till-plains** may form. Differential stresses within the ice, however, mould the till into streamlined features such as flutes and drumlins. **Flutes** vary in height from a few centimetres to several metres and may be between a few metres and many kilometres in length. Exactly how they form is not clear, but it appears that they are often initiated by boulders which protrude into the ice base and produce an elongated ice cavity. Material is squeezed laterally into this cavity by the weight of the ice to form a small ridge (Figure 23.18).

Drumlins are similarly enigmatic features, and it has long been debated whether they are a result of differential deposition or erosion of the till materials. Typically they are between 10 and 50 m in height and from 100 to 1000 m in length. Their stoss face is relatively steep and blunt, while they taper gradually in the direction of ice flow. Commonly they occur in swarms, and often seem to be associated with areas where the ice is fanning out on to a glaciated lowland (piedmont glaciers). The debate about their formation continues, but modern ideas suggest that drumlins are a product of processes similar to those responsible for flutes, particles migrating into areas of low ice pressure such as cavities. Gradual closure of the cavities by internal deformation of the ice may account for the tapered form of the drumlin.

Not all subglacial till landforms are streamlined. Some, such as so called **ribbed** or **Rogen moraines**, are aligned at right angles to flow. They occur as

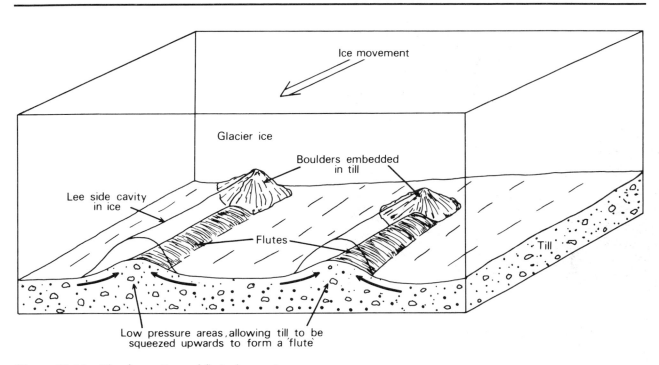

Figure 23.18 The formation of fluted moraine

long, rather low ridges, spaced at more or less regular intervals. In Quebec, for example, ridges are 10–30 m high, over 1 km in length and about 300 m apart. It has been suggested that they are associated with compressive flow just behind the glacier snout and they may result from pushing of blocks of till upwards into the ice due to the development of compressive shear planes (Figure 23.19).

Ice-marginal features
Moraines also accumulate at the ice margin. These are largely due to the flowage and slumping of debris off the ice surface, often lubricated by melt-waters. Seasonal melting also contributes englacial debris while there is some evidence that subglacial debris may be squeezed out at the ice front. As the ice margin retreats due to summer melting, these materials are left as small ridges known as **annual**, **washboard** or **de Geer moraines** (Plate 23.13). If the ice margin remains static (apart from minor seasonal variations) for a long period, these may build up into quite large **terminal moraines**. Similarly, when the ice overrides its own deposits the debris may be bulldozed into a **push moraine**. Internally, this is characterized by intense folding and shearing.

All these moraines are often **ice-cored** in that large, stagnant blocks of ice are left beneath the till. As the ice melts and the debris is let down, further deformation occurs. Over time, gradual retreat of the ice front due to climatic change, interspersed with stillstand phases or periodic readvances of the ice, may lead to sequences of moraines, decreasing in age towards the ice front. In the foreland of the Herbert Glacier in Alaska, for example, over twenty moraines can be identified marking different ice front positions over the last 200 years.

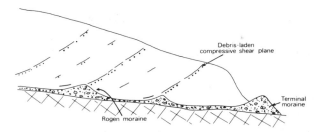

Figure 23.19 The formation of Rogen moraines. Compressive shear planes in the ice (e.g. behind an area of relatively 'dead' ice at the snout) allow particles to migrate upwards, creating a transverse ridge of debris on the glacier bed

Plate 23.14 A sandur plain, Kaldalon, NW Iceland (photo: J. D. Hansom)

Glacio-fluvial and glacio-lacustrine landforms

Many hundreds of kilometres from their sources, meltwater streams draining from glaciers are distinctive: turbid and milky white with the fine sediment they carry. Suspended sediment concentrations of as much as 3800 mg l^{-1} have been measured from the Evdalsbreen glacier in Norway. Closer to the ice margin vast spreads of coarser debris may be deposited. Glacio-fluvial processes and landforms are therefore important components of the landscape.

Not all the sediments carried by meltwaters are deposited in the glacial foreland; some accumulate

Mass-wasting features

Mass wasting of ice produces complex till deposits and a wide range of often apparently haphazard landforms. Medial and lateral moraines, for example, are let down on to the glacier bed as the ice melts, while, more generally, the ablation till on top of the ice and the englacial debris are laid down as **meltout till**. This material is highly variable in character, depending upon its origin, but is often crudely sorted due to the effect of meltwaters during wastage. In addition, **flow till** is deposited by debris slumping and flowing off the stagnant ice surface and accumulating in depressions between the decaying ice blocks. Compared to the lodgement and meltout till, this is often well-sorted and shows evidence of crude bedding.

Overall, glacier wastage tends to result in a tripartite sequence of lodgement, meltout and flow till (Figure 23.20). The surface form of these tills is frequently irregular and chaotic, with depressions (**kettle-holes**) left where buried ice blocks subsequently melted, and moraine ridges and mounds where meltout and flow tills built up to greater thicknesses. These landscapes are often called **hummocky moraine** or **dead-ice topography**.

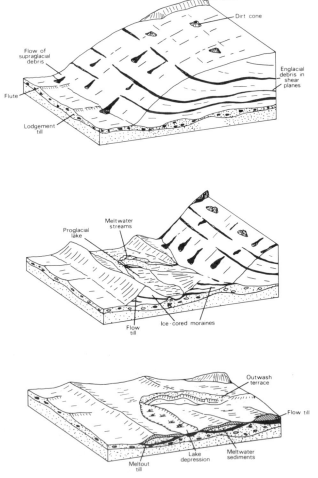

Figure 23.20 The formation of a glacial depositional landscape

within or beneath the active ice. In the former case, debris may build up in tunnels or crevasses in the ice, ultimately to be let down on to the bed as the glacier melts. In the process, the meltwater deposits may lose many of their original bedforms and are often left as relatively ill-bedded mounds or ridges of gravel and sand. Subglacial deposition occurs mainly in tunnels which are kept open by high rates of meltwater flow. The debris which builds up in these is generally coarse, the finer material being swept away.

Deposition of material in englacial or subglacial tunnels occurs mainly where flow velocity is reduced by, for example, changes in the shape of the tunnel, or where extra debris is released into the flow from a crevasse. In addition, as meltwater emerges from the ice it loses much of its tractive power and dumps its coarser sediment.

All these depositional situations are closely interrelated, so that it is not always easy to distinguish between materials formed in different environments. **Eskers**, for example, are sinuous ridges of gravel and sand, which may form subglacially, englacially and proglacially (Figure 23.21). Similarly, **kames** are variable features produced by meltwater deposition from a disintegrating ice

sheet or glacier in a wide variety of situations: depressions, cavities or lakes formed above, within, beneath or at the edge of the ice. Indeed, kames are so variable in form and genesis that the term is of doubtful validity. **Kame terraces**, however, are of rather more distinctive origin. They are produced by meltwaters flowing along the glacier margin, confined between the ice and the valley side. When the ice melts the deposits are left as more or less continuous, terrace-like features on the hillside.

In many cases, lakes also form at the glacier margin, impounded by the valley side or by moraines. Meltwaters pouring into these lakes deposit their coarse debris as deltas, while the finer sediment accumulates as lacustrine sediments. Often these are varved, for seasonal variations in sediment input result in alternate bands of relatively coarse silt (during the spring and summer) and finer clays (during the winter when the surface is frozen).

By far the majority of glacio-fluvial debris, however, is carried out of the ice on to the glacial foreland, and here the material is sorted and deposited as extensive **sandar** (singular **sandur**). Because of the erodibility of the sediments and the

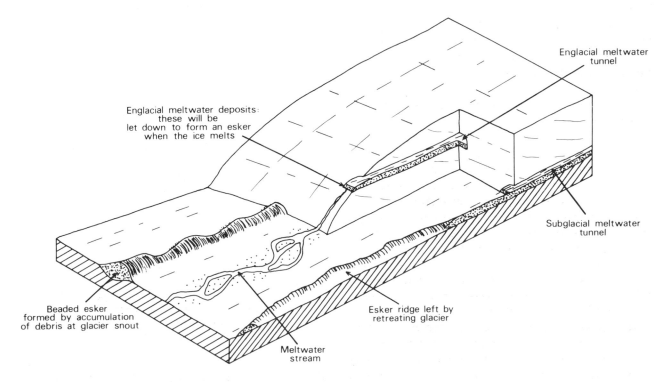

Figure 23.21   The formation of eskers

Plate 23.15 A frost-shattered boulder on Sprengisandur, Iceland. Note the way in which the rock has shattered along parallel lines of weakness (photo: D. J. Briggs)

highly variable discharge of the meltwaters, the streams tend to adopt a braided pattern, so the sandur plain is characterized by complex networks of shallow channels and low-relief bars (Plate 23.14).

Periglacial environments

The margins of an ice sheet or glacier are often dramatic: shear cliffs of ice from which emerge turbulent floods of meltwater, huge ridges of debris, and vast spreads of glacio-fluvial deposits. Beyond, however, lies a less clearly defined zone in which ice still dominates; an area typically of open, wind-swept plains and seasonally extensive snow cover. These are the periglacial regions of the world.

The periglacial areas today are found mainly in the northern hemisphere, fringing the polar ice caps in Canada, Greenland, Iceland, Scandinavia and Russia (Figure 23.1). They are areas of prolonged winters and short summers. The harsh climate restricts the vegetation to ground-hugging plants, many of which emerge and flower only briefly during the few summer weeks. The rest of the year the land is exposed and barren. Temperatures are so low that, at the surface, the

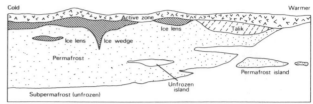

Figure 23.22 A generalized transect through a permafrost zone showing the presence of unfrozen islands and talik (thawed bodies) within the permafrost

water is frozen for much of the year, while at depth, where the sun's warmth never reaches, the ground is permanently frozen. Thus the ground within the periglacial areas consists of three general zones: at depth, the unfrozen **sub-permafrost**; above this the permanently frozen **permafrost** layer; and at the surface the seasonally frozen **active layer**. The depth of these three layers varies, depending on the regional climate and nature of the bedrocks, but the active layer is often only a few metres thick, while the permafrost layer may extend to as much as 100 m. All three layers are often discontinuous, however, varying in extent according to local conditions. Moreover, unfrozen layers may be preserved within the permafrost, while permafrost islands may occur within the unfrozen sub-permafrost or active layer (Figure 23.22).

Table 23.1 *Types of ground ice*

Type	Process of formation
Needle ice (pipkrake)	Strong radiative cooling of bare earth surfaces
Pore ice	Rapid freezing of saturated soils
Segregated ice	Slow freezing of partly saturated soils; clusters of ice form in some pores but not others
Intrusive ice	Growth of sheets and lenses of ice by adhesion as water is injected under pressure from adjacent rocks and soils
Vein ice	Freezing of water and water vapour in sub-vertical cracks produced by thermal contraction or desiccation
Extrusive ice	Overflow water from springs forced on to the surface before freezing
Buried ice	Burial by debris of snowbanks or glacier, sea, lake or river ice

After E. Derbyshire, K. J. Gregory and J. R. Hails, *Geomorphological processes*, Butterworths, 1979.

The ice which forms under these conditions is referred to as **ground ice** and takes a number of different forms, depending on the rates of freezing, the nature of the materials in which it forms and the amount of water available (Table 23.1). It is the development and, in the active layer, melting of this ice which dominate the formation of many of the landforms of the periglacial regions.

As we noted in Chapter 19, when water freezes, its volume increases by about 9 per cent, setting up considerable stresses in the surrounding materials. In addition, as the ice crystals grow, they attract to themselves more water for they create moisture gradients within the rock or soil, down which water migrates. In this way the developing ice mass 'feeds' itself, and can continue to grow. Of course, the process cannot normally proceed indefinitely, because ultimately the water supply is exhausted, but by then large ice masses may have developed and these can cause major disruption of the surrounding materials. At depth, for example, the pressures created by the expanding ice mass compress the material around it and lead to the development of hard, **indurated** layers. Examples of these still survive in many soils in northern Britain. At the surface, the pressures are relieved by uplifting of the overlying material, sometimes to heights of 50 m or more. This occurs particularly where intrusive ground ice (Table 23.1) is fed by water moving through a layer of low permeability beneath the developing ice mass. The feature which develops is known as an **open-system pingo** (Figure 23.23). Similar **closed-system pingos** may develop where unfrozen, saturated sediment is trapped between an advancing permafrost table from below and a thin surface layer of perennially frozen ground. This latter situation is often associated with old lake basins which have frozen from the surface downwards (Figure 23.23). In both cases, the material within the active layer above the ice lens may slump and wash downslope and accumulate as a bank around the pingo. When the ice lens finally melts, a crater is then formed, surrounded by a rim of slumped debris.

The development of needle ice and segregated ice also has widespread effects. Where ice crystals grow in joints and other lines of weakness in rocks, **gelifraction** may take place, the rock being broken down into angular debris (Plate 23.15). Over time, continuous gelifraction may result in the accumulation of extensive masses of this material, known as **blockfields**. Some of the best developed examples

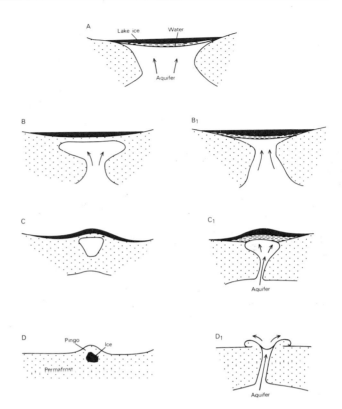

Figure 23.23 Pingo formation. Initially a broad, shallow lake forms over frost-free ground. Sediment accumulation and permafrost extension may seal off the lake from the underlying aquifer in which case a closed-system pingo develops (B–D). Alternatively, the aquifer may remain linked to the lake, with the result that water flows upwards to the surface, creating an open-system pingo (B_1–D_1). Under extreme conditions the surface may be ruptured and the trapped water may burst outwards

are seen in Labrador and Norway, in areas which have escaped full glaciation during the Pleistocene and which have thus been subjected to almost permanent periglacial activity for many millennia. Sorting of the surface sediments and soils may also take place. **Upfreezing** of blocks occurs, for example, for segregation ice develops preferentially in coarser voids which tend to be associated with larger clasts in the materials. Ice lenses forming beneath these clasts push them to the surface through the matrix of finer material. At the same time, the particles are often tilted so that they assume a vertical alignment. This is a characteristic seen today in many coarse-grained deposits which have been affected by past periglacial activity.

Plate 23.16   An ice-wedge cast in Devensian terrace gravels of the River Thames at Standlake, Oxfordshire (photo: D. J. Briggs)

During the development of ground ice, desiccation of the surrounding materials leads to shrinkage and the formation of polygonal **contraction cracks** or **frost polygons**. In the active zone, these cracks may fill with water during the period of thaw, which then freezes in the following winter. Vein ice thus develops in the cracks and, by a process of repeated growth, cracking and refilling with water this may expand into a large **ice wedge**. Adjacent to the ice wedge the ground is pushed into low mounds, while, during periods of melting, material slumps down the side of the wedge to produce down-turned bedding. When the climate improves and the permafrost melts, the wedge may be infilled by sediment from the surface to produce a fossil ice-wedge cast. Such features are common, for example, in many river terrace deposits in Britain (Plate 23.16).

Within the active zone, periods of intense freezing are followed by warmer phases during which thawing occurs. Together these two processes are important in rearranging and sorting the surface materials to produce a range of **patterned ground**

Plate 23.17   Stone polygons on Gagnheidi, Iceland. The moss-covered stones form a distinct pattern around the finer material. These polygons are exceptionally well-developed; often where the range of particles is smaller, they are much less distinct (photo: D. J. Briggs)

Plate 23.18   Thufurs, near Reykjavik, Iceland. When active, the hummocks are ice-cored. These, however, are fossil features, the ice having melted to leave mounds of moss-covered debris (photo: P. A. Smithson)

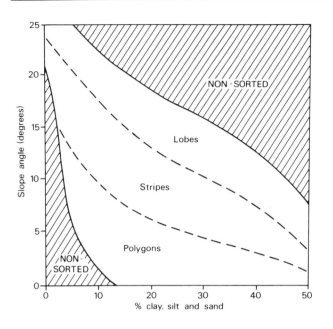

Figure 23.24 The relationship between particle size, slope angle and patterned ground formation (adapted from Goldthwait, 1976)

features. On low angle slopes, for example, **stone polygons** or **stone nets** may develop (Plate 23.17). They form, apparently, because freezing results in the formation of low mounds of material. Larger, upfrozen clasts extruded on to the surface of these mounds slump downslope during periods of thaw and collect at their edges. These zones of concentration of stones then act as lines of drainage during the thaw, and the water accentuates the sorting effects by washing out the fine material between the stones. Moreover, during refreezing, little ice growth can take place in the large cavities between the stones, so subsequent heaving is confined to the finer material in the centre of the mound. In this way the feature is maintained. On low angle slopes these sorted polygons may be exceedingly regular, but where slope angles are greater than about 5° downslope movement results in the polygons being stretched out to form **stone garlands** and **stone stripes**. On yet higher angle slopes (above about 10–25° depending on the character of the materials), patterned ground features do not develop and slope processes such as debris avalanching and

Figure 23.25 The extent of glaciation in Britain during the Quaternary. Considerable controversy surrounds the details of these limits

earthflow dominate (Figure 23.24). Similarly, sorting cannot occur in materials of uniform particle size distribution, and here features such as small hummocks (**thufurs**) and **nonsorted stone polygons** are found (Plate 23.18).

The legacy of the Pleistocene Ice Age

As we noted in Chapter 10, the last 2 million years have witnessed major climatic changes. At times, global temperatures have fallen sufficiently to allow the polar ice caps to extend far into what are now temperate areas of the world, while local ice caps developed over mountains. In Britain, for example, ice from Scandinavia reached as far as East Anglia, and extensive ice sheets grew and spread out from the mountains of Scotland, the Lake District, Wales and western Ireland (Figure 23.25). At the same time, at the margins of these vast ice sheets, periglacial conditions gripped the exposed land.

In the upland areas of western Scotland, the Lake District and Snowdonia, erosion was intense. The surface was stripped bare of weathered debris, deep corries and glacial troughs were cut, and the whole land surface greatly modified. Significantly, much of this modification seems to have taken place not during the periods of maximum glaciation, when the ice caps which covered these areas were probably cold-based and had little erosive power, but during early glacial phases, when the ice caps were developing. In other upland areas, which did not support their own, local ice caps but were affected only by overriding, cold-based ice, erosion was less intense (Figure 23.26). In the Pennines, eastern Scotland and much of south Wales, therefore, relatively little modification occurred.

In the lowlands, erosion was generally less active, for here the ice sheets were often thinner. Nevertheless, massive quantities of material were scraped off the Chalk in East Anglia and the bed of the North Sea, and off the Jurassic clays of the Midlands and eastern England. The Vale of Belvoir, for example, seems to be a product of glacial scour of the soft Kimmeridge Clay, while huge rafts of chalk, torn from the bed of the North Sea and overturned by the ice, are visible in the till cliffs around Cromer in Norfolk.

The materials removed from these areas were dumped either as tills or glacio-fluvial deposits. Most deposition occurred in the lowlands (Figure 23.27) and deep drifts, up to 100 m in thickness,

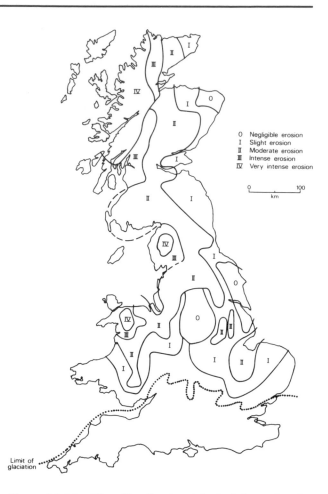

Figure 23.26 The distribution of glacial erosion in Britain (after Boulton *et al.*, 1977)

were laid down in parts of the Midlands and East Anglia. In large areas of Ireland, and more locally in England, the tills were moulded into drumlin belts. In areas where the ice sheets finally stagnated and decayed *in situ*, hummocky drift and dead-ice topography were formed. Elsewhere, lodgement tills were laid down as more or less flat till plains.

Beyond the ice margins, in the major river valleys such as the Thames, Severn and Avon, the meltwaters laid down extensive spreads of gravel which now form river terraces. In these areas, too, periglacial conditions produced widespread patterned ground features: pingos, solifluction debris (often mapped on geological maps as 'head') and loess (or 'brickearth'). As the climate ameliorated, at the end of the glacial periods, these conditions followed the retreating ice margin and affected the exposed

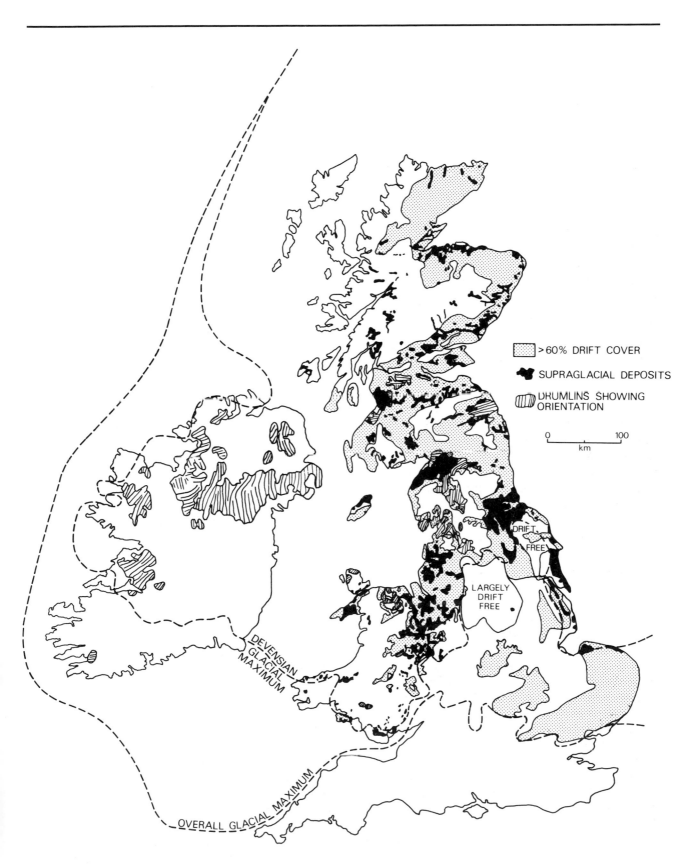

Figure 23.27 The distribution of glacial drifts in Britain (from Boulton et al., 1977)

glaciated surfaces. In addition, around the coasts, variations in sea-level, due to the combined effects of isostatic depression and recovery and eustatic processes, resulted in the development of raised beaches (such as that at Black Rock, Brighton) and submerged shorelines (e.g. in the Clyde). Clearly, there are few areas in Britain which do not bear the imprint of glacial and periglacial effects.

Aeolian systems

The work of the wind

Distribution of aeolian activity

Almost 10 per cent of the world's land surface may be described as **hot desert** (Figure 24.1). Within this area, the action of the wind is all-pervading. It plays a vital part in landscape development, carrying abrasive fragments of rock across the surface and redistributing the materials to leave vast expanses of land bare and rocky and cover others with thick layers and folds of sand. It directly influences man's activities; it constrains his use of the land; it poses barriers to his movement and communication; it enters his lungs and eyes and creates a constant health risk.

A further 10 per cent of the continents consist of **polar deserts**. Again, wind activity is important in shaping the land. As we shall see, fierce polar winds strip sediment from the river plains and carry it many kilometres across the surface. In the past, when the polar ice sheets extended far into lower latitudes, these polar deserts occurred in what are now temperate areas. They have left behind the legacy of widespread wind-blown deposits known as loess.

Wind activity is not confined to the desert areas, however. In many semi-arid and even humid temperate areas, the effects of wind erosion are of

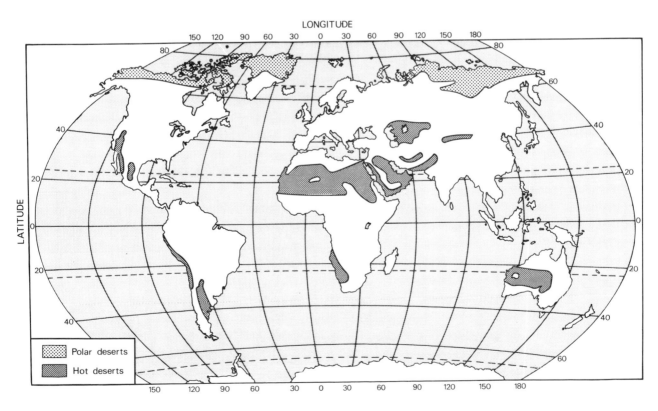

Figure 24.1   The distribution of hot and polar deserts

considerable import. Coastal areas are often subject to the action of winds from the sea, and aeolian landforms lie inland of many sandy beaches. Many agricultural areas are also prone to wind activity. The American Dust Bowl is the most oft-quoted example, but less dramatically wind erosion occurs in most intensively cultivated countries. Man's attempts to clear the land for cropping, and his repeated tillage of the soil, have often produced unprotected, powdery surfaces, susceptible to the erosive power of the wind. In Britain, wind erosion presents severe problems in the sandlands and fens of eastern England.

Aeolian processes

Entrainment

The ability of the wind directly to attack and erode the land surface is severely limited. Air has a very low fluid density; only about 1.22×10^{-3} g cm^{-3} at a temperature of about 20°C, approximately one nine-hundredth of that of water. As a result it is able to exert only a relatively small force against objects in its path. Entrainment of material by fluid drag (i.e. the direct force of the wind) is therefore restricted. On the other hand, once the air has picked up sediment, its erosive potential increases enormously, for the sand grains bouncing on the surface have a kinetic energy sufficient to move particles as much as six times their own diameter. Wind erosion is consequently a process which is subject to threshold constraints: once threshold conditions controlling the initial entrainment of material by fluid drag are reached, the rate of erosion increases swiftly.

Fluid drag

The first question which needs to be asked, therefore, is what are the factors controlling the action of erosion by fluid drag? A major determinant is clearly windspeed. In general, as windspeed rises,

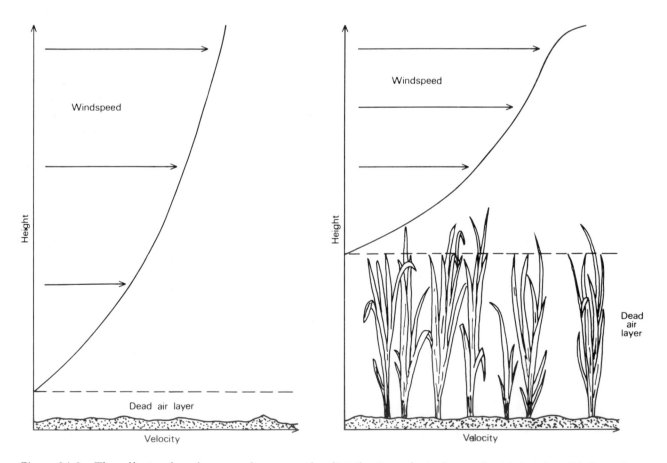

Figure 24.2 The effects of surface roughness on the distribution of wind-speed with height: (A) flow above a smooth sand bed; (B) flow above a densely vegetated surface (after Embleton and Thornes, 1979)

the entrainment potential of the wind increases. It is not, however, the mean windspeed which is most important, for over short periods – often fractions of a second – great variations in velocity may occur. It is the extreme gusts which are instrumental in entraining sediment, so the erosivity of the wind is related more closely to the duration and magnitude of these brief spells of high windspeed.

Wind conditions at the ground surface, where particle entrainment occurs, are not the same as those in the open atmosphere. The surface exerts a frictional drag against the air which reduces wind velocity and often creates a thin layer in which windspeed is zero (Figure 24.2). The depth of this **dead air layer** depends upon the surface roughness but is generally about one-thirtieth of the average height of the obstacles. Over a bare sand surface, the particles create a degree of roughness which acts to shelter grains lying in the depressions between the larger particles. Where vegetation is present, however, a much deeper dead air layer may be created which protects the whole surface from erosion. For this reason the presence of vegetation is a major constraint on erosion, and maintaining a good vegetation cover provides one of the most effective means of erosion control.

Surface roughness also helps to generate turbulence in the wind. As we saw in Chapter 20, this is important, because turbulent eddies give a vertical lift which helps to raise particles off the bed and keep them buoyant in the air. In addition, bedforms such as ripples and dunes, produced on the surface by the wind itself, modify the wind flow at ground level and turbulent eddies may be created in the lee of such features (Figure 24.3). These eddies help to keep the area between ripples or dunes free of sediment and play an important part in the dynamics of sediment transport and deposition.

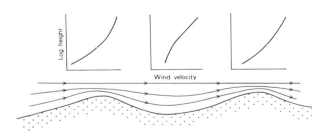

Figure 24.3 Wind flow over sand ripples

Fluid impact

As we have noted, the entrainment of particles by fluid drag represents only the first stage in the initiation of transport. Most of the material carried by the wind is entrained by impact of particles on the surface. These impacts cause individual particles to bounce upward, off neighbouring grains. Often, several grains may be set in motion at once by the impact of a single particle on a loose sand bed.

Frictional resistance

The frictional resistance of the surface materials to entrainment is largely a function of their weight. Up to a point, larger or heavier grains require higher fluid drags to initiate movement than do smaller or lighter particles. Very fine material, however, has greater resistance to entrainment because of its cohesiveness and because the particles do not protrude into the active wind flow. Thus the critical entrainment velocity for fluid drag increases below a particle diameter of about 0.1 mm (Figure 20.6).

Moisture is also important in controlling entrainment, for water films provide a source of intergranular cohesion which increases the frictional resistance of the surface materials. It is largely for this reason that wind erosion in Britain is relatively rare during the winter and very early spring. Although the soil is often bare then, and although windspeeds are often high, the land is too wet to be readily susceptible to erosion. It is not until evapotranspiration has started to dry out the soil at the start of the growing season that erosion occurs. More generally, it is the lack of moisture which is primarily responsible for the high rates of wind erosion in arid and semi-arid areas. The effect operates both directly and indirectly: it makes the soil non-cohesive and susceptible to erosion, and it inhibits vegetation development so that the surface is frequently bare and unprotected.

Transport

When grains are lifted off the surface their frictional resistance declines for they are no longer in contact with adjacent particles and they come under the influence of stronger horizontal wind flows. The horizontal forces drive the grains forward, while vertical eddies help to keep them airborne. Together, these forces act to counteract the gravitational pull which draws the particles back to the ground. In most cases, however, the vertical currents are insufficient to keep the particles aloft for

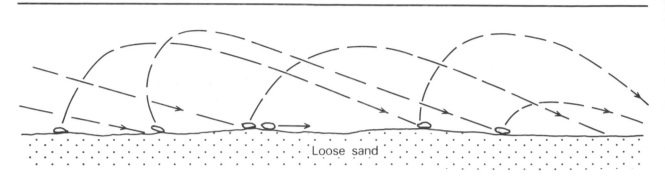

Figure 24.4   Trajectories of saltating aeolian sand grains

long and they fall slowly to the earth along a characteristic curved path (Figure 24.4). The length of this path and the height which the particles reach are governed by the relationship between wind-speed, turbulence and particle size, but under reasonably uniform conditions of sediment supply and wind velocity the particles tend to follow a fairly consistent trajectory. As we will see, this has considerable significance for the development of aeolian bedforms.

The wind is one of the most effective agents of sediment sorting. Large particles are not normally moved: the largest material that can be entrained by fluid drag is about 3 mm in diameter; particles of this size bouncing on the surface can thus dislodge grains up to about 18 mm in diameter. Larger particles can be moved – in fact there is evidence of huge boulders rolling across desert surfaces under the effect of wind – but these are exceptional and the extent of transport is extremely limited. More commonly, coarse debris is left as a lag on the surface where it helps to protect the underlying material from erosion.

The material which is entrained is sorted according to both size and shape. Larger particles (above about 1–2 mm in diameter) move almost entirely by creep and are thus subject to considerable frictional

retardation, with the result that they move relatively slowly. Material between about 0.1 mm and 1.0 mm in diameter moves largely by saltation, for it is too heavy to be retained aloft for long. Finer material, however, may be carried to heights of several hundred metres and, under convective atmospheric conditions, may be transported thousands of kilometres before deposition. Silts, transported in this way, form vast spreads of **loess** in areas such as the Gobi Desert in China and, during cold, arid phases in the Quaternary, similar silts were laid down over much of north-western Europe.

The effect of shape is more complex. Spherical particles tend to have relatively high settling velocities, and consequently do not remain airborne as long as flatter, discoidal grains. On the other hand, spherical grains tend to be thrown steeply into the air when entrained by particle impact, whereas discoidal grains rise more obliquely. As a result, spherical particles adopt somewhat steep, short trajectories when moved by saltation, or are transported preferentially by creep. Conversely, discoidal particles move mainly by saltation along shallow, flat trajectories (Figure 24.4). Because of these effects, flatter particles often move further than spherical grains, especially at low wind velocities. As windspeed increases, however, this pattern is reversed due to the tendency of more spherical grains to bounce back into the air flow after impact.

Together, these processes lead to marked patterns of sorting within aeolian deposits. At a large scale, materials become finer with increased distance from their source. At a small scale, distinct size grading occurs on individual bedforms such as ripples and dunes. Coarser particles tend to accumulate on the crest of ripples, for example, while finer material collects in the more sheltered troughs between the ripples (Figure 24.5).

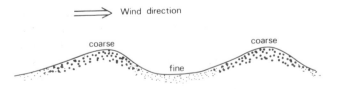

Figure 24.5 Particle size variation across sand ripples

Deposition

The sand and silt carried by the wind are ultimately deposited. Deposition occurs for a variety of reasons and in a variety of circumstances. Mass deposition of the sediment load occurs when wind-speed drops at the end of a storm. During a storm, material may be trapped against an obstacle or held by a moist patch of ground; local deposition also takes place in sheltered areas provided by small surface irregularities. Deposition against perma-nent obstacles produces **sand drifts** which are static features. Many features of aeolian deposition, however, are dynamic, steady-state forms, dependent upon a balance between rates of entrain-ment, transport and deposition.

Aeolian landforms

Landforms of hot deserts

To many people, the term 'desert' conjures up visions of limitless expanses of sand. This impression is far from generally true. Sand deserts (**ergs**) are far from ubiquitous in arid regions, and it has been estimated that about 85 per cent of aeolian sands are confined to a relatively small number of large ergs, each covering at least 32,000 km². Even within these areas, the sand cover is not con-tinuous, and between the dunes occur expanses of bare bedrock. Elsewhere, the hot deserts consist of extensive bedrock plains, spreads of gravel (**regs**), or coarse boulder pavements (**hamadas**). All these landscapes reflect the combined effects of the removal (**deflation**) and deposition of the sand by the wind. In addition, however, it is important to remember that water plays a significant part in sculpturing the desert landscape. Rainfall may be rare but when it occurs it is often intense and flash floods develop. The water cuts deep gullies and trenches, known as **wadis** and transports vast quantities of sediment. Much of this material ends up in inland **playa lakes** where the water evapor-ates to leave spreads of saline sediment.

Sources of aeolian sand

Sand deserts may not be ubiquitous, but they con-tain vast quantities of sand. The Rub al Khali erg in Arabia, alone, covers 560,000 km², with sand deposits up to 300 m in depth. Where does all this sand come from?

This is a pertinent question, for we have already noted that the wind itself is not very efficient at eroding bedrock surfaces. Most of the sand must therefore be derived from other sedimentary deposits. Much, in fact, comes from alluvial fans where streams draining the surrounding mountain areas dump their debris at the desert edge. Dry playa lakes and wadi floors also supply sand, while some is derived from *in situ* weathering of the bedrock surface. Considerable controversy still exists about the importance of this last source, for it has been argued that weathering processes in desert areas are relatively ineffective. None the less, salt weathering, heating-and-cooling and thermal contraction may all encourage rock disintegration and provide at least small quantities of material.

Sand ripples and ridges

Sand ripples are common features of almost all sand surfaces. Typically they are a few centimetres in height and are spaced at intervals of several centi-metres to metres. Their long axes run at right angles to the wind direction and they tend to migrate downwind. They may be symmetrical or asym metrical in cross-section. Their form, however, depends to a great extent on wind conditions. Over a wide range of conditions their wavelength-to-height ratio (known as the **ripple index**) is between 15 and 20, but as wind velocity increases the ripples become flatter and more widely spaced and the index rises to as much as 50 or 60.

The reasons for ripple formation are related to the dynamic interaction between wind flow and sedi-ment movement. Initiation is often due to a random factor, such as the presence of a small obstacle on the surface, a patch of moist sand, or a local vari-ation in sediment size or wind speed, which results in the accumulation of a small mound of sand. This mound creates, on its lee side, an area which is protected to some extent from the saltating sand grains, whereas the windward side tends to be subject to higher rates of particle impact (Figure 24.6). The result is that material starts to collect on the upwind side of the mound.

In time, however, the mound becomes large enough for the grains on the surface to be no longer held by whatever initially trapped the sand. There-after, as grains land on the upwind side of the mound, the impact throws new material into motion. The dislodged particles tend to move a similar distance due to their uniform saltation trajectories, and they thus create a second zone of intensified impact one 'hop' downwind. These impacts in turn set further particles moving,

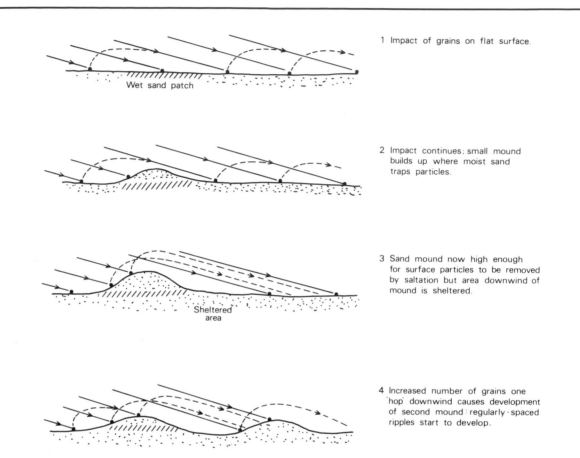

1 Impact of grains on flat surface.

2 Impact continues; small mound builds up where moist sand traps particles.

3 Sand mound now high enough for surface particles to be removed by saltation but area downwind of mound is sheltered.

4 Increased number of grains one 'hop' downwind causes development of second mound: regularly-spaced ripples start to develop.

Figure 24.6 Development of sand ripples

excavating a small depression in the process and triggering off yet further saltation downwind. In this way, ripples develop which are spaced at intervals equivalent to the average path length of the saltating particles. Ripples formed in this way are referred to as **impact** or **ballistic ripples**. Their height depends on grain size, larger ripples being associated with coarser or more poorly sorted sediments.

Under conditions of extremely poor sorting or a relatively coarse sediment relative to the wind-speed, **sand ridges** form. The wind is only able to transport the finer fraction by saltation with the result that the coarser particles are moved by creep or accumulate as a lag deposit. In time, the largest particles become concentrated at the surface, protecting the underlying sediment, and trapping particles which are being transported along the ground by creep. Gradually a ridge builds up of sediment which is too coarse to be removed by saltation (Figure 24.7). Sand ridges of this sort may reach considerable heights and appear to be essen-

tially immobile; ridges of up to 60 cm are known in the Libyan Desert, for example.

Dunes and draas

The huge dunes which characterize the sand seas of deserts such as the Libyan and Arabian deserts are some of the most spectacular aeolian landforms. They may be up to 200 m in height and up to 5 km or more in wavelength. They vary greatly in genesis and size, however, and in recent years it has become common to divide them on the basis of scale into **dunes** (with amplitudes up to 0.5 km) and **draas** (with amplitudes of 0.5–5.0 km).

In practice, the many different dune forms show immense variation and tend to merge into each other both in terms of morphology and genesis, but four main types of dune are extensively found in desert areas: seif, barchan, transverse and aklé dunes. Each seems to represent a different relationship between sand supply, wind speed and wind direction.

Seif dunes are named from the Arabic term

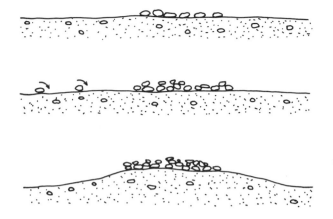

1 Coarse particles are left as a lag deposit by the wind; local concentrations of coarse material may develop.

2 Other particles moving along the surface by creep are added to the patch of coarse sand.

3 The coarse sand protects the underlying surface, and traps other particles blown from the area upwind. A sand ridge develops.

Figure 24.7   The development of sand ridges

meaning 'sword-edge' and as this implies they are long, steep-sided, sharp-crested features (Plate 24.1). They are also called in some areas **longitudinal dunes** for they are orientated parallel to the wind. They are found in many sand deserts, including the Arabian, Libyan, Namibian, Australian and Arizona deserts. In Iran, they are up to 200 m high and they are commonly 60–100 km in length. Often, seif dunes join to create so-called **tuning-fork** features (Figure 24.8).

How seif dunes form is not entirely certain. It was once thought that they were produced by cross-winds, blowing at 90° to each other. Modern ideas, however, suggest that they are due instead to the development of helical air flow between the dune ridges. Probably because of excessive heating of the sand surface, convection cells develop which result in two, parallel vortices (**rollers**) circulating in opposing directions between each pair of dunes (Figure 24.9). These act to sweep sand off the desert floor and move it obliquely up the dune face. Evidence for this type of movement is sometimes seen in the strings of debris and small-scale sand ridges running up the flanks of seif dunes.

In contrast, **barchans** often form as isolated features. They are crescentic mounds of sand, orientated with their 'horns' pointing downwind (Figure 24.10). The steep **slip face** which points downwind is typically at an angle of about 34°, the angle of repose of dry sand. Barchans vary greatly

Plate 24.1 Seif dunes in the Simpson Desert, Australia. Saline panes can be seen between the dune ridges, which are up to 100 m wide (photo: D. D. Gilbertson)

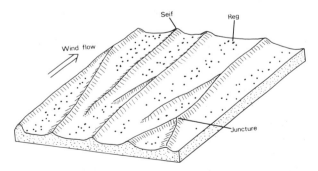

Figure 24.8   Typical system of seif dunes showing tuning-fork forms

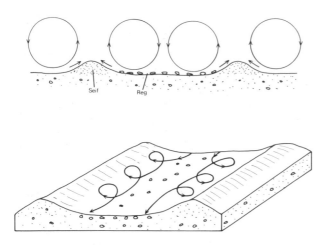

Figure 24.9   The development of seif dunes

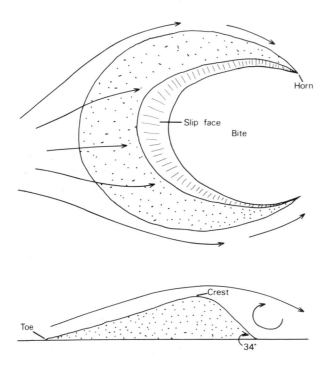

Figure 24.10   A barchan dune

in size, at their greatest reaching diameters of 400 m and heights of 30 m – though some may be as high as 150 m. The most perfect examples tend to be found as isolated dunes developed on bedrock surfaces under conditions of limited sand supply, but they commonly occur in groups as rather less well-developed forms.

Barchans are common in many sand deserts, including those of Peru, Egypt and the Imperial Valley in California, but significantly none occur in

Australia: they seem to be found only under conditions of a limited sand supply, flat bedrock surfaces and relatively uniform winds. In these situations, the barchans represent an aerodynamically stable form. As the wind passes over the dune, it divides slightly and eddies develop on the lee side. These tend to keep the 'horns' short, while within the 'bite' the air is relatively calm. As a result, sand is moved up the windward side of the dune and avalanches down the slip face. In this way, the dune gradually migrates downwind without altering its shape.

As the sand supply diminishes with increasing distance from its source, the barchans tend to disperse and break up. Conversely, where the sand supply is greater the dunes tend to coalesce and lose their crescentic form, ultimately merging into **transverse dunes**. These are long, relatively straight or somewhat serated ridges orientated at right angles to the wind direction. They are commonly arranged at regular intervals and separated by sediment-free areas or coarse gravel-strewn surfaces (Figure 24.11). They appear to be the characteristic form in areas of high sand supply in which surface heating (or other effects) are insufficient to produce the helical flow necessary for formation of seif dunes.

In practice, none of these relatively simple forms of dune are very widespread, and the majority of sand deserts are probably characterized by more complex forms known as **aklé**. These consist of sinuous dune ridges, aligned transverse to the wind direction, and composed of crescentic sections, pointing alternately upwind and downwind. The way they form is far from certain, but they appear to result from the development of complex patterns of secondary wind flow due to interactions between the bedform and the wind.

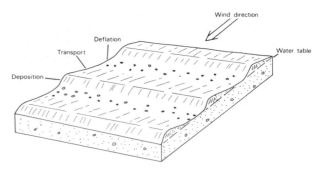

Figure   24.11   The   development   of   transverse dunes

Plate 24.2 A wind-etched 'mushroom rock' in Israel. The rock is undercut at the base because the majority of sediment movement occurs close to the surface (photo: E. Landy)

Erosional features

The sediment-laden winds of a desert sandstorm operate like a sandblasting machine, wearing away at anything in their path. In the process they produce a range of erosional landforms. Pebbles exposed on the surface, for example, are polished and faceted by the incessant abrasion, and where the winds are from a consistent direction they are shaped into **dreikanter** (German: three edges). Larger boulders and hills are smoothed and etched, weak beds being picked out to create fantastic shapes (Plate 24.2). At a larger scale, the wind may ultimately open up huge **deflation hollows** in the bedrock surface. Erosion often starts along a line of weakness (such as an area of intense jointing) or along a wadi which cuts through the cemented surface rocks, and continues until it reaches the water table. One spectacular example is Al Fugaha in Libya, an almost circular depression 2–3 km in diameter and over 60 m deep.

As we have already seen, deflation also leaves eroded desert surfaces: rock pavements and gravel plains. Indeed, it is landscapes such as these which make up the majority of many desert areas. 70 per cent of the Sahara, for example, consists of sandless rock plains, regs and hamadas. Often the pebbles and boulders in these areas are coated with so-called **desert varnish**, a smooth dark mineral layer containing iron, manganese and various trace elements. How this varnish forms is not known for certain. One theory, however, suggests that it is due to the development of a thin film of dew on the pebble surfaces which encourages ions to diffuse from the interior of the rock.

Landforms of coastal dunes

Few of us probably have the chance to witness first-hand the action of the wind in desert areas. But similar processes can operate wherever there is an adequate supply of sand, a dry and unvegetated surface and high windspeeds. Such conditions characterize many sandy coasts. Here, the source of sand is the beach: prevailing winds or local on-shore breezes are often strong, and, because of the exposure, the salinity, the coarseness of the sediments and the impacts of human trampling, vegetation growth is often limited. As a result, the winds winnow the beach material and transport it inland to form transverse dunes, running parallel to the shoreline (Plate 24.3).

The formation of coastal dunes often follows a distinct cycle in which soil formation and

Plate 24.3 Coastal dunes, Banna Strand, County Kerry, Ireland (photo: D. J. Briggs)

vegetation development play significant parts. Initially, sand is transported from the beach and accumulates a short distance inland, often where it is trapped by vegetation, in the form of transverse **foredunes**. The dune shape is governed by aerodynamic factors, and is typically relatively gentle on the seaward side and steeply sloping inland. During transport, marked sorting of the sediment by size and shape occurs and the dunes are often composed of material which is finer and more rounded than that on the beach (Figure 24.12).

Over time, the foredune may migrate inland as sand creeps up the windward slope and avalanches down the leeward side. As this happens, the area of available sand supply increases and ultimately new foredunes may develop on the seaward side. Similarly, if the sea-level falls so that the beach becomes wider, new foredunes may form. In either case, the effect is to reduce the rate of sediment supply to the older dunes, allowing them to become stabilized. Gradually, the sand weathers, soil starts to form and vegetation starts to invade. Ultimately, a whole series of parallel dunes of increasing age may form inland, separated by **swales** (Figure 24.13).

During the evolution of the stable dune, a number of additional changes may occur. Marram grass is one of the main forms of vegetation to become established because it tolerates – indeed it requires – continual burial by sand. The grass provides protection to the dune, helps to trap sediment and redu-

ces erosion. When the dune becomes isolated from the beach and sediment supply falls, however, the grass may start to die. Bare patches develop on the old dune surface and erosion may be reactivated. This, in turn, increases the supply of sediment to dunes further inland.

More drastic and localized erosion may occur where the vegetation is destroyed. These **blow-outs** often result from overgrazing of the vegetation by rabbits or sheep, or trampling by man. The material is carried inland and deposited as a longitudinal or barchan-type dune.

Landforms of cold (polar) deserts

Polar deserts are characterized by extensive, exposed surfaces with little or no vegetation. Combined with these conditions, however, there exists a climate which is dominated by low temperatures and prolonged freezing. Physical weathering of the rocks by freeze–thaw processes produces sands which may be transported by winds, but in addition the debris washed from ice sheets and glaciers often consists of fine-grained sand and silt ground up beneath the ice. This material tends to accumulate as extensive sandar, the surfaces of which are left bare and exposed to the wind. Windspeeds are high, with strong katabatic winds blowing off the

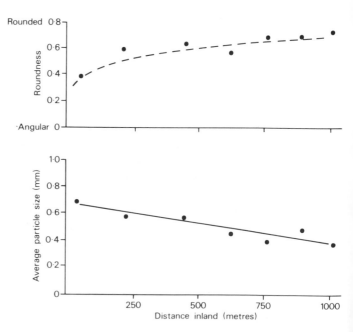

Figure 24.12 Typical variations in particle roundness and average size across a coastal dune sequence; all samples are from the crests of dune ridges

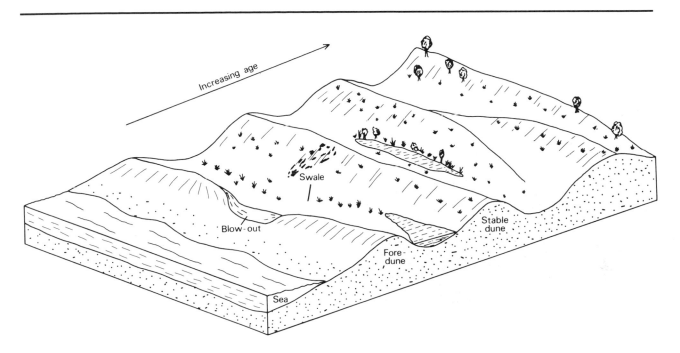

Figure 24.13 A sequence through a series of coastal dunes

icc sheels. Moreover, at low temperatures the air is relatively dense so that it exerts a greater fluid drag; the velocity necessary to move a sand grain 2 mm in diameter is about 5 m s⁻¹ less in the Antarctic than in a sub-tropical desert. Thus, sand dunes often develop at the margins of the outwash plains, while the finer material, mainly silt, is transported further afield.

These wind-blown silts are often deposited in extensive sheets of **loess**. During the Quaternary, when ice extended into what are now the cool temperate zones, widespread loess deposits were formed. These include silts up to 30 m deep in some parts of North America and northern Europe (Plate 24.4). In addition, thinner loess deposits were laid down over even wider areas and these have become incorporated into the soils to give an additional silty component. In some cases, the direction of the winds bearing the loess can be deciphered, for the silts become finer downwind. They also show a reduction in the quantities of more fragile minerals away from their source.

Aeolian processes and man

Soil erosion by wind

Now the wind grew strong and hard and it worked at the rain crust in the cornfields. Little by little the sky was darkened by the mixing dust, and the wind felt over the earth, loosened the dust, and carried it away. . . . The wind grew stronger, whisked under stones, carried up straws and old leaves, and even little clods, marking its course as it sailed across the fields. The air and the sky darkened and through them the sun shone redly. . . .

So John Steinbeck, in *The Grapes of Wrath*, describes the beginning of a dust storm in Oklahoma. After the storm, the dust settles:

All day the dust sifted down from the sky, and the next day it sifted down. An even blanket covered the earth. It settled on the corn, piled up on the tops of the fence posts, piled up on the wires; it settled on roofs, blanketed the weeds and trees.

Such were conditions time and time again throughout much of the Mid West of the USA during the 1930s. So extensive was the erosion, so widespread the damage, that the area became known as the Dust Bowl. It provided a terrible example of how the activity of the wind could affect man. It is a reminder, too, of the way man can encourage wind erosion. For while the ultimate cause of the Dust Bowl was the wind and the inherent susceptibility of the soils to erosion, it was man who triggered off the problem. For decades he had been ploughing the fragile land of the Mid West, removing the crops, leaving the soil bare and unprotected to the wind. In the 1930s, following endeavouring to understand more clearly what controls soil erosion by wind. Much of this work has been carried out by W. Chepil. He and his colleagues have analysed results from hundreds of experiments, both in the field and in artificial wind tunnels. Based on this research they have devised the so-called wind erosion equation showing the relationship between wind erosion and environmental and management factors. The equation states:

$$E = f(c, i, l, k, v, p)$$

That is, the amount of erosion (E) is a function of climate (c), soil erodibility (i), the exposed length of the field (l), the roughness of the surface (k), the vegetation cover (v) and the management practice (p).

Each of these factors can be defined and measured. The climatic factor, for example, is dependent on windspeed and soil moisture. Soil erodibility is dependent upon the proportion of the soil made up of stable aggregates greater than 0.84 mm in diameter. The equation is used to assess the amount of erosion and to help define ways of controlling soil loss. For example, the agriculturalist can calculate the effect of changing the vegetation cover (e.g. by growing a different crop), or reducing the length of field exposed to the wind (by planting hedgerows and windbreaks), or altering the surface

Plate 24.4 A section in loess at Bahlingen, Kaiserstuhl, W. Germany (photo: P. A. Smithson)

several years of dry weather, the parched soil started to erode. Crops were ripped out, seeds and fertilizer blown away, the fine fertile clays and silts lost. As it moved across the fields the dust lacerated the seedlings; as it landed, it filled ditches, blocked tracks and buried other crops.

The devastation of the Dust Bowl caused much suffering, but it was a lesson well-learned. Since then soil scientists and agriculturalists have been

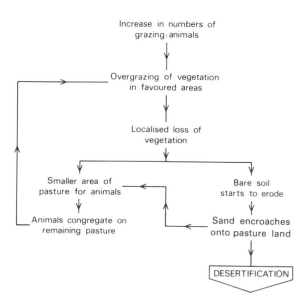

Figure 24.14 The effects of increasing grazing intensity on desertification

roughness by tillage. Based on this information he can advise the farmer on the best way of avoiding wind erosion.

Desertification

The experience of the Dust Bowl has focused attention upon the sensitivity of semi-arid areas and raised concern about long-term changes in these areas. In recent years a more general question has been asked: are the deserts advancing? Some observers certainly think so; one reported in 1975 that the desert in the Sudan had advanced 90–100 km south during the preceding seventeen years. But the rate and extent of such advances are difficult to measure. There are numerous reasons for this. The first is that desert advance is often a slow and subtle process. It rarely occurs in the form of a clearly advancing dune, but as a gradual and irregular deterioration in vegetation. The second is that from year to year marked changes take place in response to variations in rainfall, disguising the general trend. Third, desertification does not always take place at the desert margins. Often it affects sensitive areas, hundreds of kilometres from the present deserts.

What causes desertification? In part, perhaps, the process results from climatic change. More prolonged, more frequent or more intense periods of drought may prevent the vegetation re-establishing itself, so that the land becomes bare and susceptible to erosion. In addition, man undoubtedly plays a vital part, just as he did in the Dust Bowl. Overgrazing, especially during dry years, reduces the vegetation cover. Thus a series of positive feedback effects occur which lead to desertification. As the vegetation disappears the animals become concentrated on the surviving pasture; this too is overgrazed. As the soil becomes exposed, it is subject to wind erosion. Consequently the vegetation cannot regenerate so easily. The process speeds up and, in the end, land which was once pasture becomes desert (Figure 24.14). Over time 'the degraded patches, like a skin disease, link up to carry the process over extended areas'. The present problems in the Sahel area of Africa demonstrate this point vividly.

Lacustrine, coastal and marine systems

Water at the earth's surface

Water covers 70 per cent of the earth's surface. The overwhelming proportion of this area is ocean; only a fraction – less than 0.02 per cent by volume – is made up of fresh-water lakes and saline inland seas.

As we saw in Chapter 15, the movement of water through the oceans is one of the main processes by which energy is transferred from one part of the globe to another. In addition, the oceans are important components of the hydrological cycle, acting as huge stores for water running off the land and as the major sources for evaporation and the return of water back to the atmosphere. They are also vital parts of the landscape system, for much of the debris washed, blown or scraped from the continents is carried into the oceans. At the margins of the land, waves and ocean currents attack the rocks, eroding and depositing debris along the coastline. Ocean currents transport the material over long distances, along the coast and out to sea. In the still, dark depths of the oceans, the debris slowly accumulates. It is trapped there, with no ready escape. It lies there until plate movements and continental collisions force the ocean floor into mountain chains, or subsume the material into the mantle.

Lacustrine systems

Lakes as models of ocean systems

The oceans are clearly major features of the earth's surface, but we know remarkably little about them. The reason for our ignorance is simple; we can see the struggle between land and sea at the shoreline, can measure the rates of cliff erosion or beach formation and trace the movement of sediment along the coast, but in the mid-ocean areas geographic study is almost impossible. The waters are too impenetrably deep to be observed directly by divers; all we can do is to take observations with instruments lowered into the ocean, and extract samples of the sediment on the ocean floor with boring equipment mounted on ships. Otherwise we are dependent upon the clues we can pick up from geological studies of ancient sea-floor sediments now preserved in the rocks of the continents.

It would be useful, therefore, if we could gain at least a general impression of ocean systems by studying a simpler, more manageable environment – a natural model, as it were, of the oceans. And indeed we can, for many of the processes operating within these vast oceans take place microcosmically in lakes. The analogy is by no means perfect; their mode of formation is different; the magnitude of the forces acting within lakes is much less than in oceans, yet the sediments may be of a similar size. Nevertheless, they form a fascinating introduction to ocean systems, and they provide a 'model' which can be studied almost anywhere.

The formation of lakes

The word **lake** conjures up different images to different people; the vast water bodies of the Great Lakes or Lake Victoria, the mountain tarns and pools found in glaciated uplands, the spectacular caldera-rimmed lakes such as the Lac d'Issarlès in France. As this indicates, lakes are formed by a variety of processes: by subsidence of the land, volcanic action, warping and folding of the surface, differential erosion, or by damming of river valleys with sediment. These and many other processes of lake formation are illustrated in Plate 25.1 and Figure 25.1.

Inputs and outputs of lacustrine systems

Rivers draining the surrounding land provide the main inputs to lakes; they supply water and sediment and dissolved substances. Rainfall, too, introduces material; not only water but also dust, gases and solutes washed from the atmosphere. In addition small quantities of water may seep into the

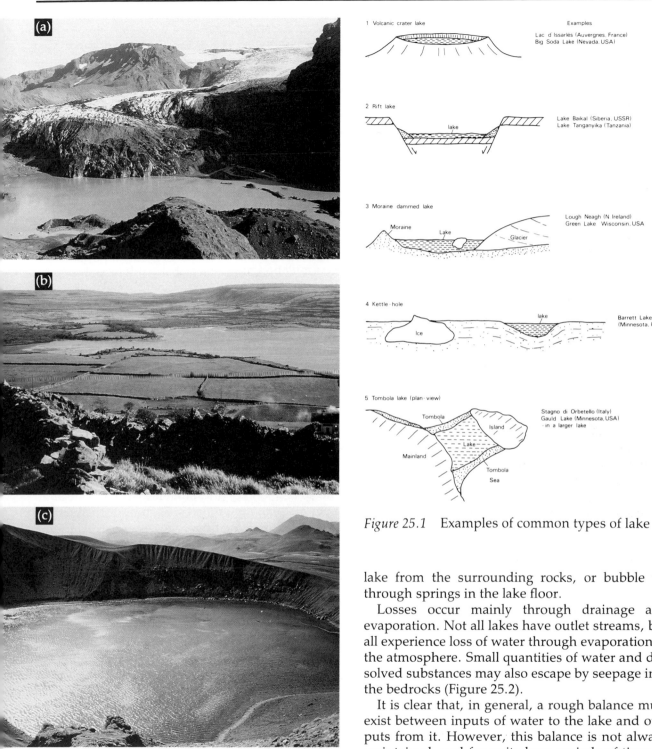

Plate 25.1 Examples of lakes: (a) an ice-marginal lake, Steinholtsjökull, Iceland (photo: A. J. R. Briggs); (b) solution lake in limestone, the Burren, Co. Clare, Ireland (photo: P. A. Smithson); (c) volcanic crater lake, Krafla, Iceland (photo: D. J. Briggs)

Figure 25.1 Examples of common types of lake

lake from the surrounding rocks, or bubble up through springs in the lake floor.

Losses occur mainly through drainage and evaporation. Not all lakes have outlet streams, but all experience loss of water through evaporation to the atmosphere. Small quantities of water and dissolved substances may also escape by seepage into the bedrocks (Figure 25.2).

It is clear that, in general, a rough balance must exist between inputs of water to the lake and outputs from it. However, this balance is not always maintained, and for quite long periods of time an imbalance may exist. As in any system, this leads to changes in the storage within the lake; in other words, the amount of water held in the lake changes and the lake level rises or falls. When inputs exceed outputs, the lake level rises; when outputs exceed inputs the lake level falls.

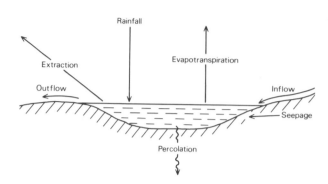

Figure 25.2   The water budget of a lake

This has happened on a large scale. In many parts of the southern USA and Africa, for example, the Quaternary Period saw dramatic fluctuations in lake levels due to variations in climatic conditions. At times, areas which are now desert or semi-desert were much more humid, for disruption of atmospheric circulation by changes in the extent of the polar ice caps brought rain-bearing winds to these regions. Rainfall increased, and the small desert playas were transformed into vast water-bodies (Figure 25.3). These periods were known as **pluvials** (from the Latin **pluvius** meaning rain). In between times, in the **interpluvials**, arid conditions returned; losses by evaporation may have increased while inputs from rainfall fell. The lakes shrank or disappeared.

Disparities between inputs and outputs also occur in the much shorter term, due to yearly or seasonal variations in rainfall and evaporation. In many temperate regions there is a tendency for lake levels to fall during the summer months, when rainfall declines and evaporation increases. Some lakes even vanish completely, for the water table, too, may fall, so that the lake drains away into the underlying, permeable bedrock (e.g. the Devil's Punch Bowl in the Brecklands of East Anglia).

Short-term climatic oscillations also produce variations in lake levels. The level of Lake Victoria, for example, fluctuates in close harmony with rainfall inputs. For a long time these variations coincided with sunspot activity, and it was postulated that this controlled rainfall in the area – and therefore lake level. But more recent studies have shown that the fluctuations are in fact much more complex, although still related to rainfall (Figure 25.4). Lake Eyre in Australia shows a similar pattern of variation. At times, within quite short periods, its size and depth may change dramatically (Figure 25.5).

Sedimentation in lakes

The waters flowing into lakes often carry vast amounts of debris swept from the surrounding land. So long as the water is flowing rapidly, this material is carried in suspension, but as the waters enter the lake their velocity declines, the sediments start to settle out and material accumulates on the lake floor.

Sedimentation produces a variety of features. At the mouth of the inlet stream, **deltas** develop as coarser debris is deposited. As we saw when considering fluvial processes, the material builds up as dipping layers of sediment, each new flood tending to erode debris from the top of the delta which is deposited at the delta front (Figure 25.6). The rate of accumulation may be rapid. The Rhine is estimated to add 2,790,000 m³ of sediment each year to the Lake Constance delta.

Further away from the stream mouth, the finer materials collect: clays and silts which sink slowly to the bottom of the lake and produce layer upon layer of new sediment. Often, where the waters of the lake are still and quiet, and where seasonal variations in the input of material occur, these form laminated clays or varves. As we saw in Chapter 23, varves are formed in pro-glacial lakes. Each spring, as the snow and ice melts, water pours into the lake, bringing with it coarse debris which collects on the lake floor. Each autumn and winter, as the land refreezes and stream discharges decline, finer material accumulates. Each year, therefore, a pair of **laminae** are deposited, one coarse and one fine. Over the years a rhythmic sequence of these varves is formed. So regular are these sequences that varved clays and silts have been used to date the age of the lake; simply by counting the number of annual varves in the sediments. Moreover, changes in the thickness and particle size of the varves can give clues to the climate at the time of deposition. Rather like tree-rings, these sediments preserve a record of annual changes in the environment.

The material swept into the lake is not always allowed to accumulate in peace. Often it is reworked and redistributed, in the process forming a range of depositional features. It accumulates at the lake margin as **beaches**, and it is carried down the shoreline by currents and deposited as **spits** and **bars** (Figure 25.7). At the same time, the waves and currents attack the shoreline, dislodging sediment and cutting **cliffs** and **wave-cut notches** in the more resistant rocks.

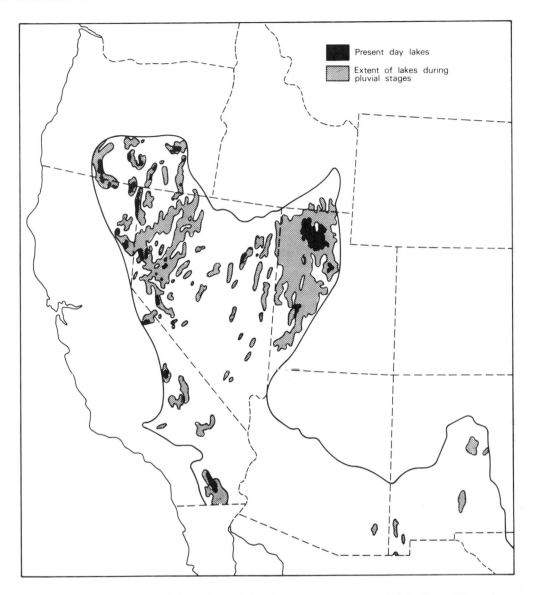

Figure 25.3 Changes in the extent of the lakes of the Great Basin region, USA (from Thornbury, 1954)

We mentioned earlier that a rough balance tends to exist between inputs and outputs of water in the lake system. But what about the sediment? What happens to this material?

The answer is that it remains trapped within the lake. Unlike the water it has no escape. Over time it accumulates in the lake; the storage of sediment continually increases. We can see the process happening in our local pond or pool; we know that it happens in reservoirs. Eventually, the sediment builds up, encroaching from the margins, reducing the capacity of the lake until it is nothing but a boggy, muddy depression. As it does so, as land is produced where once there was water, vegetation becomes established, and the old lake becomes no more than a memory. In the once glaciated lowlands of Britain and Ireland the evidence of these processes is common. Old kettle-holes, produced by melting of stagnant ice during the glacial period, were once lakes. In the centuries since they were first formed they have been slowly infilled by sediment and peat growth and invaded by vegetation.

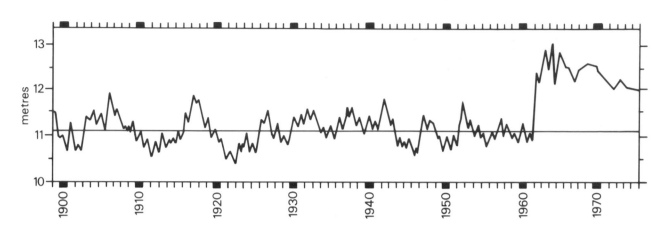

Figure 25.4   Variations in the level of Lake Victoria, 1899–1975 (from Lamb, 1977)

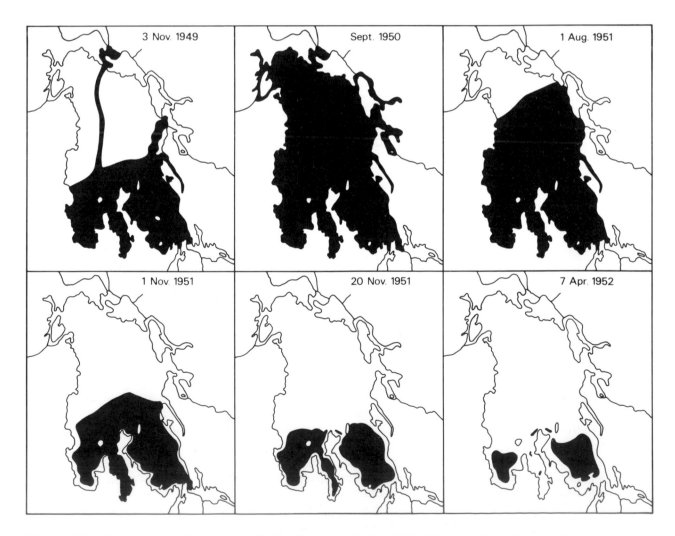

Figure 25.5   Variations in the extent of Lake Eyre, Australia, 1949–52 (from Bonython and Mason, 1953)

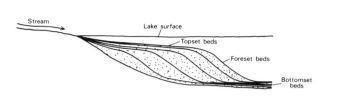

Figure 25.6 The structure of a simple delta

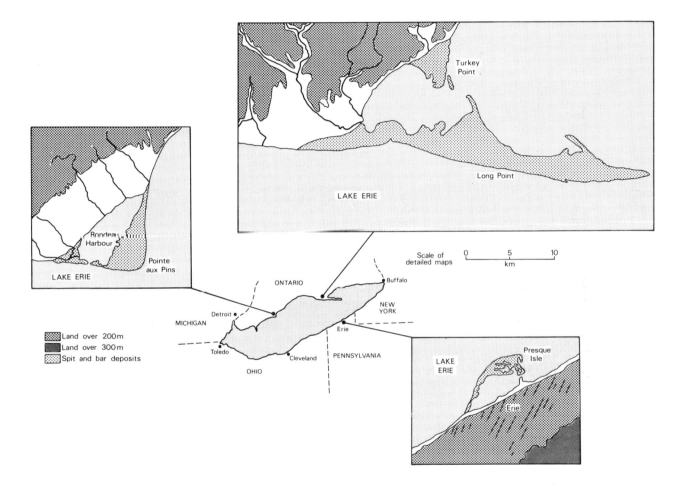

Figure 25.7 Lacustrine spits and bars, Lake Erie, USA

The world's oceans

We have a picture, now, of the processes operating within lakes. We can represent the broad structure of the system as in Figure 25.8. To what extent does that picture provide a model of marine systems? Can we now apply it to that 70 per cent of the earth's surface made up of ocean?

Clearly the scale is rather different. Lake Superior is a mere 83,300 km² in area; the huge inland sea of the Caspian 436,400 km². The Pacific, on the other hand, is an estimated 165,384,000 km², and the Atlantic about half that size (Table 25.1). The depths of the oceans, too, are orders of magnitude greater. The deepest known lake, Lake Baikal in Russia, reaches a depth of about 1740 m; the Pacific approaches depths some six times that value (Table 25.1). Different processes also operate within the world's oceans. As we have seen, crustal formation is occurring in the mid-ocean ridges as magma wells

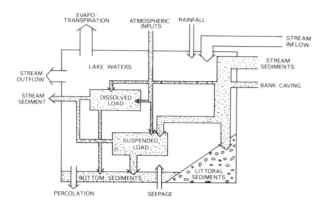

Figure 25.8　A model of the lacustrine system

changes and expansion and contraction of the polar ice caps caused fluctuations in the water budget of the oceans. During cold, glacial periods the ice caps grew and locked up vast volumes of water. The sea-level throughout the world fell. As the climate ameliorated, and the ice melted, the sea-levels rose again. These variations, however, were super-imposed upon a longer-term decline in sea-level related to a gradual increase in the volume of the ocean basins (Figure 25.9).

The structure of the oceans

The oceans can be subdivided into three main zones (see Figure 15.2, p. 232). In the open ocean areas occur the **ocean basins**. These are many kilometres deep – much of them between 3 and 6 km – and they are underlain by crustal material which is both shallower and younger than that beneath the continents. In these areas the processes of sediment accumulation are slow and poorly understood. Over the last 2–3 million years, depths of no more than a few metres of sediment have built up in many places. The main exceptions are at the margins of these zones and the adjacent continental shelves, where vast accumulations have occurred. These zones act as the final resting place for much of the sediment swept off the continents.

The **continental shelves** are in most areas only a few kilometres in width; the mean width is about 75 km. They are also relatively shallow. On average they are about 130 m deep at their outer margins. They occupy about 8 per cent of the ocean area. Most of the sediment carried into the ocean collects initially on the continental shelves; a proportion remains there almost indefinitely, until continental collisions and crumpling force it up into mountain chains; but a proportion is stored there before being

to the surface and as the ocean floor spreads as a result of plate movement. Simultaneously, in the deep marine trenches, the crust is subsiding and the lithospheric plates are being consumed. Inputs of sediment by wind and inputs of both debris and water by the calving of ice sheets are also more important in the oceans, while the interconnections between the main oceans mean that major transfers of water occur between them. It is also inconceivable, of course, that oceans should suffer infilling as do the smaller lakes. There are, therefore, limits to the validity of our model.

Nevertheless, the general principles remain valid. Precipitation, streamflow and seepage provide inputs in the same way; evaporation and percolation into the crustal rocks cause losses of water. Over time, as well, the same balance between inputs and outputs is important, and during periods when there is an imbalance the sea-level changes. It did so in the Quaternary, when climatic

Table 25.1　*The size of the world's lakes and oceans*

Name	Origin	Area (km²)	Maximum depth (m)
Pacific Ocean	–	165,384,000	11,730
Atlantic Ocean	–	81,484,800	9,225
Indian Ocean	–	75,110,400	7,450
Caspian Sea	tectonic	436,400	946
Lake Superior	glacial erosion	83,300	307
Lake Victoria	tectonic	68,800	79
Lake Baikal	tectonic	31,500	1,740
Crater Lake	volcanic	55	608
Lake Windermere	glacial erosion	15	66

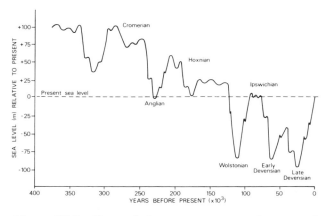

Figure 25.9 One of the most commonly quoted reconstructions of sea-levels during the Quaternary, developed by Fairbridge (1961). Different reconstructions have been made by other researchers, and there is considerable controversy about many details of the sequence

carried over the **continental slope** and into the ocean basins.

The continental shelves fringe the continents, and it is on the inner margins of the shelf that most activity probably occurs. Here are the main dumping grounds for the streams which enter the sea; here the oceans and the land interact. And it is in the coastal zone that we can see landform processes at work.

Coastal form and process

The nature of coastal systems

Coastlines, whether in lakes or in oceans, vary considerably in structure and character. They include the steeply shelving margins of mountainous regions and the gentle sweep of the aggrading, lowland areas. They include cliffs and bays cut into resistant rocks (Plate 25.2) and the extensive salt marsh and mud flats which characterize coastlines such as the Wash and Morecambe Bay (Plate 25.3). Generalizations about the structure of coastal systems must therefore be made with caution.

In many coastal areas, however, it is possible to identify three main zones. At some distance from the coastline there occurs a **deep-water zone**, where the action of the waves does not reach to the sea floor. This gives way landward to what is often called the **offshore zone** – the region in which the waves start to interact with the bottom. Within this zone a mutual reaction occurs; the waves affect the sea floor, which in turn modifies the waves. Eventually, the waves start to break and enter a third zone: the **nearshore zone** (Figure 25.10).

Processes in the deep-water zone

The coastal areas of the world are dominated by the progress of energy landward from the open sea. It is this energy which transports material to and along

Plate 25.2 A hard-rock coastline on the Dingle Peninsula, Co. Kerry, Ireland (photo: D. J. Briggs)

Plate 25.3 The degraded edge of a saltmarsh at Silverdale in Morecambe Bay, Lancashire (photo: D. J. Briggs)

411

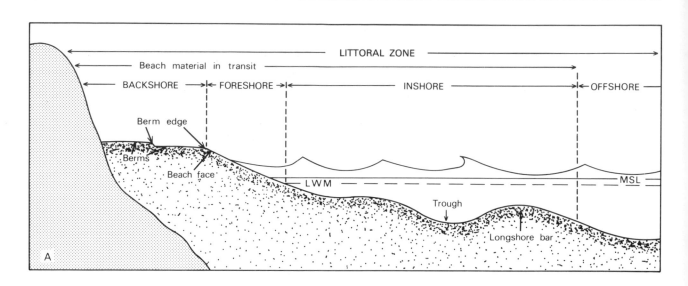

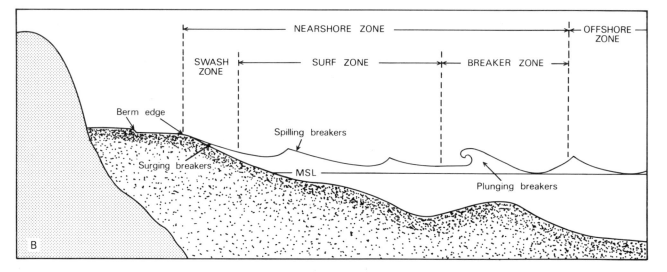

Figure 25.10 The zonation of coastal systems: (A) according to beach profile; (B) according to wave action (from Derbyshire *et al.*, 1979)

the coastline, and it is this energy which attacks the land and erodes the coastal rocks. It is within the deep-water zone that this energy gathers.

The energy is derived mainly from the development of **waves**. These are generated in turn by wind. The friction and pressure of the moving atmosphere set up an oscillation within the water, by which individual water particles move in a circular manner. It is important to appreciate that the water particles themselves do not travel forward with the wave. The apparent motion of the wave is produced by transmission of energy through the water, from one particle to another

(Figure 25.11). The orbital movement of the water declines with depth, the diameter of the orbit being halved for every increase in depth of about 0.11 L (where L is the wavelength).

The character of the waves can be described by their wavelength, height and rate of progression. The **wavelength** is largely a function of the wind speed and the **fetch** (uninterrupted travel distance) of the waves. **Wave height** is also controlled by wind speed and fetch, as well as the gradient of the water surface. This develops because of the piling up of water in the deep-water zone by tidal currents. In general, wave height gives a good indication of the wave energy, the two being related as follows:

$$e \propto h^2$$

where e is the energy and h the mean height of the waves.

The oscillation of a water particle takes place within a single wave. It rises to the wave **crest** and sinks within the **trough** (Figure 25.11). The speed of the wave thus depends upon the rate at which this orbit occurs. In general, wave speed is closely and positively related to wavelength, although there is a tendency for higher waves to travel faster than low waves.

It is this wave motion which is responsible for the main energy transmission through the deep-water zone and into the shoaling zone. There are, however, other processes operating within the deep-water zone which may supplement this energy. **Tidal currents** may be important, although their force is generally rather limited. In many cases they travel at little more than 0.2 m s⁻¹, and they affect only the surface of the ocean. Locally, however, more active tidal currents develop, particularly where they are constricted in some way. Thus, off the New South Wales coast of Australia, a southerly current reaches 2.0 m s⁻¹, while tidal currents sweep through Hell Gate in New York's East River at over 2.5 m s⁻¹.

Tidal currents may therefore act as a means of energy transmission into the offshore zone, and,

where they are deep, they may disturb and entrain sediments on the floor of the deep-water zone. Certainly within the English Channel there is evidence of marked and consistent movements of sediment on the sea floor, well beyond the reach of surface waves. The tides have an additional effect, however, for by piling up water in the coastal zone they extend the influence of coastal processes over a considerable height range. The **tidal range** is generally no more than a few metres, but where the tides are channelled into a narrow, exposed inlet, much greater tidal ranges are found. In the Bristol Channel, in England, for example, the tidal range is in the order of 12 metres. At the head of the Bay of Fundy, Nova Scotia, Canada, the tidal range reaches 15.4 m.

Processes in the offshore zone

Within the offshore zone, the interaction between sea floor and wave action takes effect. Significant disturbance of the floor only occurs when the depth of water is less than half the wavelength. At this point, the horizontal motion of the water particles at the bottom starts to be retarded by friction with the bed. This produces a marked difference in the velocity of water movement between the water surface and the sea floor which results in the wave *breaking*. The first signs of the wave breaking occur when the water is about twice the depth of the wave height; at this point the waves become asymmetrical and peaked. At a depth of about 1.3 times wave height, they break (Figure 25.10). At the same time, as the wave becomes increasingly influenced by the sea floor, the wavelength decreases and increased disturbance of the bottom sediments occurs.

It is within this zone, therefore, that waves are able to entrain the bottom sediment and carry it landward. Finer material only is moved in this zone, because the energy available to carry the sediment is relatively low at the bottom of the water, and because only little turbulence occurs. Removal of the finer particles from the bottom deposits leaves a coarse-grained **lag** material (Figure 25.12).

The nearshore zone

Breaking of the incoming wave represents the boundary between the offshore and nearshore zones. In the nearshore zone the oscillatory motion of the particles gives way to a forward movement, creating a **wave of translation**. This runs landward, expending its energy by friction with the rising

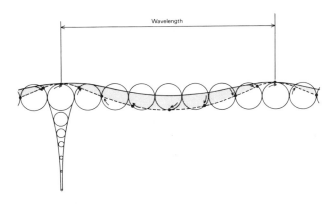

Figure 25.11 Water movement in an ocean wave. Each circle represents the orbit of individual water particles. Note that the diameter of the orbit declines with depth such that at a depth equivalent to half the wavelength the diameter of orbit is only 4 per cent of that at the surface

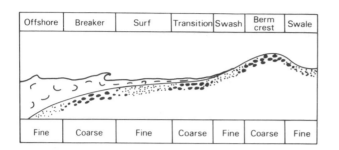

Offshore	Breaker	Surf	Transition	Swash	Berm crest	Swale
Fine	Coarse	Fine	Coarse	Fine	Coarse	Fine

Figure 25.12   Variations in sediment size across a sand beach

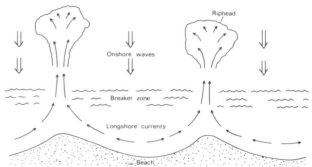

Figure 25.13   The development of rip currents

slope of the bottom and by collision with the backwash from earlier waves.

The nearshore zone is often divided into a number of more specific areas related to the character of this water movement and its effects upon beach morphology. At its outer margins, where the waves break, is the **breaker zone**. This grades landward into the **surf zone**. Here, the action of the incoming waves is often accompanied by longshore currents, and by seaward movement of the water. Particularly important are **rip currents**. These are localized streams resulting from the build up of water on the beach until it bursts back in the form of fast-flowing narrow currents. They may cut through the breakers and attain speeds of over 1 m s^{-1} for periods of several minutes. They represent a form of output from the landward zones of the coastal system by which water and sediment are returned to the open sea (Figure 25.13).

Landward of the surf zone is the **swash zone**, within which the water runs up the beach as an increasingly shallow film of turbulent water. We only have to watch this process on the beach to see that, again, there is a reverse process in operation; the **backwash**. As the incoming wave loses energy in the swash zone the water starts to run back down the slope. Much of it percolates into the beach, but some flows back into the surf zone where it collides with the next, incoming wave. This zone of collision is often known as the **transition zone** (Figure 25.12).

As ever, processes within the foreshore zone must be in some form of equilibrium, such that the rate of input of water to the beach is balanced by the return of water to the sea. As we have seen, backwash and the action of rip currents perform this function to some extent. But not all the water returns directly. Often there is a local build up of water at the coast which results in a lateral flow

down-coast. In this way **longshore currents** are created. In addition, tidal currents may be channelled along the coast, particularly in the narrow zone landward of the breaker zone; here they may reach velocities of several metres per second.

One of the most common causes of longshore movement, however, is wave **refraction**. The incoming waves are slowed by friction with the sea floor, and as a result they become progressively slower as the water becomes shallower. Where the incoming waves approach the shore obliquely, this leads to a bending of the waves parallel to the coast (Figure 25.14). Even so, the waves still tend to reach the shore at an angle, and the swash runs obliquely up the beach. The backwash, on the other hand, runs directly down the beach slope, to give a net longshore movement to material transported by the water (Figure 25.15). The process is known as **longshore drift**.

Wherever waves approach the coast obliquely wave refraction and longshore drift occur. These processes are common, therefore, around head-

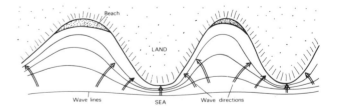

Figure 25.14   Wave refraction along a headland and bay coastline. The effect of refraction concentrates wave energy on the headlands and results in erosion of the cliff; in the bay, wave energy is less and deposition tends to occur

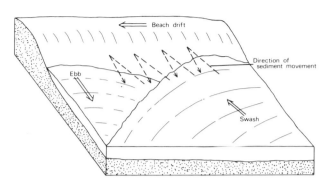

Figure 25.15 Sediment movement by longshore drift

lands and in bays (Plate 25.4), where the waves are bent by friction with the changing coastline. They are also common where on-shore winds blow obliquely on to the coast.

The movement of material by longshore drift may be remarkably rapid. On coasts in California, the effects of wave refraction give rise to longshore movements of material at rates of around 30 cm s⁻¹, and rates of 1.2 m s⁻¹ have been recorded. The pathway followed by these particles tends to be irregular, however, so that not all the movement represents longshore progress. Even so, considerable volumes of material may be moved annually by longshore drift: at Santa Barbara on the Californian coast, for example, over 200,000 m³ of sediment may be transported in a single year; 40 million m³ of sediment are estimated to move northwards along the coast of the Netherlands each year.

The coastal system

It should be clear that the three zones or subsystems which make up the coastal system are closely interactive. Outputs from one represent the inputs to the next. It is also clear that, over time, a general balance tends to exist between them. The progress of water and sediment landward through the action of on-shore waves and currents should be balanced by an equal return of water and sediment if the system is not to change. In reality, of course, change does occur. Sediment accumulation may take place at the coast to create aggradational beaches; or removal of sediment may exceed supply with the effect that the coast is eroded and the shoreline retreats. Moreover, as we have seen, movement of both sediment and water occurs not only in an onshore–offshore direction, but also laterally. Longshore currents result in the circulation of water and sediment down-coast. It is the

interaction of all these processes that determine the shape and the landforms of the coastline.

Coastal landforms

Landforms of deposition

Most of us are familiar with **sandy beaches**. We may not have visited such exotic and spectacular examples as Long Beach in California or Daytona Beach in Florida, but we are likely to have seen more everyday examples. Beaches of this type form in relatively low energy environments where there is an ample supply of fine, transportable material. Where the material is coarser, and the wave energy higher, **pebble beaches** tend to form. Both sand and pebble beaches consist in general of a sloping area which is covered by water at high tide and one or more almost parallel ridges or **berms**.

Both the shape and the composition of beaches are products of the action of waves and currents operating within the foreshore zone, and both tend to change over time as the balance between deposition and erosion alters. If we were to monitor these changes, by repeatedly surveying the **beach profile** or analysing the character of the sediments, we would discover that beaches are highly dynamic features (Figure 25.16). Detailed studies would also show that there is a close interaction between the beach profile, the size and shape of the sediments and the processes operating at any point on the beach.

Plate 25.4 Wave refraction in a marine embayment (photo: J. D. Hansom)

415

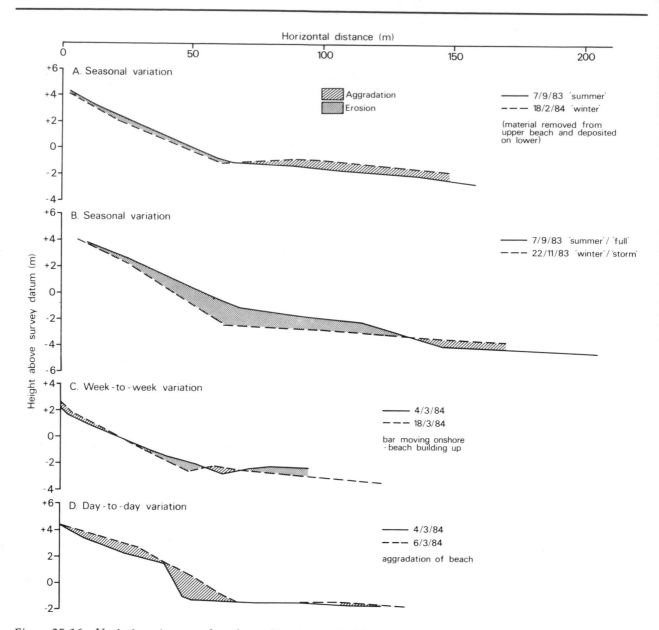

Figure 25.16 Variations in cross-beach profiles on the Holderness coast of eastern England (data provided by Susan Mason)

At the outer margin of the surf zone, for example, where the waves are breaking, there is a relatively high energy area in which there is considerable turbulence. This allows the water to pick up material from the bottom, and all but the coarsest particles are carried landward. The result is the development of a coarse lag material in the breaker zone (Figure 25.12). As the water moves inland, the friction with the bed, and the gradual shallowing of the water, results in a reduction in wave competence, and the progressive deposition of the coar-

ser material. The sediments therefore become finer.

Near the landward margin of the surf zone – in the transition zone – the collision of swash and backwash again creates a high-energy, turbulent zone, and coarser materials accumulate. Beyond, in the swash zone, the waters run relatively gently on to the beach, losing energy rapidly as the beach becomes steeper, and transporting only the finer sediment.

The berms are created mainly from material thrown on to the beach by storm waves. They are, therefore, composed of coarse material which has been transported at times when wave energy is high. Often two berms occur, a lower **summer berm** and a higher **winter berm**, reflecting the more intense storms of winter. In addition, on sandy coasts, wind action may redistribute the berm materials and create a series of **sand dunes** inland (Chapter 24).

All these processes interact to produce the characteristic profile of the beach and the sequence of sediments shown in Figure 25.12. Superimposed upon this general pattern, however, are more local features. **Beach cusps** form due to the fan-like intrusion of waves on to the beach (Plate 25.5); accumulations of shell material may occur at the upper tidal limit. **Ripples** may also form.

Material moved by longshore drift tends to accumulate in areas where the current is impeded. This may occur for a variety of reasons: due to the occurrence of shallower water, or to the intrusion of a headland into the flow, or to the meeting of two, opposing currents. The sand and silt which are deposited as a result often form a spit or sand bar. **Spits** develop through the accumulation of sediment outwards from the shore. They tend to be self-producing to some extent, since once deposition has started, the effect of the spit is to reduce current velocity further and encourage further deposition (Figure 25.17; Plate 25.6a).

Plate 25.5 Beach cusps on Anniestown Beach, Co. Waterford, Ireland (photo: J. D. Hansom)

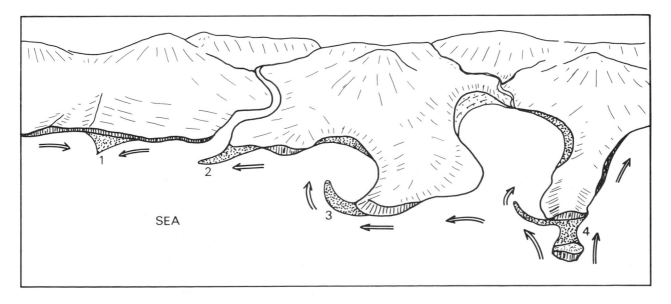

Figure 25.17 Formation of coastal spits and bars: (1) cuspate bar; (2) bayhead bar; (3) spit; (4) tombolo and spit

Plate 25.6 Coastal spits and bars: (a) cuspate spit at Ferry Point in Youghal Bay, Co. Waterford, Ireland (photo: J. D. Hansom); (b) shingle bar and lagoon at Slapton, Devon (photo: D. J. Briggs); (c) tombolo on Ardley Island, South Shetland Isles, Antarctica (photo: J. D. Hansom)

Sand bars take a variety of forms. They often develop across bays or at the mouths of rivers (Plate 25.6b) and a specific type – the **tombolo** – may be formed between the mainland and an island (Plate 25.6c). Great and Little Orme in North Wales are examples. Coastal bars are not only formed of sand; gravel may also accumulate in the same way, as at Chesil Beach in Dorset. Such features do not always protrude above the water, and submerged sand bars are a traditional hazard to shipping.

Elsewhere, the accumulation of fine sediments in very sheltered inlets and estuaries, or behind spits, results in the formation of **mudflats**. In time, as these build up, vegetation may become established to produce **salt marshes**. The plants which invade these areas need to be highly tolerant of salt and of occasional flooding, but once established they act to trap the sediment and accelerate deposition. Eventually, the salt marsh builds up until it stands above all but the highest tides (Plate 25.3).

Landforms of erosion

Rocky coastlines such as those of Cornwall and western Ireland (Plate 25.2) present an exciting contrast to the calm of many sandy beaches. As the roar of storm waves on the rocks indicates, these coastlines develop in relatively high energy environments – those exposed to waves travelling over long distances, and facing strong onshore winds.

One of the most dramatic features of these coasts is the **marine cliff**. This is formed by the constant attack of the sea, particularly against resistant, massive rocks. It is not only the power of the breaking waves which erodes the cliff, but also the sediment carried by the water. Boulders and large pebbles may grind incessantly against the foot of the cliff, while, during periods of storm, coarse material may be thrown against the rock face and chip material from the surface. These processes are supplemented by more subtle effects. Wetting and drying of the rock face, growth of salt crystals in minute cracks and joints in the rock, and even the boring of **marine molluscs** may help to erode the cliff. As the cliff face is undermined, and as the debris falls into the sea, the waters wear the material down and carry away the fragments; they then serve as material with which to continue the attack upon the cliff.

The vigour of this attack is often greatest just below the water surface, and thus erosion of the cliff is at a maximum here. Over time, a **wave-cut**

notch may form at the base of the cliff. In time, as well, the cliff tends to retreat as erosion at the base undermines the rock face and causes slumping and rockfalls. Cliff retreat leaves behind a level or gently sloping **shore platform** (Plate 25.7). This platform acts to reduce the power of the incoming waves in much the same way as a beach does, so it eventually protects the cliff from further erosion. For this reason, there is a limit to the width which most shore platforms reach. This is a case of negative feedback; features develop which damp down the initial changes.

The shore platform may itself be attacked. When the tide is out, erosion of its outer margins may occur, while the exposed inner parts are subject to wetting and drying and salt weathering. Marine molluscs are also particularly effective in the inter-tidal zone, and it is possible to identify a sequence of erosional features along many rocky coasts, reflecting the interaction of biological and physical weathering and erosion.

Variations in the intensity of wave attack, or in the resistance of the rocks, in a long-shore direction result in differential erosion. A **headland-and-bay** coastline therefore develops. The softer rocks are cut back, leaving the harder rocks as headlands. Again it is possible to detect in the evolution of these coastlines a form of negative feedback, for the embayments are eventually sheltered from the incoming waves by the headlands, and longshore drift tends to carry material into the bays, slowly silting them up. At the same time, the exposure of the headlands to active wave attack results in their slow destruction. Thus, a form of dynamic equilibrium is established.

Coral coastlines and barrier reefs

Where the coastal waters are clear and unpolluted, where they are warm and rich in nutrients, **coral reefs** may develop. The association of corals and algae result in the secretion of calcium carbonate in the form of intricate growths, attached to the sea floor. Colonies of corals produce extensive deposits, which build outwards into the sea in the form of a **fringing reef** (Figure 25.18). The coral is attacked by the waves and broken down into coral-line sand which accumulates as beach material, and is often cemented into **beach rock**. The waves carry clean water with abundant food materials which feed new coral growth. A balance between the growth and destruction of the reef is therefore attained.

Plate 25.7 A shore platform exposed at low tide, Quilty, Co. Clare, Ireland. Note the way in which erosion has picked out the rock structure (photo: J. D. Hansom)

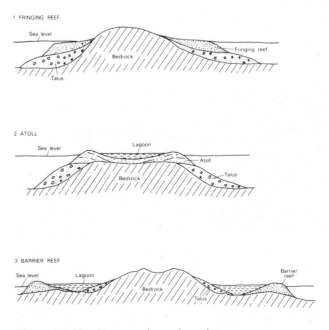

Figure 25.18 Types of coral reef

419

The growth of coral reefs is slow, and at intervals changes in sea-level may occur which affect the development of the reef. A slow rise in sea-level results in the drowning of the old, dead inner reef and the formation of a **lagoon** behind the actively growing fringe. In this way a **barrier reef** is formed. The lagoon is linked to the open sea by channels cutting through the barrier reef. Possibly the most famous examples are found along the eastern coast of Australia – the Great Barrier Reef.

Small barrier reefs have formed around isolated volcanic islands in some parts of the world, particularly the Pacific. These islands are themselves interesting features and their existence has for long puzzled geomorphologists. Today it is believed that they are a result of volcanic activity along the mid-ocean ridges. Fringing reefs have developed at their shorelines and, over time, the sea-level has risen, possibly due to the melting of the polar ice caps and the vast ice sheets of the Pleistocene. The coral has grown to keep pace with the rising waters, but ultimately the island itself has been drowned. An **atoll** is therefore created: an almost circular reef enclosing an empty lagoon.

As we have mentioned, coral development is closely dependent upon coastal conditions. Temperature is particularly critical, and coral growth is only active where the water temperatures are above 20°C. In the geological past these conditions have been widespread, and rocks produced by coral development – coralline limestones – are extensive. Today, however, the distribution of active growth is much more restricted, and in many places it is being further inhibited by the action of man. Coastal pollution and the attack of the coral by marine organisms accidentally introduced or encouraged by man, is posing a threat to several reef areas.

Coasts and man

For centuries, man has watched as the sea has eaten away his coastline. Until quite recently he has been powerless to do much about it. The East Coast is an example. Near Withernsea, the cliffs were eroded back 130 m in the twenty-five years from 1852 to 1876; and since Roman times it is estimated that the Holderness coast has retreated some 4–5 km, devouring many villages and large areas of agricultural land in the process (see Plate 20.4, p. 318). In East Anglia, at Covehithe, a storm surge in 1953 caused 12 m of erosion in a single day.

Over the last 200 years, however, man has attempted to arrest the action of the sea and protect the coastline from erosion. In many areas **groynes** have been constructed to catch the material which is being swept away by longshore drift (Plate 25.8). The groynes are normally arranged in series, often of different length to give a tapering sequence. Unfortunately, there is considerable disagreement about which way the taper should run – whether the groynes should become shorter up-current or down-current, so they are not always as effective as intended. Moreover, trapping of the sediment in one place may encourage erosion elsewhere, for the areas down-coast of the groynes are deprived of material which once protected them. In other words, the coastline acts as an integrated unit; man cannot change one part of it without also affecting other parts.

It is not only coastal erosion which threatens man's use of the coast. Sedimentation is also a problem. In the past, ports of considerable commercial importance have been left stranded by the incessant encroachment of the land into the sea. Often, these processes have been exacerbated by human activity, sometimes far removed from the coastline. In the Persian Gulf, for example, the delta of the Tigris River has advanced over 100 km during the last 2500 years as silts washed down the rivers have accumulated along the coast. Vegeta-

Plate 25.8 Groynes in Aberdeen Bay, Scotland. The direction of longshore drift at the time of photography is towards the camera; hence sediment is accumulating on the far side of the groynes (photo: J. D. Hansom)

tion clearance, canal-building and drainage in the Tigris and Euphrates valleys were largely responsible for this encroachment.

Similar effects are produced today by river management. The building of dams, the diversion of streams and the control of flooding may all disrupt the sediment supply to the coast and upset the balance between erosion and deposition. In the Nile Delta, major changes in coastal conditions occurred because of the building of the High Aswan Dam. The reduction in the amount of sediment entering the Mediterranean resulted in the destruction of offshore bars which had been important in keeping the coastal waters brackish. When they were destroyed, salt water invaded the coastal lagoons and, among other things, killed off the fish which were the basis of Egypt's sardine industry.

Siltation of coastal waters is often controlled by **dredging**. This is necessary to keep channels open and prevent flooding in the river valleys. The sediment is normally taken out to sea and dumped there, but the procedure is not always successful. Tidal currents may return much of the material within a short time. In the Humber Estuary in eastern England, the movement of sediment was monitored using radioactive tracers. It was discovered that the material was being washed back into the estuary within a matter of days.

The story of man's intervention in coastal systems is not always one of defeat and error, of course. In many cases erosion has been successfully controlled, and in some instances some of the land which has been lost has been regained by **coastal reclamation**. One of the most remarkable examples is in The Netherlands, where reclamation has been going on since the Middle Ages. Despite the natural tendency for the area to be drowned due to the slow rise in sea level and the settling and subsidence of the land, the Dutch have managed to reclaim something in the region of half a million hectares. The work has not been without its problems; in 1953 violent coastal storms led to the loss of almost two-thirds of the reclaimed area. But almost all this land has again been salvaged, and, protected behind the extensive Zuider Zee dam, the sediment of the Ijssel and Vecht are being trapped in huge lakes. As these materials build up, and as the lakes are pumped dry, huge **polders** are created. Once drained and treated with freshwater to remove the salt, the land provides invaluable fertile soil (Figure 25.19).

Increasingly, man is becoming aware of the need

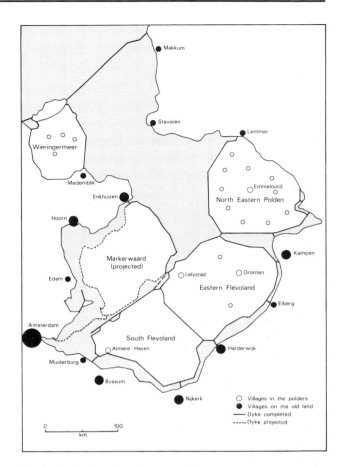

Figure 25.19   The Dutch polders

to manage coastal areas as a whole, taking account of the interactions within the system. One of the most intriguing examples of this new philosophy is the California Coastal Plan. This is aimed not only at controlling coastal erosion and sedimentation along the whole of the Californian coast, but also at managing wildlife resources, agriculture, recreation and industry in a co-ordinated fashion. Plans of this type are exciting, but it is important to remember one thing: they will only succeed if man's understanding of the system he is trying to manage is reasonably accurate. And, at that scale, failure could be highly expensive.

The continental shelves and ocean basins

The continental shelves
Just beyond the edge of the continents, the sea is relatively shallow (*c.* 130 m) and the sea-bed forms

a bench-like feature. This is the continental shelf. It forms an important zone, for it collects much of the sediment washed into the sea, and processes operating upon the shelf control what happens to this material in the long term.

Sediment is carried on to the continental shelves by the action of waves and tidal currents. Most of the material is fine, of course, for the coarser debris tends to be deposited close to the coast. However, in the past, glacial deposits have also accumulated on the shelf areas, for the Pleistocene ice sheets extended across these surfaces as they were exposed by the falling sea-level. It is estimated that sea-levels may have fallen by up to 100 m, sufficient to expose most of the earth's shelf areas. During these periods, weathering and erosion must have modified the shelves, and they must have been partially buried by sediments swept over them by the action of wind and water. Deltas and stream alluvium must have accumulated, as the rivers extended towards the retreating sea; while in some areas the ice sheets themselves plastered glacial deposits across their surface.

During periods of rising sea-level, transgression of the sea across these areas introduced coastal deposits. We may visualize the slow encroachment of the rising sea over the continental shelf; in the process all the activities we associate with modern coastlines must have acted upon the surface. Gravel and sandy beaches, cliffs – with the debris eroded from them lying at the base – and numerous other coastal features must have formed. The materials on the continental shelves are consequently varied.

Processes acting on the continental shelves

Today, three main processes affect the continental shelves. Deep tidal currents carry sediment across their surface. These are particularly active around the mouths of rivers where quite rapid currents may extend from the estuaries some way out to sea. They carry with them silts and clays washed from the land. Slower-moving currents associated with larger aspects of oceanic circulation may also be important (Chapter 15). Cool polar water moving into lower latitudes at depth may transport fine shelf sediments, while currents such as the Gulf Stream may be active on the continental shelf; around Britain a distinctive pattern of sediment movement seems to occur due to the circulation of the North Atlantic Drift through the Irish Sea and southwards down the North Sea (Figure 25.20).

The outer margins of the continental shelves

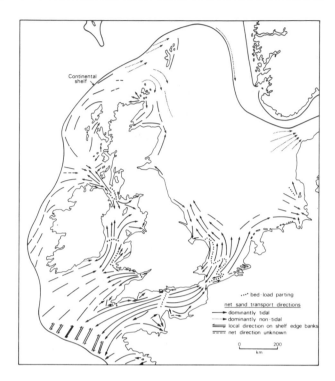

Figure 25.20 Sediment movement around the British Isles. Arrows show directions of net sediment transfer (from Stride, 1982)

merge into the continental slopes, which dip gently (mean slope 4°) for a kilometre or so, before plunging more steeply to the **abyssal plain** (Figure 15.2, p. 232). Within this area there occur many **submarine canyons**, some of immense size. The processes responsible for the formation of these canyons have been the source of much argument, but it is now believed that they were formed by the movement of sediment-laden seawater down the continental slope. Here, therefore, a third, and most important, process is in operation: the action of **turbidity currents**.

Turbidity currents are composed of dense masses of silt and mud moving over the ocean floor under the effect of gravity. As the materials roll and sludge down the continental slope they gather speed. The particles erode the surface of the slope as they bounce along, pulling more material into motion. As the surface flattens out, the turbidity current starts to slow down and the material is deposited. Deposition occurs mainly at the base of the continental slope, and produces a fan-shaped feature, similar to an alluvial fan.

The initiation of turbidity currents is only poorly

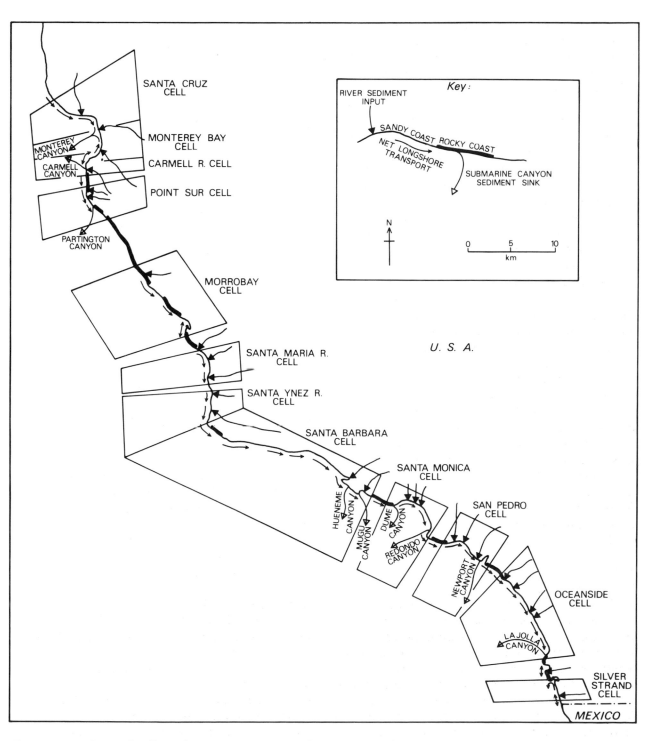

Figure 25.21 Littoral cells, submarine canyons and rivers in southern California (adapted by J. D. Hansom from Department of Navigation and Ocean Developments, 1977)

understood. Clearly it is difficult to study these processes in action, and much of our knowledge is derived from laboratory simulations and experiments. It appears, however, that convergence of deep ocean currents may cause localized erosion of the shelf surface, resulting in an increase in the concentration of sediment in the bottom waters. As we have seen, once the sediment is entrained in this way, further erosion is encouraged; the turbidity current is produced.

Cold, and often sediment-laden waters from major rivers may also initiate turbidity currents, and it is notable that many submarine canyons coincide with the mouths of large rivers (Figure 25.21); the Ganges, Indus, Nile, Niger, Rhône and Mississippi are all associated with submarine canyons. More commonly, perhaps, turbidity currents arise from the effects of localized slumping of the sediment on the continental slope. This may be due to oversteepening of the deposits laid down by submarine sedimentation, or by the trigger action of earth tremors or severe storms.

The effect of turbidity currents is illustrated by just such an example. In 1929 an earthquake is known to have triggered off a major slump in the Grand Banks area south of Newfoundland. This became a turbidity current which swept down the continental slope and severed the trans-Atlantic

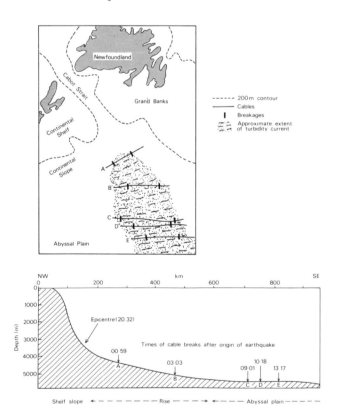

Figure 25.22 The effects of the Grand Banks turbidity current in 1929 on telegraph cables in the north Atlantic: (A) sketch map showing the positions of some of the cables and the locations of the breakages; (B) section showing the timing of breakages (after Holmes, 1965)

telephone cables which lay in its path. Over a period of $12\frac{1}{4}$ hours the current travelled almost 1000 km and cut five cables (Figure 25.22).

The ocean basins

The deepest parts of the ocean basins – the abyssal plains – are probably the least known parts of the earth. The processes which go on there are largely a matter of speculation. Only at their margins – where they receive inputs of sediment from the continental slopes – and in the vicinity of the mid-ocean ridges where they are subject to volcanic activity are there likely to be any great depths of sediment; elsewhere, they must be dark, silent and still. All that happens is the slow, perpetual accumulation of fine detritus which settles from the water above: microscopic remains of marine organisms (radiolaria) and volcanic debris and dust carried by winds from the distant land. These materials comprise what is known as the **marine ooze**.

Close to the mid-oceanic ridges, where the ocean floor is still young, the ooze is thin. Further away on the older rocks, it has been gathering for hundreds of thousands of years, more or less undisturbed. Here it reaches thicknesses of several metres and contains a record of recent earth history: of the periods of volcanic activity when the rain of dust was more intense, and of the fluctuations in climate during the Quaternary Period, when the marine organisms changed in response to variations in temperature and salinity. As we saw in Chapter 10, by analysing the foraminifera preserved in these sediments, we can reconstruct changes in global temperatures over the last 2–3 million years.

These are not the only materials on the ocean floor, however. In addition, there are rich deposits of ores which have precipitated out from the ocean waters. Manganese is particularly abundant, mainly in the form of oxide nodules, crusts and pavements at depths of 3500 to 4500 metres. Some estimates suggest that there may be as much as 4000×10^9 tonnes of manganese on the sea floors in this form. Many of the nodules also contain copper, nickel and cobalt, while copper and zinc are also locally abundant in the areas where hot, salt-rich brine has leached minerals from the volcanic rocks beneath. At present we cannot exploit these extensive reserves, but one day, perhaps when economic conditions are appropriate and the necessary technology further developed, they may provide vital sources of ore.

Conclusions

We still have much to learn about the processes operating on the sea floor and about the distribution of sediments by tides, ocean currents and turbidity currents. What is clear, however, is that the movement of sediment into and through the oceans represents one of the last stages in the debris cascade, for once on the ocean floor the material tends to be trapped. Nevertheless, this is not the end of the story. As we saw in Chapter 17, earth building and plate tectonics may result in the uplift of the ocean sediments into mountain chains or their consumption by plate subduction. In either case, the materials are recycled: when the mountains are again lifted, or when the magmas from the earth's interior are spewed again on to the surface, the whole process of weathering, erosion and deposition is started once more.

Ecosystems

General structure of the ecosystem

Man and the ecosystem

Man derives his food from the plants and animals which occupy the earth. He also tries to manage them, to control them in order to safeguard his food supply, or to increase the quantities of food he can obtain from a given area of land, or for a given amount of effort. In all these cases, he is operating within a complex system; a system founded upon the interactions between living organisms. It is referred to as the **ecosystem**.

Man's role in the ecosystem is a unique one. More than any other component he tends to control it. He acts as a **regulator**. Because of his influence on the system it is, of course, vital that he understands it, although the lesson of history is that he does not always do so. Through his ignorance he often damages the ecosystem, disrupts it with eventual cost to his own well-being. The consequences, such as famine and disease, have been seen in all parts of the world, on innumerable occasions.

What, then, is the nature of this ecosystem, on which man so clearly depends? What are the principles which govern its function?

Components of the ecosystem

We can best answer these questions by considering a simple example. Let us examine the relationships and processes involved in a small isolated woodland.

The wood consists of several different components: the trees, the understorey of ground vegetation and the animals which live within the undergrowth and in the trees. These are all closely associated (Figure 26.1). Many of the animals derive their food from the plants; others are predators that live off fellow animals. Some of the plants are parasites that live off other plants.

The relationships extend further. The plants themselves obtain their energy from the soil, the sun and from the atmosphere. Water and nutrients are taken up from the soil; vital elements are extracted from the atmosphere; the whole process, as we might expect, is powered by energy from the sun. The network of relationships therefore includes the processes operating within the soil and the atmosphere of the woodland. In the soil there are countless organisms which take part in the conversion and transfer of material to the plants; they also break down the debris which accumulates on the ground as the plants die or lose their leaves. In the atmosphere, there are similar cycling processes which remove products from the vegetation and return vital materials to the plants. Overall, therefore, we must consider the woodland as an

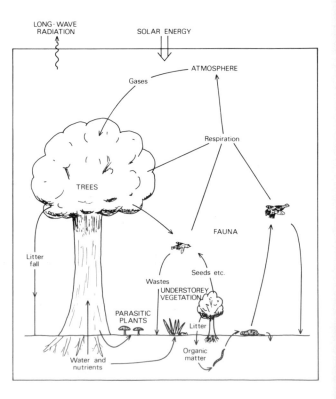

Figure 26.1 The general structure of the ecosystem

ecosystem including not only the trees and the vegetation and the surface animals, but also the soil and all the organisms it contains, and the immediate envelope of the atmosphere with which the woodland interacts (Figure 26.1).

These, typically, are the basic components of any ecosystem: the organisms and the environment with which they interact. At the small scale, we can recognize detailed ecosystems including only a few organisms operating within a minute area; at the large scale we can consider the whole of the earth's surface as a huge and complex ecosystem.

At whatever scale we examine an ecosystem, it is clear that one of the fundamental features is the cycling of matter and energy between its various components. Plants obtain their foodstuffs from the soil and atmosphere; they are eaten by animals which may be hunted by other animals. Man possibly hunts them; in many cases he is the ultimate predator. The remains of these animals and of the plants return to the soil. They are attacked and decomposed and mixed into the soil. Ultimately, the residual materials are converted to a form which can, again, be taken up by plants.

Inputs and outputs of the ecosystems

Of course the processes operating within the ecosystem do not function in isolation. We have already mentioned that they are powered by energy from the sun, and thus solar radiation represents one of the basic inputs into the ecosystem. There are, however, many other inputs. Rainfall provides a supply of water and minerals to the ecosystem; much of the incoming water is stored in the soil and is eventually used by the plants and animals. Rock materials are broken down by weathering to supply inputs to the soil. The organisms also may, in part, enter the ecosystem as inputs from outside. Migrating animals may move into the area; winds and streams may carry seeds which germinate into plants. Man, too, may introduce various inputs. He may bring in seeds or seedlings to replenish the vegetation; he may also introduce animals, often accidentally. In addition, he may apply fertilizers to encourage growth, or pesticides to control the woodland pests. All these represent inputs from outside the ecosystem (Figure 26.2).

By the same token, the ecosystem loses outputs

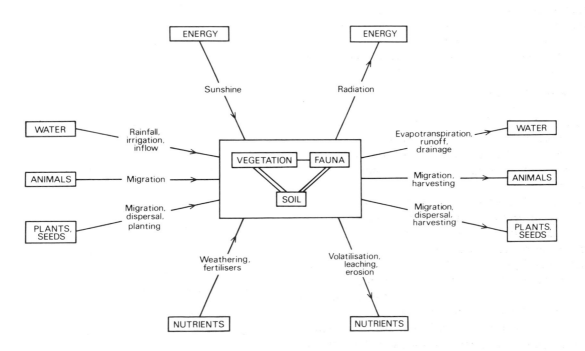

Figure 26.2 Inputs and outputs of the ecosystem

(Figure 26.2). Water drains through the soil and flows out of the area; animals migrate away; the wind picks up and carries the dry leaves; man removes timber and the bodies of the animals he has hunted; gases and heat escape back to the atmosphere. The ecosystem is consequently an open system. It receives and loses both energy and matter. And, as with all open systems, it is the balance between these inputs and outputs which governs the state of the system. Again, we can see this by considering our woodland example. If man removes more timber than he, or other processes, replace, the woodland slowly degenerates; if he brings in and plants more saplings, the woodland becomes denser. In addition, either action is likely to lead to fundamental changes within the woodland, for the internal character of the ecosystem depends upon the interaction between its components. If these components are disturbed or altered, the effects may spread throughout the ecosystem. Removing certain trees for timber, therefore, may have implications for many other aspects of the woodland; changes in the density of the tree cover may trigger off adjustments in the ground vegetation, and will almost certainly lead to changes in the animal populations. In addition, the amount of material being returned to the soil will be reduced, so similar adjustments are likely to occur in the soil. As ever, the integration of the system makes it respond as a whole to changes in the external conditions.

It is also apparent that the inputs to and outputs from any ecosystem represent a link with other parts of the environment. The ecosystems themselves are associated in this way, for some of the outputs from one area become inputs into adjacent areas. For example, animals migrating from our woodland will become immigrants in another area. In addition, ecosystems are linked to other environmental systems. The debris from the forest floor may be carried by streams into fluvial systems, while the gaseous losses and evaporation from the vegetation become inputs to atmospheric systems. In the same way, ecosystems often act as controls upon other processes in the environment. The vegetation, as we have seen (Chapter 14), intercepts rainfall and controls inputs of water into the soil and streams; it acts as a control within the hydrological cycle. Vegetation also protects the land surface from erosion, and thereby acts as a control on many geomorphological processes. Indeed, it is partly because of these interactions that ecosystems represent such an important feature of man's environment – why he needs to manage them with care and understanding.

Cycling processes in ecosystems

Ecosystem structure and cycling

Both energy and matter are cycled through ecosystems. Most of the energy moves in the form of chemical energy, while the matter which is cycled includes water and a wide range of mineral substances. These latter are referred to as **nutrients**, and their transfers from one component of the ecosystem to another makes up the **nutrient** or **biogeochemical cycle**.

These transfers of energy and nutrients do not occur haphazardly; they tend to follow reasonably distinct and consistent pathways. This is because the ecosystem, despite its awesome complexity, is structured and organized. Thus, the relationships within it follow certain clear patterns.

One of the fundamental patterns which we can see in any ecosystem is represented by the **food chain**. This illustrates the links between the different organisms as shown by the energy transfers. Each organism tends to fulfil a specific role within the ecosystem, and to maintain specific relationships with its neighbours. Thus energy passing through the ecosystem does so along certain, well-defined routeways.

We will look at the details of these routeways later; here it is useful to sketch in the broad picture of food chains. All organisms may be generally classified into one of two groups: **autotrophs** and **heterotrophs**. The autotrophs are those organisms which manufacture their own food, using energy from the sun, carbon dioxide from the air, water and minerals from the soil. By far the most important members of this group are the **green plants**, although many micro-organisms, including **algae**, are capable of the same function. Autotrophs are often referred to as **producers**.

The heterotrophs are those organisms which derive their energy from the autotrophs. That is to say they either eat the producers (**herbivores**) or each other (**carnivores**) or they attack the dead tissues of other organisms (including other heterotrophs) and decompose the material. Thus, they are considered to include **primary consumers**, **secondary consumers** and **decomposers**.

As we can see, this classification of organisms

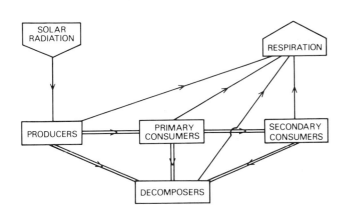

Figure 26.3 The flow of energy through the ecosystem

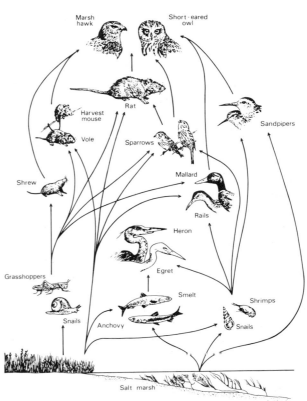

Figure 26.4 A food web: salt marsh, San Francisco Bay, USA (after Smith, 1966)

provides a framework for understanding the flow of energy through the ecosystem. Producers (auto-trophs) convert solar energy to chemical energy and act as the foodstuff for the primary consumers. These are in turn devoured by the secondary consumers, while all three groups are attacked by the decomposers. Energy therefore flows through the system as in Figure 26.3.

If we consider specific ecosystems we find that particular organisms often fill the roles we have identified in this diagram. The linkages between them therefore represent a food chain; put very simply, food flows from one link to the next in the chain. In fact, the term chain is a little misleading. It is better to think of it as a **food web**, for the links are not unidirectional; many interactions occur between the various members (Figure 26.4).

Energy flows in the ecosystem

The food web therefore provides a simple picture of the energy flow within the ecosystem. The inputs are derived from solar energy, but most of the transfers occur in the form of chemical energy. The energy is cycled from one **trophic level** to another.

The transfer of energy in this way is far from perfectly efficient. Considerable losses in energy occur during each transfer. Much of this energy is expended by the living animal in the form of body heat and respiration; much is also lost because of the incomplete breakdown of the compounds which are eaten by each organism. Thus leakage of energy from the system occurs. Some of this energy becomes outputs from the ecosystem, in the form of heat lost to the atmosphere or as kinetic energy and chemical energy when organisms leave the

ecosystem. In addition, some of the energy becomes concentrated in forms which are unavailable to the organisms. It is therefore temporarily withdrawn from the energy cycle.

Two processes are of particular importance in relation to energy cycling. These are **photosynthesis** and **respiration**.

Photosynthesis is the process by which the auto-trophs convert solar energy, in the form of sunlight, to chemical energy. It is a highly complex process, but in general terms can be seen as a reaction involving carbon dioxide, water and light energy. The carbon and water combine to form **carbohydrate**, one of the basic components of living organisms, and oxygen is given off as a by-product. The energy involved in the reaction is stored within the carbohydrate:

$$H_2O + CO_2 + \text{sunlight energy} \rightarrow -CHOH- + O_2.$$

(water) (carbon dioxide) (carbohydrate) (oxygen gas)

In this way plants can retain solar energy for later use.

The energy is released through the process of respiration, during which the carbohydrate is decomposed to water and carbon dioxide. This requires the combination of the carbohydrate with oxygen, and thus involves the intake of oxygen by the plant. As we will see, most of the oxygen is absorbed from the soils by plant roots. The subsequent process of respiration is in reality an oxidation process in which the carbohydrate breaks down to simpler molecules, releasing energy in the process:

$$-CHOH- + O_2 \rightarrow H_2O + CO_2 + \text{chemical energy.}$$

The energy in this case is in the form of chemical energy, which is stored within specific energy-carrying molecules in the plant. It is subsequently used to produce the other cellular compounds required by the plant, such as amino acids. These require the supply of mineral nutrients (e.g. nitrogen) and can therefore be considered within the context of nutrient cycling (Chapter 29).

By the time animals eat the plants, a large part of the initial energy input has been lost. Typically, little more than 10 per cent of the energy is passed on to the next trophic level, although in somewhat special circumstances as much as 50 per cent may be transferred. Thus the efficiency of energy transfers can be estimated by comparing the proportion of organic matter in each level of the ecosystem (Figure 26.5). As we can see, it is normally low. Moreover, as this figure also shows, it is rare for more than four trophic levels to exist within an ecosystem, since the loss of energy during each transfer is such that more levels could not be sustained.

The inefficiency of these transfers has a further implication, for it means that each successive level within the ecosystem is by necessity much smaller. It requires a very large population of producers to support the total pyramid. As a consequence, if we plot **trophic pyramids** for different ecosystems we find a very consistent pattern (Figure 26.6); in almost all cases, the total weight (**biomass**) of the trophic levels declines rapidly upward.

As we will see later, this fact has been used as evidence for the inefficiency of many of man's agricultural systems. It is argued that meat production by cattle involves an unnecessary loss of energy; far more energy could be made available if the food chain was shortened further and man ate the plants rather than the primary consumers.

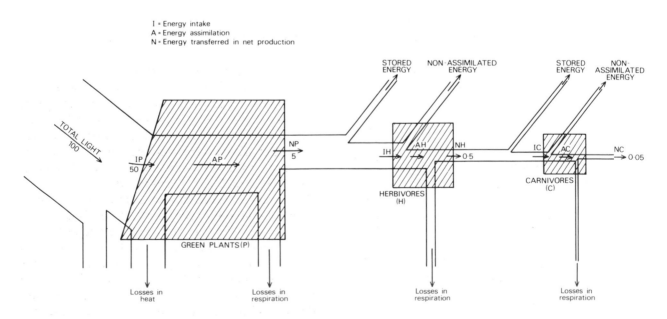

Figure 26.5 Energy transfers through a typical woodland ecosystem. Thicknesses of the bars are generally proportional to the amount of energy flowing along the specified route; size of boxes relates to the amount of energy stored in the specified component. Figures indicate the percentage of incoming energy at each stage

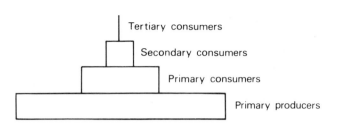

Figure 26.6   A typical pyramid of numbers

Biogeochemical cycling

The consumption and transfer of materials through the ecosystem involve, in addition to a cycling of energy, a movement of mineral nutrients. Organisms require these nutrients for their healthy growth. They take part, for example, in the formation of new cell material and the production of compounds such as proteins and amino acids. Many nutrients are involved in these processes; among them the most important are compounds of carbon, oxygen, nitrogen, potassium and phosphorus.

The inputs of nutrients to the ecosystem occur in a number of ways. Weathering of rocks releases inorganic compounds into the soil; rainfall introduces material from the atmosphere; soil organisms and plants absorb substances from the air; man applies materials such as fertilizers and pesticides. Within the ecosystem, these compounds are altered and transferred, passing through food webs, and breaking down through chemical reactions in the soil.

Within the food webs it is common to recognize two main cycles. One relates to what is known as the **grazing food chain**, and involves the plants, the herbivores which feed off them and the carnivores which predate on the primary consumers. This operates mainly in the vegetation layer (Figure 26.7). It forms a basis, as we will see later, for examining biogeochemical cycling in the vegetation and fauna. The other main component of the food chain is the **detritus food chain**. This is dominated by the heterotrophic decomposers; the organisms which break down dead organic material and release the nutrients contained within it. They often convert the organic compounds back to inorganic form through a process of **mineralization**. It is a process which operates mainly within the soil, and it will form the basis for our examination of biogeochemical cycling within the soil (Figure 26.7).

The cycling of nutrients is not only a biological phenomenon. As the term 'biogeochemical' implies, various physical and chemical processes also take part. Organic debris may be decomposed purely by physical and chemical activity; nutrients within the soil may be stored by physico-chemical forces operating on the surfaces of the minute colloidal particles (clay and microscopic organic fragments). Water percolating through the soil and moving both laterally and vertically under the effect of capillary forces also transports nutrients through the soil. Overall, however, these cycles are dominated by biological and chemical processes.

World ecosystems

Factors affecting ecosystems

It is apparent that the character of the world's ecosystems varies considerably. The range of organisms involved, and their relationship with their immediate environment, differ from one area to another. These variations arise mainly from the difference in the external factors which control ecosystem structure and behaviour.

Three main factors may be mentioned. Climate is clearly of major importance, since this directly controls many of the inputs to the ecosystem (in particular solar energy and rainfall), and indirectly influences many other inputs and the rates of many ecosystem processes. The rate of weathering, for example, is related to climatic conditions, and thus climate influences the inputs of nutrients. The operation of many chemical and biological reactions is also climatically dependent; temperature and water availability, for example, influence the rate of

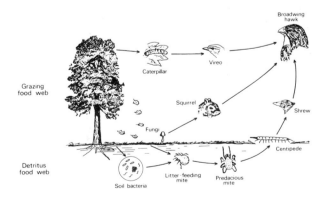

Figure 26.7   Examples of grazing and detritus food webs from a temperate deciduous forest, USA (from Smith, 1966)

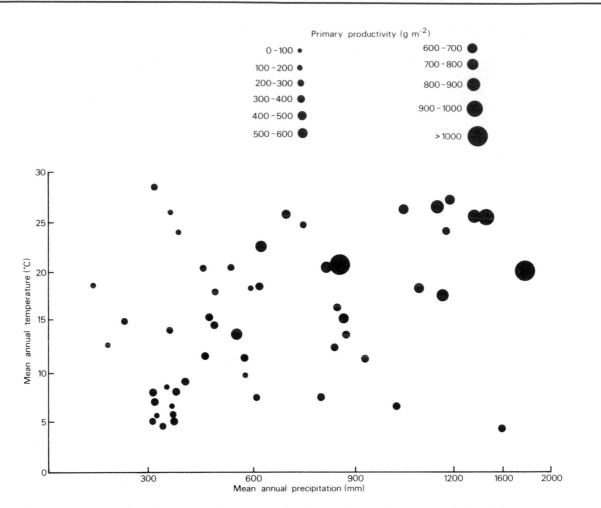

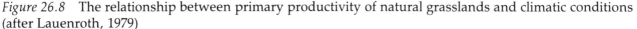

Figure 26.8   The relationship between primary productivity of natural grasslands and climatic conditions (after Lauenroth, 1979)

primary production (i.e. organic material formation) in the ecosystem (Figure 26.8). These effects are also evident in the global distribution of biological productivity (Figure 26.9).

Geology also is important, since this tends to influence the inputs of nutrients. In general, ecosystems are more productive on rocks which weather rapidly and release large quantities of nutrients than on resistant, nutrient-deficient rocks.

Finally, man plays a vital role, since he controls the structure of many ecosystems through his management of the environment. At one extreme, he creates totally artificial ecosystems as part of his agricultural exploitation of the environment; at the other extreme he accidentally and subtly interferes

with natural ecosystems through his production of pollutants and his effects on climate.

Terrestrial ecosystems

We can compare ecosystems on the basis of their net annual primary productivity (the quantity of new organic matter produced each year). Three main terrestrial ecosystems may be mentioned here. We will discuss these and others in more detail later (Table 26.1).

Woodland (or **forest**) **ecosystems** represent the natural state of much of the earth's land surface. That such ecosystems today only occupy about one-quarter of that surface is to a large degree a result of man's interference with the environment, mainly in the pursuit of agricultural production. These

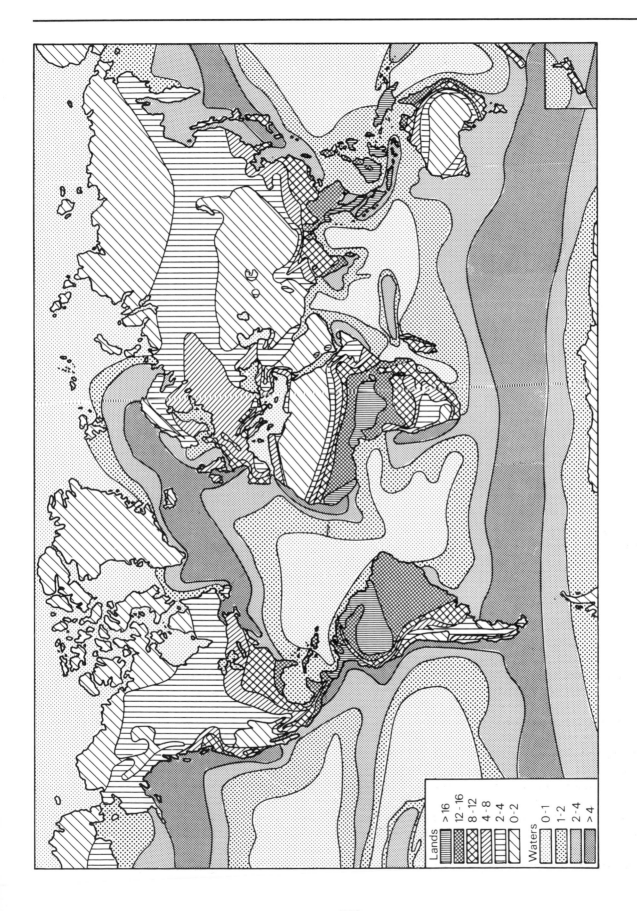

Figure 26.9 The primary productivity of terrestrial and oceanic systems. Units are t ha⁻¹ y⁻¹ (after Lieth, 1965)

Table 26.1 *Net primary productivity and total biomass of selected ecosystems*

Ecosystem	Net primary productivity (t ha⁻¹y⁻¹)		Biomass (t ha⁻¹)		Area
	Range	Mean	Range	Mean	10⁶ km²
Tropical rain forest	10–35	22	60–800	450	17
Temperate deciduous forest	6–25	12	60–600	300	7
Temperate evergreen forest	6–25	13	60–2000	350	5
Boreal forest	4–20	8	60–400	200	12
Savanna grassland	2–20	9	2–150	40	15
Temperate grassland	2–15	6	2–50	16	9
Tundra	0.1–4	1.4	1–30	6	8
Semi-desert	0.1–2.5	0.9	1–40	7	18
Desert	0–0.1	0.03	0–2	0.2	24
Arable land	1–40	6.5	4–120	10	14
Lakes and streams	1–15	4	0–1	0.2	2
Estuaries	2–40	15	0.1–40	10	1.4
Upwelling ocean zones	4–10	5	0.05–1	0.2	0.4
Continental shelves	2–6	3.6	0.01–0.4	0.01	26.6
Open ocean	0.02–4	1.25	0–0.05	0.03	332

Data From R. H. Whittaker and G. E. Likens, 'The biosphere and Man', in H. Lieth and R. H. Whittaker (eds.), *Primary productivity of the biosphere*, New York: Springer Verlag, 1975, pp. 305–28.

systems differ markedly in their detailed structure, but on the whole they represent the most complex terrestrial ecosystems. They contain many different structural levels (Figure 26.10), and tend to have a very high rate of primary production. In tropical areas, for example, the rain forests have a net annual production of about 2000 t km⁻². Even in mid-latitudes a mixed oak forest is capable of producing as much as 1500 t km⁻² of organic matter each year.

In addition to their high annual rate of primary production, forest ecosystems are distinctive in that they store a disproportionate quantity of nutrients and organic matter in the vegetation. As we will see, this is a feature which causes some difficulty when man comes to exploit these areas, for in clearing the vegetation he removes most of the nutrients within the system; the system which is left is therefore relatively unproductive.

In contrast, **natural grassland ecosystems** have a relatively low annual productivity, with average rates of about 500 t km⁻². This is somewhat less than the average achieved by many agricultural systems. Cereal farming, for example, can attain productivities of several thousand tonnes per year, while the average in mid-latitudes is probably between 500 and 1000 t km⁻² yr⁻¹. In the case of **agricultural ecosystems**, of course, man's influence

is of fundamental importance. He controls and maintains the system, often by using relatively high levels of inputs. Thus, primary net production figures may be misleading; they may be achieved at the cost of considerable inputs from other systems. In addition, agricultural ecosystems are often highly simplified. The food chains are deliberately shortened and unwanted components such as wildlife and competitors are eliminated. Thus they tend to be dominated by a very small range of species, sustained by the application of fertilizers and pesticides, and exploited through the use of machinery requiring fossil fuels. Moreover, as we will see later, the outputs from the system are much greater than those experienced in natural ecosystems, and it is not always clear that the system is being maintained in a condition of long-term stability; exhaustion of nutrient reserves, and damage to the soil, may threaten many of the more intensive agricultural ecosystems.

Aquatic ecosystems
We have concentrated until now on **terrestrial ecosystems**, but these are far from the only ones on the earth. **Aquatic ecosystems** make up more than two-thirds of the earth's surface, and although they characteristically have a relatively low annual productivity, they are clearly important (Table 26.1).

Figure 26.10   The vertical structure of a forest ecosystem

Within the oceans, the character of the system tends to vary according to water depth and the proximity to land. In general, shallow, littoral waters are more productive than deep mid-ocean waters, largely because of the greater food supply from the land. Estuarine areas, for example, may be as productive as the most productive terrestrial ecosystems, for they receive vast inputs of nutrients from the rivers. The continental shelves are less productive, but are nevertheless important for man as the source of most of his fishery products. Open ocean areas, which cover about 60 per cent of the globe, are generally the least productive, save where convergence of the deep currents results in upwelling cool waters in otherwise warm areas. This produces active plankton growth which gives rise in turn to increased productivity of other marine animals. Many of the important fishing areas, such as Peru and south-west Africa, are based upon this phenomenon. In addition, the stimulus it gives to the food chain results in the formation of large deposits of **guano**, the waste excretions of the sea birds which feed on the fish. This is rich in phosphate and is therefore exploited as a fertilizer.

Lacustrine ecosystems may also be productive, again mainly due to the high inputs of nutrients from the surrounding land. These inputs may in fact become a problem, for excessive quantities of nitrogen and phosphorus compounds from farmland and human wastes often accumulate in inland waters, leading to a process of **eutrophication**. The nutrients stimulate such intense algal growth that the waters become depleted in oxygen and ultimately are left sterile. Indeed, one of the problems of lacustrine ecosystems is their isolation; they act almost as closed systems, so that they cannot pass on the effects of changes in inputs. Instead, the inputs accumulate and ultimately choke the system.

An approach to ecosystems

One way of tackling the study of ecosystems is to

concentrate upon the biogeochemical cycles. This is the approach we will adopt in the following chapters. As we have seen, two main components to these cycles may be identified: those operating within the soil, and related primarily to the detritus food chain, and those operating within the vegetated layer, and related largely to the grazing food chain. Later, therefore, we will trace the flow of nutrients through these main subsystems before looking at specific examples of ecosystems in the world.

We should also be aware, however, that the cascade of nutrients through the ecosystem is associated with a morphological development of the soil and vegetation. Over time, the processes operating at the surface of the earth act to create specific types of soil and forms of vegetation. Thus, we need also to analyse the development of ecosystem morphology. In the next chapter we will consider one aspect of this: the nature of soil formation. We can then use this as a starting point for our study of nutrient cycles.

27

Soil formation

Introduction

The nature and function of soils

When rocks are exposed to the atmosphere, they undergo weathering. The weathered debris collects to form a veneer of material known as regolith. In many cases, the regolith is able to support plants and animal life, and in time organic remains accumulate and become mixed with the mineral matter. In this way soil is formed. **Soil** is the mixture of rock debris and organic material which develops at the earth's surface.

However, the soil is not a totally disorganized mixture of material; it has an internal organization and arrangement. It is, in the words of the early Russian soil scientist, Dokuchaev, 'an organized natural body'. We can see the truth of this if we dig a pit and look at the soil (Plate 27.1). Normally, we find that it consists of layers (horizons) of material of slightly different colour, composition and structure. These horizons are not normally depositional features – they have not formed by the sequential accumulation of sediments – but are a result of processes operating within the soil itself. They form what is called the **soil profile**, and they reflect the combined action of many different and often conflicting processes of soil formation. In this chapter we will consider the nature of these processes and the way they influence the development of the soil profile.

Before we do so, it is important to stress the role of soils within ecosystems. They are, in fact, a fundamental component of almost all terrestrial ecosystems. They provide a substrate for plants, a medium within which their roots can take anchorage, and a source of water and nutrients for the vegetation. They also act as a habitat for many of the organisms involved in the cycling of nutrients through the ecosystem. In addition, outputs from the soil often form inputs to aquatic ecosystems: lakes, rivers and coastal waters derive many of their nutrients from the soils of neighbouring areas. The character of the soil therefore exerts a major control upon ecosystem processes and, more particularly, upon the productivity of the ecosystem.

Composition of the soil

In addition to the solid material – the organic and inorganic matter derived from weathering and plant decay – the soil consists of two further components: water and air. It is thus referred to as a **three-phase medium**, implying not only that it consists of three main components, but also that significant transfers of matter and energy take place

Plate 27.1 A podzol profile, showing horizonation due to the operation of soil forming processes. Note the dark topsoil (Ah horizon) which contains organic matter, and the lighter, leached (Ea) horizon below. Beneath that there is a layer which is enriched with organic matter washed from above (the Bh horizon), and finally a mineral horizon enriched with iron and aluminium oxides (Bs horizon) (photo: D. J. Briggs)

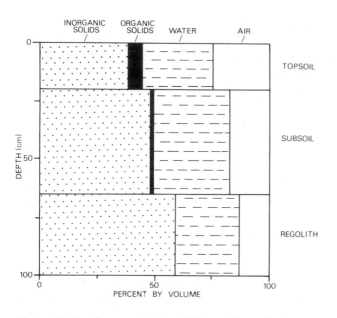

Figure 27.1 Composition of the soil: typical proportions of inorganic and organic solids, water and air in the soil profile

between them. The relative proportions of the three phases vary considerably, although in many soils the solid phase makes up about 50 per cent of the total volume, and the air and water make up the remainder (Figure 27.1). Air and water compete for a place in the voids between the solid particles, however, so their relative quantities change markedly over time in response to rainfall inputs and losses by drainage and evaporation.

The composition of the three components also varies to some extent. On the whole, the air within the soil is similar to that of the open atmosphere, but enriched with carbon dioxide and deficient in oxygen (Table 27.1). This is due to the effects of plant respiration. The composition changes over time, depending upon the rate of organic activity and the ease with which the gases can diffuse through the soil into the open atmosphere. The soil

water also differs somewhat from the water we find in lakes or rivers. It tends to be much richer in dissolved substances washed from the soil and vegetation and, because it comes under the attraction of the soil particles, is much less free to move.

Both the organic and inorganic matter that make up the solid phase of the soil vary considerably in character. The inorganic fraction consists in the main of partially weathered rock fragments and minerals, including in particular the more resistant residues such as quartz, feldspars, clay minerals and compounds of iron and aluminium. The composition of the inorganic material depends upon the nature of the rocks from which they have been weathered, the climate and the time available for weathering so many other minerals may also be present. These materials vary considerably in their form, ranging from large stones or boulders to minute clay minerals. The finer particles – the sand, silt and clay fractions of the soil – rarely occur as individual grains but are bound together by cohesion and various cementing agents into **aggregates**. The organic fraction of the soil is similarly variable. It consists of the living and dead cells of animals and plants and the organic acids (e.g. **fulvic** and **humic acid**) which are formed by decomposition of plant materials. In some cases, these compounds occur as discrete horizons or masses of organic matter; in **peat**, for example, almost the whole soil is composed of plant debris. But, in other cases, the organic matter is intimately mixed with the mineral material so that we cannot see it, save perhaps by the dark colour it gives to the soil. Nevertheless, even in this state it plays an important part in the soil, helping to cement the particles together and acting as a vital store for plant nutrients.

Although much of the organic matter in the soil is relatively transient, soon decomposing to form organic acids or being liberated into the atmosphere as gases, some of it – the portion normally called **humus** – is persistent. This persistent fraction, together with the inorganic solid fraction, provides the main constituents of the soil profile, and it is differences in the character, arrangement and proportions of these components that give rise to differences in soil profiles.

The soil-forming system

Inputs to the soil system

If we think of the soil profile as a system, we can

Table 27.1 *Composition of the soil air relative to the open atmosphere*

| | Per cent by volume | |
	Soil air	Open atmosphere
Nitrogen	79.0	79.01
Oxygen	18.0–20.8	20.96
Carbon dioxide	0.15–0.65	0.03

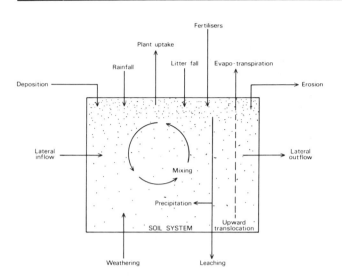

Figure 27.2   The soil system

readily appreciate that, like other systems, its form depends upon three groups of processes: the inputs of raw materials to the soil, the loss of materials from the profile, and the internal transfers and reorganization of these materials within the profile

(Figure 27.2).

The main inputs of matter come from the bedrocks or parent materials of the soil. These are released from the rocks by weathering at the base of the soil, and contribute to the lower layers of the profile. We can often see evidence of this input at the bottom of the profile: joints are opened up and rock fragments occur in a state of partial detachment (Plate 30.13). The rate at which these materials are released depends upon the rate of bedrock weathering, itself a function of geology and climate. On massive, resistant rocks, for example, the rate of weathering is often slow, with the result that soils are thin. On softer, less resistant rocks much more rapid weathering may occur and larger quantities of material are released, leading to deeper soils. Similarly, where climatic conditions favour active weathering, deeper soils develop. As we saw in Chapter 19, rates of chemical weathering tend to be greatest under conditions of high temperature and humidity, and, not surprisingly, we find that soils in tropical areas are often several metres deep, while those in polar regions are shallow and poorly developed (Figure 27.3).

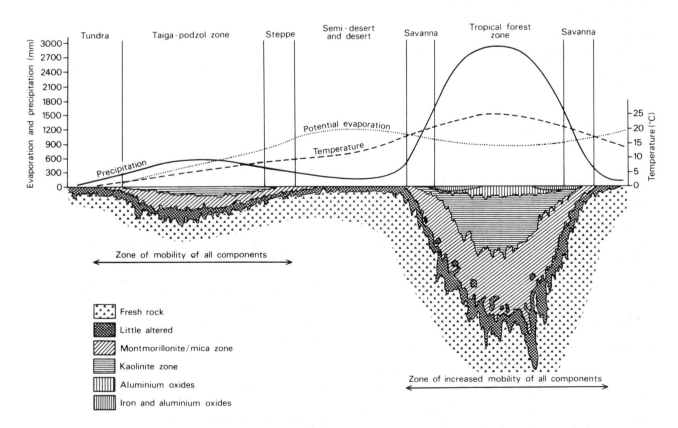

Figure 27.3   Variations in soil composition and depth in relation to climate (after Strakhov, 1967)

439

In addition to the input of material from bedrock weathering, significant inputs of material come from surface accumulation. This often results in the formation of **buried soils**. One of the most important situations in which this occurs is in alluvial valleys, where flood waters deposit silt on the floodplain surface. Indeed, it has been largely because of this process that many early civilizations became established in alluvial valleys such as the Indus, Nile and Ganges, for the annual floods helped to replenish the soil and renew the nutrients removed by agriculture. Less welcome accumulations also occur on soils in the desert fringes, for, as the deserts expand, winds carry sand on to the surrounding agricultural lands, resulting in gradual encroachment of the deserts. During the Quaternary Period, widespread surface accumulations also took place in the northern hemisphere as winds blew fine-grained glacial debris across the frozen ground, forming extensive loess sheets. At times, when wind activity declined and the surface became more stable, these were weathered and reorganized to form soils, so that the deposits now consist of alternating sequences of loess and soil. On a more local scale, surface accumulation occurs in many footslope areas, for mass movement and surface runoff carry material downslope. Here it accumulates to give buried soils (Plates 27.2 and 27.3).

As we have seen, organic inputs to the soil are also important. These are derived mainly from the vegetation, which provides an annual input of debris in the form of leaves, plant litter and dead roots, which accumulate at the surface. The character of this material clearly depends mainly upon the nature of the vegetation. Beneath coniferous forests, the organic debris is typically acid and resistant to decomposition, with the result that it accumulates as a distinct layer of humus. Debris from deciduous trees is often much less resistant, and is rapidly broken down and mixed with the soil.

Plate 27.2 A buried soil in hillwash deposits, Stanage Edge, Derbyshire. The dark, buried A horizon is clearly visible just above the end of the tape; beneath it is the leached Ea horizon of the original soil. The profile has been buried by soil washed from the hillside, possibly as a result of over-grazing or vegetation clearance. A thin Ah layer is now developing at the surface (photo: D. J. Briggs)

Plate 27.3 A Late Devensian (Allerød) soil (shown by trowel) buried beneath hillwash at Glen Wyllin, Isle of Man. Burial probably occurred in the cold zone III stage, when sands and gravels were washed off the surrounding hills (photo: P. A. Smithson)

The rate of organic inputs also depends upon the density of the vegetation, and thus is related to climatic conditions. In the tropics, the annual input of plant debris may total as much as 2 t km^{-2}, while in the sparsely vegetated desert margins as little as 0.1 t km^{-2} of organic carbon may accumulate each year.

Outputs from the soil system

The main losses from the soil occur through erosion and leaching. Erosion is often the more dramatic process, since it acts at the surface and has direct effects upon soil fertility and thus upon man. Moreover, it is often encouraged by agriculture, which leaves the soil bare and unprotected from the wind and rain. Thus, rates of soil loss by erosion from arable soils may be many times greater than those from similar grassland or woodland plots (Table 27.2). The susceptibility of the soil to erosion is a function of many factors, including climate, soil conditions, vegetation cover, topography and management practices. We saw the way these factors interact to influence wind erosion in Chapter 24, and similar interactions affect erosion by rainfall (Figure 27.4). In both cases it is clear, however, that man plays a dominant role in causing and controlling losses by erosion.

Leaching is a more insidious process, for it is less readily visible and often less catastrophic in its

Table 27.2 *Rates of erosion from soils under different vegetation covers*

Vegetation cover	Rate of soil loss (kg ha^{-1})	Runoff (per cent of rainfall)
Woodland	0.82	0.09
Grassland	4.10	0.29
Rotation crops	4,440.8	8.8
Cotton	10,320.9	10.5
Bare soil	26,604.1	29.1

Data from H. H. Bennett, *Soil conservation*, New York: McGraw Hill, 1939.

effect. It involves the loss of soil materials in solution, and is most active under conditions of high rainfall, rapid drainage and acid vegetation (e.g. pine trees and heather). These conditions favour the development of abundant hydrogen ions in the soil water which displace other nutrient elements from the soil and take place in weathering reactions such as hydrolysis. The percolating waters carry soluble substances downwards through the soil profile, depositing some in the lower layers, but removing the most soluble entirely.

Numerous other processes also remove material from the soil profile. In some cases man has a direct effect, for example by extracting peat for fuel; in the Republic of Ireland almost one-third of the

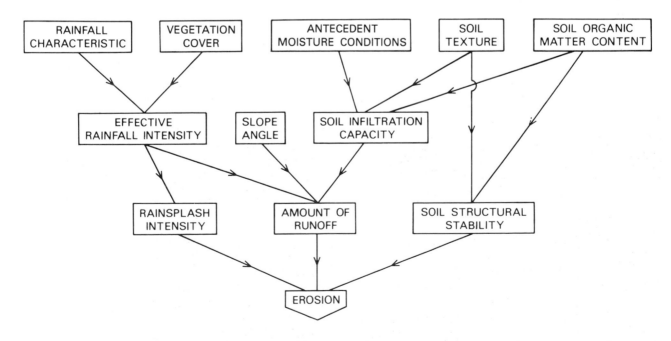

Figure 27.4 Factors affecting soil erosion by rainfall

Table 27.3 *Soil organic matter content of soils under different vegetation covers*

Vegetation cover	Organic matter content (per cent)
Woodland	5–15
Grassland	8–10
Grass-arable rotation	2–5
Continuous arable	1–3

country's energy is still derived from peat. Plants, too, remove substances, while significant losses may occur in the form of gases. Organic matter, in particular, is lost in gaseous form, as we will see in Chapter 29, for it decomposes in the soil to carbon dioxide, hydrogen and oxygen which escape to the atmosphere. Once more it is a process which is encouraged by cultivation of the soil, for during tillage the soil is loosened and aerated so that organic matter decomposition is speeded up and the rate of gaseous diffusion is increased. As a result, arable soils often have markedly lower organic matter contents than equivalent grassland or woodland soils (Table 27.3).

Soil profile development

The relative magnitudes of inputs to and outputs from the soil determine whether or not the soil profile gets deeper. During the early stages of soil

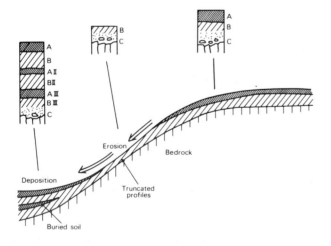

Figure 27.5 Profile truncation and the formation of buried soils on steep slopes. A, B and C are distinct horizons in the soil profile.

development, inputs characteristically exceed outputs and the soil depth increases. Under conditions of excessive erosion, as for example when steep slopes are cultivated for arable crops, outputs may exceed inputs and the soil profile is **truncated** (Figure 27.5). The internal character of the profile, however, depends upon the processes of reorganization within the soil. These involve transfers of soil material by water, ice, soil organisms and man. They may involve either upwards or downwards movement of material, some of the processes acting to segregate the material and favouring the development of horizons, while others cause mixing of the soil and destroy horizons.

Lessivation

Water is one of the main agents of soil profile development, and the downward percolation of rainwater through the soil transports material in both suspension and solution. Movement of material in suspension is common under many conditions. Clay movement (**lessivation** or **clay translocation**, as it is known) involves the detachment of individual clay particles or the dissolution of clay minerals in the upper layers of the soil and their transport downwards through the profile. The process is most active in porous, slightly acid soils containing moderate quantities of kaolinite clay. The clay is washed down the root channels, worm tunnels, fissures and pore spaces and is deposited or reprecipitated in the subsoil. Here it tends to form thin veneers or skins (**cutans**) around larger particles and along the walls of the pores (Plate 27.4). Many soils in the humid temperate regions have been affected by this process, particularly under conditions of high rainfall and acid vegetation.

Not only clay, but silt, iron oxides and organic matter all move in the same way. Translocation of organic matter is particularly significant. The exact processes involved are not clearly understood, and it seems that, in some cases at least, the organic matter moves in solution and is reprecipitated at depth. It also seems that deposition of the organic matter lower down in the profile is controlled by small changes in the acidity, wetness and aeration of the soil, so that distinct layers of humus may accumulate in the subsoil (Plate 27.1).

Leaching and cheluviation

Chemical compounds are also washed downwards

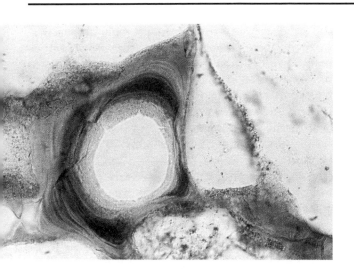

Plate 27.4 A thin section of soil showing cutans in the pore spaces. The laminated appearance of the clay cutan is due to the repeated precipitation of clay in the pore spaces (photo: J. Rose)

through the profile in solution. We have already mentioned the process of leaching, and this is of major significance in soils throughout the temperate and humid tropical world. Rainwater picks up substances from the atmosphere and vegetation which makes it slightly acid; in many cases the rainfall is a weak carbonic acid due to the presence of carbon dioxide dissolved from the atmosphere (Chapter 19). When this water enters the soil, it is able to dissolve soluble substances, such as calcium carbonate, and carry the calcium into the subsoil. It also removes nutrients attached to the surfaces of clay and organic matter by a process of cation exchange (see Chapter 29) and washes these downwards.

Cheluviation is a related process. Rainwater, picking up organic compounds from the vegetation or the surface layers of the soil, is converted to a weak organic acid. The organic compounds in the soil are then able to combine with mineral elements to form **organo-mineral complexes**. This process is known as **chelation**, from the Greek word meaning 'to claw'. These complexes of organic and mineral materials are highly mobile and are readily washed down through the soil to accumulate at depth. As with leaching it is a process encouraged by acid vegetation.

Both leaching and cheluviation therefore lead to the removal of the more soluble elements from the surface layer of the soil. Under extreme conditions this may leave the topsoil acid and relatively

infertile. The most mobile compounds may also be removed from the profile, while the less soluble materials are precipitated at depth. The depth at which accumulation occurs depends upon the relative solubility of the substances, the amount of water passing through the soil (and hence the rainfall) and specific chemical conditions within the soil profile. Calcium carbonate, for example, can be seen to be precipitated at differing depths in relation to rainfall (Figure 27.6), while aluminium and iron carried downwards by cheluviation under acid conditions seem to accumulate largely as a response to changes in the degree of aeration and the quantity of organic substances lower down in the profile. Thus, under certain conditions, the iron may be precipitated in the form of a distinct layer (an **ironpan**) within the soil.

Leaching and its related processes operate to some extent in most soils. They are responsible for much of the loss of fertilizer nutrients, particularly nitrogen and calcium, from agricultural soils, and thus contribute to the process of **eutrophication**;

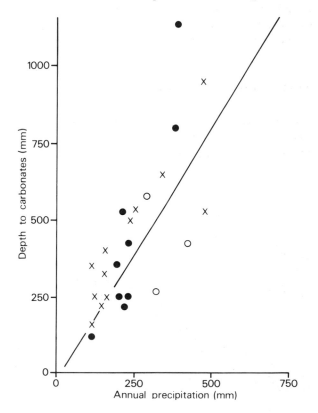

Figure 27.6 The relationship between depth to calcium carbonate and annual precipitation (after Arkley, 1963); × = sandy soils; ● = clay loams; ○ = clays

443

the nitrogen is washed into rivers and becomes concentrated in lakes. It is also, of course, a major output from the soil.

Upward translocation

Net movement of compounds upward through the soil by water occurs mainly in arid and semi-arid areas, where evapotranspiration exceeds rainfall. This results in a tendency for water to be drawn to the surface of the soil along the capillary pores (Chapter 13), and then for the salts it contains to be precipitated at or near the surface as the water evaporates. Sodium compounds are particularly affected by this process, and they give rise to **salinization**; salt crusts develop in the surface layers causing considerable problems for farming. Calcium, iron and even silica may also be affected in this way, however, and the process can then lead to the development of calcretes, ferricretes and silcretes. Under tropical conditions in which marked seasonal variations in rainfall occur, the process is particularly active. The vegetation, the high temperatures and the heavy rainfall during the wetter months encourage the solution of iron and silica within the soil; then, during the drier months, water is drawn to the surface and the iron is precipitated in the upper layers. The process is therefore implicated in the formation of what are known as laterite soils, though, as we will see later, other mechanisms are also important.

Biological processes

Various biological processes take part in the redistribution of soil materials within the profile. Plants, in particular, cause upward translocation of compounds, for they extract water from the lower levels in the soil and thereby draw the dissolved substances in the water to the surface. Since the plant debris and nutrients are returned to the soil surface, this leads to a net upward movement of material.

Many other organisms are also active in disturbing and rearranging the soil. To some extent they tend to act as agents of mixing, for the burrowing of rodents, the tunnelling of earthworms, and the smaller scale activity of micro-organisms, all overturn the soil.

Physical mixing

Mixing of the soil is also helped by the development of ice crystals. Due to freeze–thaw activity, large ice crystals (pipkrakes) may develop in the soil, and these break down the soil aggregates and cause the soil to heave. As the ice melts, finer material is washed into the space, so that the soil is reorganized. Under intense conditions, as found in tundra regions, the effect is to sort the soil into patterned ground features (see Chapter 23); where it is less intense the main effect is to mix the soil and destroy the horizonation.

Man, too, causes physical mixing of the soil. Ploughing of the soil, to create a seedbed for his crops and to remove weeds and bury stubble, disturbs the soil to a depth of 20 cm or more, eradicating the original horizons and creating an evenly mixed, ploughed layer. Similarly, disturbance caused by root growth and uprooting of trees (e.g. by wind or animals) helps to mix the soil.

Gleying

One further process might be mentioned, although this involves little real redistribution of the soil constituents. It is the process of **gleying**, and it results from the effects of oxidation and reduction. During periods of waterlogging the soil experiences **anaerobic** conditions, for the pore spaces become filled with water and air cannot enter the soil. Thus, the oxygen that normally acts as a sink for electrons released by plants and organisms in the soil is lacking, and other compounds, such as iron and aluminium oxides, accept the electrons. As we saw in Chapter 19, the result is chemical reduction of these compounds.

Ferric iron is converted by this process to the ferrous form:

$$Fe^{+++} + electron \rightarrow Fe^{++}$$

in which state it forms a ferrous oxide:

$$2Fe_2O_3 \rightarrow 4FeO + O_2.$$

This is characteristically grey or green in colour. Aluminium, sulphides, sulphates, nitrates and many other compounds may be reduced in the same way. The soil, as a result, becomes duller in colour, and many of the reduced compounds, which are typically more soluble, may be washed from the soil.

During drier periods, air enters the soil and a reversal of this process occurs. Oxygen accepts the electrons and combines with the reduced substances to form higher valency compounds. The ferrous iron is converted back to the ferric state, which is typically red or yellow in colour. The soil thus becomes brighter in colour.

Alternate phases of reduction and oxidation are

common in soils due to fluctuations in the water table, and this results in the creation of varied soil colours, known as **mottles**. Oxidation is often seen to be most marked along the larger pores, such as root and earthworm channels, due to the better access of air along these.

Soil formation and soil properties

During the course of soil formation the original raw materials – the rock debris and organic matter – are redistributed and altered. In the process, the physical, chemical and biological properties of the soil develop, resulting in the creation of distinct horizons. As is shown in Appendix II, it is on the basis of these features that we recognize and describe soils and can deduce both the nature of their formation and their potential use. The number of properties we can consider is almost infinite, and each can tell us something different about the soil's history or management. Let us look, therefore, at some of the main morphological properties of the soil profile and examine the ways in which they develop.

Soil texture

If we examine a soil sample closely – or better still if we are willing to get our hands dirty by moistening it and rubbing it between our fingers – we can discern one of the fundamental properties of the soil: its texture. Strictly speaking, soil texture refers to the physical feel of the soil, and as such is affected by a large number of factors, including the size and shape of the individual particles, the moisture content, the amount of organic matter and the soil structure. By far the most important aspect of texture, however, is particle size, and soil surveyors are able to use the feel of the soil to determine the size distribution of the constituent particles.

The material in the soil ranges in size from large boulders to minute clay minerals too small to be discerned with the naked eye. For the purpose of description, the soil particles are often divided into size classes. Numerous systems of defining particle size have been developed, but the method used in Britain defines six size grades: cobbles, stones, gravel, sand, silt and clay on the basis of particle diameter (see Table 17.2, p. 251). The most important of these in many ways are the sand, silt and clay fractions for it is these which are chemically most active and are able to store both nutrients and water

in the soil. Together they are referred to as the **fine earth fraction**, and they contribute vitally to soil fertility. Consequently, it is with the fine earth that the soil scientist is mainly concerned when he analyses the soil, and to characterize the soil texture he assesses the relative proportions of sand, silt and clay. On the basis of these values the soil is then referred to a textural class (Figure 27.7). Each textural class represents a combination of particle sizes that not only feels different but also acts differently under management.

It is clear from what we said earlier that the texture of the soil depends to a great extent upon the character of the original parent material; that is, on geological factors. It is also influenced by the degree of weathering. More intense or more prolonged weathering results in finer-grained soils. In addition, texture is affected by the processes of soil horizon development, and lessivage, for example, may reduce clay contents in the topsoil and enrich them in the subsoil (Figure 27.8). By examining the vertical pattern of texture in a soil profile, therefore, we get an indication of the nature of soil formation.

Clay mineralogy

Of all the constituents of the inorganic fraction of the soil, probably the most important is the clay component. This has a fundamental influence on both the physical and chemical character of the soil; it helps to give the soil a structure, it acts to hold water in the soil, and it provides the capacity to store and exchange plant nutrients.

These vital attributes arise from the physical and chemical structure of the clay particles. Clay comprises particles less than 0.002 mm in diameter. The small size of the particles means that they have a very large surface area relative to their volume and this makes them highly reactive. It also means that some of the clay particles (those less than about 0.001 mm) can behave as **colloidal materials**, with the ability to disperse spontaneously in a solvent such as water. It is this property which, in part, accounts for the tendency for clay particles to be lessivated by acid, percolating waters.

In addition, many clay minerals are composed of aluminium and silica, arranged in a lattice-like structure which is to some extent flexible and plastic. This allows the clay particles to shrink and swell, and it also permits the clay to absorb water and other substances into their structure. Moreover, clay particles have a negative electrical charge, giving them the capacity to attract and hold

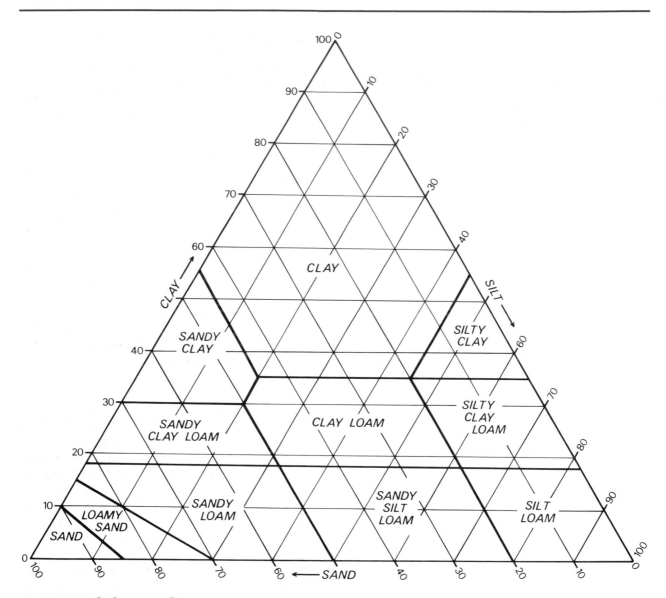

Figure 27.7 Soil texture diagram

(**adsorb**) positively charged ions, and to repel other negatively charged surfaces.

Together, these properties of clays are of major significance to both soil formation and fertility. But how do they develop? To answer this question we need to consider the structure of clays and the way in which they form in rather more detail.

The structure of clays

Most clay minerals are silicates. Originally, they were believed to be amorphous, with no regular structure, but since it has been possible to examine them with X-rays and electron microscopes we

have come to realize that they do, in fact, have a very definite crystalline structure.

The basic building blocks of this structure are atoms of silica, aluminium, oxygen and hydrogen. Silica and oxygen are arranged in the form of a single Si^{++++} atom surrounded by four O^{--} atoms, giving a four-sided molecular structure. This is known as the **silica tetrahedron** (Figure 27.9). Adjacent tetrahedra are linked by shared oxygen ions to produce a sheetlike layer. Conversely, each aluminium atom is surrounded by six oxygen or hydroxyl (OH^-) ions to give an eight-sided molecular structure, called the **aluminium octahedron**

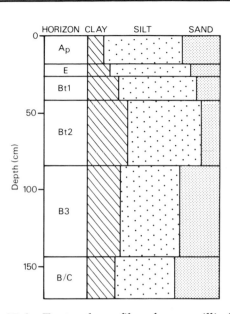

Figure 27.8 Textural profile of an argillic brown earth (Hamble Series), showing the increased clay content in the subsoil as a result of clay translocation (data from Jarvis, 1968)

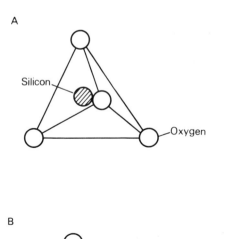

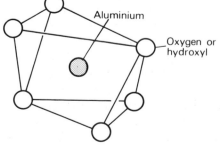

Figure 27.9 The structure of the silica tetrahedron (A) and aluminium octahedron (B)

(Figure 27.9). Again, these are linked by shared oxygen atoms, producing an octahedral sheet.

The silicate clays are built from various combinations of these two sheets. Two common arrangements are found. Alternating sheets of silica tetrahedra and aluminium octahedra produce what are called **1:1 clays**, the most widespread example of which is **kaolinite**. Each pair of sheets is held together by hydrogen ions, making a relatively rigid and stable structure. In the **2:1 clays**, the aluminium octahedra are sandwiched between two silica sheets. In minerals such as **montmorillonite** each sandwich is, in turn, linked by weak oxygen bonds. As a result, these 2:1 clays are relatively flexible and are able to expand. Moreover, their internal surfaces are open to weathering reactions, making them highly reactive. Not all 2:1 clays are of the same structure, however, and in **illite**, for example, the crystal units (the 'sandwiches') are held much more rigidly by potassium ions. This produces a more stable, non-expanding clay.

Clay formation

As we saw in Chapter 19, clays are produced by the weathering of alumino-silicate minerals. Two dominant processes occur: recrystallization and alteration. **Recrystallization** involves the precipitation of new, secondary minerals from solutions containing the dissolved weathering products of the primary alumino-silicates. A wide range of clay minerals may form, depending upon the chemical composition of the solutions and the weathering environment.

Alteration involves the selective removal and substitution of certain ions from the primary minerals and a gradual loosening of the crystal structure. **Isomorphous replacement** plays a major role in this process. Ions of silica and aluminium are replaced by other ions, of similar size but lower electrical charge. It is this process which gives the clay its negative charge. For example, the silica ion has four unbalanced protons and thus has a valency (i.e. a charge) of four. If silica is replaced by a divalent ion, such as magnesium (Mg^{++}), then the clay has effectively lost two positive charges. This must leave two unsatisfied negative charges in the clay. Similarly, if the magnesium is replaced by potassium, which is monovalent (K^+), another unbalanced negative charge is produced. Because of this, the magnitude of the electrical charge developed on the clays depends on how many and which ions are available to take part in the substitution. In

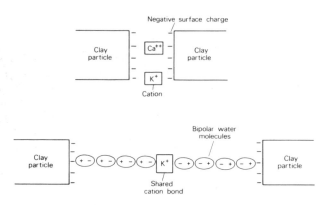

Figure 27.10 Cohesion due to (A) cation bonding, and (B) development of chains of bipolar water molecules

alkaline environments, where there is abundant potassium, magnesium and other bases, clays with a high charge form. In more acid environments, clays with a lower charge tend to develop. On the whole, the 2:1 clays such as montmorillonite and vermiculite have a higher negative charge than the 1:1 clays such as kaolinite. It should also be noted that some non-crystalline (i.e. amorphous) clays do occur. How they form remains uncertain, but they tend to have a lower electrical charge, and to be less plastic and less cohesive than the crystalline clays.

Soil structure

We only have to look at the surface of a ploughed field to appreciate that the soil rarely consists of a collection of loose particles. Instead, the particles are arranged into aggregates. In their natural state the aggregates are referred to as **peds** and they occur through the effect of a number of forces. The initial impetus for aggregation comes from the cohesive forces in the soil. Cohesion is partly a result of the processes of clay formation and organic matter decomposition. As we have seen, these lead to the development of small colloidal particles with minute negative electrical charges at their surface. When these particles are drawn very close together these electrical charges form a mutual attraction with positive charges inside the clay structure. In addition, water molecules and positively charged ions, such as calcium (Ca^{++}) or potassium (K^+) may form bridges between the colloids and create small aggregates (Figure 27.10). The colloidal particles may also form 'domains' which envelop and bind together larger sand and silt grains (Figure 27.11).

The effects of cohesion play an important role in generating soil structure, but they are not permanent, and when the soil is wetted, for instance, water enters the pore spaces between the particles and may push them apart. The cohesive forces are then reduced. Stability is given to the aggregates by the action of cements in the soil. These include both chemical compounds, such as calcium carbonate and iron oxides, and organic compounds formed during the decomposition of plant materials. Complex organic polymers called **polysaccharides** are particularly important in this respect since they are water repellent. They therefore act like a waterproof coating to the aggregates, and make the soil resistant to structural disintegration by wetting. In addition, fungi and plant roots may enmesh the particles and bind them together.

It is clear that the character of the soil structure depends to a great extent upon the texture and the amount and type of organic activity in the soil. In this context earthworms play a major part for they help to decompose the plant debris and mix it with the other soil materials. Consequently, structure is best developed in soils with an active earthworm population. Different types of soil structure are shown in Figure 27.12.

Soil pH

The ability of the soil to support plant life depends upon a wide range of chemical properties. One of the most critical is the **soil reaction** or **pH**. This relates to the concentration of free hydrogen ions in the soil. Hydrogen ions are provided by the dissociation of water, by root activity, and by many weathering processes. As the concentration of

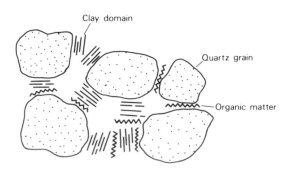

Figure 27.11 The binding of quartz sand grains by clay domains and organic matter. Each 'domain' is composed of a group of tightly linked clay particles

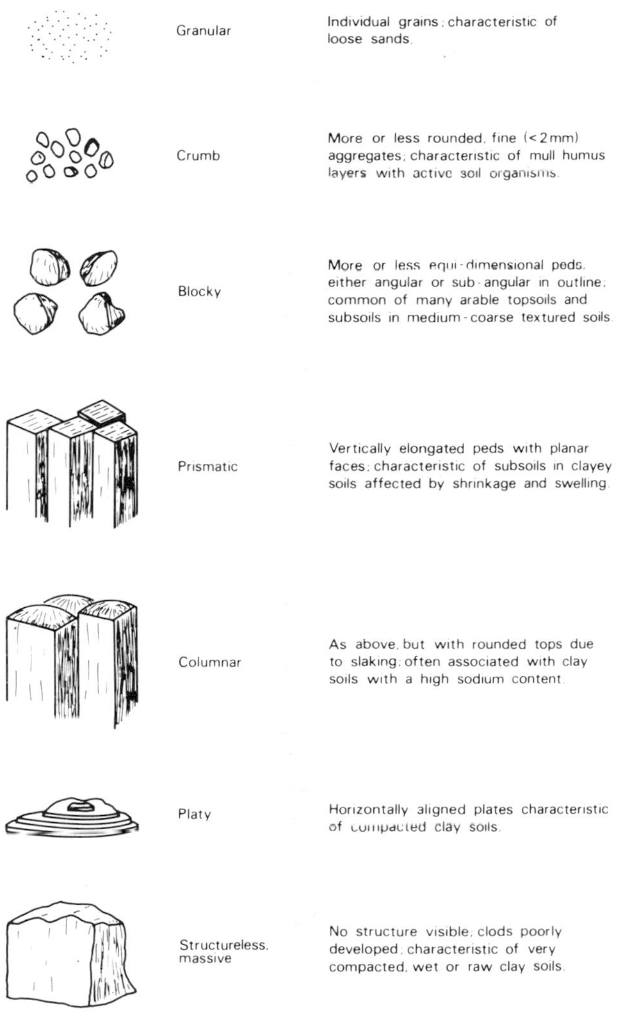
Granular
Individual grains; characteristic of loose sands.
Crumb
More or less rounded, fine (<2mm) aggregates; characteristic of mull humus layers with active soil organisms.
Blocky
More or less equi-dimensional peds, either angular or sub-angular in outline; common of many arable topsoils and subsoils in medium-coarse textured soils.
Prismatic
Vertically elongated peds with planar faces; characteristic of subsoils in clayey soils affected by shrinkage and swelling.
Columnar
As above, but with rounded tops due to slaking; often associated with clay soils with a high sodium content.
Platy
Horizontally aligned plates characteristic of compacted clay soils.
Structureless, massive
No structure visible, clods poorly developed, characteristic of very compacted, wet or raw clay soils.

Figure 27.12 Types of soil structure

hydrogen ions increases, the soil becomes more acid.

Acidity is measured on the pH scale. This, simply, measures the concentration of active hydrogen ions, thus:

$$pH = \log_{10}\frac{1}{[H]}$$

that is to say: the pH is the logarithm of the reciprocal of the molar concentration of the hydrogen ions in the soil solution. What this means in reality is indicated in Table 27.4. As this shows, the soil is considered to be acid when the pH is below 7.0; it is alkaline when the pH is above 7.0. Note two things: that acidity increases with a reduction in pH; and that the scale is logarithmic so that each time the hydrogen ion concentration shows a tenfold change, the pH alters by one unit.

Table 27.4 *The pH scale*

pH	*Hydrogen ion concentration (moles/litre)*	*Hydroxyl ion concentration (moles/litre)*	*Description*
3.0	0.001	0.00000000001	Excessively acid
4.0	0.0001	0.0000000001	Strongly acid
5.0	0.00001	0.000000001	Moderately acid
6.0	0.000001	0.00000001	Slightly acid
7.0	0.0000001	0.0000001	Neutral
8.0	0.00000001	0.000001	Alkaline
9.0	0.000000001	0.00001	Strongly alkaline
10.0	0.0000000001	0.0001	Excessively alkaline

The importance of pH to soil processes and fertility is mainly that it influences the solubility of many nutrients. At low pH, there is a tendency for

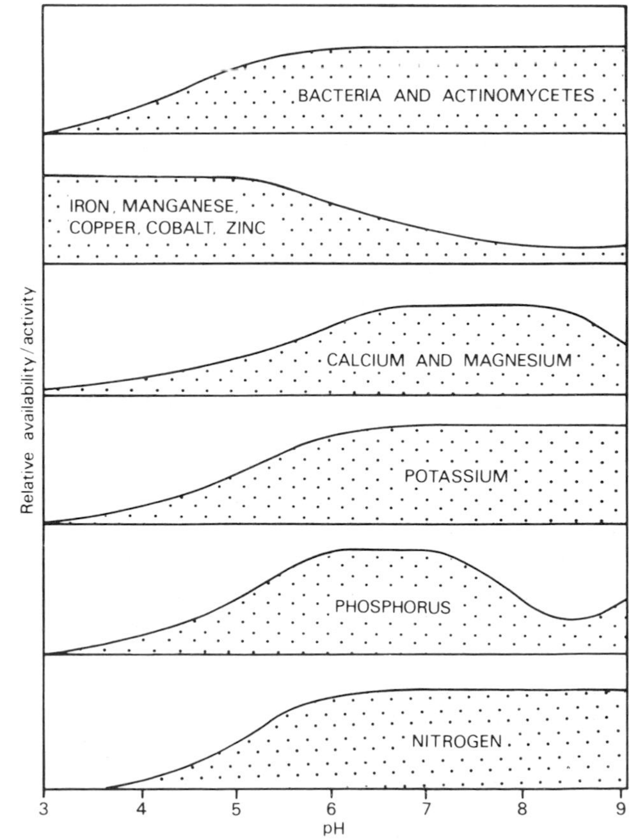
BACTERIA AND ACTINOMYCETES
IRON, MANGANESE, COPPER, COBALT, ZINC
CALCIUM AND MAGNESIUM
POTASSIUM
PHOSPHORUS
NITROGEN
Relative availability/activity
3
4
5
6
7
8
9
pH

Figure 27.13 The effects of pH on nutrient availability and biological activity most nutrients to be very soluble; so much so that they are readily washed from the soil and become unavailable to plants. At high pH the nutrients are often insoluble, and plants cannot extract them from the soil. As this indicates, the optimum conditions for plant growth lie somewhere between (at a pH of about 6.0 to 7.0), and it is the aim of the farmer to keep his soil pH within this range in most cases. pH also affects the operation of some soil organisms, however. In this way it influences biological cycling of nutrients, and has a further effect upon plant growth (Figure 27.13).

Soil pH is particularly sensitive to soil-forming processes. In calcareous soils, pH is high because the calcium carbonate dissociates to form free carbonate ions:

$$CaCO_3 \rightarrow Ca^{++} + CO_{3-}$$

calcium carbonate / calcium ion / carbonate ion

The carbonate then combines with free hydrogen ions to form hydrogen carbonate which is highly soluble and is removed from the soil in drainage waters:

$$CO_3^- + H^+ \rightarrow HCO_3^-$$

carbonate ion / hydrogen ion / hydrogen carbonate ion

Leaching reduces soil pH (makes it more acid) because it removes calcium and leaves the hydrogen ions free in the soil.

Calcium carbonate content

As this indicates, the calcium carbonate content is an important chemical property of the soil, affecting pH and therefore many other soil properties. In particular, it influences soil structure, for calcium ions act as bridges between the colloids, and calcium carbonate is a strong cementing agent. As with pH, the content of calcium carbonate is influenced by leaching, and as the intensity of leaching increases the depth to calcium carbonate in the soil increases also. We noted earlier, for example, that moving eastwards across the Mid West of the USA into areas of higher rainfall, and therefore greater leaching, we find a progressive increase in the depth to calcium carbonate accumulation (see Figure 27.6).

Organic matter content

One of the most important and diagnostic properties of the soil is its organic matter content. As we have seen, organic debris enters the soil from plants and is then broken down to form humus. In the absence of soil organisms – for example, in very wet or acid conditions – the organic debris breaks down only slowly and accumulates on the surface. In acid environments, such as heathlands and moorlands, it leads to the development of a **mor** humus layer, consisting of a clear sequence of fresh **plant litter**, partially decomposed organic matter (the **fermentation layer**) and fully decomposed humus. Where soil organisms are active, however, the organic matter is mixed in with the soil and becomes intimately bound up with the mineral particles. In this state it is referred to as **mull** humus and it acts as a cementing agent which stabilizes the soil structure.

The organic matter content of the soil is naturally a reflection of the quantity and character of inputs from vegetation. It is also closely associated with the activity of soil organisms. On the one hand the organisms are the major agents of organic matter decomposition; on the other hand the plant debris provides the basic foodstuff for the fauna. It is also notable that the organic matter is affected by climatic conditions, for rates of decomposition are greatest under high temperatures and low moisture content, whereas accumulation of organic matter tends to occur in cool, wet or acid environments (Figure 27.14).

Soil formation and soil type

As any farmer knows, soils vary markedly over even quite short distances. At a broader scale, we can identify more fundamental variations in soil type – from the deep, red lateritic soils of tropical areas, for example, to the ashen grey, podsolized soils of cool temperate latitudes. What causes these variations?

The immediate answer is differences in the soil-forming processes operating in different places: in the inputs of material by weathering and deposition; in the redistribution of the material within the soil by leaching and cheluviation, lessivation, upward translocation, gleying and mixing; in the losses of material by leaching, erosion and plant uptake. The soil, like any other component of the environment, can be seen as a process–response system, the form of which is a function of the processes acting within the system (Figure 27.15).

This, however, merely raises another question: what controls the intensity of these processes? One

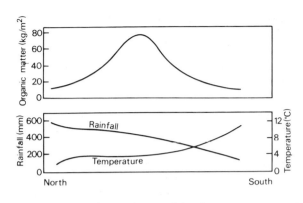

Figure 27.14 The relationship between organic matter content and climate in a transect across the USSR (based on Kononova, 1966)

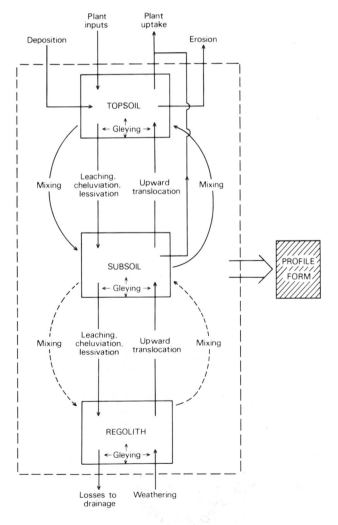

Figure 27.15 The soil as a process–response system

attempt to answer this question was provided by the American soil scientist H. Jenny, in the 1940s. He argued that the soil profile is a product of a number of soil-forming factors, all operating over time. He expressed this relationship in the form of an equation, as follows:

$$S = f(C,G,V,R)T$$

where S is the soil condition, C is climate, G is geology (or parent material), V is vegetation, R is relief and T is time.

To some extent, Jenny's equation is useful, for it indicates the main factors influencing processes of soil formation, but it is far too generalized to be of much value in trying to explain the detailed distribution of soil types. There is no way we can solve this equation as it stands. What we need to do is to consider rather more closely the relationships between the individual processes of soil formation and the external environmental conditions, so that we can see how, under any given set of environmental conditions, the different processes interact.

Some indication of these relationships is given in Figure 27.16. As this shows, a major control on soil

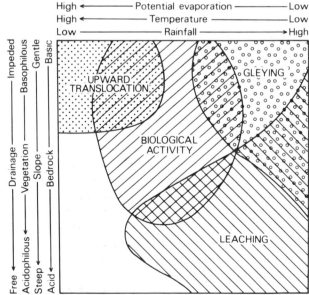

Figure 27.16 The environmental domains of four main sets of soil forming processes. The diagram shows the zones in which each set of processes is at an optimum in relation to climatic (horizontal axis) and vegetational, geological and topographical (vertical axis) conditions

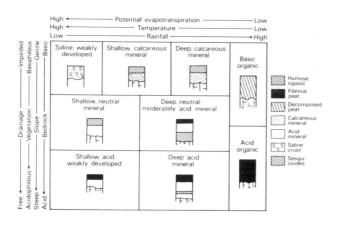

Figure 27.17 Relationships between environmental conditions and general soil types

formation is the availability of water for this is involved in processes such as leaching, cheluviation, lessivation and erosion. This, in turn, depends upon the balance between rainfall and evapotranspiration. Similarly, the rate at which water moves through the soil is important, and this is influenced by factors such as bedrock permeability, slope angle and slope position. Leaching and cheluviation, for example, are inhibited where water flow is restricted, but gleying and organic matter accumulation are encouraged. Water quality is also important, and rates of chemical weathering as well as many processes of redistribution are enhanced by increased acidity. This depends among other things on the character of the vegetation.

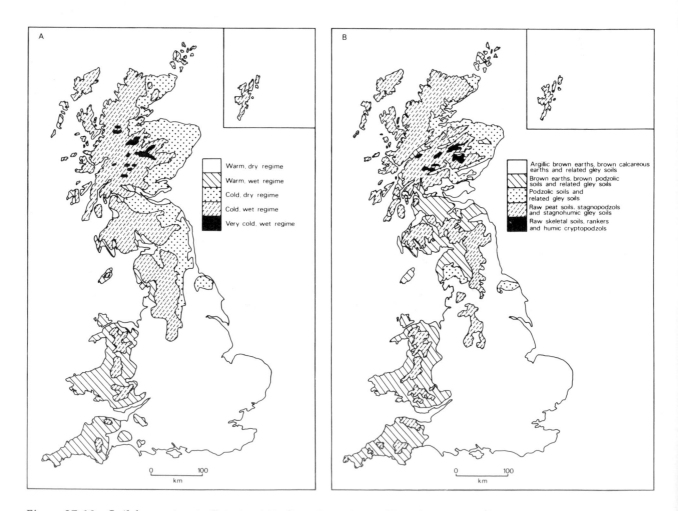

Figure 27.18 Soil formation in Britain: (A) climatic regions; (B) soil regions (after Burnham, 1970)

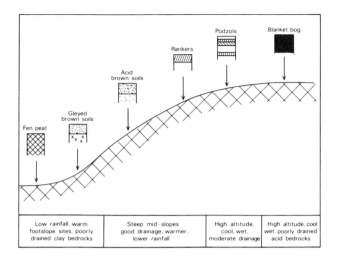

Figure 27.19 Soil variations in relation to slope, altitude, geology and climate: a generalized transect down the eastern slope of the Pennines.

All these relationships are reflected in the character of the soil. It is therefore possible to develop a general model of soil formation as shown in Figure 27.17. We would expect to find, for example, intensively leached, acid soils associated with areas characterized by some combination of high rainfall relative to evapotranspiration, acidic vegetation cover, acid bedrocks, low temperatures and high rates of water flow. Alkaline and saline soils, on the other hand, tend to occur where the R : ET ratio is low, where temperatures are high and/or where the vegetation and bedrock are base-rich. Organic soils are associated with wet or acid conditions; shallow, stony soils with dry and cold environments and hard, resistant parent materials.

We will see the implications of these relationships at a global scale in Chapter 30, when we consider the distribution and character of the world's major biomes. An indication of the general validity (and limitations!) of the model is shown in Figure 27.18, which illustrates the broad distribution of soils and environmental conditions in Britain. At an even more detailed level we can see similar relationships affecting the pattern of soils within a single valley or an individual slope: soil conditions vary in sympathy with differences in slope angle, geology, aspect (and therefore microclimate) and vegetation (Figure 27.19).

Conclusions

As we have seen, the soil is a complex system. Because it lies at the interface between the lithosphere, the atmosphere and the biosphere it is influenced by a wide range of different factors and it represents the products of the continual conflict and adjustments operating at this boundary. We cannot consider soil formation independently of geology, topography, climate and vegetation.

The soil, however, is not only a product of the processes operating at this interface. It also affects the processes and influences the other components of the environment. It influences the rate of weathering and the processes of sediment movement and deposition, for example, and thereby affects landscape-forming processes. It influences the absorption and reflection of energy and the release of water into the atmosphere, and thus has an effect on microclimate. It affects the pathway taken by water as it flows into streams and seeps into the bedrock, and thus it is an important component of the hydrological cycle. Above all, the soil exerts a fundamental control on processes of plant growth and thereby affects the development of the vegetation layer.

28

Development of the vegetation layer

The structure of the vegetation layer

Vegetation

Outside the desert and glacial areas, most of the land surface is covered for at least part of the year with vegetation. The character of the vegetation varies considerably. It may consist only of simple, ground-hugging plants such as lichen and mosses or include herbaceous plants which temporarily invade the surface in periglacial areas during the brief spring; it may include grasses and shrubs; it may be a complex forest.

Within the vegetation, however, there typically exists several layers or **strata**. The roots represent the lowest level, penetrating into the soil, and acting as the main link between the soil and vegetation systems. Immediately above the ground surface occur the low-lying plants, often only a few millimetres in height. They are overshadowed by taller herbs and grasses, which are in turn overlooked by shrubs. Then comes the tree layers; the young seedlings at the base of this zone, the secondary or understorey trees and finally the dominant or overstorey trees (see Figure 26.10, p. 435). The whole structure represents a vertical horizonation; it is a structure which has significance not only in relation to the visual character of the vegetation but also to the processes operating within the vegetation. We have seen earlier (Chapter 9) that specific climatic conditions develop within the vegetated layers; in addition, competition between plants is associated with this structure, for the light reaching plants lower down in the vegetation hierarchy are controlled by the plants above. Moreover, the structure of the vegetation provides a range of different habitats for the fauna.

Of course, not all vegetation consists of all these layers. As we have already mentioned, in some extreme environments only the lowest layers are present. Where the growth of plants is favoured, however, as in many moist tropical areas, even more strata may exist (Figure 28.1).

It is clear that this layering of the vegetation represents a distinct structure of plant types; each layer tends to be dominated by specific plants. Within any one layer, however, there may be considerable diversity. Thus, within the overstorey layer it is common to find three or four main species of tree in temperate areas, while tropical forests may contain a dozen or more.

The vegetation, therefore, represents an **assemblage** of plants, each with a particular **niche** within the overall structure. One of the fascinating aspects of the vegetation is the way these different plants, each acting within their own specific position, are interdependent.

The fauna

Within the vegetation there frequently exists a varied and active wildlife. This, too, shows a vertical structure related to the character of the vegetation. Thus we may discern animals that live in the canopy of the trees, animals which concentrate mainly in the understorey and within the stem zone, and animals which live on the ground. Again, these animals show a distinct spatial variation. Often, they inhabit specific territories; almost invariably they have a particular niche. Furthermore, like the plants, they are closely associated; through their search for food, and their relationship of predator to prey and prey to scavenger, they exhibit a network of interactions.

Changes in the biosphere

The evidence for change

The vegetated layer of the earth – the **biosphere** – is a particularly dynamic system. Both fauna and vegetation are constantly changing. We can readily see some of the short-term changes if we wander through a wood. Trees die and fall and decay, animals migrate into or out of the area, the wood itself may be encroaching on to neighbouring

454

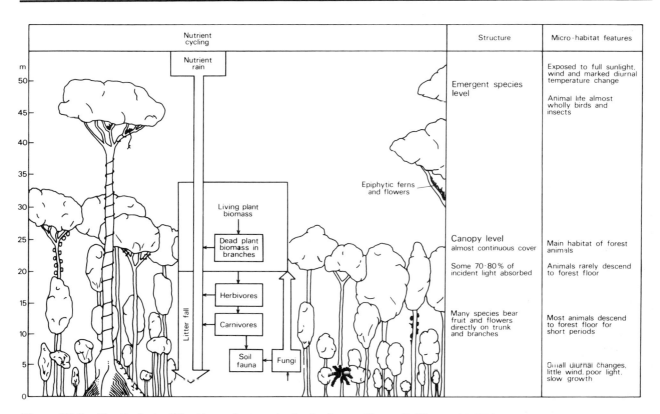

Figure 28.1   Vertical stratification of a tropical rain forest (from Collinson, 1977)

abandoned farmland. In the long term the whole structure of the vegetation may change as a result of changes in external conditions. We know, for example, that the climatic variations during the Quaternary Period led to marked adjustments in vegetation; the evidence is preserved in the pollen records of peat bogs and lakes (Figure 28.2). We know, too, that man has modified the vegetation by forest clearance, cultivation and pollution. Even landscape changes are important. Irrespective of any climatic influence, the invasion of vast areas by ice sheets, and the consequent removal or burial of the original soils, led to major adjustments in the character of the vegetation and fauna. The plants and animals which invaded the area as the ice withdrew had to tolerate very different soil conditions. Mountain uplift, erosion and deposition have the same general effects. As the landscape rises, the climatic and soil conditions alter; as erosion strips away the rock it alters the soil; as deposition builds new features it creates new substrates for soil and vegetation development. The salt marshes and sand dunes that we discussed in Chapter 24 are clear examples. Thus, all the time that the

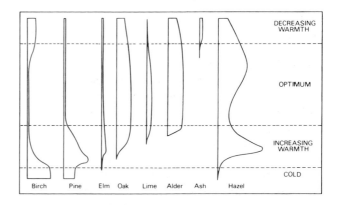

Figure 28.2   Generalized pollen diagram showing changes in vegetation during the transition from a cold (glacial) to warm (interglacial) phase. Changes like this occurred following the last (Devensian) glaciation in Britain (from Godwin, 1975)

environment is changing, so too is the vegetation and the fauna.

Processes of change

Plants and animals respond, therefore, to their environment. Their adjustments are yet another example of an equilibrium response; they attempt to maintain an equilibrium with the climate and the soil. However, these environmental relationships are not the only factors affecting the processes of change. For change in any ecosystem invariably means adjustments in the inputs to the ecosystem. New animals and plants start to appear, and thus the interactions between these individuals become important. **Competition**, both between individuals of the same species and between different species attempting to fill the same niche within the environment, acts as an important control upon the process of ecosystem development. Clearly, if one species is more competitive than another, it will tend to oust its rivals and become established. This is the basis of Darwinian thought: survival of the fittest.

In addition, other aspects of the plants or animals must be considered. For an individual to become established in any area it is not just sufficient for it to be a good competitor; it must also be able to get there. Thus, its means of **dispersal** or **migration** are important. All else being equal, we can readily appreciate that the plants and animals which can most rapidly get to an area are most likely to become established.

We can identify three main factors affecting the development of the biota, therefore:

1   The environmental relationships of individual species.
2   The interactions between the plants and animals themselves.
3   The means of dispersal and migration.

We must consider these separately before looking at the more general question of ecosystem development.

Environmental relationships

Environmental factors

Plants and animals respond to two main **environmental factors**: to the climate and to the soil. The prevailing state of these two factors determines to a great extent whether an individual is likely to survive. In general, individuals are able

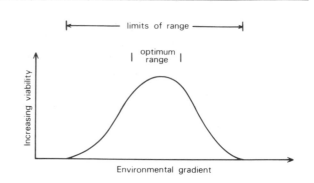

Figure 28.3   The viability of an individual plant in relation to changing conditions along an environmental gradient. The gradient shown may represent, for example, increasing pH (see Figure 28.8)

to tolerate a range of conditions. At the extremes of this range their physiological processes become less effective and they consequently survive less well; within the middle of this range they find their optimum conditions (Figure 28.3).

Similar relationships can be seen between whole species and environmental factors. In this case, however, the individuals which make up the species are variable. Thus there is a tendency for the range of tolerance to be extended slightly (Figure 28.4). If one individual is not able to survive, another member of the species might.

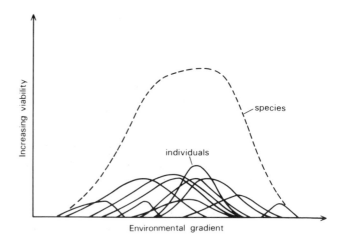

Figure 28.4   The relationship between the viability of a species and the individual members of that species along an environmental gradient. Each individual has a slightly different range of tolerance but most individuals have optimum viability near the centre of the species range

In both cases, the plants and animals are not responding to the general nature of the climate or the soil, but to specific aspects of them. In the case of the climate, the important factors are the rainfall, the temperature, the wind and the humidity. These factors between them tend to control the period when the plant is able to grow (i.e. its **growing season**), the rate of transpiration (and hence of nutrient uptake) and the rate of photosynthesis. They similarly govern the bodily functions of animals. Here, one of the main factors is the balance between heat loss and energy assimilation. The same is, of course, true of the soil. It is not simply the general character of the soil which is critical, but the specific conditions, such as soil depth, nutrient reserves, pH, water retention and drainage and soil structure.

The law of the minimum

One principle which follows from what we have just said is that inadequacies in relation to any of the climatic or soil properties are able to limit the growth of plants and the survival of animals. It matters little, for example, that rainfall is adequate and the duration of sunlight ideal if the temperature of an area is too low to allow plant growth. In this instance temperature acts as a **limiting factor**. This principle is embodied in what is known as the **Law of the Minimum**. It was developed in the nineteenth century by the German agricultural chemist Justus von Liebig, and it may be illustrated by the example in Figure 28.5. In simple terms, the Law of the Minimum states that it is the most limiting factor that controls the response of the individual.

This idea seems very simple and logical, but we must beware of some of its hidden implications. One implication is that, for any individual, changes in any but the limiting factors have no real effect upon its growth. Thus, if we see that the plant in our plant pot is dying, it will have no substantial effect to give it fertilizer, sun-ray lamps and encouraging words, while its real need is for more water. Another implication is that if we remove the effects of one limiting factor, another will come into operation.

Nevertheless there are reasons to doubt the total validity of this principle when extreme conditions do not exist, for to some extent the factors governing growth of plants and animals are substitutable. Within fairly narrow limits it is possible to compensate for inadequacies in one factor by improve-

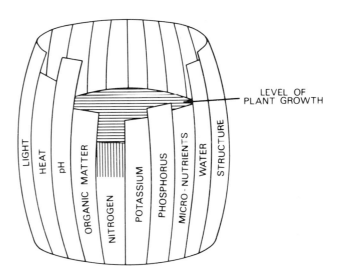

Figure 28.5 The law of the minimum (after Brady, 1973)

ments in another. Thus, an increase in the supply of certain nutrients may enable the plant to compensate for low temperatures; increasing the food supply of an animal may enable it to withstand harsh climatic conditions.

Another aspect which must be remembered in this context is that many of the factors controlling plant and animals responses are closely related. This is part of the explanation behind the substitutability of these factors. It also operates in a rather more indirect fashion, however, for it means that if one of the factors is changed there may develop complex reciprocal effects with other factors. The truth of this can readily be seen. If the climate becomes warmer, not only do temperatures rise, but also evaporation and water loss; soil conditions too may alter, for the increased temperatures may stimulate more vigorous organic activity and faster nutrient cycling. Many of these responses may counteract each other. As we have seen on numerous occasions it is often a question of the balance between opposing feedback reactions that determines the response of the system. But, either way, an individual factor acts as a trigger mechanism for many other responses.

Environmental gradients

We need to take these aspects into account when we consider the environmental relationships of plants and animals. However, they do not invalidate the basic principle – that individuals respond to their environmental conditions. All else being equal, changes in one of the controlling environmental factors leads to a response in the plant or creature. If the change is too great, the individual may die or leave.

The relationship between the response of an individual and a single environmental factor is, as we saw in Figure 28.3, represented by an approximately bell-shaped curve. The optimum response typically occurs somewhere in the central part of the range. In the case of a whole species, the overall response is built up of the cumulative responses of many individuals (Figure 28.4). Thus, given a particular set of environmental conditions, whether or not an individual will survive at any point in space depends upon whether one of the individuals which makes up the species is able to tolerate the conditions. There follows from this the fact that plant and animal species tend to react to **environmental gradients**. More individuals will survive within the range of optimum conditions; less will survive at the extremes. Thus, if we imagined an area in which one, critical environmental property varied from one extreme to another, we should find that the number of an individual plant species was small at the extremes and large in the central optimum area. This, in a simplified form, is the concept of environmental gradients. The activity of the species responds to the condition of the main limiting factors. Since, in the real world, these factors vary markedly from one area to another, changes in the environmental conditions lead to differences in the vegetation and animals.

Competition

Plant and animal assemblages

As we have noted, plants and animals do not act entirely as individuals; they do not exist in isolation. Thus they are forced to compete with one another for a position in the world. This involves competition between individuals of the same species, and between different species. It also includes competition between plants and animals.

It is important here to appreciate that, in the real world, both the vegetation and the fauna are made up of a number of different species which exist together as an assemblage. There is a tendency for a specific set of environmental conditions to be associated with a specific plant and animal assemblage,

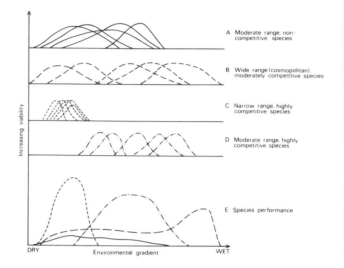

Figure 28.6 The viability of individual members of four species of differing competitive ability (A–D) in relation to soil moisture and the effects on species distribution (E)

so that it is possible to talk about particular vegetation types occurring in certain environmental conditions. We talk, for example, about tropical rain forests and savanna grassland. In neither case will the vegetation be the same everywhere, but in general terms the same species will be represented in similar relative quantities.

This pattern arises partly from the competitive advantage that certain individuals and species have under specific conditions. It is clear that if we consider a simple environmental gradient, and the responses to it of individual members of different species (Figure 28.6), there will be situations in which one species has a competitive advantage (that is, it responds more actively and favourably to the conditions), and other situations in which another species has the advantage. Thus, over the small scale, different individuals from different species may become established. Overall, of course, it is possible that the individuals of one species are better adapted to the environment than those of another. Thus, the more competitive individual will tend to dominate the environment and it will be found in larger numbers within the assemblage (Figure 28.6).

Competition between individuals and species therefore relates to their degree of adaptation to the specific environmental conditions. It is not, normally, an absolute thing, and as conditions change so individuals that were once more competitive become less so.

The nature of competition

The way in which this competition is expressed, however, varies. One of the main means of competition is through light. More competitive plants may grow more rapidly and shade out their rivals. The shaded plants may thus find their growth further restricted; they will be less able to take up water and nutrients and this will retard their growth even more. A process of positive feedback may develop which results in the poorer rival being eliminated (Figure 28.7). One of the critical factors in competition for light, of course, is the rate of upward growth, and there is a tendency for fast growing, tall plants to dominate their smaller competitors. In addition, the changes in the environment brought about by this competition for light may create conditions highly suitable for other plants. Beneath the canopy of dense, light-restricting trees, for example, there may develop substrata of more shade-tolerant plants.

Competition does not occur only for light. Competition for water and nutrients is also important. In the competition for water, root density, root depth and transpiration rate are clearly important. Plants which draw their water from similar levels may be in severe competition, and those with the greatest root density may be most successful. In areas of water shortage, plants which have longer roots may be more competitive than rivals that obtain water from near the surface. Competition for nutrients also relates to these factors, but in addition various microbial relationships become important. **Mycorrhizal** associations with fungi and bacteria may give one plant an advantage over another; thus under conditions of low phosphorus content, plants with the necessary bacterial associations tend to be more competitive. In addition, some plants exude substances or encourage microbial activity which specifically inhibits others. An intriguing example of this was provided by the apparent competition between the lowly ling plant (*Calluna vulgaris*) and the Scots pine (*Pinus sylvestris*). It was found almost impossible to grow pine trees upon ling-covered

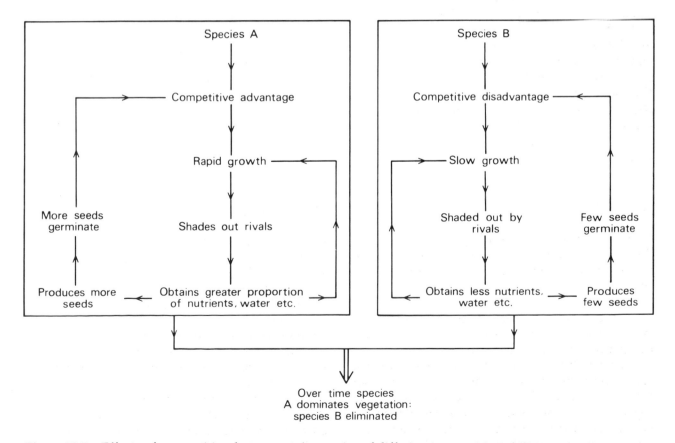

Figure 28.7 Effects of competition between two species of differing competitive ability

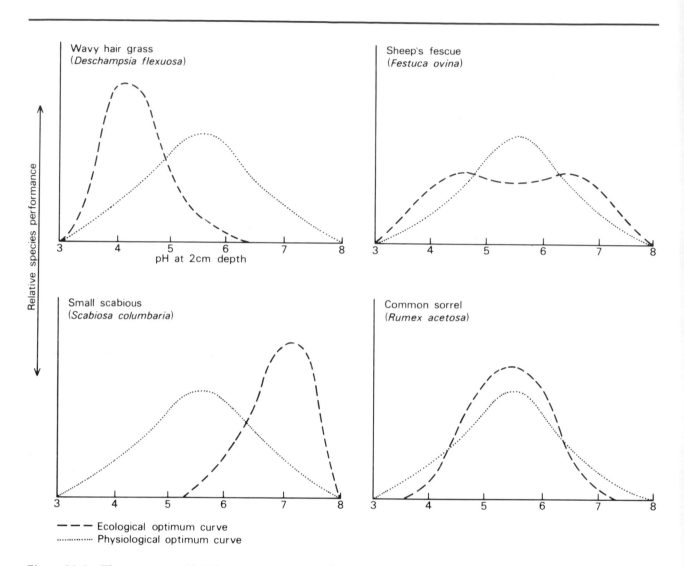

Figure 28.8 The response of different species to soil pH in pastures in the Sheffield area. The physiological optimum curve shows their relative performance under non-competing conditions; the ecological optimum curve shows their frequency under conditions of competition in the field, as determined from studies of 340 1 m² quadrats (from Rorison, 1969). Note that wavy hair grass is confined to more acid sites than it would otherwise favour; small scabious inhabits more alkaline sites; and only common sorrel (the most competitive species) occupies its physiological optimum range

moorland, and many possible reasons were explored. In the end it was discovered that the *Calluna* roots were associated with specific mycorrhiza that inhibited the fungi that normally grew in association with the roots of the pine trees. Lacking these fungi the pines could not obtain sufficient nutrients for their growth.

Competition between individuals tends to be keenest between members of the same species, for they clearly have almost identical requirements. In terms of the assemblage as a whole, however, it does not matter much which individual of the same species wins the competition; in each case the composition of the assemblage will be the same. More important is competition between members of different species. Again this is most intense where the species have similar requirements (Figure 28.8). Where this occurs less competitive species may be forced out of their optimal areas and may instead occupy sub-optimal niches where competition is less severe. In some instances, competition between plants trying to fill the same niche results

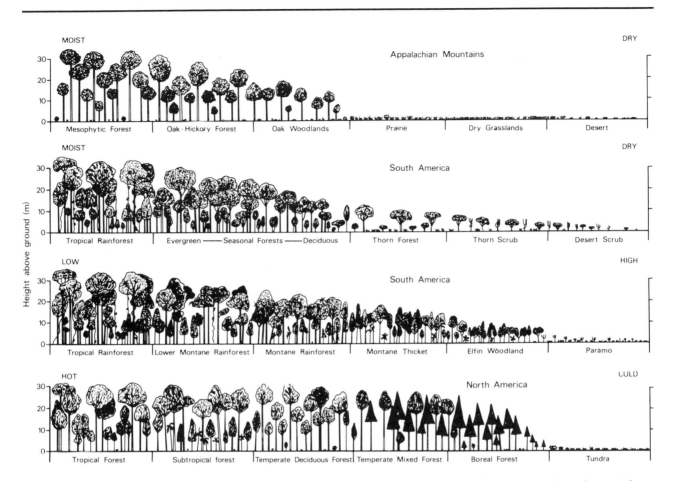

Figure 28.9 Changes in the structure of the vegetation along selected environmental gradients (from Whittaker, 1975)

in an almost invariable success for the same species. In this case the less competitive species may only survive in areas which are, in reality, sub-optimal for its growth, but in which it nevertheless has an advantage over its rivals. At the larger scale these differences in the ability of plants to compete successfully under different environmental conditions account for the variations in vegetation type we see across the world. Some examples are illustrated in Figure 28.9.

Dispersal and migration

Processes of migration

In the case of animals, movement from one area to another occurs through physical migration. An individual wanders into a new area, forsaking its old home and finding a new one.

The reasons for migration vary. In many cases minor and subtle changes in environmental conditions may be important, such as the development of a minute imbalance between predator and prey in the area; thus, the predators may migrate to new areas in search of their food. In other cases, it is a result of changes in the supply of plant foods; the loss of food due to overgrazing of the grasslands may cause herbivores to move to new pastures. In both these cases, however, even more subtle and more general changes may be involved. The cause of overgrazing or the movement of the prey may be a response to gradual climatic changes. Vast migrations of this sort must have happened during the Quaternary as a reaction to the fluctuating environmental conditions. For any individual, the change may have been abrupt and specific; for the populations as a whole, however, migration was almost certainly slow and imperceptible.

As these comments have indicated, migration is often a process which has numerous implications.

461

If one part of the foodchain moves, other parts are affected. Thus imbalances are set up and, in time, the whole ecosystem may change. Alternatively, other, rival animals may move in to fill the vacant niche. In recent times, of course, man has played an important part in these processes of animal migration, deliberately and accidentally introducing new species to areas of the world which once lay well beyond the reach of the animals concerned. In several cases, the migrants found a ready niche. In some cases, too, their arrival has created grave imbalances within existing ecosystems and led to major and occasionally catastrophic adjustments: the introduction of rabbits to the Brecklands, for example, caused dramatic changes in the heathland vegetation.

Plant dispersal

The ability of plants to adjust to changes in environmental conditions is more restricted. To a small degree some plants may be able to adapt their life forms to new conditions, but in general they cannot respond so directly as do animals. Instead, they live, die and regenerate as individuals, only the vigour with which they perform these functions changing as the environment alters. A pattern to these responses, in the form of an apparent migration of the plants, arises mainly through the preferential pattern of survival which results from those environmental conditions. Thus, improvement in one part of their habitat does not lead to a physical movement of plants towards it, but instead to the better survival of seeds or young plants in that area. Conversely, an adverse change in conditions results in greater mortality.

One of the critical factors in the development of vegetation, therefore, is the dispersal mechanism of the plant. Plants which are able to disperse their seeds over a wide area are able to respond to environmental changes more readily than plants which distribute their seeds locally or reproduce by vegetative propagation. It is symptomatic that many of the common weeds which infest large areas of the world are dispersed by wind; the seeds are characteristically light, produced frequently and in great abundance, and often with a high degree of genetic variability. This latter factor allows them to find a niche far more widely; the range of conditions that the species can tolerate is very large.

In the case of seed-bearing plants the means and efficiency of dispersal vary considerably. Some plants produce heavy, robust seeds which tend to travel only short distances, but to survive well due to their internal food supply. Others produce seeds which are transported not by the wind but by animals, especially birds. Man, too, often acts as a major means of plant dispersal; possibly he is the most effective means of dispersal that the world has known. The tomato provides an interesting example, for its seeds are very robust and they have been widely dispersed in sewage and domestic refuse.

Vegetation succession

The basis of vegetation succession

It is apparent that as a new area of land becomes available to plants and animals, the effects of dispersal and migration are to bring individuals to the area. Then, the effects of environmental conditions and competition determine whether individuals, with specific tolerances and competitive characteristics, are able to survive. The vegetation that develops, and the fauna that it supports, depend upon the integration of these three factors.

The same is true as environmental conditions change. The loss of vigour of one species, as the conditions become less favourable for it, allow others, with slightly different environmental requirements, to become established. In both cases, a vegetation change occurs.

Typically, these changes in the vegetation are not instant, but are slow and progressive. Moreover, as the vegetation changes, it causes related adjustments in the external environment – in the soil and the micro-climate – so that complex feedback mechanisms start to operate. For this reason, **vegetation succession** is often a complex and controversial topic.

Developments due to environmental changes

Probably two main reasons for environmental changes account for the major changes in vegetation patterns over the last 2 million years. One is the fluctuation in climate associated with the growth and decay of the polar ice caps; the other is the effect of man, related, in particular, to his management of the land for agriculture.

Climatic changes during and following the glacial periods of the Pleistocene are a fascinating topic in themselves (Chapter 10). Their effects upon the vegetation can only be sketched here; clearly these effects varied according to the specific part of the

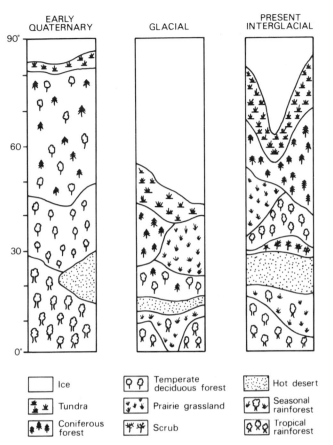

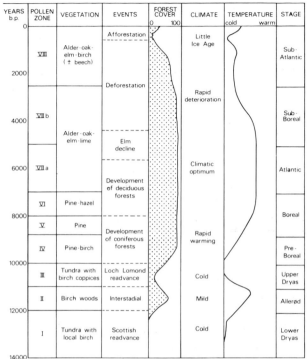

Figure 28.11 Changes in the vegetation of Britain during the Late- and Post-glacial periods

Figure 28.10 Variations in the distribution of vegetation zones during the Quaternary Period

world under consideration. Nevertheless, broad patterns of change can be seen which show that the cold, glacial phases were associated with a much more restricted pattern of vegetation zones, and the virtual disappearance of certain zones (Figure 28.10). It seems, for example, that the tropical rain forests were all but eliminated throughout much of the equatorial region. Only isolated remnants remained, and it was from these that the vegetation dispersed as climatic conditions ameliorated. Interestingly, this is possibly one reason for the relative diversity of the flora and fauna of these areas today; the periods of isolation allowed the development of different sub-species in different areas – sub-species which subsequently became intermingled in the expanding tropical forests.

The evidence for vegetation changes in the earlier parts of the Quaternary are limited, however, and the most extensive evidence relates to the post-glacial developments. These are outlined in Figure 28.11. As this indicates, there was a period in

modern temperate regions when the vegetation developed from a sub-arctic tundra type to a rich mixed oak forest. The peak of this development was reached in the Atlantic period, about 6000 years ago. There followed a period when temperatures fell and ice sheets readvanced. Many of the deciduous forests retreated southwards, and coniferous forests dominated the northern latitudes during the Sub-Boreal period. Subsequently, slight renewed warming occurred, and the deciduous forests again advanced northwards.

During the last 5000 years, however, the effects of climatic change have been overshadowed by the activities of man. He has cleared large areas of forest, burned the natural vegetation, introduced new plant species, eliminated competition, and restricted the range of species present over large areas of the world. We still see these changes operating today.

Vegetation succession and the climax concept

The changes which have accompanied climatic fluctuations have themselves been accompanied by developments in soil and microclimatic conditions. Similar relationships occur whenever the

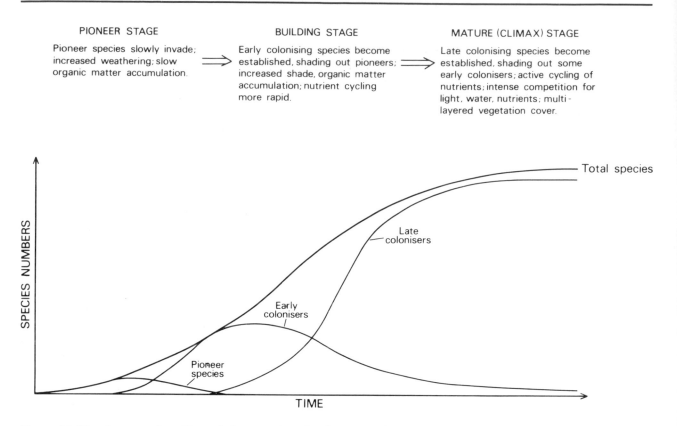

PIONEER STAGE

Pioneer species slowly invade; increased weathering; slow organic matter accumulation.

BUILDING STAGE

Early colonising species become established, shading out pioneers; increased shade, organic matter accumulation; nutrient cycling more rapid.

MATURE (CLIMAX) STAGE

Late colonising species become established, shading out some early colonisers; active cycling of nutrients; intense competition for light, water, nutrients; multi-layered vegetation cover.

Figure 28.12 A general outline of changes involved in a seral succession

vegetation changes, for the plants exert an influence on their surrounding environment. We can see this most clearly if we consider events following the invasion of a virgin area of land, irrespective of any climatic change. This is known as a **seral succession**.

During the early stages of plant invasion, as we have seen, it is likely to be the most readily dispersed plants which arrive first. These will include locally derived species and long-distance, but efficiently dispersed plants. The survival of these initial pioneers depends, however, upon their degree of adaptation to the prevailing conditions. Often these are relatively harsh, with exposed, infertile soils and lack of water. Thus, a restricted range of more hardy pioneers becomes established.

In time, these provide shelter for other species; they help to weather the soil and release nutrients, and they allow the establishment of other plants. Indeed, by altering their environment they create the conditions for their own elimination, for subsequent arrivals include plants which are relatively more competitive in these new conditions (Figure 28.12).

Over time, the process is repeated. Lichen and mosses may be replaced by grasses and herbs, these are replaced by shrubs, and these, eventually, by woodland. Throughout the sequence the soil becomes deeper; it is more fertile and it retains water better. The microclimate within the vegetation becomes more distinctive, with greater shading, lower temperature extremes and higher humidity. All these changes promote the development of the next stage in the succession.

It has been argued that this process moves towards an ultimate and characteristic form of vegetation, known as the **climatic climax vegetation**. The composition of this depends mainly upon gross climatic conditions; soils and other initial environmental factors become less important as they are increasingly modified by the vegetation. Thus, irrespective of the original rock type, a deep, well-weathered soil ultimately develops. In many instances it is argued that the final climax vegetation consists of woodland: tropical rain forest in equatorial areas, mixed oak woodland in many warm temperate areas, northern softwood (coniferous) forest in the cooler temperate zones.

The validity of this concept is now disputed. There are many reasons for doubting its practical significance, not least the fact that the climatic climax vegetation develops so slowly that few areas remain stable enough for the length of time needed to allow it to be achieved. Throughout the Quaternary, at least, climatic fluctuations have occurred too rapidly for the climax to be reached over large parts of the globe. Moreover, it is no longer accepted that climate plays such a dominating role on the vegetation.

Nevertheless, the principle of vegetation succession remains valid. Given the initial creation of a new substrate for vegetation development – whether it be a sand dune, a bare rock surface, a spoil tip or an area reclaimed from the sea – a sequence of soil and vegetation changes is bound to occur, during which increasingly luxuriant vegetation tends to become established.

Ecosystem dynamics

The example of the lithosere

In order to summarize these processes of ecosystem change, it is useful to consider two or three different examples of seral successions, and examine the factors and relationships involved in each. We might take as a starting point the sequence that characterizes ecological successions on a bare rock surface, such as that left by the retreat of ice from the uplands of Scotland following the last glacial phase or which can be seen more recently in Alaska (Figure 28.13). This sequence is often known as a **lithosere.**

The initial conditions in this case are the bare, unweathered surface. The lack of nutrients, and the total exposure of the surface, mean that few plants can survive. Consequently, it is mainly the epiphytes, which obtain their water and nutrients from the atmosphere, and particularly hardy species which are able to extract nutrients from the unweathered rock, that become established. These pioneers include mainly lichens and mosses. These provide a substrate on which other plants can root, and subsequently perennial herbs (e.g. thyme, stonecrop) may become established. As they die and decay they add to the organic litter and the humus, and help to weather the underlying rocks. Animals and micro-organisms slowly build up, including herbivores which graze on the plants and decomposers that break down the plant and animal residues. In time, dwarf shrubs appear (e.g. rest harrow) and these, in turn, provide a basis for taller shrubs and ultimately forest. Thus, gorse, broom and hazel give way to birch, beech and eventually to mixed oak woodland.

The example of the psammosere

A generally similar sequence is shown by the development of vegetation on sand dunes (a **psammosere**). As we saw in Chapter 24 the young

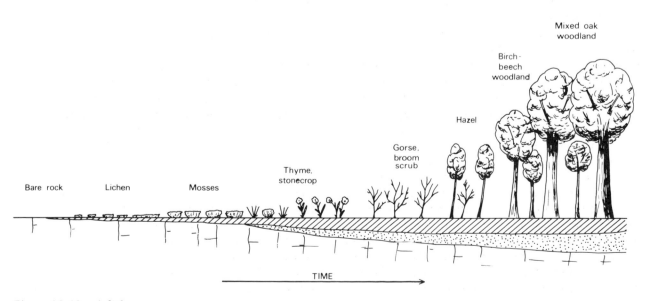

Figure 28.13 A lithosere

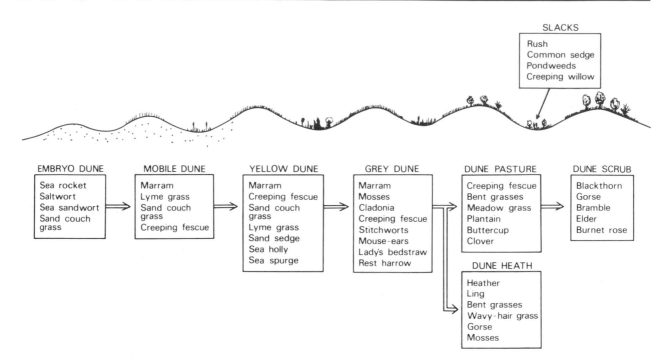

Figure 28.14 A psammosere sequence showing some of the plants typical of each stage

dune surface is unweathered and subject to frequent burial by new sand. Both water and nutrients are lacking, and thus few plants can become established. Only drought-resisting plants such as marram grass and ragwort can survive; these are tolerant of burial and are able to trap and stabilize the surface. This allows weathering to proceed. Lichens and mosses can then invade the surface, together with ground-hugging shrubs such as buckthorn. These create a suitable environment for other species, including **calcicoles** (plants tolerant of alkaline conditions) such as bloody cranesbill and burnet rose (Figure 28.14).

The build-up of humus, the stabilization of the sand and the opportunity for weathering results in the slow formation of soil on the surface, and a reduction in pH. A wider range of shrubs and grasses becomes established, as the dunes become more acid. Ultimately scrub and woodland develop, with holly and pine dominating the vegetation and shading out the light-demanding ground flora.

The example of the hydrosere

Both the lithosere and the psammosere are examples of dryland successions, known as **xeroseres**. In many cases, however, the develop-

Plate 28.1 A small lake formed by solution of evaporites beneath a till cover, near Harrogate, North Yorkshire. The vegetation in the lake shows a clear hydrosere succession, from floating water plants to carr sedge (photo: D. J. Briggs)

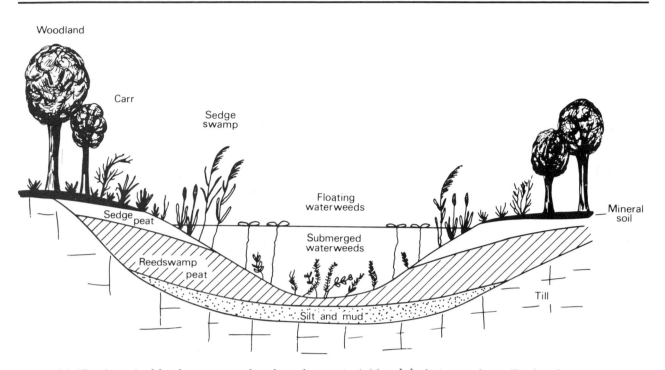

Figure 28.15 A typical hydrosere, as developed in a glacial kettlehole in northern England

ment of vegetation occurs upon a wet or even flooded surface, such as a lake which gradually infills. In this case the succession is known as a **hydrosere** (Plate 28.1).

The example of the development of many kettlehole lakes formed during deglaciation of the lowlands provides an ideal example of this process (Figure 28.15). Here, the first stage in the evolution was the establishment of submerged aquatic plants. As sediment enters the lake and as the waters become more nutrient rich, these are replaced by surface aquatic plants. Continued infilling, especially around the margins of the lake, leads to the appearance of sedges and grasses. All the time the remains of these plants are accumulating in the lake, and trapping material introduced by slope wash or wind. Continually, therefore, the lake is infilled, and the hydrophytic sedges and grasses are able to migrate towards the centre of the lake. At this stage two rather different sequences may be followed.

If the prevailing conditions are acidic – that is, if the inflowing water carries few nutrients – the vegetation tends to develop towards a **damp heath** or **acid bog**. Plants such as sphagnum moss dominate, obtaining most of their nutrients from the atmosphere and keeping the surface moist. Con-versely, if conditions are base-rich – if nutrients are being supplied by the inflowing waters – then grasses and ferns and ultimately trees such as willow, ash and beech or oak may develop. In this case the trend is towards a **eutrophic fen bog**.

Conclusions

The changes that take place during the development of the vegetation represent a natural progression towards equilibrium: an adjustment towards a state of balance with environmental conditions. As in all such developments, the rate of change tends to be most rapid at the start and to fall off with time, as equilibrium is approached. One of the main problems in trying to manage the vegetation is to predict the final state to which it will change following interference and how fast it will get there. After burning of rangeland or moorland, for example, we need to know how quickly the vegetation will recover, and what its new composition will be. Or if we remove animals from an area of pasture we need to assess the speed and character of the changes in the vegetation. Sometimes it may revert to its original state; often when we interfere

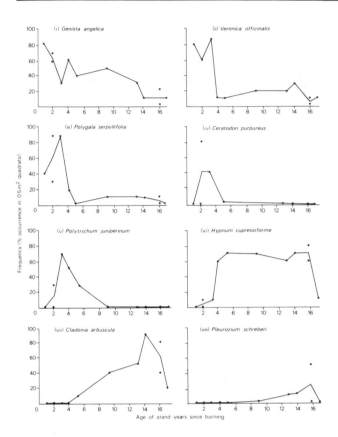

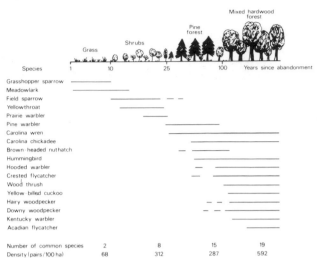

Figure 28.17 Changes in bird populations during a succession on abandoned farmland in south-east USA (data from Johnston and Odum, 1956) (Note: not all species are listed)

Figure 28.16 Changes in the vegetation of a heathland following a deliberate burn (from Gimingham, 1972)

with conditions we initiate a whole new sequence of development (Figure 28.16).

We must also realize that it is not only the vegetation that changes. Vegetation and fauna are intimately related, so that as one changes the other

does also. As the vegetation progresses from pioneer stage to climax forest, the fauna develops accordingly (Figure 28.17). The soil, too, shows a parallel succession. The vegetation, and the accumulation of organic debris, for example, may encourage weathering and leaching, leading to deeper soils. Leaching, in turn, may make the soils more acid, so that calcicole species are unable to survive. In time, as a forest cover develops, the increased interception of rainfall may reduce leaching and changes in the soil may be reversed. Soils, vegetation and animals are intimately related in the ecosystem, therefore, each affecting the other, each developing in harmony.

29

Biogeochemical cycling

Nutrients in the biosphere

The development of the concept of nutrient cycles

During the seventeenth and eighteenth centuries, at a time when modern science was slowly being born, a great controversy raged among agricultural chemists: what was the food of plants? Many ideas were put forward. The chemist and diarist, Evelyn, suggested that there was some 'nitrous spirit' in the air from which they fed, though Jethro Tull (a famous agriculturalist) complained that since this 'nitrous spirit' could seemingly rust iron bars he could not conceive that it could be the food of delicate plants. At about the same time, von Helmont carried out an experiment, in which he grew a willow tree in a sealed tub of soil, to which nothing but water was applied. By weighing both soil and tree before and after the experiment he showed that while the tree gained in weight from only 5½ lbs to 165 lbs, the soil lost no more than a few ounces of weight. He concluded that plants must live off water. Tull, himself, claimed that plants devoured the soil particles and soundly attacked a Dr Woodward who disagreed with him, saying that Woodward was like the old woman who could see a needle on a barn but could not see the barn. It was all furious and bitter stuff!

Only in the mid nineteenth century did the truth begin to emerge, and then thanks to the work of chemists like Sir Humphrey Davy and Justus von Liebig, and agriculturalists like William Lawes, the secrets of plant nutrients were discovered. Following years of careful experiments these and other scientists showed that plants required a range of nutrients, including nitrogen, phosphorus, potassium and calcium. Many of these the plants obtained from the soil, others from the atmosphere.

These discoveries had widespread implications. From them arose the fertilizer industry and the modern approach to agriculture. From them, too, developed the concept of biogeochemical cycles.

The nutrients required by plants, it was realized, flowed through the ecosystem along far-reaching and complex pathways; from soil to plants, from plants to fauna, from fauna back to soil (Figure 29.1). It was probably the German scientist Ebermayer who first established the concept of the nutrient cycle, in 1876, but it seems to have been a Russian ecologist, Vernadskii, who coined the term 'biogeochemical cycles'. Since then the concept has become a fundamental part of ecological studies.

The nutrients

It is time, perhaps, to ask a fundamental question. What exactly are nutrients? Almost any chemical element may, in fact, act as a nutrient; the term is simply applied to elements or compounds which are required by plants to fulfil their growth functions. Eighteen essential nutrients are commonly recognized, and these are split into two main groups: macro-nutrients which are required in large quantities, and micro-nutrients which are used in much smaller proportions. The main macro-nutrients are carbon, oxygen, nitrogen, calcium, potassium, phosphorus, sulphur and magnesium. The main micro-nutrients are manganese, iron, silica, sodium and chlorine, together with several trace elements such as boron, copper, zinc and

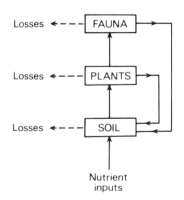

Figure 29.1 The general structure of the biogeochemical cycle of an ecosystem

Table 29.1 *Plant nutrients*

Nutrient	Amount needed (mg/litre)	Function in plants	Sources
Oxygen	–	Respiratory processes	Atmosphere
Carbon	–	Carbon used in photosynthesis	Atmosphere
Nitrogen	15	Protein synthesis, formation of nucleic acids, amino acids, vitamins	Soil, as ammonium, nitrite or nitrate; often fixed by bacteria in plant roots
Sulphur	1	Protein and vitamin synthesis	Soil, as sulphates, gypsum or pyrites; as hydrogen sulphide in waterlogged soils
Calcium	3	Metabolic processes; vital constituent of cell membranes	Soil, from limestone, feldspars, augite, gypsum
Potassium	5	Protein synthesis and transphosphorylation (conversion of sugar to phosphate	Soil, from feldspars, micas and clay minerals
Magnesium	1	Constituent of chlorophyll; enzymic reactions	Soil, from dolomitic limestone, montmorillonite clays, biotite, augite, hornblende
Phosphorus	2	Component of many organic molecules; major source of energy through conversion of ATP to ADP	Iron, aluminium and calcium phosphates in soil; dissolved phosphates in soil solution
Iron	0.1	Oxidation-reduction processes in respiration	Iron oxides, sulphates and silicates; often chelated with organic acids
Manganese	0.01	Small quantities used in enzymic reactions	Iron-magnesian minerals
Copper	0.0003	Respiratory metabolism	Igneous vein minerals
Zinc	0.001	Enzymic reactions	Igneous vein minerals
Boron	0.05	Cell division during growth	Soluble borates (mainly from marine sources)
Molybdenum	0.0001	Nitrogen fixation and assimilation	Igneous vein minerals
Cobalt	0.00001	N fixation in root nodules	Igneous vein minerals
Sodium	0.05	Unknown	Sodium chlorides; sea spray
Silica	0.0001	Unknown	Silicate minerals
Chloride	0.05	Regulates osmotic pressure; balances cation concentration in cells and sap	Dissolved in rainwater entering soil

molybdenum (Table 29.1).

All these nutrients are derived, ultimately, from rocks or from the atmosphere. But they enter ecosystems by a variety of processes: weathering; in rainfall; biological processes (including human activity); and by deposition. They may also be lost from ecosystems. Erosion may remove nutrients from the soil, man removes them when he harvests the crop, waters percolating through the soil wash away essential elements. As in all systems, therefore, the flow of nutrients depends upon the relative magnitudes of inputs and outputs.

Nutrient inputs to ecosystems

Weathering

One of the main, long-term inputs of nutrients comes from rock weathering. The process is slow, and the quantities of nutrients released into the soil in this way are often small, but it is nevertheless vital to the maintenance of the biogeochemical cycle. Numerous processes are involved. Mechanical weathering breaks up the rocks, and chemical processes such as hydration, and carbonation bring the substances into solution. Possibly the most important processes, however, are hydrolysis and oxidation. Together, these alter primary rock minerals to secondary, clay minerals. As we saw in Chapter 27, isomorphous substitution then results in the release of certain atoms from the internal structure of the clays. Silica, aluminium, manganese, potassium and sodium are all made available as nutrients for plants in this way.

Table 29.2 *Chemical composition of rainfall*

Nutrient	Rothamsted, UK	Saxmundham, UK	Concentration (mg l⁻¹) Hubbard Brook, New England, USA	Czechoslovakia	Nova Scotia, Canada
Sodium	1.9	4.4	0.06	0.18	0.08
Potassium	0.7	0.6	0.07	0.08	0.06
Calcium	1.8	1.6	0.16	0.62	0.13
Magnesium	0.4	0.7	0.03	0.07	0.10
NH_4-Nitrogen	1.7	1.8	0.21	0.85	0.01
NO_3-Nitrogen	1.1	1.2	0.22	0.68	0.18
Phosphorus	0.13	0.02	0.003	0.006	n.d.
Chloride	5.9	8.8	0.45	0.41	1.44
Sulphur	3.3	3.3	0.90	1.72	0.47

Atmospheric inputs

Significant quantities of nutrients are derived from the atmosphere, either in rainfall or by biological processes. The fauna, for example, obtains its oxygen for respiration from the open atmosphere, while plants absorb both carbon dioxide and nitrogen through their leaves. In addition, large quantities of atmospheric nitrogen may be fixed in the soil by bacteria and algae. We will examine this process more fully later in this chapter, but its importance is illustrated by the fact that these atmospheric inputs account for about one half of the nitrogen circulating in agricultural systems.

Rainfall, too, acts as a major source of nutrients. Nitrogen, calcium, phosphorus and sodium are all washed into the soil in this way (Table 29.2). The amounts involved are often small, but in low-productivity ecosystems, such as acid bogs, they may be vital to plant growth. Epiphytes (plants which derive their moisture from the atmosphere) like *sphagnum* moss, for example, are wholly dependent upon atmospheric sources for their nutrients.

Fertilizer inputs

In our modern world, man governs many ecosystems and significantly interferes with others. A major way in which he does so is by the application of fertilizers. Most intensive farming systems in the western world are today based upon high inputs of fertilizer nutrients – in particular, of nitrogen, phosphorus and potassium. In addition large quantities of calcium are applied as a liming agent to control soil pH. Total world consumption of fertilizers was estimated as 112 million tonnes in 1980 (Table 29.3), and in some ecosystems (e.g. in arable farming systems) this may account for 80 per cent or more of the total inputs. Even where fertilizers are not used directly, they may act as a major source of nutrients, for a significant proportion of the nutrients washed into the soil from the atmosphere is derived from agricultural fertilizers – either as dust blown away during fertilizer application or gases released into the air during chemical decomposition of the substances.

Biological inputs

Man introduces nutrients to ecosystems in many other ways. Young livestock, seeds and seedlings, for example, act as important inputs of nutrients, while imported feedstuffs – such as hay or corn bran for cattle – can be a major source of nutrients. In intensive grazing systems in the USA, it is common for 25 per cent of nitrogen and 10 per cent of potassium to be derived from these sources; in the Netherlands, where dairy farming is even more intensive, the values may reach 60 per cent and 40 per cent respectively.

Natural processes of migration may also provide inputs. Animals wandering into a new area carry

Table 29.3 *World consumption of fertilizers (1939–80)*

Year	Consumption (t × 10⁶) Nitrogen	Phosphate (P_2O_5)	Potassium (K_2O)
1939	2.6	3.6	2.8
1960	9.7	9.7	8.6
1970	28.7	18.8	15.5
1980	57.2	31.1	23.5

Data from G. W. Cooke, *Fertilising for Maximum Yields*, Granada, 1983.

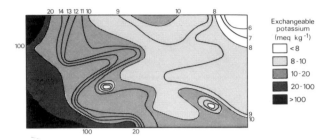

Figure 29.2 The distribution of potassium across a field following grazing by sheep (based on Hilder, 1964)

with them nutrients which they release either in their faeces or when they die and their bodies decay. We see evidence of this on a small scale on the farm, for cattle and sheep tend to graze in certain parts of the range or field and deposit their faeces elsewhere. This results in a gradual transfer of nutrients from one part of the system to another (Figure 29.2). On a larger scale, the mass migration of animals either seasonally, or in response to longer term climatic change, may lead to quite marked inputs of nutrients in some areas.

Nutrient outputs from the ecosystem

If the ecosystem is in equilibrium, the rate of nutrient inputs is balanced by the outputs, so that the total quantity of nutrients in circulation remains the same. This is not always the case of course, and in many agricultural systems outputs may exceed inputs so that nutrient reserves fall and the system degenerates. In other cases, the reverse may be true; lakes, for example, may act as nutrient sinks – the nutrients that enter them may be unable to escape, with the result that the substances build up in the waters, eventually choking them. This is the process of eutrophication. In most situations, however, nutrients are lost through a number of processes.

Erosion

One of the most important means by which nutrients are lost is soil erosion. The process is particularly active in agricultural systems, where cultivation leaves the soil bare and unprotected, so that wind and water can carry away the finer materials. It is these finer fractions – the clays and organic compounds especially – that store many of

the nutrients in the soil, so when the particles are lost the nutrients go with them. In this way, erosion results in the selective removal of soil nutrients; in particular, phosphorus, potassium and nitrogen.

Leaching

Leaching is another important process of nutrient loss. As we saw in Chapter 27, water percolating through the soil carries with it nutrients in solution. Many of these are ultimately washed out of the soil profile and carried into streams and rivers, where they contribute to water pollution; so the outputs from one system become the inputs to another (Figure 29.3). It is a process that is encouraged by high rainfall and acid vegetation (e.g. pine), but leaching losses are often at their greatest in arable farming systems, for arable soils contain large quantities of readily soluble fertilizer nutrients.

There is, however, an interesting dichotomy with leaching. If the nutrients in the soil are to be available to plants, they must be soluble, for plants obtain most of their essential elements as solutes in the water they take up. On the other hand, if the nutrients are soluble they are also highly susceptible to leaching. Thus, increasing the availability of nutrients to plants may increase leaching losses.

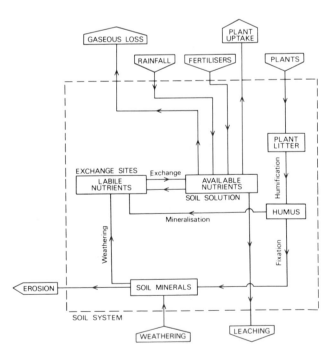

Figure 29.3 The nutrient cycle in the soil showing the main inputs, outputs and pathways for nutrient transfers

Table 29.4 *The effects of seven days' waterlogging on crop yields in Hungary*

Crop	*J*	*F*	*M*	*A*	*M*	*J*	*J*	*A*	*S*	*O*	*N*	*D*
				Per cent reduction in yield due to waterlogging in month								
Grass	–	–	–	10	15	20	20	10	–	–	–	–
Sugar beet	–	–	50	50	50	40	40	40	40	10	–	–
Potatoes	–	–	80	80	90	100	100	100	40	–	–	–
Winter wheat	5	5	15	25	40	50	–	–	–	–	–	–

Data from B. D. Trafford, 'Field drainage', *Journal of the Royal Agricultural Society of England*, **131** (1970), pp. 129–52.

Gaseous losses

Significant losses of nutrients often occur in gaseous form. Especially when the soil is wet and air is lacking, many compounds are reduced to gas which can then diffuse through the water and through the soil pores into the open atmosphere. Nitrogen, in particular, is lost in this way, and it has been estimated in the Netherlands that as much as 80 per cent of the nitrogen applied to the soil as fertilizer may be lost by this process of denitrification. This represents a significant waste of fertilizer, of course, but other effects may also be felt. Some of the gases produced by reduction of carbohydrates, for example, are highly toxic. Ethylene, which is formed by the decomposition of organic compounds under waterlogged (anaerobic) conditions, may damage roots with long-term consequences for plant growth. As a result, even a few days waterlogging at critical times of the year may greatly reduce crop yields (Table 29.4).

The plants themselves contribute to the gaseous loss of nutrients. During respiration, plants release carbon dioxide through their roots. Carbon dioxide is highly soluble and diffuses through the water in the soil and escapes to the atmosphere. In addition, oxygen is lost through the leaves of growing plants.

Emigration and harvesting

Just as material may be introduced to ecosystems by migration, so too may it be lost. The emigration of animals, and the removal of vegetation by man are both processes by which outputs occur from the system.

Man plays a particularly important part in these processes, since he uses much of the land for agriculture or forestry, and consequently removes large quantities of the material during harvesting. Crops are cut, animals are taken for slaughter; milk is taken from cows, wool from sheep. The nutrients in these are thus removed (Table 29.5).

In addition, many of man's agricultural activities cause significant losses. In many agricultural systems he burns the vegetation, either to clear the land for tillage or to remove crop residues and weeds. Considerable quantities of plant nutrients are lost in the smoke, and the remainder in the ash. Not all these nutrients are even retained in the ecosystem as a whole. The losses to the atmosphere may be blown out of the area; the losses to the soil may be removed rapidly by leaching. Thus, repeated burning can lead to a decline in the fertility of the ecosystem.

Biogeochemical cycling in the soil

The biogeochemical cycle in the soil is complex. The main inputs to the soil come from weathering, rainfall, fertilizers, atmospheric sources and plants. Under natural conditions, inputs from plants are the most important, including not only nutrients released by organic matter decay, but also substances

Table 29.5 *Nutrients retained in crops and livestock*

Crop	Yield (t ha^{-1})	N	P	K	Ca	Mg	S
		Nutrient content (kg ha^{-1})					
Wheat	6	120	25	80	20	15	25
Barley	5	100	18	60	15	8	20
Potatoes	50	180	25	200	10	20	25
Grass	10	250	30	250	70	20	20
Rice	2.2	26	8	8	2	4	–
Tea	1.3	60	5	30	6	3	–
Cattle	2.5/ha	50	30	15	25	10	–

Data from C. W. Cooke, *Fertilising for Maximum Yields*, Granada 1983.

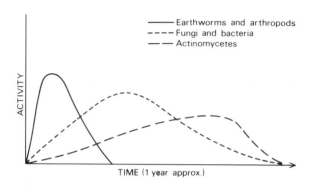

Figure 29.4 The changing activity of different groups of soil organisms during decomposition of organic matter (from Briggs, 1977)

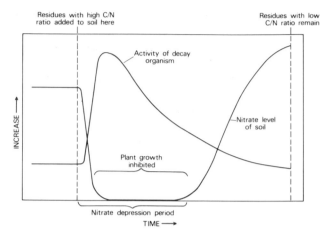

Figure 29.5 The effects of raw organic residues on soil biological activity, nitrate levels and plant growth (after Brady, 1973)

washed in from the plant leaves. Losses are by leaching, erosion, gaseous loss and plant uptake. Within the soil, nutrients are stored on the particles or in chemical compounds, and circulated both in the water and in the organisms.

Organic matter decomposition

The input of nutrients through the decomposition of organic matter is dependent upon the action of soil organisms. It is, therefore, a process which is closely associated with the detritus food chain. The initial attack on the plant debris often occurs before the material reaches the soil. The leaves of trees, for example, may be partially decomposed by fungi while still on the tree. This process releases some of the more unstable compounds such as the sugars and rainfall washes them into the soil. Once the leaf falls to the ground, larger soil organisms attack it. Earthworms, arthropods (such as ants, beetles and termites) and gastropods (slugs and snails) macerate (chew up) the leaf material and make it more susceptible to attack by the smaller, micro-organisms. General purpose organisms – mainly fungi and heterotrophic bacteria – tend to wage the next attack on the debris, and during this stage more resistant carbohydrate compounds (e.g. polysaccharides and starches) are decomposed. Finally, **actinomycetes** and various more specific fungi and bacteria attack the remaining substances: mainly **cellulose** and **lignin** (Figure 29.4). At the end of the process, the resistant fulvic and humic acids remain. It is these which comprise the humus of the soil.

The whole procedure may take only a few months; the leaves of many deciduous trees, for example, rot within nine months. In tropical areas the process may be completed within a matter of weeks, for the voracious appetite of termites and the thriving population of micro-organisms allow rapid breakdown. On the other hand, the leaves of pine needles, in cool temperate areas, may take nine years or more to decompose fully.

The remaining humus decomposes even more slowly for it breaks down mainly by chemical oxidation. Only in tropical conditions, where soil temperatures are sufficiently high to speed up the chemical processes, does the humus disappear at all rapidly; in temperate areas dating of the material by radiocarbon methods has shown that some of

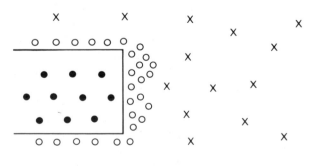

● Ions held in colloidal particles (unavailable)
○ Ions adsorbed on colloidal surfaces (exchangeable)
X Ions in solution (available)

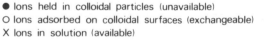

Figure 29.6 The distributions of cations in relation to colloidal particles (from Briggs, 1977)

the humus in the soil may be as much as 1000 years old.

Organic matter decomposition releases nutrients into the soil, and is a process of considerable significance to the nutrient cycle. The nutrients are not always released immediately in an available form, however, for many are bound up within the bodies of the consumers and decomposers. They are made available only when these organisms themselves decay and the organic compounds are mineralized into inorganic forms. This effect of organic assimilation of the nutrients may be important, for during active decay the populations of the organisms may increase markedly, and in the process they may take up nutrients which would otherwise have been available to plants. For a short period, therefore, nutrient deficiencies may occur, and plant growth be inhibited (Figure 29.5).

The compounds which are released in an available form are not all stored; some may be lost from the soil before they can become attached to the soil particles. Carbon, oxygen and hydrogen may all be lost in gaseous form; calcium and magnesium may be dissolved in the soil water and lost by leaching. Water itself is another common decomposition product. Thus, the input of nutrients to the soil nutrient store is far from totally efficient; considerable losses may occur in the process.

The nature of soil nutrients

Nutrients are stored within the soil in a variety of forms. Most occur as ions; that is as atoms or groups of atoms bearing an electrical charge. Positively charged ions are known as cations (e.g. calcium, Ca^{++}, or sodium, Na^+); negatively charged ions are anions (e.g. nitrate, NO_3^-, or phosphate, PO_4^-). Some are bound up within the structure of the soil particles – for example, within the clay minerals. Others, such as phosphorus are held in chemical compounds (e.g. calcium phosphate). These nutrients are thus unavailable to plants unless the minerals are weathered and the nutrients are released. They are called labile nutrients. Many of the cations are also held in an exchangeable form; they are bound loosely to the surface of colloidal clay or organic particles. Although they are not immediately available to plants, they may be released relatively easily by the process of cation exchange. For this reason they are referred to as exchangeable nutrients. Finally, there are the available nutrients – those which are dissolved in the soil water (or soil solution as it is commonly

known) and which may be removed directly by plants.

The exchangeable nutrient reserves in the soil have particular importance. If growth is to be maintained, the nutrients removed from the soil solution by plants must be replenished by release of labile or exchangeable ions. Release from labile forms is slow and for this reason nutrients such as phosphate, which are held mainly in this form, cannot be replaced rapidly. On the other hand, most of the cations in the soil are held in exchangeable form, and these may be released quickly, ensuring that deficiencies of these nutrients rarely occur. Let us look at the processes involved.

Cation adsorption

The ability of soils to store nutrients in an exchangeable form arises from the colloidal properties of the clay and organic materials. As we have seen, isomorphous substitution during clay formation results in the replacement of certain atoms in the clay structure by others of similar size. In most cases, this process involves atoms of decreasing valency. For example, silicon, which has a valency of four (Si^{++++}), is replaced by aluminium, with a valency of three (Al^{+++}). This, in turn, may be replaced by divalent magnesium (Mg^{++}). In this

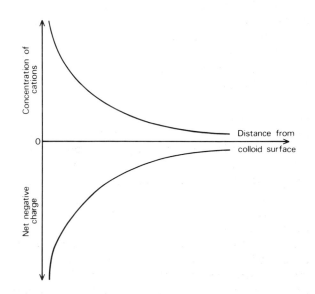

Figure 29.7 The distribution of adsorbed cations and the strength of the negative charge with distance from a colloidal surface. The whole graph relates to a distance of only a few molecular layers

way, the clays progressively lose positive charges, and this leaves unsatisfied negative charges within the mineral. Similar unsatisfied negative charges may also be created by breakage of the clay minerals during weathering. In both cases, the result is to give the particles a net negative charge.

Generally similar processes occur during the decomposition of organic matter.

During the oxidation of organic matter, the hydrogen and carbon compounds tend to be released, leaving unsatisfied negative charges associated with the oxygen atoms exposed at the surface of the material. This negative charge is similar to that produced in clay minerals.

As we know from our own experiments with two magnets, opposite poles attract each other. Thus, the negative charges on the colloidal particles attract positively charged cations. In this way the colloids collect around themselves a swarm of cations (Figure 29.6). These cations are said to be **adsorbed** to the colloids. The strength of adsorption depends upon a number of factors, but is greatest close to the surface of the particle and declines with distance; within the equivalent of a few molecular layers it is negligible.

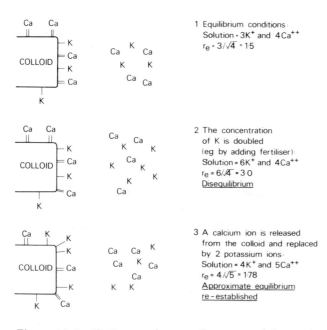

Figure 29.8 Cation exchange. Because of the ratio law a change in the absolute concentrations of two cations of different valency upsets the chemical equilibrium of the soil solution and causes cations to be released from the colloids

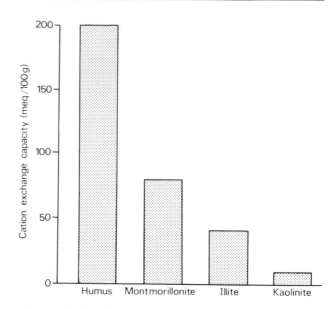

Figure 29.9 The cation exchange capacity of different soil colloids

Cation exchange

Release of the cations from the particles occurs mainly because of a disequilibrium between the concentration of ions in the soil water which surrounds the colloid and the concentration on the colloidal surface. The equilibrium conditions are to some extent a function of the valency of the particular nutrient cation under consideration (Figure 29.7). However, in general, if the concentration of ions in the soil water is increased, a net migration of those ions on to the colloidal particles occurs, with the consequent release of others; if the concentration of ions in the soil water is reduced, a net movement of those ions away from the colloids occurs (Figure 29.8).

It is through this process that plants obtain many of their nutrients. Roots of plants tend to release hydrogen ions, which thus become dissolved in the soil water. Over time, therefore, there is a tendency for the concentration of hydrogen ions in the water to increase; this is counteracted by the adsorption of hydrogen on to the colloids and the release of other ions into solution. The ions which are released may be taken up by the plants. This is the process of **cation exchange**.

Cation-exchange capacity

Colloidal particles vary considerably in the density of the negative charge which develops on their surfaces. In general the densest charges occur in

476

organic compounds, although these are not always stable, for in time the organic matter decomposes further and is lost. In the case of the clay minerals, the charges are greatest on the 2:1 clays such as montmorillonite (see Chapter 27). This is because these minerals contain relatively weak bonds within their lattice structure which are able to expand to some extent and allow ions to enter within the lattice. Thus, these clays have both a surface and an internal charge which is available to the nutrient ions. In other words, in relation to their volume, they have a very large active surface area. The lowest charges are found on the 1:1 clays of the kaolinite group. These are much more rigid, they have a smaller surface area, and they are consequently less active.

The charge developed by the colloids is measured in terms of **milli-equivalents** per 100 g. This, simply, is the weight of hydrogen ions that could become attached to the surface of 100 g of soil if all the negative charges were occupied by hydrogen. The range of the charges developed on different colloids is indicated in Figure 29.9. These values are often referred to as the **cation-exchange capacity**.

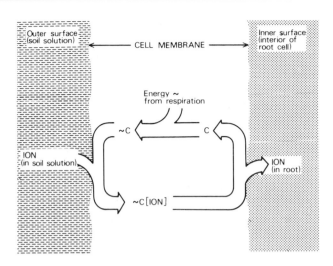

Figure 29.10   The role of enzymes in the uptake of ions by plant roots. The enzymes act as carriers (C) specific to particular groups of ions. Energy is derived from respiration, allowing the ion to be taken up by the enzyme which then transports it across the cell membrane. Inside the root cell the ion is released, the enzyme obtains new energy from respiration and the process is repeated (after Brady, 1973)

Nutrient cycling in the vegetation layer

Plant uptake

Plants require nutrients for a variety of functions. Nitrogen is a vital component in amino acids, proteins and chlorophyll. It thus acts as a major control upon photosynthesis and energy production. Phosphorus is an important component of all living cells; potassium plays a role, which is not fully understood, in cell production and growth, and thereby influences the uptake of other elements and the processes of respiration and transpiration; calcium is used in the cell walls and is an important control upon the growth of the meristem – the cells in the root which allow it to extend into the soil.

Most of the nutrients, as we have seen, are derived from the soil, and thus a good point at which to enter the nutrient cycle within the vegetation is the uptake from the soil. This involves a variety of processes, not all of them perfectly understood. Certainly mass flow represents one of the main means by which the plants obtain their nutrients; the plants transpire moisture from their leaves, creating a moisture gradient within the plant. Moisture then moves up the plant, generating a tension within the roots, and water migrates into the roots from the soil. In the process, the nutrients are transported into the roots.

This process is supplemented by **diffusion**. Nutrients tend to move through the soil solution in response to concentration gradients; that is, they move from areas of high concentration to areas of low concentration. As plants remove nutrients from the soil water around their roots, new nutrient ions move in to take their place. In this way a slow flow of ions occurs through the soil water.

Both mass flow and diffusion operate over relatively small distances and plants seem to be dependent upon root extension for locating both the water and the nutrients. They also obtain nutrients through direct contact with the colloidal particles; ions – in particular hydrogen – are then released from the roots and replaced by others from the colloid.

The exact process by which nutrients pass through the cell wall of the roots is not clear. It has been suggested that it is a process aided by **enzymes** which act as carriers for the nutrient ions (Figure 29.10). In addition, soil organisms, in particular bacteria and fungi, may transport the

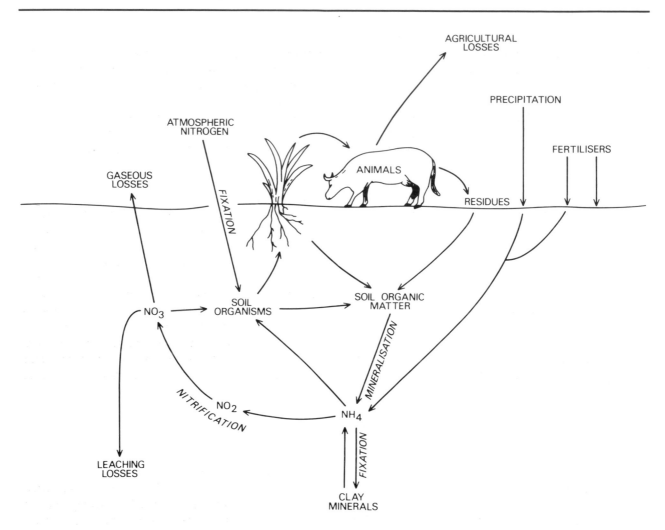

Figure 29.11   The nitrogen cycle

nutrients into the plants. As we will see later, these **rhizosphere** associations are particularly important in relation to nitrogen and phosphorus.

Within the plants, the nutrients enter the **xylem** – the cells within the plant along which the solutions from the soil move. The nutrients thus rise through the plant by capillary action. This is essentially a process of mass flow. It leads to a concentration of nutrients in the younger parts of the growing plant. This arises largely because, as the solutions evaporate from the leaf surface, the dissolved substances are left within and on the leaf. Rain is then able to wash the nutrients from the leaves, and when they die the leaves return the nutrients to the soil surface. Only a small proportion of the nutrient content of the vegetation is therefore held for any length of time in the plants. This storage takes place mainly in the boles and branches of trees and shrubs which may persist throughout the life of the plants.

The rate at which nutrients move through the plant has been measured by 'labelling' nutrients with radioactive substances. These are monitored as they emerge from the plant leaves in the rainfall, as they are released by organic matter decomposition of the plant residues, and as they are taken up and further released by other plants in the ecosystem. From studies of this sort it can be shown that nutrients such as phosphorus, injected into the roots of trees, may emerge from the leaves within 28 hours and be taken up and released by other plants in about 3 days. Clearly not all the nutrients are moving at this rate; some are retained and released by the much slower decomposition of the leaves during the autumn. Nevertheless, it is clear that the nutrient cycle within the plant may be very rapid.

The return of nutrients to the soil, therefore, occurs partly through rainwash and partly through organic matter decomposition.

Cycling within the fauna

The plants, as we have indicated, are the main producers in most ecosystems. They convert solar energy to chemical forms and extract water and nutrients from the soil to form plant tissue. They also act as the foodstuff for the consumers and decomposers; they stand at the base of the food chain. From the plants, nutrients are transferred through the fauna, mainly via the grazing food chain.

Animals typically browse or graze off the younger parts of the plants and thus use the more nutritious components. In fact, in some agricultural systems, specific attempts are made to encourage the growth of young plant material so that animal nutrition is improved. Thus, in grasslands, the animals are allowed to eat the young grass; as the plants get older and start to flower their nutritional value declines, for they have higher carbohydrate contents (i.e. more green plant) but lower concentrations of nutrients and proteins. Similarly, moorlands may be burned to encourage heather regeneration; this creates a young and more nutritious vegetation cover.

The nutrients taken in by the primary consumers are, to some extent, stored within their bodies. A large proportion, however, are lost in the excretion products. These are attacked by the decomposers – including various flies, beetles and micro-organisms – and the nutrients either released into the soil or atmosphere, or stored within the bodies of the decomposers. At the same time, the primary consumers may be hunted and eaten by secondary consumers, which in turn may be devoured by tertiary consumers (often man). The detritus from the prey at each stage in this process is eaten by further decomposers, and thus enters the detritus food chain.

If it were possible to monitor the movement of nutrients through a natural ecosystem in this way, it would be found that the food web created by these organisms is typically complex. Even within very simple situations, a wide range of animals may be involved (see Figure 26.4, p. 429). It is also apparent, however, that each step in the process involves considerable losses of energy and nutrients from the system, so that it is rare for more than four trophic levels to exist. Energy losses occur in the form of respiration; nutrient losses occur mainly through excretion.

Nutrient cycles

The nitrogen cycle

The nitrogen cycle represents one of the most interesting nutrient cycles (Figure 29.11). Nitrogen (N) is possibly the most important, and often most

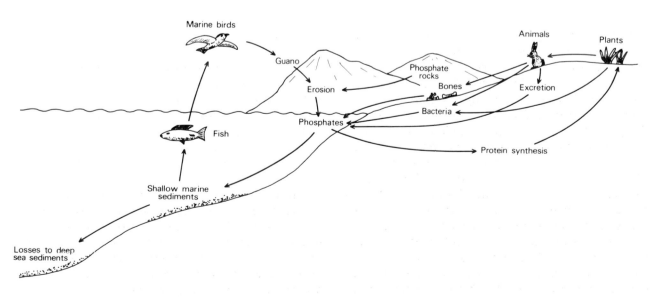

Figure 29.12 The phosphorus cycle

limiting, nutrient for plant growth. Moreover, it follows a cycle in which soil organisms play a fundamental role.

Most nitrogen is derived from the atmosphere. Small proportions enter the soil in rainfall or through the effects of lightning. The majority, however, is fixed within the soil by micro-organisms. These are of two general sorts. Some live freely (non-symbiotically) in the soil and build up nitrogen from the atmosphere in their bodies, releasing it as they die and decay. A wide range of fungi, bacteria and algae take part in this **non-symbiotic fixation**. The other group consists of the **symbiotic nitrogen fixers**. These are mainly bacteria which live in close and mutually beneficial association with plant roots. They irritate the roots and produce small **nodules** in which they live. They absorb nitrogen from the air and pass it into the plant. These symbiotic relationships are restricted to specific plants (in particular legumes), but they represent a major source of nitrogen in the soil.

In the case of the symbiotically fixed nitrogen the nutrient enters the soil store mainly following death and decay of the plant, or as a result of return to the soil in animal wastes. The non-symbiotic fixation of nitrogen requires the death and decomposition of the soil organisms before the nitrogen is released. In both cases, decomposers within the soil attack the organic compounds and the nitrogen is converted from organic ammonia (NH_3^+) to inorganic ammonium salts (NH_4^+). This process is known as **mineralization** and it is carried out by a variety of bacteria, actinomycetes and fungi.

In the form of ammonium the nitrogen may be adsorbed on to the clays, and thus enters the exchangeable nutrient store. When it is released through cation exchange, it may be altered by a small range of autotrophic bacteria into **nitrite**. *Nitrosomonas* is one of the main bacteria to take part in this process:

$$2NH_4^+ \; + \; 3\,O_2 \; \rightarrow \; 2NO_2^- \; + \; 4H^+ \; + \; energy$$
$$\text{ammonium} \quad \text{oxygen} \quad \text{nitrite} \quad \text{hydrogen} \quad \text{ions}$$

The process is essentially one of oxidation.

Further alteration by a restricted range of bacteria including *Nitrobacter* oxidizes the nitrite to nitrate, a process known as **nitrification**:

$$2NO_2^- \; + \; O_2 \; \rightarrow \; 2NO_3^- \; + \; energy.$$
$$\text{nitrite} \quad \text{oxygen} \quad \text{nitrate}$$

In the form of nitrate the nitrogen is available to plants and it may be lost through root absorption. However, it is also very soluble and may be removed by leaching. In addition, it may be reduced to gaseous nitrogen and then lost to the atmosphere, or again taken up by soil organisms and recycled. It is mainly in the form of nitrate, therefore, that nitrogen losses from the system occur. Nitrogen taken up by the plants is returned, of course, in the plant debris and released into the soil during organic matter decomposition.

The phosphorus cycle

Phosphorus (P), like nitrogen, occurs mainly as an anion (e.g. PO_3^-, PO_4^-). It is derived not from the atmosphere, however, but from the weathering of phosphatic minerals, such as apatite and many clay minerals. It occurs within the soil in three main forms: as inorganic phosphate in the original minerals or as compounds coating mineral particles; as organic phosphate bound up within the plant debris and soil fauna; and as soluble phosphate within the soil solution. On the whole phosphorus is highly insoluble, so the last of these is quantitatively the least important. Nevertheless, it is mainly from the soluble component that plants obtain their phosphates. Thus, it is essential that there is a constant turnover of phosphorus between the other two forms and the soluble phosphate.

The importance of this can be appreciated when it is seen that the total phosphorus content of the soil solution would be exhausted within a few hours if constant replenishment by weathering did not occur.

Phosphorus is taken up by plants in the form of phosphate. The process is often aided by soil organisms, including fungi and bacteria. From the plants, the phosphorus may return directly to the soil, mainly through leaf fall and organic matter decomposition, or it may be passed along food webs. In the latter case much of the phosphorus is returned in animal wastes, particularly dung. Considerable proportions of phosphate may be fixed within the animal bones, however, and this is only returned to the soil when the animal dies, and then but slowly. Indeed, in the past bones were a major source of phosphate 'manure'; on one occasion English agriculturalists were accused of robbing the graveyards of Europe in their search for bone manure!

As this implies, phosphates are also returned to the soil in considerable quantities as fertilizers. In addition to bone manure, 'artificial' fertilizers are

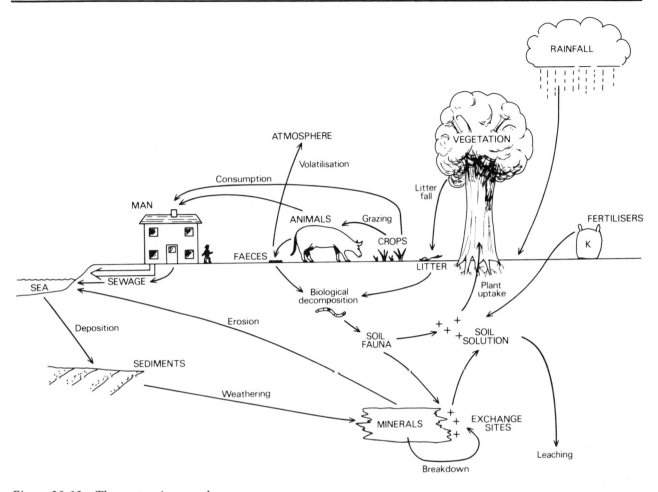

Figure 29.13 The potassium cycle

used, often derived from phosphate-rich rocks. **Guano** is another source, and this illustrates a further aspect of the phosphorus cycle (Figure 29.12), for the nutrients which are washed from the land, enter streams, are carried to the sea, absorbed by fish, eaten by seabirds and eventually deposited as guano. Man then comes along and collects it for fertilizer.

The phosphorus that is returned to the soil in organic form is slowly released during organic matter decomposition. Bacteria play a vital part in this process, mineralizing the phosphorus and converting it to inorganic forms. It may then be again taken up by plants or may combine with other elements, such as iron and calcium to form the insoluble phosphate compounds in which it is stored. Small amounts are also adsorbed on to colloids.

The potassium cycle

Potassium (K) is the third essential element and,

like phosphorus and nitrogen, is commonly applied as a fertilizer in order to ensure adequate plant nutrition in agricultural systems. Unlike these other two nutrients, however, potassium is a cation and thus is subject to rather different processes of cycling.

The main natural sources of potassium are potassic minerals such as K feldspars and micas. Together these account for between 90 and 98 per cent of the total potassium in the ecosystem. These primary minerals slowly weather, producing potassium-containing clay minerals such as illite and vermiculite, and releasing potassium into the soil. Muscovite, for example, weathers to illite as follows:

$$K_2Al_4(Al_2Si_6)O_{20}(OH)_4 + Si^{++++} \rightarrow$$
Muscovite
$$K_{0.2}(K_{0.8})Al_4(AlSi_7)O_{20}(OH)_4 + K^+ + Al^{+++}.$$
Illite

481

Most of the potassium held in the clay minerals (i.e. that shown as $K_{0.8}$ in the above equation) is not readily available to plants, but is held within the clay lattice. In the long term, it is released by gradual breakdown of the clay. A small proportion (that shown as $K_{0.2}$ above), however, is held in an exchangeable form on the surface of the clay minerals. This is released by exchange with other cations (e.g. H^+) in the soil solution and is thus available for plant uptake. These exchangeable reserves are themselves replenished by potassium from within the clay lattice. Thus, a three-way transfer of potassium takes place:

non-exchangeable $K \rightleftharpoons$ exchangeable $K \rightleftharpoons$ soil solution K

The processes of transfer are inter-dependent so that these three reserves of potassium are kept in more or less constant ratios.

At any one time, about 10 per cent of the readily available potassium in the soil is dissolved in the soil solution and is thus directly accessible to plants. This potassium is also available for leaching, however, and a small proportion is consequently carried away in drainage waters, much of it ultimately reaching the sea. Losses by leaching are greatly dependent upon pH, and almost four times more potassium may be lost from an acid soil (pH c. 4.5–5.0) than from a neutral soil (pH 7.0). In addition, soil erosion results in the loss of mineral and exchangeable potassium held on the soil minerals.

The potassium taken up by plants is used in photosynthesis, starch formation and translocation of sugars. Some is subsequently removed in harvested material and stored in human bodies from where it passes into the sewage system. Much, however, is returned to the soil in plant residues, animal wastes and manures. Together with inputs from inorganic fertilizers, these provide the main source of soil potassium. Within the soil, these inputs are released by biological decomposition and chemical reactions, then recycled, leached, adsorbed, or fixed within the clays by incorporation into the mineral structure (Figure 29.13).

Conclusions

The importance of biogeochemical cycles within the ecosystem is clear. It is through this circulation of nutrients that life-giving substances are transferred. It is a process which ties the components of the ecosystem intimately together. The significance of the process extends far beyond the ecosystem, however, for the movement of nutrients also involves the hydrological and landscape systems. Water is one of the main agents of nutrient cycling; it carries the nutrients to the plants, it washes them from the leaves and animal wastes, it takes part in the decomposition of organic matter. Water is also a fundamental agent in the initial weathering and release of the nutrients from rocks, and it carries nutrients out of the system by leaching and erosion. For this reason, many of the outputs from the ecosystem become inputs to the surface hydrological system, as pollution of streams and lakes by fertilizer residues shows.

In the same way, there is a close relationship between biogeochemical cycling and landscape processes. The majority of the nutrients are released by weathering of rock materials; the nutrient inputs to the soil and vegetation are thus the outputs from landscape processes. Similarly, losses of nutrients by wind erosion and rainfall erosion are a result of geomorphological processes. As ever, nature recognizes no clear boundaries between the systems we try to identify.

30

World ecosystems

Soil and vegetation patterns

The classification of soils and vegetation

As we journey across the world – even over short distances – we may become aware of quite marked changes in our ecosystem. We pass from forest to grassland, from agricultural land to semi-natural or natural areas. We may also notice variations in the soils. Often we can see slight changes in the colour of the topsoil, or we can see differences in their surface expression; some create large, cloddy fields, others give fine, smooth surfaces; some show evidence of erosion; some look peaty and fibrous. We can see these changes, and to some extent can categorize what we see, but if we are to be more specific and analytical about the different types of ecosystem, we need to classify these features much more definitively.

In fact, a great deal of effort has gone into methods of classifying the world's soils and vegetation. In the case of soils, the reason is largely to aid agriculture; by knowing more about the type of soil in any area it is possible to develop better plans for its agricultural management. This is particularly true where man is trying to open up or develop new areas for agriculture. A survey of the soil, to show what conditions the farmers must contend with, is an essential step. In addition, soil surveys provide a means of collecting general information on one of the world's most important resources and of developing an inventory of the resource. Soil surveys and classification also have scientific value since they facilitate the transfer of information. Using brief descriptions of the soils it is possible to convey a great deal of information from one soil scientist to another.

The same general principles are true of vegetation surveys and classifications. The aim again is to collect information which may aid man's understanding and management of the vegetation, particularly with regard to his conservation and protection of the ecosystem. Again this is achieved by surveys, normally carried out in the field, by trained field workers.

Working in the field, it is clearly possible – although it would take an inordinate length of time – to map the location and type of each plant. The same approach is not possible with soils; the soil does not consist of individuals but is a **continuum**. We can dig holes and look at the soil profile, but each hole is merely a section through a laterally continuous and variable medium. In both cases, therefore, man is faced with a major problem of classification. He must be able, if he is to produce maps of the soil and vegetation (and produce them within a realistic time scale), to classify the vegetation into general types, so that he can say that the whole of this area is of one type of vegetation, while the adjacent area is of another. Similarly, he must be able to define areas in which the soil is approximately uniform, so that farmers or other users know that they can treat it as a single, constant unit.

There are two major problems facing both these tasks. The first, as we have already hinted, is that we are trying to classify and map a continuum. On our map we must draw lines separating one soil or vegetation type from another, but of course, in reality, this line does not exist. Often, indeed, it is a transitional zone. The second problem is that in trying to identify different soil or vegetation types we are dealing with what are called **multivariate phenomena**. In other words, there are a myriad of different properties which we might use to classify them. Colour, texture, structure, depth, mineralogy, chemistry, organic properties – these and many more in the case of the soil; the occurrence and relative frequencies of an almost infinite range of species of plant in the case of the vegetation. Clearly, any system of classification is likely to be imperfect.

We should note this conclusion carefully. All too often it is assumed that the classification method we are familiar with is the only one, and the optimum one. This is not so. In Appendix II of this book, for

		COOL CLIMATES	WARM CLIMATES	
MOIST CLIMATES	PEDALFERS	PODZOLIC	TUNDRA SOILS		INCREASED LEACHING
			PODZOLS	
			BROWN FOREST SOILS	RED & YELLOW PODZOLIC SOILS
			PRAIRIE SOILS	
		LATERITIC		TROPICAL RED SOILS
				SOILS WITH LATERITE HORIZONS
DRY CLIMATES	PEDOCALS		CHERNOZEMS		INCREASED LEACHING
			CHESTNUT-BROWN SOILS	VARIOUS TROPICAL PEDOCALS
			BROWN SOILS	
			SIEROZEMS	

Figure 30.1   The zonal classification of soils according to Marbut (from Eyre, 1968)

example, we outline two methods used to classify soils; remember that they are but examples from a very long list of similar approaches.

World soils

At a world scale, it was once common to relate soils to climatic conditions. The Russians – in particular V. V. Dokuchaev and N. Sibertzev – were among the first to devise a systematic classification of soils in this way, late in the last century, and C. F. Marbut, in the 1930s, extended their approach to the United States. It was an approach which subsequently became widely used in many other parts of the world, including Europe, and it was based upon the principle that the soil was a product of its prevailing environment, within which climate and vegetation were dominant influences.

This approach involved the recognition of zonal, azonal and intrazonal soils. **Zonal soils** were those that are in close accord with climatic and vegetational conditions, and included most soils developed under conditions of good drainage. **Azonal soils** were those which were not wholly in equilibrium with their climate, because other factors, such as slope or insufficient time, had limited their formation. **Intrazonal soils** were those associated with poor drainage or salinity or calcareous parent materials.

Each of these broad classes of soil were subdivided on a hierarchical basis into suborders and great soil groups and so on (Figure 30.1). However,

several problems were encountered in using the approach. In particular, recognition of a particular soil type depended upon correct interpretation of the soil formation; this is not always possible with any certainty. Second, as new information on soil types became available, it was clear that many of the world's soils did not fit happily into the scheme.

For these reasons new approaches were sought, and in the United States a series of provisional classifications were introduced, leading up to the definition of a working method in 1960, known as the **Seventh Approximation**. It was given this name because it was the seventh attempt to produce a final version; it was recognized, however, that it was still not definitive. Since then further development has taken place, and the method has become refined into a system known as the **Comprehensive Soil Classification System**. Many other countries have produced classifications which are broadly similar in structure, including the system used by the Soil Survey of England and Wales. The principle behind the Comprehensive Soil Classification System is the recognition of soil types on the basis of measured properties of the profile. It is, like most soil classifications, a hierarchical approach, such that there are six levels of classification; soil classes

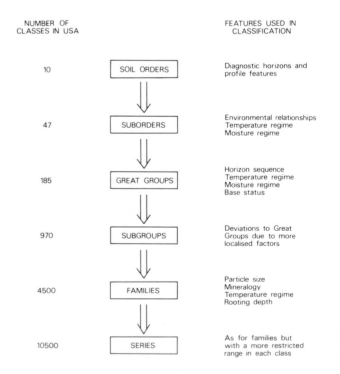

Figure 30.2   The Comprehensive Soil Classification System

recognized at one level are composed of amalgamations of classes defined at lower levels (Figure 30.2). As can be seen this allows a large number of locally significant soil series to be grouped into a few much broader groups, suborders or orders for use at a more general scale.

This method is briefly outlined in Appendix II. It is not an easy method to apply without experience, and there is little point in learning the sometimes unmemorable names of the classification merely for the sake of it. What is more important is to appreciate the principles involved in the approach and some of the problems it encounters. Among the problems, two specific ones may be mentioned. The first is that because the method is based upon strict quantitative definitions of many properties, slight differences in key criteria may lead to separation of soils which are similar in many other respects. The second is more philosophical. Like most classifications it is hierarchical in structure, yet soils are not really arranged in that way. Thus, one cannot think of groups of soils of different morphological character being part of some 'family' of soils; each soil profile develops, as we have seen, through the interaction of specific sets of environmental processes. We cannot think of soils being related, as, for example, we do animals and plants which have evolved from a common ancestry.

Nevertheless, the method has much to commend it, and it introduces a number of significant principles. We might note three in particular. The method was one of the first to confine itself to measurable properties of the soil and to avoid classification on the basis of assumptions about genesis or environmental relationships. It puts much weight upon the hydrological properties of the soils, a factor which is most certainly of great importance to agriculture. It is also based upon the concept of the pedon. A **pedon** is considered to be a three-dimensional soil column (often considered to be hexagonal) extending to the base of the soil. This is the smallest unit of soil that can be recognized and it is considered to be uniform in character. Over space, soils can be arranged into **polypedons**. These are groups of similar pedons which are contiguous. Inevitably, such pedons vary slightly, and 'rogue' pedons become included within the polypedon. Thus polypedons are rarely totally uniform. Nevertheless, the polypedon represents an area of soil that can be mapped as a more or less uniform unit (Figure 30.3).

On the basis of this classification, general maps of

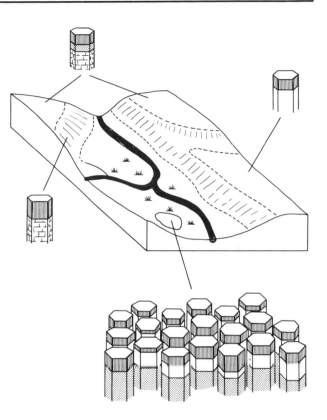

Figure 30.3 Soil distributions on a landscape, showing representative profiles (pedons) and the way in which a number of contiguous pedons can be grouped to form a polypedon

the soils of the world can be produced (Figure 30.4). At this scale, it is possible to show only the suborders, but it is apparent that at this level climatic aspects, including temperature and moisture regime are important. None the less, classification at this scale has only the most general value.

Vegetation of the world

The classification of vegetation presents rather different problems to that of soils. Again we are confronted by the need to subdivide a natural continuum into artificial classes, but now we are dealing with a continuum composed of an almost infinite number of discrete individuals. Thus, classification depends upon the spatial arrangement of these individuals into recognizable and repetitive associations or communities.

The basis of many vegetation classifications is the principle that these assemblages are distinctive due to the response of the constituent plants to the

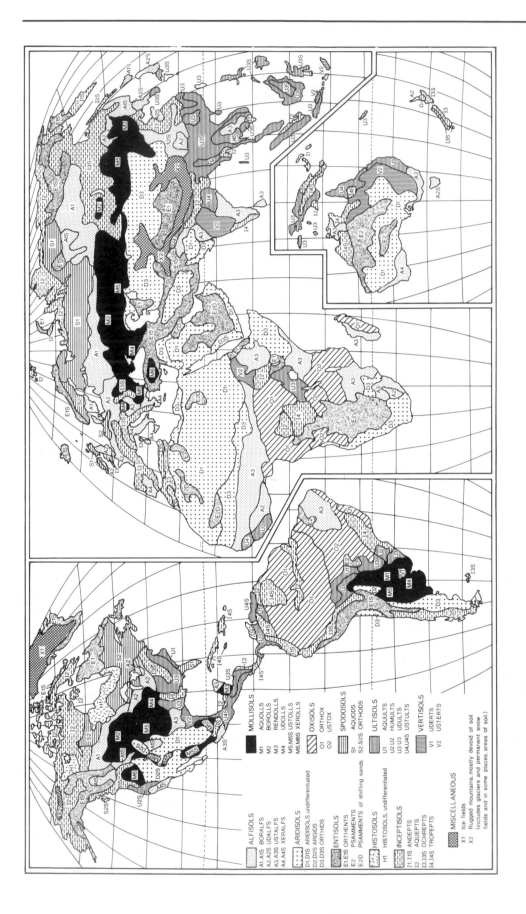

Figure 30.4　　Soil map of the world, according to the Comprehensive Soil Classification System (after USDA, from Brady, 1973)

prevailing environmental conditions. French biogeographers, for example, have attempted to classify vegetation on the basis of characteristic or faithful species, arguing that certain plants provide a definitive indication of the ecological conditions. In other cases, classification is more quantitative, based upon the overall similarity of the plants within the assemblage. This approach often requires the use of complex statistical procedures and computers to analyse the vast array of data on the abundance or distribution of all the constituent species within the vegetation.

These approaches are useful when examining vegetation at a local level, but they are not really suitable for identifying the broader picture. At this scale it is common to classify vegetation in a much more intuitive way on the basis of the appearance of the plant communities. One of the most successful examples of this approaches was that developed by Küchler in the United States. This has been tested in many parts of the world and found to be reliable.

The **physiognomic approach**, as it is known, recognizes **formations** – vegetation types with distinctive and uniform structure, appearance and composition. These are typically grouped into larger classes called **formation-types**. Figure 30.5 shows natural formation-types at a world scale.

Ecological relationships and the biome

The nature of soil–vegetation relationships

It is apparent if we compare the maps of world soil and vegetation distribution that there is a broad similarity between them. This reflects the overall climatic influences, together with the major physiographic constraints imposed by the high mountain chains. However, this comparison is to some extent misleading, for, as we have seen, each of the main soil and vegetation types recognized at this scale represents very variable and complex units. Even at a local scale, however, we can recognize a relationship between the distribution of soils and vegetation (Figure 30.6).

This should not surprise us. We have seen in earlier chapters the ways in which the two are associated within the ecosystem. In particular the biogeochemical cycles by which plants gain and lose nutrients represent a major link. Moreover, the vegetation tends to influence soil development and the two evolve in harmony. To some extent, therefore, we find that it is possible to identify ecologically integrated units within which both soil and vegetation are reasonably uniform.

Relationships with animals

Animals also are intimately related to the soil and vegetation. This is apparent at the local scale, where the structure of the food chain means that consumers tend to be tied closely to their food source; thus there is a strong link from vegetation to herbivore to carnivore. Many of the decomposers are also specific to particular types of organic debris and they therefore are associated with specific plant and animal assemblages.

These relationships are obviously less valid at a general level; as ever, numerous complications and exceptions occur once we start to generalize. Nevertheless, it is within this context that the concept of the **biome** has developed. A biome is considered to be an area which is ecologically integrated and uniform. It represents the soil, the vegetation, the animals and their physical environment. An indication of the main terrestrial biomes of the world is given in Table 30.1 and can be interpreted from Figure 30.7.

In the next sections we will consider some of the more extensive biomes and their constituent formation-types, and analyse the character and, more importantly, the processes related to each.

Tropical forests

General character

Tropical rain forests cover about 17 per cent of the land surface of the earth, a total of about 400×10^4 km^2 (Table 30.1). They are, of course, variable in composition at a detailed level, and they include areas in which there is no distinct seasonality as well as more marginal regions with a seasonal pattern of climate (Plates 30.1 and 30.2).

The climate of these areas is dominated by the high temperatures and rainfall, and the consequent high humidity. The precipitation is thus in excess of the potential evapotranspiration and this allows the development of a luxuriant vegetation.

The vegetation is typically diverse both in composition and structure. Whereas temperate forests may contain only three or four tree species per hectare, the tropical forests often include as many as 100 different tree species in the same area. Most are evergreen, and those that are leafless for any period shed their leaves at irregular intervals. Thus

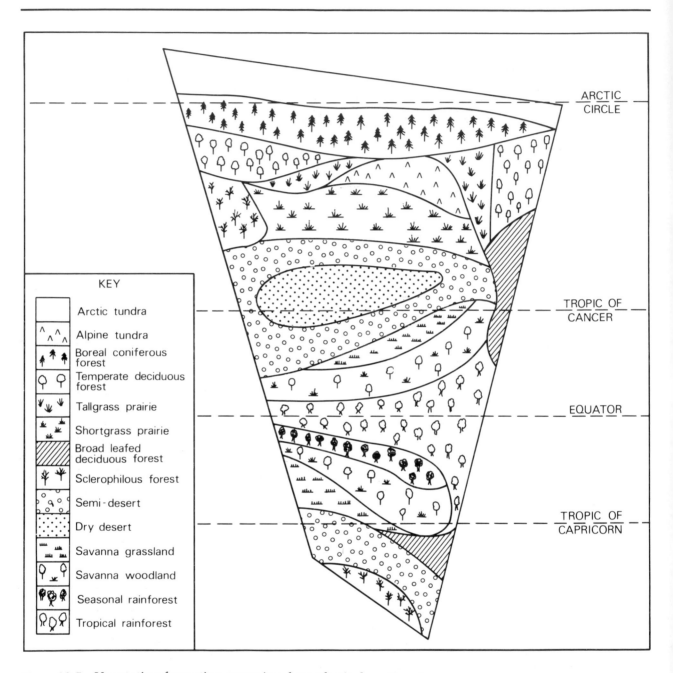

KEY

	Arctic tundra
^ ^ ^	Alpine tundra
♠ ♠	Boreal coniferous forest
♀ ♀	Temperate deciduous forest
♈ ♈	Tallgrass prairie
♠ ♠	Shortgrass prairie
⫽⫽	Broad leafed deciduous forest
♣ ♣	Sclerophilous forest
°₀°	Semi-desert
⋯	Dry desert
⊥⊥	Savanna grassland
♀ ♀	Savanna woodland
♠♠♠	Seasonal rainforest
♀♀♀	Tropical rainforest

Figure 30.5 Vegetation formation-types in a hypothetical continent

there is no autumn in the sense that we know it in more temperate areas. Moreover, the intense competition and the diversity of the vegetation leads to a distinct structuring of the forest, with five or more strata recognizable (see Figure 28.1, p. 455).

The soils associated with these forests are characterized by intense and perhaps prolonged weathering, and by active leaching. Organic matter breakdown is rapid, so despite the high inputs of plant debris the soils rarely develop a distinct organic surface layer. Moreover, the intense weathering and leaching means that the more soluble constituents are totally removed, iron and even silica may be mobilized and the soils have a low cation exchange capacity and a limited supply of bases, such as calcium and potassium (i.e. a low base status). Using the USDA classification described in Appendix II, the soils are mainly oxisols and

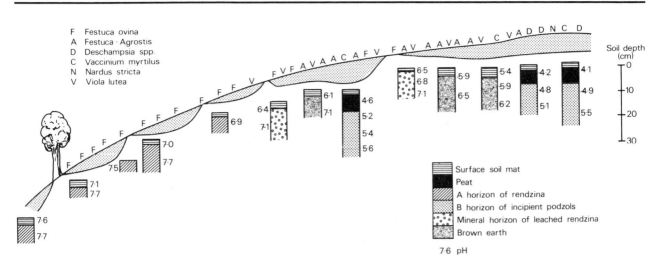

Figure 30.6 The relationship between vegetation and soil conditions on a limestone slope (after Balme, 1953)

ultisols. They show marked oxidation which gives rise to reddening of the soil colours and include the ferricretes and laterites which are frequently defined as tropical forest soils (Plate 30.3).

The fauna, like the vegetation, is varied. Animals with many different life-forms are present, taking advantage of the diverse niches provided by the vegetation. A large majority of the animals live within the trees, particularly at the canopy level. These rarely descend to the floor. Animals dwelling at lower levels in the trees – the middle-zone fauna – come to the ground more frequently. In addition there are both large and small animals occurring on the ground. Only the subterranean animals are rare compared to other environments.

Environmental relationships and variations

As we have noted before, the history of the rain forests is probably more complex than was once thought. It is no longer viable to think of them as an ancient, almost unchanged formation-type which has survived since the Tertiary. Instead, it seems that they experienced quite drastic fluctuations in fortune during the Quaternary, and probably owe much of their present diversity to the periods of isolation they experienced at that time.

Table 30.1 *Major biomes and formation classes and their environmental relationships*

Biome	Formation-type	Climate*	Soils†
Forest	Tropical rain forest	Tropical rainy (Af)	Oxisols, ultisols
	Seasonal rain forest	Tropical monsoon (Am)	Oxisols, ultisols, ustalfs, vertisols
	Broad-leafed evergreen	Moist warm temperate (Cfa)	Udalfs, udults
	Sclerophyllous	Mediterranean (Csa)	Xeralfs, xerolls, xeralts
	Temperate deciduous	Warm temperate (C)	Udalfs, boralfs, udolls
	Boreal coniferous	Cold boreal (Dfb, Dfc)	Spodosols, boralfs, histosols, cryaquepts
Savanna	Savanna woodland	Seasonal tropical (Aw)	Ustalfs, ultisols, oxisols, vertisols
	Savanna grassland	Dry steppe (BSh)	Ustalfs, ultisols, oxisols, vertisols
Grassland	Tall-grass prairie	Moist warm temperate (Cfa)	Udolls
	Short-grass prairie (steppe)	Cold steppe (Bsk)	Borolls, ustolls, xerolls, aridisols
Desert	Semi-desert scrub	Hot dry desert (BWh)	Aridisols, psamments
	Dry desert	Hot dry desert (BWh)	Aridisols, psamments
Tundra	Alpine tundra	Mountain (H)	Mountain soils
	Arctic tundra	Tundra (E)	Cryaquepts, cryorthents

Notes: * See Appendix I.   † See Appendix II.

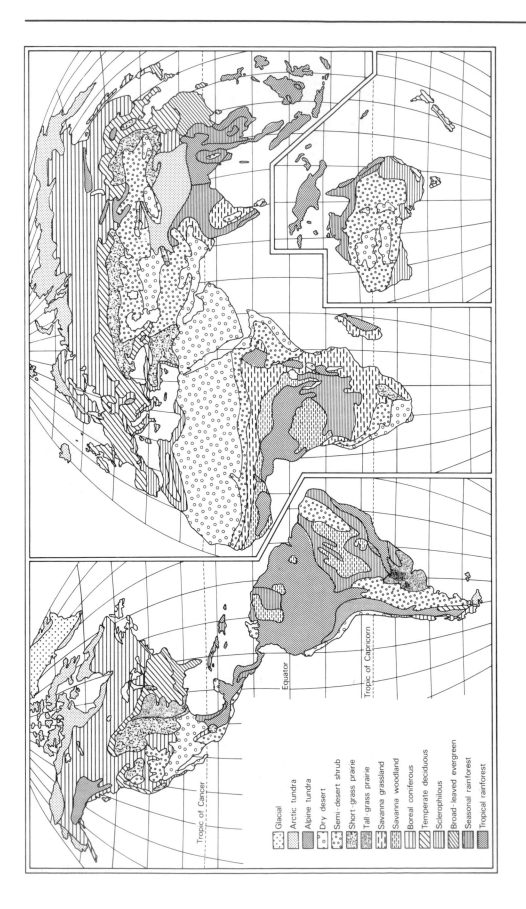

Figure 30.7 Formation-types of the world

Glacial
Arctic tundra
Alpine tundra
Dry desert
Semi-desert shrub
Short-grass prairie
Tall-grass prairie
Savanna grassland
Savanna woodland
Boreal coniferous
Temperate deciduous
Sclerophilous
Broad-leaved evergreen
Seasonal rainforest
Tropical rainforest

Tropic of Cancer
Equator
Tropic of Capricorn

Plate 30.1 Tropical rain forest, Borneo (photo: J. Rose)

Plate 30.2 Seasonal rain forest, Tibet (photo: J. Ulmanis)

It is also clear that within this formation-type, quite distinct variations occur, related mainly to climatic and geological factors. An idealized picture of these patterns is shown in Figure 30.8. Moreover, away from the equator more seasonal forests occur.

Biogeochemical cycling

One of the main features of tropical forests is the huge concentration of nutrients within the vegetation. The soil has a low storage capacity for nutrients, and it is mainly in the dense and lush vegetation that nutrients are retained. This has considerable significance, for when the forests are cleared, as they may be by man and as they are by natural processes in the seasonal forests, the bulk of the nutrient store is lost; the remaining soil is very infertile by comparison and regeneration is consequently less effective than might be imagined.

Under undisturbed conditions, the cycling of nutrients and energy through the tropical forest occurs rapidly and in large quantities. It has been estimated that the annual biomass in a central Amazon rain forest near Manaos is in the order of 1100 t ha^{-1}. A high proportion of this annual production is associated with the green parts of the plants – the leaves, fruits and flowers of the trees, ground flora and epiphytes – which typically account for 4–9 per cent of the total biomass (compared with 1–2 per cent in temperate forests), and the contrastingly low biomass of the fauna (only 0.02 per cent in the Amazon example). The annual litter fall is also high, but the rapid decomposition and leaching mean that little surface accumulation occurs. Most of the decomposition is carried out by fungi, so that soil animals, instead of feeding upon the organic matter itself, tend to feed on the fungi. Earthworms are therefore confined to the upper rooting zone within the soil, and there is little mixing of the soil by animals. Consequently, the

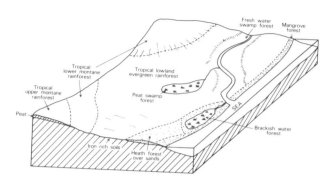

Figure 30.8 Relationships between tropical rain forest formations and environmental conditions (based on Collinson, 1977)

soil horizons are distinct.

In the seasonal forests, the rate of biomass production is considerably less and an average of just over 200 t ha^{-1} seems typical. Moreover, almost all the processes we have just described operate more slowly, so that leaf litter can accumulate at the surface, and more direct relationships between litter and soil organisms are found. Fire is also an important component of these areas. This releases nutrients to the soil and reduces storage in the vegetation; however, the less marked leaching means that more of the nutrients are retained in the soil.

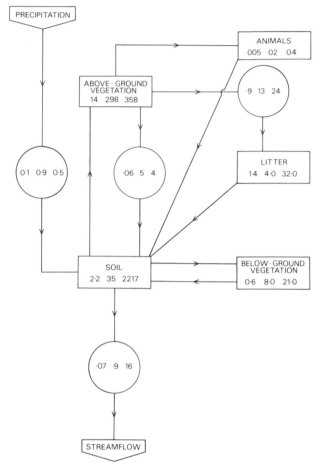

Figure 30.9 The nutrient flow of a tropical rain forest, Panama (data from Golley, 1978). Figures in boxes represent total stores of phosphorus, potassium and calcium in g m^{-2}; figures in circles represent annual flows of P, K and Ca in g m^{-2} y^{-1}

Plate 30.3 A lateritic soil, showing corestones, Bega Valley, NSW, Australia. This soil would be classified as an ustox according to the USDA Soil Taxonomy (photo: J. C. Dixon)

The nutrient budgets of tropical rain forests are interesting in that they show much higher annual turnovers than almost any other biome. The rate of cycling is perhaps three to four times that in temperate forests, and large quantities of nutrients are returned every year to the soil (Figure 30.9). Leaching is active, and the loss of these nutrients from the soil to the streams is intense.

Savanna

General character

Savanna biomes cover some 20 per cent of the land surface of the earth. They include many of the sub-

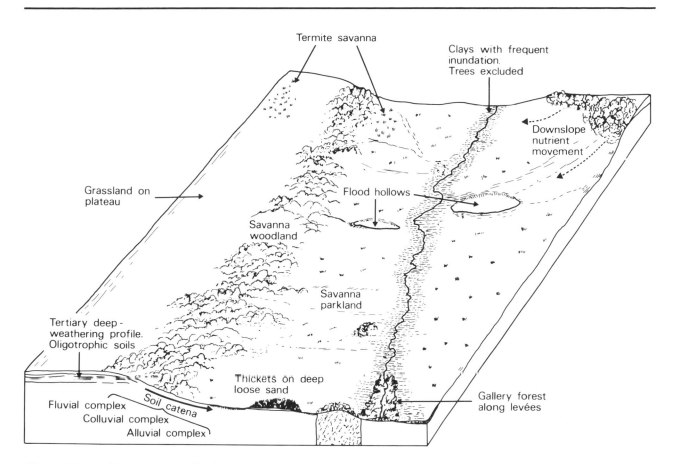

Figure 30.10 Environmental relationships in a savanna area (after Collinson, 1977)

tropical regions fringing the rain forests; in particular in Africa and South America. Although it is often referred to as a grassland biome, it is an open woodland in many cases, with widely spaced and rather scrubby trees (Plate 30.4).

The formation of the savanna areas is one of considerable interest and dispute. It seems that climatic factors alone cannot account for the character of these areas, for although they experience a distinct dry season and have a variable precipitation, in many cases the climate could support a much more luxuriant and diverse flora than it does. One possible reason for this disparity is that the savanna represents a form of **plagioclimax**, one which has been severely curtailed by the activities of man. Fire, in particular, has played a major part in the development of these areas, and many of the trees that now occur are fire resistant. The action of fire, and the voracious appetite of termites, mean that seeds rarely survive, and thus the trees produce enormous numbers of seeds each year. *Acacia karoo*, for example, releases as many as 20,000, of

Plate 30.4 Savanna, South Africa (photo: D. J. Briggs)

493

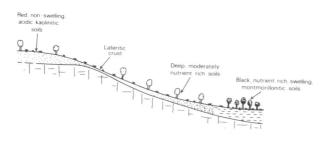

Figure 30.11   Catena sequence in a savanna area

which about 90 per cent are typically fertile. Few survive to grow into trees, however, as the scattered nature of the arboreal vegetation shows (Plate 30.4).

Many of the trees are xerophytes, and are therefore adapted to survive the dry period. They are also deep rooting, and have flattened crowns. Some shed their leaves during the dry season in order to reduce transpiration. They are often stunted, and may be overtopped by the tall grasses. Browsing of the trees by the savanna animals is a major constraint on their growth and survival, and overgrazing represents one of the main causes of degradation. Some plants have developed thorns as a protection against browsing.

The herbaceous plants are dominated by a few species. African elephant grass is sometimes abundant, and may reach heights of several metres. The density of trees relative to grass is to some extent climatically controlled, and trees become scarcer in the drier margins of the savanna. In addition, however, there are often subtle local variations in the vegetation related to the topography and drainage (Figure 30.10). In the drier areas, in particular, the vegetation tends to be attuned to the shorter growing season, and rapid growth of the ground flora takes place once the rains come. An interesting phenomenon of plant uptake of water is indicated here, for it seems that although the soils are relatively dry at the end of the arid season, there is no need for the rain to 'top up' the soil moisture store before plants can extract the water efficiently; instead they transpire at their full rate immediately. Thus, much of the rainfall seems to be directly used by the plants before it can be moved into the finer pore spaces in the soil.

The soils of the savanna area are variable. They include oxisols, ultisols, vertisols and ustalfs (Appendix II), their distribution being related to climatic, geological and geomorphological conditions. Slope processes are active, for the vegetation is often insufficient to prevent erosion and downwashing of nutrients. Thus marked catena sequences develop on the hillslopes, with a gradation from shallow stony soils to deeper, less well-drained, base-rich alluvial soils (Figure 30.11).

Compared to the tropical forests the fauna of savanna areas is impoverished, in that few species occur. On the other hand, large numbers of individuals of these species are found. Many of the larger game animals are familiar to us all (although not through personal contact fortunately). Rather surprisingly, perhaps, the competition between these seems to be limited, for each fills a separate niche within the ecosystem. For the most part the food chains are relatively short, with few secondary consumers; most carnivores prey directly upon herbivores, as illustrated by the lion which attacks mainly zebra, wildebeest, antelope and giraffe. In addition, however, there are large numbers of scavengers and decomposers, both among the

Plate 30.5   A termite mound built around a thorn bush, South Africa (photo: D. J. Briggs)

mammals and the insects. Termites are particularly abundant, and their mounds form a major feature of the savanna landscape (Plate 30.5). They attack the plant debris and macerate it, making it more readily available for decomposition by other organisms. They also eat growing plants, especially during periods of drought.

Biogeochemical cycles in savanna areas

Relatively little is known about the nutrient cycles of savanna areas. It is clear that the rate of biomass production is much less than that of the tropical rain forests. An annual total of 2 to 20 t ha^{-1} seems typical, but the quantities vary considerably according to climate and tree cover.

Nutrient cycling is also relatively rapid due to the speedy breakdown of the organic residues by the soil organisms. The high temperatures encourage chemical activity, and silica appears to be particularly soluble in these areas, often making up a major part of the total nutrient budget. Indeed, plant uptake of silica is so great that clots of amorphous silica may form in the leaves.

It is also apparent that the nutrients which are returned to the soil are less readily lost by leaching (Figure 30.12). This is for two related reasons. The lower rainfall means that weathering and leaching are less intense, while the soils contain clays with a greater cation exchange capacity than the kaolinite-dominated soils of the rain forests. The main source of nutrient loss in many of these areas, in fact, seems to be soil erosion. This is one of the major problems facing agricultural utilization of the savanna, for overgrazing and tillage leave the soil unprotected; the sudden intense storms which are characteristic of the savanna then result in intense rainsplash and surface erosion.

Desert and semi-desert biomes

General character

The layman's picture of the desert areas as vast expanses of barren, shifting sand is false for all but a relatively small part of the arid world. Elsewhere a widespread, although often relatively sparse, vegetation occurs which supports a distinctive wildlife.

The arid and semi-arid lands possibly cover as much as one-third of the land surface of the globe. Of this total, perhaps no more than one-fifth (that is, about 6 per cent of the land surface) is true desert. The remainder varies from steppe grassland to thorny scrub. The common feature in all cases, however, is a climate in which potential evapotranspiration greatly exceeds rainfall (Figure 13.6, p. 203). This condition arises for a variety of reasons. In coastal areas bordering cold upwelling waters it is due to the cool on-shore winds being warmed and thus causing evaporation as they blow on to the land; this occurs, for example, in Peru and the Baja Californian desert. Rainshadow deserts occur to the lee of mountain barriers (e.g. the Mojave and Patagonian deserts). The sub-tropical deserts of the Sinai and Sahara are a result of the stable anticyclonic atmospheric circulation with which they are associated, while the continental interiors support deserts because the winds, blowing inland from the sea, deposit most of their moisture *en route*: this is typical of the central Asian and Arizona deserts.

The vegetation of these areas consists for the most part of short perennial grasses and thorn scrub (Plate 30.6). Only in extreme cases, such as the rocky hamadas and reg and the shifting sand dunes and sand seas of the Sahara is vegetation totally lacking. Even in these areas, locally developed lines of vegetation occur along wadis, while more lush growth occurs around oases. In all cases, the vegetation has to be able to withstand periods of drought, however, and thus xerophytic plants predominate. The adaptation of plants to

Plate 30.6 Hot desert, near Aggeneys, South Africa (photo: P. A. Smithson)

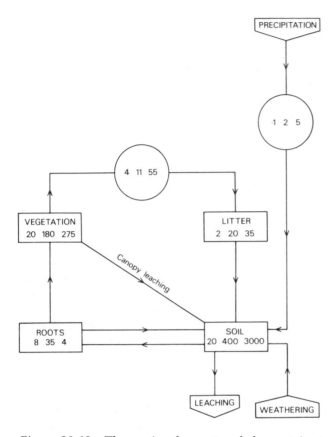

Figure 30.12 The main elements of the nutrient cycle in a savanna area. Figures in boxes represent amounts of phosphorus, potassium and calcium in kg ha⁻¹; figures in circles represent annual flows of P, K and Ca in kg ha⁻¹ y⁻¹. Data derived from Nye and Greenland (1960) and Rodin and Bazilevich (1965)

these conditions varies: in the case of the saguaro cactus, it is through the development of a widely-spreading root system; the mesquite has deep roots (sometimes over 50 m); while many other plants, including various cacti and agaves, store water in their roots, stems and leaves. In addition, water loss through evaporation is reduced by control of the stomata. Other plants have long dormant periods, growing and flowering briefly and irregularly when moisture is available.

The soils associated with these conditions are typically little weathered, and lacking in humus. In the most extreme cases no true soil exists, and, even where sufficient vegetation does occur to give surface accumulation of plant debris and a food base

for the soil fauna, the lack of leaching and chemical breakdown means that the soils are relatively infertile. Salinity is also a major problem, for the constant evaporation from the surface tends to draw water from the lower layers of the soil and leads to the accumulation of salts in the upper horizons (see Plate 13.1, p. 207). This salt is derived in part from the groundwaters; seepage of salty seawater into the aquifers occurs in many coastal areas. Alternatively, winds blowing from the sea introduce the salt in rainfall; practically no leaching occurs, so even though the salt is soluble it accumulates in the soil.

The fauna of these desert areas is, perhaps, more diverse than often imagined. Nevertheless, relatively few large mammals can survive in these conditions, and insects and arachnida predominate. Flies, scorpions and orthoptera (crickets, locusts and grasshoppers) are among the most common. In addition, however, several vertebrates are found. The ostrich, the sand-grouse and various rodents are among the more abundant.

Biogeochemical cycles

Again there is not a great deal of information on which to base generalizations of nutrient cycles in the arid regions of the world. It is apparent that cycling is slow and involves very low quantities of nutrients, for the total biomass is small. The annual rate of productivity is closely related to rainfall, and probably ranges from about 50 to 200 kg ha⁻¹. About 80 per cent of the organic material is underground, however, and cycling mainly occurs through the decay of root material. Leaching losses are negligible, but gaseous losses associated with the chemical oxidation of plant materials are likely to be significant. In addition, soil erosion possibly represents a significant loss. Due to the low rainfall and the limited weathering, the more insoluble compounds are present only in small quantities. Sodium and calcium tend to dominate the nutrient budgets (Figure 30.13).

Temperate grassland biomes

General character

Steppe and **prairie grasslands** represent one of the most important agricultural zones in the world, and many of the original semi-natural grasslands have been taken over for intensive farming. Only the

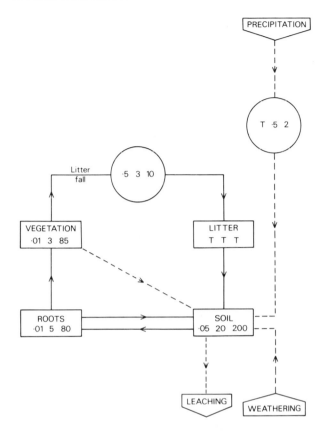

Figure 30.13 The nutrient cycle of a hot desert. Data on such biomes are rare and only very general estimates can be given. Figures in boxes represent the amounts of stored phosphorus, potassium and calcium in kg ha^{-1}; figures in circles represent annual flows of P, K and Ca in kg ha^{-1} y^{-1} (T = trace). Dashed lines show minor nutrient flows

marginal areas remain in a semi-natural state, and these merge into the desert biomes that we have just discussed. Characteristically, these areas experience a significant soil moisture deficit due to a long period of drought during the late summer and autumn. Compared to the savanna areas, climatic conditions are somewhat unfavourable for vegetation growth in many cases, and the American steppe tends to be dominated by short grasses such as buffalo grass (*Buchloe dactyloides*) and bluestem wheatgrass (*Agropyron smithii*). In the areas of higher rainfall, e.g. the more eastern states of the Mid West (Iowa and Missouri), occur tall-grass prairie, in which big bluestem (*Andropogon gerardii*) and little bluestem (*A. scoparius*) grasses predominate.

Three main soil types occur in the steppe and prairie areas. In the former, the more arid conditions give rise to aridisols, but over much of the area more fertile mollisols occur. These are deep, unleached soils, with high organic matter contents and a high nutrient supply. The pH ranges from neutral in the wetter areas to slightly alkaline (7–8) in the drier regions, and organic activity is intense. Thus, in their undisturbed state, the soils retain large quantities of nutrients. When the soils are tilled, however, they are prone to erosion.

In terms of the fauna, these temperate grasslands are undistinctive. Many of the animals which inhabit the forests and deserts on their margins seem to merge here, and it thus represents a transitional zone between these two more extreme biomes. In the past, of course, the American prairies and steppe were dominated by the bison (*Bison bison*) and the pronghorn (*Antilipcapra americana*), but the ravages of the early settlers and the irresistible spread of cattle and sheep resulted in the virtual extinction of these animals.

Biogeochemical cycles

The productivity of the prairies is relatively high, even under semi-natural conditions, and the total biomass of these grasslands probably exceeds 100 t ha^{-1}, with an annual production of about 15 t ha^{-1}. In the drier, short-grass steppe, the quantities are lower. In some of the better steppe land in Russia the total plant biomass was estimated as 20–50 t ha^{-1}, and generally it seems that the annual production of new material averages about 5 t ha^{-1}.

The cycling of nutrients within this ecosystem is dominated by the close interaction between the grass and the soil. Uptake of nutrients by the grass is balanced by high rates of return both from the roots and from the above-ground parts of the vegetation. Litter accumulates at the surface for the rate of decomposition is rather slow, partly because of the long cold winters which give a short active season for the soil fauna. As much as 40 per cent of the carbon content of these areas may occur within the litter layer, and this has significant effects upon the soil processes. It provides a protection from erosion, it absorbs heat and it holds moisture, often preventing deep percolation. This helps to restrict leaching, but it also creates problems of drought for the water cannot be stored at depth, away from the effects of intense solar radiation.

Temperate forests

General character

Within the **temperate forests** can be seen four main formation-types: the **temperate deciduous forests**; the **cold temperate (boreal) coniferous forests**; the **broad-leaved evergreen forests**; and the **sclerophyllous forests**. These represent responses to slightly different climatic conditions. The coniferous forests occur mainly in cooler areas, while the broad-leaved forest is concentrated in the warmer and more humid zones. Sclerophyllous forests are associated with areas of Mediterranean climate.

The environmental relationships and general composition of these forests are shown in Figures 28.9 and 30.14. The deciduous forests (Plate 30.7) are typically dominated by oak, beech and hickory, with smaller quantities of understorey trees such as birch, hazel, sycamore and maple. The coniferous forests (Plate 30.8) include mainly pine, while the evergreen forests comprise oak, magnolias, hollies and sabal palms. Sclerophyllous forests are characterized by drought-tolerant species such as olive, sessile oak and Aleppo pine. Where man has disturbed these Mediterranean forests, however, by overgrazing and clearance, they tend to have degraded into scrub. This is called **maquis** where the shrubs are thorny (Plate 30.9), or **garigue** where the vegetation is dominated by waxy-leaved shrubs such as juniper and broom.

Like the tropical forests, these formation classes, to some extent, show a distinct vertical structure. Only three main layers can normally be recognized, however, and in the case of the coniferous forests the understorey of trees may be almost absent, for the canopy may cut out as much as 99 per cent of the incoming sunlight. The competition for light is thus one of the main factors involved in these areas because solar radiation inputs are limited, particularly during the winter months. Consequently, there is a marked seasonality related to temperature and sunlight, and one of the major factors controlling the forests (especially in the more northerly zones of these areas) is the length of the growing season. Deciduous trees lose their leaves during the winter months and thereby reduce their rate of transpiration; this limits their energy requirements in the winter. Coniferous trees, on the other hand, have needle-like leaves which cause less transpiration. They are therefore able to survive without losing their leaves during the coldest months. In the more extreme conditions of the boreal forests or taiga, where the temperate forests merge into the tundra areas, the deciduous softwood larch (*Larix* spp.) tends to prevail. This combines the advantage of both needle-leaf and deciduous trees: a low rate of transpiration and an ability to reduce it further during periods of low energy inputs.

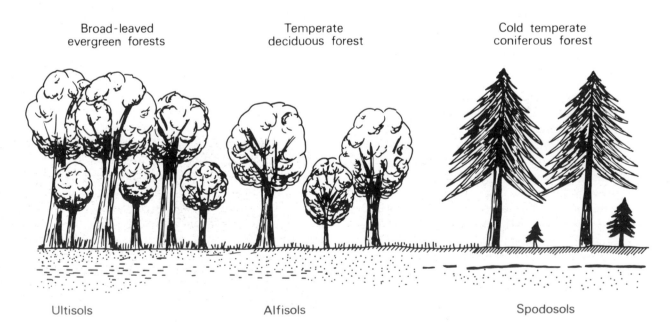

Broad-leaved evergreen forests · Temperate deciduous forest · Cold temperate coniferous forest

Ultisols · Alfisols · Spodosols

Figure 30.14 Temperate forest formations and soil type along a generalized transect from warm to cool temperate zones

The soils beneath the temperate forests range from ultisols and alfisols in the case of the evergreens to histosols and spodosols in the more northerly coniferous forests (Figure 30.14). This sequence is related to the climatic pattern in part, but it is also a response to the vegetation itself (Plates 30.10 and 30.11). As we have noted, the litter produced by the coniferous trees is relatively acid and resistant to decomposition. It also provides organic acids which encourage cheluviation. Thus, in the coniferous forests there is a tendency for increased nutrient loss and active translocation of iron, clay, and organic matter. In contrast, the moist, humid conditions of the evergreen forests are more akin to those of the tropics, so soils of low base status with rapid rates of organic matter decomposition occur, due to the high temperatures and intensified chemical processes.

The fauna of the forests shows a similar range, although it is far more restricted in terms of species diversity than that found in the tropical forests. Numerous tree-dwelling animals such as squirrels and chipmunks are found, while the ground fauna includes deer (e.g. *Cervus elephus* and *Capreolus* sp.), foxes, wild cats, lynx and various mice. Soil organisms are abundant in the deciduous woodlands, although less so in the coniferous forests,

Plate 30.8 Boreal coniferous forest, in the eastern foothills of the Canadian Rockies, at an altitude of about 1000 m a.s.l. The species in the foreground is Western red cedar (*Thuja plicata*); Engelmann spruce (*Picea engelmanni*) occurs extensively in the background and the ground flora is dominated by willow scrub and *Vaccinium* species (photo: N. V. Pears)

Plate 30.7 Temperate deciduous forest, Ecclesall Woods, Sheffield. Like most woodlands in Britain, these are largely 'man-made'. Species present include birch, sycamore, beech and oak. At this time of the year (in winter) the trees are bare and the canopy open; last autumn's leaves are decomposing on the floor (photo: J. Owen)

Plate 30.9 Maquis, Majorca. Species present include juniper, wild olives, asparagus, artemisia and anthyilis (photo: D. J. Briggs)

Plate 30.10   Acid brown earth (alfisol, cf. boralf) in an area of temperate deciduous forest, southern England (photo: D. J. Briggs)

Plate 30.11   A podzol (Hamod) under coniferous forest, near Penrith, Cumbria (photo: D. J. Briggs)

largely because of the less nutritious plant materials and the more acid conditions in the latter.

Biogeochemical cycles

Many studies have been carried out of nutrient and energy cycling in forests of the temperate latitudes, and for once there is no shortage of relevant data. It is known, for example, that the annual biomass of these forests varies both spatially and over time to a considerable degree. The total biomass of temperate forests is shown in Table 26.1. As this indicates, the biomass of evergreen forests may reach about 2000 t ha^{-1} in some cases, though the average for temperate deciduous forests is only 300 t ha. There are also marked changes over time, however, and the ecosystem biomass seems to reach a maximum at an age of about 200 years (Figure 30.15). The rate of annual assimilation of organic matter similarly varies.

Monitoring of nutrient cycles in temperate forests has shown that considerable quantities of nutrients may be stored within the vegetation, although most of it is retained in the leaves and young shoots (Figure 30.16). Thus, the annual turnover of nutrients is large. Considerable quantities of nutrients are also retained in the soil, except where intense leaching leads to rapid removal. One of the controls upon this nutrient cycle, however, is the release of mineralized material from the organic compounds. Organic matter breakdown in many of the cooler region forests is slow, with the result that uptake by the trees is reduced. Nitrogen and phosphorus, in particular, may be limited in these situations.

The distribution of nutrients in these forest systems has some importance, for man often attempts to utilize them, with drastic consequences upon the ecosystem (see Chapter 31).

Tundra biomes

General character

We have already discussed the character of the world's **tundra** regions from the point of view of climate and topography. As this showed, they are areas characterized by intensely cold conditions and the development of a layer of permafrost at depth in the soil. Only for a brief period in the summer do temperatures rise above freezing point and allow plant growth to occur. Then, with incredible rapidity, the plants complete their life cycle

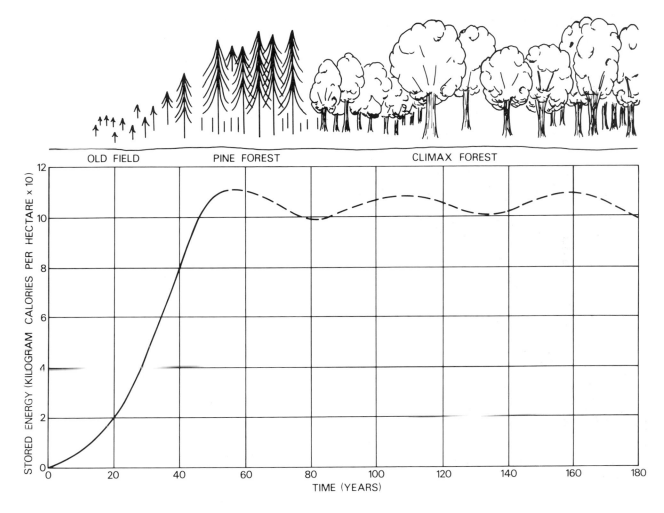

Figure 30.15   Changes in total energy storage of a forest over time (from Woodwell, 1963)

before the next cold season curtails their activity (Plate 30.12).

The vegetation of this zone is composed mainly of a few hardy species of dwarf shrubs, lichens and mosses. Trees tend to exist only in low-lying forms such as the dwarf birch (*Betula nana*) and dwarf willow (*Salix herbacea*). Saxifrage species are also common, but by far the most abundant plants are mosses such as sphagnum and simple lichens which grow on the bare rock.

The soils of these areas are poorly developed, a product mainly of physical weathering with little chemical breakdown (Plate 30.13). As a result nutrients are lacking. In addition, the effect of the permafrost layer is to prevent downward movement of moisture, so waterlogging occurs, resulting in gleyed soils. On the higher, gravel ridges better

drained and stony soils develop, while peats fill depressions.

The fauna, like the flora, is limited. Animals have to be adapted to the cold temperatures. This adaptation occurs in a variety of ways. Most animals are large so that their body area is small in relation to their weight; this reduces heat loss during the cold periods; some animals, such as the Arctic stoat, change their colour seasonally, becoming dark-coated during the summer in order to gain as much heat as possible from the brief period of intense solar radiation, and also as a means of camouflage. Few animals either migrate or hibernate in these areas. The migrations that do occur, in mammals such as the reindeer, are over relatively small distances and provide little relief from the harsh climate.

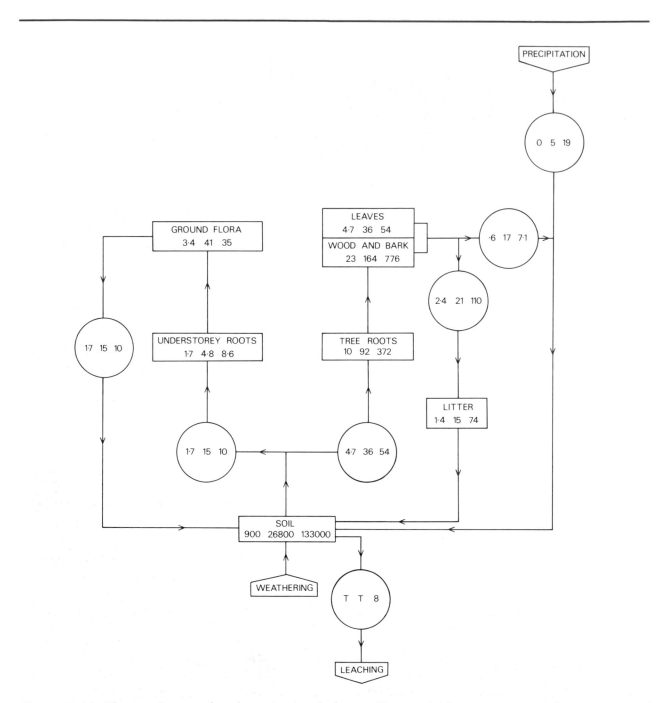

Figure 30.16 The nutrient cycle of a mixed oak forest. Figures in boxes represent the amounts of phosphorus, potassium and calcium stored in the mature vegetation and soil in kg ha^{-1}; figures in circles show the annual flows of P, K and Ca in kg ha^{-1} y^{-1} (Data from Duvigneaud and Denaeyer-De Smet, 1970)

Biogeochemical cycles

As can be imagined, there have been few attempts to monitor nutrient cycles in these areas. It is clear, however, that the total productivity of tundra ecosystems is small and that the rate of nutrient cycling is slow. The very slow rate of organic decomposition and weathering of the rock materials means that the quantity of available nutrients is small. Nitrogen dominates the nutrient cycles and is assimilated mainly through the

Plate 30.12 Arctic tundra, Norway. The main species are dwarf willow and coarse moor grasses (photo: J. Rose)

Plate 30.13 Shallow ranker-like soil (entisol) typical of many tundra areas (photo: D. J. Briggs)

activity of the lichens and shrubs, most of which are nitrogen-fixers.

Where measurements of the total biomass of tundra biomes have been made, the paucity of the environment is illustrated. Assessments in the Arctic indicate a total of 14.3 t ha^{-1} at a latitude of 63°N, but only 3.4 t ha^{-1} at a latitude of 75°N. The majority of this organic material lies below the ground, and the annual productivity probably averages about 0.5 t ha^{-1}; values typically range from 0.2 t to 1.8 t ha^{-1} y^{-1}.

Conclusions

It is apparent that we are still far from the ultimate state of understanding fully the processes operating within the world's main ecosystems. This ignorance is to some extent understandable, for, as we noted in earlier chapters, the problems of obtaining the necessary data are formidable, particularly in the inhospitable conditions of tropical rain forest, desert or tundra. None the less, it is an ignorance which we should not condone, for many of man's attempts to exploit and manage his ecological resources have run into difficulties because of his incomplete understanding of the systems he was dealing with. Already, man is exploiting large areas of tropical rain forest and is extending his influence into even the remote desert and tundra regions. We will see in Chapter 31 how far-reaching his effects can be. It is uncertain whether we are yet able to deal with the consequences of our interference with these remaining, and often fragile, natural ecosystems.

31

Man and the ecosystem

History of man's activities

Agriculture and ecosystems

Throughout most of the temperate world it is difficult to find landscapes that do not bear the imprint of man's agricultural activities either now or in the past. In much of the tropical and sub-tropical world the situation is similar. We have noted already that the savanna grasslands possibly owe their existence at least in part to man's activities; even the tropical rain forests have been widely affected by clearance and cultivation. We saw, too, that man has left his mark upon the deserts, and the present extent of the arid scrubland and desert owes much to the activities of past farmers. Only the more remote environments – the tundra regions, the central deserts (such as the Australian Desert), the interior of the larger and more impenetrable tropical rain forests, and the more inaccessible mountain areas – have escaped the distinct impact of man. Even in those areas, of course, man is not entirely unknown. He does, however, live in a closer harmony with the natural ecosystem; he has altered it less.

The history of forest clearance

The character of man's impact upon these diverse environments varies. Moreover, it has often been operating for many centuries. We can recognize three early stages in man's history when he became a tool-maker and user, and from this time he probably started to shape his ecosystem to suit his own purposes; he started to act as a control upon the system. In the earliest stage, the **Palaeolithic** (Old Stone Age), his effect was probably limited, for his technology was crude and his numbers small. As he advanced technologically, however, and as his population grew, his influence became more far-reaching. Thus, during the **Mesolithic** (Middle Stone Age) he began to settle in more permanent camps, and to clear and manage the surrounding area for grazing and fuel. In the last phase of the

Stone Age, the **Neolithic**, his influence grew further.

Throughout the early periods of his development, man therefore affected the ecosystem partly as a hunter and gatherer, and partly as a settler and farmer. In the former role he possibly influenced the populations of certain animals, hastening their extinction in some areas; in the latter role he had a more fundamental impact, for he cleared large areas of woodland, and set in motion marked changes in the ecosystem. There is evidence, for example, that Mesolithic forest clearance in upland Britain led to the decline of certain tree species, particularly lime (*Tilia*), and the opening up of the forest canopy resulted in major changes in soil development. Many of the upland peats may have been produced by these human influences (Figure 31.1).

Later forest clearance extended and intensified these effects. During the Neolithic, and even more so during the **Iron Age** and **Bronze Age**, the development of a relatively sophisticated and widespread form of settled agriculture saw the removal of vast areas of woodland in Europe. About 4000 years ago, during the Bronze Age, there was a marked decline in elm (*Ulmus*), and at about the same time the evidence of fossil pollen from peat bogs shows the increase in weed and cereal plants within the ecosystem (Figure 31.2). Thus, as well as removing the natural vegetation, man was enabling new species to become established and introducing others for his own use.

These episodes of human activity are not synchronous everywhere. We know, for example, that man has progressed at different rates in different parts of the world, so that even today we can recognize civilizations which are little different from the stone-age communities of Africa some 10,000 years ago. The spread of cultural innovations is slow and often it was many centuries before new techniques were adopted in the areas remote from their discovery. Thus, in the Middle East, settled

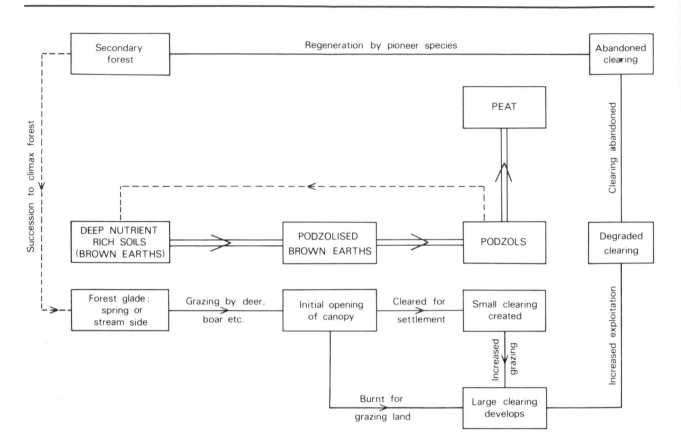

Figure 31.1   The effects of prehistoric man on forest ecosystems (based on Simmons, 1964)

agriculture, with tillage and systematic cropping, became established about 10,000 years ago, probably in Iran. In subsequent millennia, agriculture spread through diffusion of the ideas, and it was almost certainly reinvented, quite independently, by different peoples. Thus, in Britain, it emerged about 5000 years BP; in the Americas, it was invented independently at about the same time. There is evidence from Africa that agriculture evolved by 7000 years BP, and possibly earlier, while the scanty clues obtained from the tropical parts of Asia hint at an even earlier history; possibly before 12,000 years BP.

Whenever settled farming started, the implication is clear. Seed growing required land free from the natural vegetation; it involved clearance and tillage of the soil. With the growth of settled populations, and with the steady progress in agricultural techniques, the impact upon the environment increased. Not only did the area of cleared land expand, but so did the intensity of the changes. Many of the initial agricultural systems were probably based upon a principle of **slash-and-burn**.

Areas were cleared by burning, then cultivated for a few years until yields declined and the natural vegetation started to reinvade; the plot was then abandoned in favour of a new site. So long as population densities were low, this system could be maintained, for the natural vegetation could regenerate. Once the frequency with which an individual plot was cleared increased, however, the natural vegetation could not recover, and a spiral of decline in the ecosystem occurred.

These developments illustrate two interesting examples of positive feedback. In terms of human development, the concentration of the population into more or less permanent communities initiated a cycle of technological and cultural advancement which increased man's ability to manage his environment (Figure 31.3). At the same time, the tendency for more settled farming, or for more rapid cycles of slash-and-burn, led to a progressive decline in the fertility of soils and the energy status of the ecosystem (Figure 31.4).

In the last four or five centuries, man's agricultural progress has speeded up considerably. The

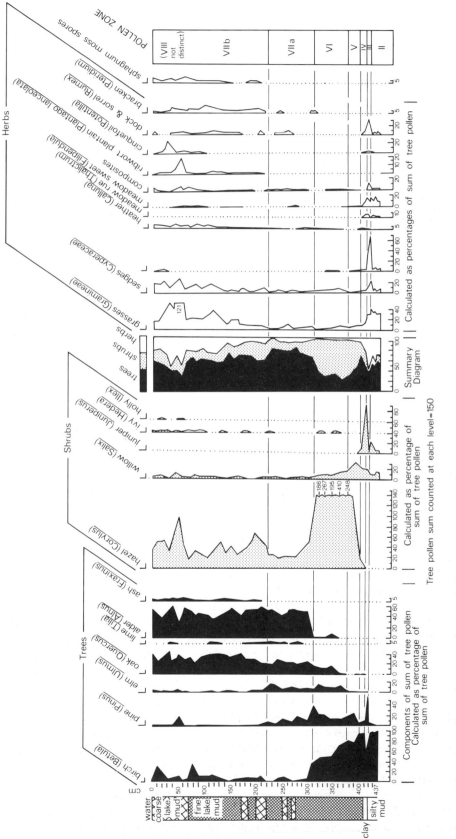

Figure 31.2 A pollen diagram from Blelham Tarn, English Lake District. Not all pollen types are represented. Note the decline in elm and the increase in herb pollen at the zone VIIa-VIIb boundary (from Evans, 1970)

Figure 31.3 The social and technological effects of concentration of population into permanent farming communities

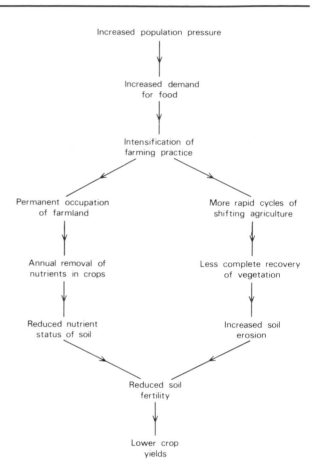

Figure 31.4 The long-term effects of increased population levels on soil fertility and crop yield

development of sophisticated ploughs and seed drills by men such as Jethro Tull; the generation of new seeds and new strains of cereal and animals; the discovery of the principles of plant nutrition and the consequent evolution of the fertilizer industry; the increased use of machinery in agriculture; and the recent upsurge in the use of pesticides – all these have fundamentally altered the agricultural ecosystem.

Industrialization and ecosystems

Of course man has progressed not only through the development of his agricultural technology. He has also created a complex and sometimes frightening array of other technologies; industry and arms, in particular, have been mirrors of this progress. Both have had a major impact upon the world's ecosystems.

It is common to regard the development of industry as a recent phenomenon, started in western Europe during the nineteenth century; we speak of the **Industrial Revolution**. Since then, industry based upon the widespread use of fossil fuels, particularly coal and, more recently, oil, has become one of the main consumers of the earth's resources. The extraction and exploitation of these resources has affected the landscape, and it has also disrupted the natural ecosystems (Plate 31.1). These effects are partly direct: the creation of large holes in the ground and the dumping of waste spoil material on the surrounding land is clearly a significant cause of disruption at a local level. But it is also more indirect and more pervasive. In particular, pollutants – waste products released during the exploitation of these resources – have been spread through most of the world. Airborne dust and gases, waterborne substances, accumulations of waste materials in the soil have all affected the natural vegetation and the fauna.

The development of nuclear weapons reveals a

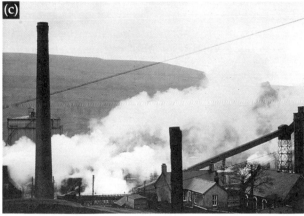

Plate 31.1 The effects of industrialization on the environment: (a) dumping of China Clay waste, Lee Moor, Dartmoor (photo: D. J. Briggs); (b) remnants of an algal bloom in an urban stream, Nottingham (photo: B. Pyatt); (c) air pollution, Calderdale, West Yorkshire (photo: D. J. Briggs)

similar story. Radioactive fallout from weapons testing during the 1960s led to widespread dispersal of substances such as strontium-90 and caesium-137. The latter has been found in Arctic food chains; it is absorbed by lichen, passed to the caribou and thence assimilated by eskimoes. Other predators such as wolves and foxes have also developed concentrations of caesium-137 in these areas.

Clearly, man's impact upon the ecosystem is significant. Sometimes this impact is deliberate, and the consequences are calculated. Often it is not. The history of man's activities indicates that he needs to understand the ecosystems he is dealing with if he is not to cause damage, not only to them but also, ultimately, to himself.

Agricultural ecosystems: cereal cropping

The general nature of cereal cropping

One of the most intensive and extensive forms of agriculture is cereal cropping. The development of cereal cultivation goes back to the early periods of farming history, when, by some chance of nature presumably, the wild wheat crossed with a natural goat grass to form a hybrid wheat plant. The result was a plumper, more nutritious grass known as **emmer**. Later, a further cross-fertilization with another goat grass occurred, and **bread wheat** developed. This has been the foundation of many of the cereal-growing cultures of the world. Meanwhile, domestication of maize probably occurred in the New World (perhaps in Mexico), and this has become the focus for cereal growing in other areas.

Today, deliberate cross-breeding of cereals by man has led to the development of many different types of maize and wheat, but the basic principles

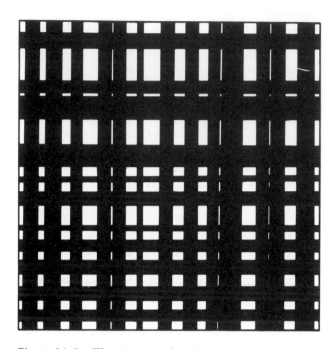

Figure 31.5 The extent of soil surface coverage by tractor wheelings during a typical annual sequence of cultivation

of their cultivation are similar, and the effects upon the ecosystem are comparable.

In most cases, cereal cropping involves the creation of a seedbed within which the crop is sown by artificial means. In all but the more primitive cereal-growing systems, this involves the use of tractors and seed drills. Sowing is therefore preceded by various tillage practices: ploughing, harrowing and rotovating, to break up the soil aggregates, to remove weeds and bury or chop up the residues of previous crops. Following sowing, the land may be rolled and harrowed to bury the seed and create a more compact soil environment for germination. During growth, fertilizers and pesticides may be applied and when the crop is mature, harvesting is often carried out by combines.

All these practices have considerable effects upon the soil. Repeated passage of vehicles over the land may mean that as much as 90 per cent of the surface is directly affected by vehicle wheels (Figure 31.5). If the soil is wet, wheel slippage and compaction may damage the soil structure (Plate 31.2). In addition, tillage alters the internal soil processes. It may kill soil organisms (particularly earthworms); it may encourage chemical oxidation of the humus by opening up the soil and improving aeration; it may,

as a result, reduce nutrient retention and lead to instability of the soil structure.

As agriculture becomes more intensive – both in the attempt to feed the growing world population and, more importantly in many western countries, to provide the population of industrialized countries with a higher standard of living – there is a tendency for cereal growing to move towards a system of monoculture. The same crops are grown repeatedly. Even where rotations of different crops are used, they may be very similar in their requirements. Over time, therefore, marked changes in the soil may occur (Figure 31.6).

Many of these changes relate to the effects of the tillage operation, and in recent years attempts have been made to combat these by introducing systems of cereal cropping which involve less frequent tillage. Thus, in the United States, and more recently elsewhere, **zero-tillage** and **minimal cultivation**

Plate 31.2 Structural damage to the soil caused by harvesting kale late in the year when the land was too wet to bear the heavy machinery (photo: D. J. Briggs)

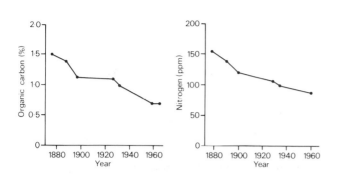

Figure 31.6 Changes in soil conditions under long-term intensive arable farming (after Mattingly *et al.*, 1975). Plots had received NPK fertilizer and had been under wheat and barley continuously since 1876

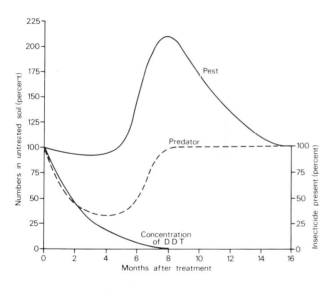

Figure 31.7 Effects of DDT on predator and prey organisms in the soil (after Edwards, 1969)

soil numerous detrimental side effects may occur. Many insecticides, for example, kill off both pests and beneficial soil organisms. More active organisms are particularly vulnerable to the toxic effects of pesticides, for they move through a wider area of the soil and come into contact with greater quantities of the chemical. Since, in general, it is the predatory organisms which are more active, the result may be to kill off preferentially the very creatures that naturally control the population of parasitic organisms and agricultural pests.

The delicate balance of the soil fauna can be disturbed so easily that the effect may be totally the opposite to that desired (Figure 31.7). In time, admittedly, the fauna may recover, but where repeated applications of pesticides are made, permanent suppression of certain organisms – many of them beneficial – may occur. This may have a number of consequences. It may lead to increased numbers of pests; it may disrupt the detrital food chain so that the whole structure of the soil ecosystem is altered; and it may inhibit biological decomposition of organic matter and nutrient cycling.

Two critical factors in the effect of pesticides in the soil are their persistence and toxicity. **Persistence** is measured in terms of the **half-life** of the substance: the time it takes for its concentration to be halved. Examples of the half-lives of common pesticides are shown in Table 31.1. It can be seen that on the whole insecticides have longer half-lives – and are therefore more persistent – than

have been introduced. In these, ploughing and seedbed preparation are reduced to a minimum or totally eliminated. The seed may be sown directly into the undisturbed soil, weeds and crop residues are controlled by **herbicides**, and insect pests – which may build up due to the lack of soil disturbance – are killed with **insecticides**.

The accumulation of pesticides in the soil and in runoff water has been worrying in some cases. It has been shown that as much as 50 per cent of compounds sprayed on to the vegetation may be lost by leaching and runoff. Moreover, within the

Table 31.1 *The persistence and toxicity to animals of common pesticides*

Pesticides	Half-life (months)	Lethal dose (mg kg^{-1})
DDT	27.5	113
Lindane	21.2	125
Dieldrin	22.0	90
Aldrin	8.3	60
Parathion	1.5	3.5
Phorate	0.3	3.7
Simazine[1]	1.8	5000
Atrazine[1]	2.7	3080
2,4,5-T[1]	2.5	300–500
MCPA[1]	0.7	700

Notes
[1] = herbicide; all others = insecticides

511

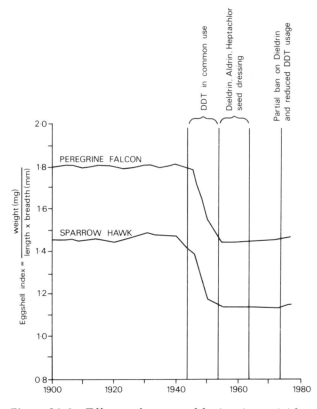

Figure 31.8 Effects of organochlorine insecticides on eggshell thickness of peregrine falcons and sparrow hawks in Britain (from Brown, 1976)

herbicides. **Toxicity** is measured by a parameter known as LD_{50}. This is the lethal dose necessary to kill off 50 per cent of the target population; again examples are given in Table 31.1.

In the past, there was a tendency to use very persistent pesticides of the **organochlorine** type. DDT, dieldrin and lindane were widely used forms. These broke down only slowly in the soil, became concentrated within the food chain, and led to unacceptable pollution not only in the soil but also in higher animals. DDT, for example, was responsible for the near extinction of the sparrow hawk in Britain during the 1950s; the accumulation of the chemical in the birds resulted in the thinning of their eggs with the consequence that relatively few survived to hatch (Figure 31.8). More recently, many of the organochlorine compounds have been banned or strictly controlled, and **organophosphate** pesticides are now used instead. These are more volatile and therefore decompose by chemical and biological processes much more rapidly. Their half-lives are thus measured in terms of weeks or

months instead of years (Table 31.1). They are also more toxic, so smaller quantities can be applied.

Breakdown of pesticides occurs by four main processes. **Volatilization** involves the reduction of the compound to gaseous form, whence it escapes to the atmosphere. **Biological decomposition** occurs through the action of certain soil organisms which convert the pesticides to different forms. In most cases the result of this conversion is a less harmful compound, although in some instances it may produce a more lethal substance. DDT, for example, may break down to produce DDD, which is highly toxic. **Chemical decomposition** may also take place, and some pesticides are subject to **photodecomposition** (breakdown under the effects of sunlight); this occurs either at the soil surface or on the leaves of the plants. Pesticides may also be adsorbed on to colloidal particles in the soil and stored there, in much the same way as plant nutrients. Erosion of the soil may lead to their removal. Finally, leaching may carry the soluble pesticide compounds out of the soil and into streams. As may be imagined, this removes the pesticide from the soil, but it provides an input of pollution to aquatic ecosystems.

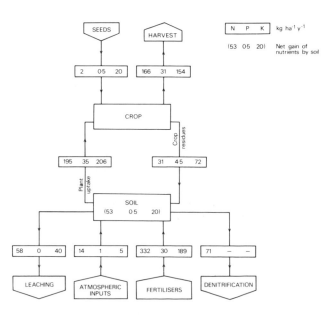

Figure 31.9 The nutrient cycle of an intensive arable farm in which crop residues are removed from the land: potatoes–wheat–sugar beet rotation on clay soil in the Netherlands (data from Henkens, in Frissel, 1978). Figures show flows of nitrogen, phosphorus and potassium in kg ha^{-1} y^{-1}

It is perhaps easy to become too concerned about the effects of pesticides on the environment. There is certainly cause for concern. In Australia, for example, there is evidence that organisms may become immune to pesticides; originally DDT was applied at rates of 1 kg ha^{-1} to control pests; by the late 1960s doses of 100 kg ha^{-1} were necessary. On the other hand, man has a lot to be grateful for. In some parts of the world, pesticides, more than any other factor, have been responsible for huge increases in agricultural output, and in a hungry world that is important. They have also helped to rid vast areas of disease; malaria has been eradicated from many thousands of square kilometres through the use of DDT. The critical need is to weigh up the advantages and disadvantages before pesticides are used; this, as ever, requires an understanding of the system with which we are dealing.

Nutrient cycling in cereal growing

One of the most important aspects of cereal-growing systems is the frequent removal of the crop in the harvest. The loss of this material results in the loss of a major part of the total biomass and nutrient reserves of the system. It is for this reason that man needs to replenish the soil fertility by applying fertilizers, for continued cereal cropping leads to a loss of organic matter from the soil (Figure 31.6) and possibly to soil exhaustion.

These two features – the removal of the crop and the input of fertilizers – represent vital components of cereal-growing systems, and ones which distinguish them from most other natural or agricultural systems. The quantities of nutrients introduced by fertilizers far exceed the inputs from either rainfall or weathering, while losses in the crop are normally greater than those from leaching or erosion (Figure 31.9).

Within the system, cycling of nutrients is often rapid, for although there tends to be less soil organisms than under grassland (Table 31.2), the action of tillage encourages oxidation and the mineralization of the nutrients. It is also apparent that the cereal ecosystem involves a very limited range of species – often only one – for competitors are ruthlessly removed by the use of herbicides. In addition, there is practically no fauna outside the soil. Thus the internal food-chain is very simple; apart from a few parasitic animals which may attack the growing crop, only decomposers are at work. Man, or the livestock he feeds with his crops, represent the sole consumers of any significance.

Table 31.2 *Earthworm numbers in arable and grassland soils*

Land use	Earthworm nos. ('000 ha^{-1})	Earthworm weight (kg ha^{-1})
Permanent pasture	700–1100	500–700
Ley grass	550–800	475–700
Young arable	575–825	400–825
Old arable	175–300	70–150

Data from E. W. Russell, *Soil Conditions and Plant Growth*, Longman 1974.

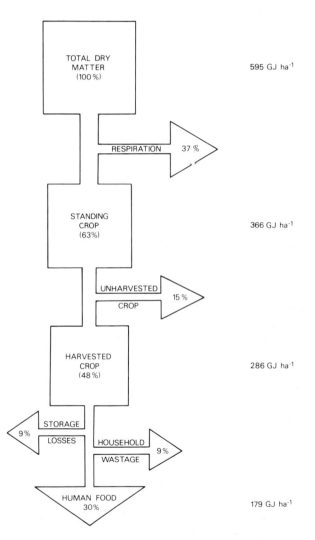

Figure 31.10 Energy efficiency of a typical intensive arable farming system (potato crop). The total dry matter produced in the system is about 595 GJ ha^{-1} y^{-1}, which is about 0.7 per cent of solar radiation inputs. The human food produced is only 0.22 per cent of solar energy received (data from Duckham and Masefield, 1970)

Table 31.3 *Losses of nutrients by soil erosion*

| | Nutrient losses (kg ha⁻¹ y⁻¹) | | | |
	N	P	K	Ca
Erosion from continuous maize	74	20	678	246
Erosion from rotational crops	29	9	240	95
Crop removal	134	25	112	45

Data from N. C. Brady, *Nature and properties of soils*, New York: Macmillan, 1973

Indeed, this is one of the main characteristics of all agricultural systems, the attempt to shorten food chains and thereby avoid waste. Nevertheless, it is significant that the total energy absorbed by man through potato growing probably averages no more than 0.2 per cent of the initial solar inputs (Figure 31.10).

Under some conditions, cereal cultivation may lead to a marked increase in erosion losses. Tillage and crop removal leaves the soil unprotected and prone to the winnowing effect of the wind (Plate 31.3); compaction by machinery, and channelling of runoff waters by plough furrows, may encourage gullying. In these circumstances, nutrient losses through erosion may be serious (Table 31.3). These processes are exacerbated by the loss of organic matter from the soil. As we have seen, the humus acts to bind the soil particles into stable aggregates. When the humus is lost – as it is due to increased oxidation and reduced organic matter inputs – the structural stability of the soil may decline. Damage by machinery becomes a greater risk, as does erosion.

It should also be noted that leaching losses in cereal-growing systems may be considerable. Although they may not represent a significant loss in relation to the large inputs of nutrients in fertilizers, they do represent an important contribution of nutrients to streams. Nitrogen, in particular, is lost in this way, for it is highly soluble, especially where it is not retained in organic form in the humus. As we will see later, one of the consequences is pollution of aquatic environments. This process is mirrored by pesticides. We have noted that large quantities may be applied to reduce competition from both weeds and animal pests; a major part is commonly lost by leaching to the streams.

Grazing systems

General nature of grazing systems

Man's use of grasslands is widespread. Utilized grasslands possibly cover as much as 30 per cent of the world's land surface, ranging in character from very extensively grazed **range** and **moorland**, to intensively stocked, artificially **sown pasture**.

In the more extensive grassland systems, the role of man and his livestock is limited; it represents little more than a regularization of the natural grazing that would occur by the wildlife of the area. Often, however, this is accompanied by simple rangeland management practices such as burning of the vegetation to encourage new plant growth. This may have severe effects upon the ecosystem. It

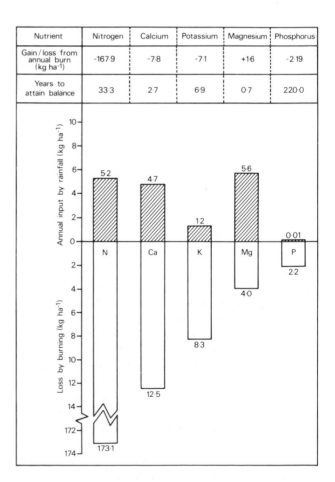

Nutrient	Nitrogen	Calcium	Potassium	Magnesium	Phosphorus
Gain/loss from annual burn (kg ha⁻¹)	-167.9	-7.8	-7.1	+1.6	-2.19
Years to attain balance	33.3	2.7	6.9	0.7	220.0

Figure 31.11 Changes in soil nutrient contents of burned moorland soils and the minimum interval between burning necessary to maintain nutrient levels (data from Chapman, 1967)

Plate 31.3 Wind erosion in the Vale of York, England. Cultivation of the field on the right of the photograph for cereal crops left the sandy soil bare during the spring. Strong winds caused rapid erosion and total loss of the crop (photo: D. J. Briggs)

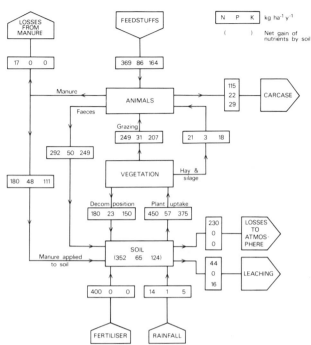

Figure 31.12 The nutrient cycle of an intensive grazing system: a Dutch dairy farm on a clay soil (data from Henkens, in Frissel, 1978). Figures show flows of nitrogen, phosphorus and potassium in kg ha^{-1} y^{-1}

kills or drives out many of the natural fauna, and it may lead to marked changes in species composition of the grass sward and shrubs. It also affects the soil (Figure 31.11).

Of course, in addition to the character of the grass, one of the most important attributes of grazing systems is the animal. Sheep, cattle and goats are the main animals in many parts of the world. Cattle include beef and dairy cattle; sheep may be reared for either wool or meat. In all cases, where the agricultural system is reasonably sophisticated, specific varieties of animal are used. The merino sheep, for example, is the basis for much of the world's wool production. Hereford cattle are common as beef animals while dairy production uses a range of animals, including the Fresian and Charolais. The importance of cross-breeding to find the ideal strain for the conditions is illustrated by the experience in Israel. Initially, using the local cattle, milk yields were in the order of 300 l yr^{-1}; introduction of Lebanese and Syrian cross-bred cattle produced an increased output to about 1000–

2000 l yr^{-1}. Finally, cross-breeding with Fresian cows resulted in yields of as much as 6000–8000 l yr^{-1}.

Nutrient cycling in grazing systems

In cereal-growing systems, the main process of nutrient cycling is carried out by decomposition of plant debris directly by soil organisms and chemical oxidation. In grazing systems this plant nutrient cycle is supplemented by an animal nutrient cycle (Figure 31.12). The vegetation is eaten by the grazing animal, and the nutrients are either retained in the animal's body or returned to the soil in the form of dung and urine. Relatively small proportions of the total nutrient reserves are normally retained in the animals (Table 31.4) and even smaller quantities are lost either in their milk or by annual removal of the beasts for slaughter. By far the greatest proportion of the total nutrient uptake is returned.

The nature of the return varies from one nutrient to another. Phosphorus and potassium are returned mainly in the dung; calcium and nitrogen

Table 31.4 *Nutrient retention and returns by cattle*

	Ca	Mg	P	K	Na
			Nutrient		
Total intake (kg)	120	95	85	135	30
Retained in body (per cent)	10	6	9	3	7
Retained in milk (per cent)	12	5	29	6	6
Returned in dung (per cent)	75	79	62	11	30
Returned in urine (per cent)	3	10	0	80	57

Data   from G. B. Davies, D. E. Hogg and H. C. Hopwell, 'Extent of return of nutrient elements by dairy cattle: possible leaching losses', Joint Meeting of the International Soil Science, Comm. IV and V, New Zealand (1962), pp. 715–20.

in the urine. In both cases, the nutrients which are contained in these waste products are readily available to the plants; they are easily soluble. This is particularly true of the nutrients in the urine. Thus, the rate of nutrient cycling tends to be speeded up, and the total quantity of nutrients in circulation increases. Moreover, the abundant supply of readily digestible organic matter, together with the general stability of the environment (there is less regular ploughing to disturb the soil fauna), means that the populations of earthworms and other soil organisms rise (Table 31.2).

The solubility of the nutrients in the animal wastes has further implications, however, for like fertilizers they are susceptible to leaching. Again, pollution of water bodies may be a problem where grazing is very intensive. Another implication of the nutrient cycle illustrated in Figure 31.12 is that the total efficiency of the system in terms of the energy supply to man is relatively small. Perhaps no more than 0.02 per cent of the initial energy inputs is made available to man, while in the less intensive rangeland systems the figure may be as low as 0.004 per cent. On the other hand, intensive grass growing is potentially very productive. In many cases the yields of the grass are far closer to the potential possible than is achieved in cereal growing. The losses come mainly in the steps from the grass to the animal and from the animal to man (Figure 31.13).

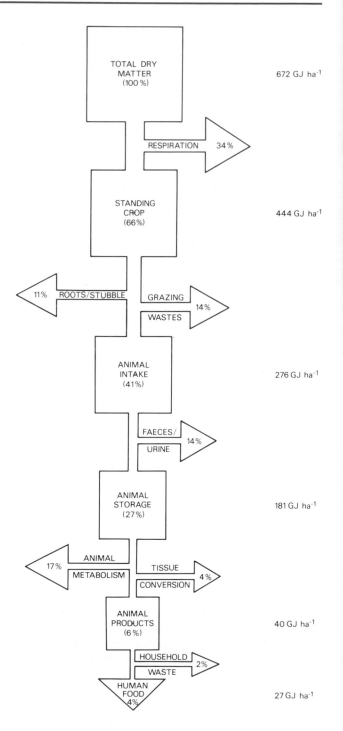

Figure 31.13 Energy efficiency of a typical intensive grazing system. The total dry matter produced in the system is about 672 G J ha^{-1} y^{-1}, which is only 0.5 per cent of solar radiation inputs. The human food produced is only about 0.02 per cent of solar energy received (data from Duckham and Masefield, 1970)

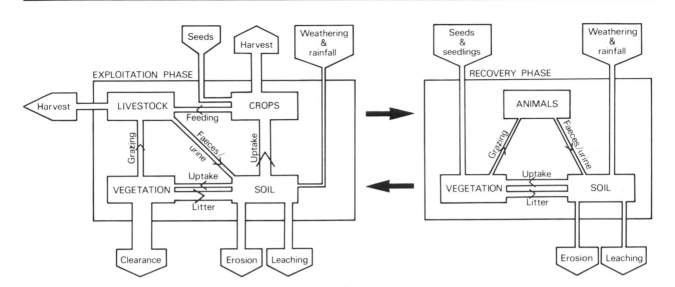

Figure 31.14 Nutrient cycles during exploitation and recovery phases under shifting cultivation. The width of the arrows shows the approximate relative magnitude of the nutrient flows

Shifting cultivation

General nature of shifting cultivation

Over wide areas of the tropical and sub-tropical world, man carries out a system of cultivation which has changed little in principle from the methods used by prehistoric man. Inputs to the system are few, and he relies upon natural replenishment of the soil by rainwater and weathering to maintain the fertility which is tapped by crop removal. Natural or semi-natural woodland and scrub is cleared, cultivated and cropped for a few years, then abandoned. The crops grown vary from one area to another, but typically include sorghum, millet, maize and cassava.

The length of the plot rotation in many of these **shifting cultivation** systems is related to the population density. While the population is low and there is ample land available, the rotation may be long; each plot may be cropped for no more than two or three years, then abandoned for twenty or more years to allow regeneration. Under these conditions regeneration is effective and the ecosystem shows no major decline in fertility. On the other hand, as the population density increases, and less land is available (something often exacerbated by the extension of plantations on to the more fertile forest soils), the rotation becomes shorter (Figure 31.4). Often less than seven years may be left for regeneration, while the plots may be cropped continuously for almost as long. Soil exhaus-

tion, erosion and an overall decline in ecosystem fertility then occur.

Nutrient cycling in shifting cultivation

The nutrient cycle under shifting cultivation cannot be considered on a short-term basis like that of cereal or grassland systems. Instead, relatively long periods of exploitation and net nutrient loss alternate with even longer periods of rest and nutrient accumulation. The long-term character of the system depends upon the balance between these two processes (Figure 31.14).

The period of loss is generally initiated by felling of the trees and burning of the residues, a practice known as slash-and-burn. This has a severe effect upon nutrient reserves, for considerable quantities of nutrients are lost in the smoke. Potassium, in particular, is removed in this way, for it is released mainly in the fine ash particles and as gases which are swept into the atmosphere. Other nutrients are returned to the soil in the ash which collects on the surface. These nutrients are readily soluble, however, and may be leached rapidly from the soil. Leaching is encouraged during this stage because the removal of the forest vegetation means that there is less interception of rainfall and increased rainfall inputs to the soil. The combined effects of burning and subsequent leaching are thus to deplete the total nutrient reserves of the ecosystem. Since, as we have noted, the majority of the nutrients of many of these tropical forests are stored within the vegetation, this may represent a serious loss of fertility (Figure 31.14).

The bare surface is also prone to erosion, so that additional losses occur in runoff. Gullying and soil exhaustion may therefore be initiated even before the land has been cropped.

During the following years of cropping, further depletion of the nutrient reserves takes place due to the combined effects of erosion, leaching and crop removal. Leaching is reduced to some extent, for the crops are often able to provide a reasonably complete and continuous cover. They also give some protection against erosion, but the nutrients retained in the soil following burning are taken up by the plants, and considerable losses may occur in the harvest. The use of fertilizers is almost unknown, and only small quantities of animal wastes are supplied to the soil. Thus, relatively quickly, yields start to decline as fertility falls. The diminishing yields are often due to increased weed competition as much as soil exhaustion.

When the land is abandoned, one of two processes may result. In the case of small plots, surrounded by lush vegetation, regeneration occurs quite rapidly. Nutrients are taken up from the soil and held within the vegetation, so that total soil reserves may decline for a time. On the other hand, the reduced leaching and erosion help to stabilize the system. Nutrients from the rainfall and weathering are stored, and in time a slow recovery of the overall nutrient reserves may occur.

In the case of larger plots, regeneration may be so slow that the bare surface becomes eroded and fertility is further reduced. If this happens, the opportunity for revegetation may be lost and a spiral of decline may commence (Figure 31.4).

The development of these systems therefore depends upon the subtle balances between nutrient removal and soil erosion during periods of utilization, and the rates of recovery in the rest periods. Over time, the overall nutrient reserves of the system may be maintained, albeit with marked short-term fluctuations, so long as the farmers do not return to the area too soon. If the rotation is short, however, in comparison to the rate of recovery, a gradual downward trend in the nutrient reserves may occur (Figure 31.15). Moreover, if the slash-and-burn process is too widespread and effective, rapid decline may occur due to erosion and leaching which prevents regeneration.

Managed forests

General nature of managed forests

Increasingly, man is attempting to make use of forests as a resource in their own right. He requires the timber they produce for a variety of purposes: for paper making, for building, for fuel and for chemical industries. Some indication of these uses is given by the estimates made by J. D. Ovington in 1965: 42 per cent of the world's harvested woodlands were used for fuel; 37 per cent for building; 11 per cent for paper and pulp and 4 per cent for pit props.

Man obtains these timber products from a diverse range of forests, however, and manages them at very different levels of intensity. At one extreme, he merely harvests natural and semi-natural forests and leaves the trees to regenerate naturally. At the other extreme he plants forests specifically for timber production. In this latter case he creates an ecosystem which is in some ways similar to that of cereal cropping; one comprising stands of one or, at the most, a few species of similar age. In the same way he may fertilize and manage the stand until it is mature, then harvest it by a process of clear-felling.

The main management procedures in these instances are site preparation, sowing, thinning and felling. Site preparation often involves clearance of the original vegetation, or the previous forest crop. Large ploughs and tractors are often used at this stage, and, as can be imagined, the

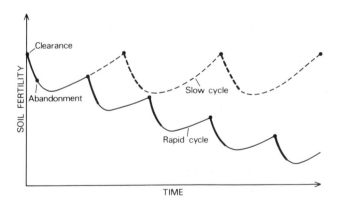

Figure 31.15   The effects of shifting cultivation on soil fertility. Heavy lines represent clearance–exploitation phases; light lines represent abandonment–recovery phases. Note that the soil fertility continues to fall for a short while after abandonment until the vegetation cover is re-established

disruption of the soil is considerable. Drainage may also be carried out, and often the soil is ridged to allow planting on slightly higher ground above the wetter furrows. The young trees are sown close together to give mutual support and shelter, and sometimes fertilizers are added. As the plants grow, they start to compete for light and nutrients, so that one or more thinnings are carried out to reduce competition and to encourage growth of the better trees. **Wind-throw** is sometimes a problem at this stage, for as the tree canopy is opened up the turbulence created by the tree-tops and the channelling effect of the wind along open areas may cause considerable pressures during gales. Trees are often relatively shallow rooting and thus are unable to withstand these pressures.

When the trees reach maturity, the stand is **clear-felled**, and the procedure starts again. The age at which felling takes place depends upon the tree species, its potential use and the environmental conditions. In the case of the Sitka spruce trees in Britain, felling often occurs at about fifty years, but in the USA pines may be grown to ages of 100 years or more.

In many parts of the world forest management is less intensive, partly because much greater areas of forest are available to be exploited. Felling is more selective, and regeneration is either allowed to occur naturally, or takes place by planting of small, irregular plots of trees. In these cases the tree stands are not of a single age, and thus the trees reach maturity at different times. Nevertheless clear-felling has occurred in many of these forests, and the lack of attention in the past to revegetation has meant that the world's forest reserves have shrunk dramatically over recent years. This is particularly true of many tropical forests; but it is also a feature of the woodlands of Canada and the United States to some extent. Clear-felling, as we will see, has a considerable impact upon the ecosystem.

Nutrient cycles in managed forests

Except in the more intensively managed plantations such as those that characterize the commercial woodlands in Britain, the nutrient cycles in managed forests are similar in broad terms to those of natural forests. There is, however, one major exception, and that is the removal of the crop following harvesting. This leads to a loss of biomass and of nutrients that does not occur under natural conditions. The loss is less than might often be expected, however, since only the boles of the trees

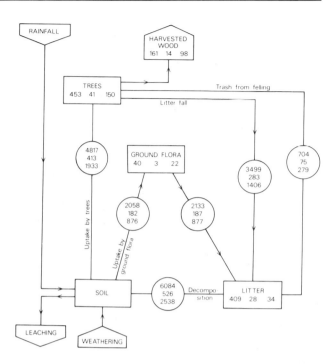

Figure 31.16 The nutrient cycle of a fifty-five-year-old managed pine forest (*Pinus sylvestris*). Figures in circles show flows of nitrogen, phosphorus and potassium totalled over the fifty-five years of growth; figures in boxes show the total amounts retained at the end of the period. Note that inputs by rainfall and weathering, losses by leaching and nutrients retained in the soil have not been measured (data from Ovington, 1962). Units are kg ha^{-1}.

are normally removed, and these contain a relatively small proportion of the total nutrient reserves. The leaves and branches, which are more nutrient rich, are normally left on the ground (Figure 31.16). They may be burned, in which case there are significant losses in smoke and by leaching, or they may be allowed to rot *in situ*. In this latter case, decomposition is slow and nutrients are released gradually.

The most dramatic losses of nutrients occur when the forests are clear-felled. Some idea of the effects of clear-felling on nutrient cycles is shown by the work of Likens and Bormann (1972) in the United States. In one experiment in Hubbard Brook they clear-felled an experimental catchment and prevented regeneration of the forest by applying herbicides. They then compared nutrient losses from the cleared catchment with those from a nearby

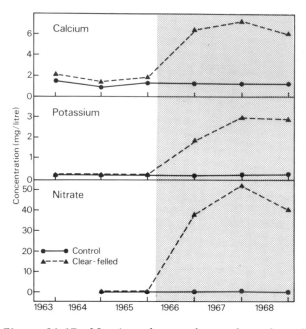

Figure 31.17 Nutrient losses from cleared and undisturbed forests at Hubbard Brook. The shaded area represents the period when the cleared forest was devegetated (from Bormann and Likens, 1979)

forested catchment. The results clearly showed the increased losses by leaching following clear-felling (Figure 31.17).

These losses are a product of several related processes. Removal of the trees resulted in less rainfall interception and transpiration, greater inputs of water to the soil and increased leaching. The decomposition of the organic debris left after clear-felling released new nutrients into the soil. Erosion also increased, so that sedimentation occurred in the rivers. Indeed, as can be seen, a complex series of responses was initiated (Figure 31.18). This had implications not just for the immediate area but also for streams and reservoirs in neighbouring localities.

Man's impact on aquatic ecosystems

Pollution of aquatic ecosystems

As this last example shows, man's use of the land often has significant effects upon aquatic ecosystems. This is true not only of his agricultural and silvicultural activities, but also of many of his industrial processes. Waste products tend to be washed into streams and to become concentrated within lakes and aquatic organisms.

The pollutants which affect these ecosystems are therefore varied. Possibly the most important in large areas of the western world are the pesticides related to agriculture. Nitrogen and phosphorus from fertilizers are also significant, however, while a wide range of industrial materials, including heavy metals such as cadmium and zinc, and radio-active substances may accumulate in aquatic systems.

The impact that these pollutants have is magnified by the distinctive structure and operation of many aquatic ecosystems. Bottom-dwelling plants may be in close contact with the pollutants which accumulate in the basal muds of the river or lake, and they absorb large quantities of these compounds. Fish swimming through the waters pass large quantities of water through their bodies, and in the process absorb many of the compounds in solution. In addition, the relationship between predator and prey, with each predator needing to eat a large number of prey, means that at each step up the food chain a considerable concentration of the pollutants tends to occur (Figure 31.19). Even greater concentrations occur in the seabirds which feed off the larger aquatic animals.

Effects of pollutants on lacustrine ecosystems

These effects tend to be most marked in lacustrine environments, for these act as sinks for many of the substances carried into the rivers; the lack of mixing and of dilution results in high concentrations of pollutants and serious impacts upon the biota.

One of the most dramatic examples of this process is shown by the Great Lakes, particularly Lake Erie and Lake Ontario. Industrial pollutants from the urban complexes along the lake shore, agricultural pollutants from the surrounding farmland and sewage waste have combined to alter both the chemical and biological composition of the lakes. The concentrations of calcium, sulphate, chloride, sodium and potassium have all increased markedly (Figure 31.20). In Lake Ontario these changes are related to the growth of cities such as Toronto, Hamilton and Rochester and to the intensification of industry in the upper Niagara River. Developments in the Detroit area have probably affected Lake Erie.

The chemical changes have been reflected by significant modifications to the fauna of the lakes. The populations of important fish species in Lake

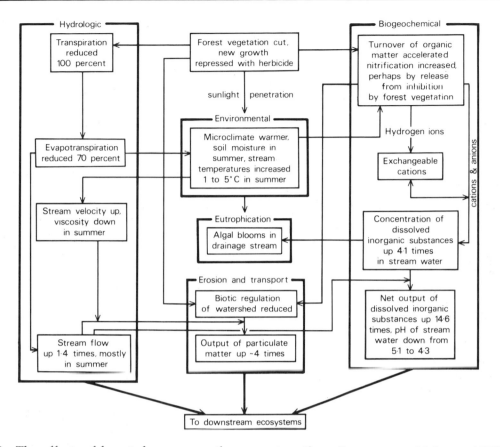

Figure 31.18   The effects of forest clearance on the ecosystem (from Bormann and Likens, 1979)

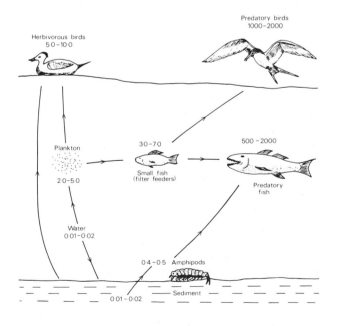

Figure 31.19   DDT concentration (parts per million) along a food-chain: e.g. Clear Lake, California

Erie, such as the lake herring, walleye, blue pike and whitefish, have all been drastically reduced. Commercial catches of the lake herring, for example, were over 20,000 t y⁻¹ prior to 1925; by 1925 they had fallen to 7600 t and in 1962 the catch was only 32 t. Catches of the blue pike show a similar trend. Early in the century they averaged about 6800 t y⁻¹; in 1962 they were only 0.45 t. Not all species have declined, of course. Some, such as the sea lamprey, have found conditions more favourable and have increased markedly in numbers; lack of competition for food and breeding grounds and increased supplies of nutrients in the inflowing river waters have probably contributed to these changes.

It is also apparent that the effects of chemical pollution in these lacustrine ecosystems are growing over time. Monitoring of nitrogen in the river waters of the Stour in England have shown the gradual rise in pollution levels (Figure 31.21), and where such waters enter enclosed lake basins the effect is clearly important. The Norfolk Broads, a

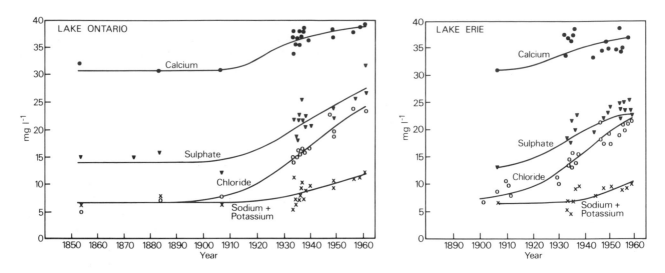

Figure 31.20 Changes in concentrations of selected nutrient elements in Lakes Ontario and Erie (after Beeton, 1971)

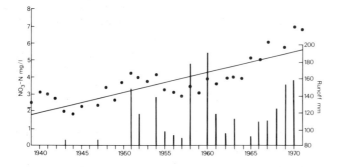

Figure 31.21 Nitrate levels in the River Stour, Essex, 1939–71 (after Edwards, and Thornes, 1973)

popular tourist area in eastern England, have also been affected. Phosphates are a major problem here, and they are leading to increased eutrophication of the waters and the destruction of the wildlife. Both fertilizers and household wastes probably contribute to pollution in this case.

Acid rain

More recently, the effects of another source of pollution have been recognized: acid rain. Pollutants given off by combustion of fossil fuels and by various industrial processes break down in the air and combine with water in the atmosphere to produce acids. Sulphur dioxide (SO_2), for example, is released in large quantities from power stations, heavy industry and vehicles. New York alone produces an estimated 2 million tonnes of SO_2 each year, while 6 million tonnes are generated in Britain. Since 1950 the quantities have risen rapidly (Figure 31.22).

In the atmosphere, the sulphur dioxide is gradually broken down, and within a period of about 43 days is converted to SO_3. During this time, however, it may be transported considerable distances, especially under conditions of limited turbulence. Thus, eventual deposition of the pollutants may occur many hundreds, or even several thousands, of kilometres from their source. It is claimed, for example, that Scotland suffers from acid rain effects derived from the USA, while sulphur dioxide produced in Britain is said to be affecting large areas of Scandinavia and West Germany.

Deposition occurs when the SO_3 is dissolved in rainwater to form a dilute solution of sulphuric acid. The effects are wide ranging. The acid may accumulate on plant leaves and inhibit photosynthesis, it may accumulate in the soil and reduce the pH, or it may wash into streams and lakes and acidify the waters. In Germany, for example, there is evidence to suggest that acid rain is inhibiting the regeneration of coniferous forest, lowering the soil pH and encouraging the leaching of soil nutrients such as calcium and aluminium. In Scandinavia, many streams and lakes have apparently been affected. A survey of 21 lakes in Sweden between 1966 and 1979 showed that all had experienced a fall in pH of between 0.9 and 2.2 units, while sulphate

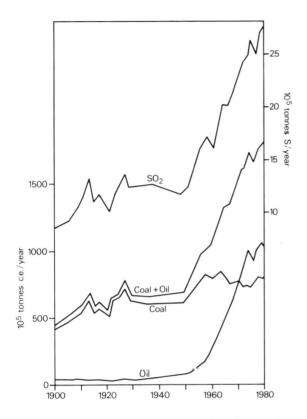

Figure 31.22   The consumption of coal and oil and estimated emissions of sulphur dioxide in Europe, 1900–80 (from Semb, 1978)

levels had increased by 6–13 mg l^{-1}. The effects may not stop there, however, for the increased acidity helps to mobilize iron and aluminium and heavy metals such as cadmium, zinc and copper. These may then build up in the waters to toxic levels, seriously affecting fish populations. One study in southern Norway suggested that 1771 lakes had totally lost their fish populations as a result of acid rain, while a further 941 had suffered a major decrease in fish numbers.

In reality, it is still not certain that all these effects are due to acid rain. Other factors, such as the natural ageing of lakes, soil acidification by fertilizers and conifers, and marginal climatic changes may also be significant. None the less, the problem is being taken seriously enough for it to be a major political as well as research topic and much government money is now being devoted to investigating the causes of acid rain. Over the next few years we can expect it to remain an issue of considerable importance.

Conclusions

Man is, of course, a part of his own ecosystem. He is also a much more adept and omniscient member of the ecosystem than any other organism. Consequently, he has tended to act as a control upon ecosystem processes, manipulating them for his own good. In the process he has deliberately created new, artificial ecosystems, which are maintained only by his own actions. He has also accidentally affected many natural ecosystems. Sometimes, these effects have serious repercussions for the ecosystems concerned, and sometimes man himself suffers.

The problems of not understanding the ecosystems he tries to create or manage are shown by the fateful ground nut scheme of the late 1940s. A project was launched to develop the agricultural economy of a part of east Africa by growing ground nuts in the savanna. Lack of local geographical and ecological knowledge led to both poor crop yield and inevitable damage to the ecosystem (inevitable to us with our knowledge of hindsight, that is). The scheme failed.

Unfortunately, the effect of man's activities are not always so clear, nor so immediate. The consequences may develop slowly and incrementally; man himself may respond to the first, subtle consequence in a totally inappropriate way, with the result that he ultimately increases the problem. Thus, when faced with soil exhaustion and low yields, the reaction of the farmer may be to counteract these effects by intensifying his agriculture even further; the consequence is likely to be even more rapid soil exhaustion.

Man's effects upon ecosystems have thus been felt throughout the world. Indeed, in many areas it is inappropriate to talk of natural ecosystems. Clearly, the ability of man to understand and manage these ecosystems effectively is critical to his future survival on this planet.

Appendix I
Köppen–Geiger system of climatic classification

Köppen's attempts at classifying climates into recognizable groups began in 1884, but his first classification system to achieve popularity was not published until 1918. Several modifications were made to the 1918 methods by Köppen and subsequently by his co-workers Geiger and Pohl. Unfortunately this has led to some confusion as climatologists have used each modification, claiming them to be based on Köppen without always specifying which classification.

The version presented here was published in 1953 following revision by Geiger and Pohl.

The five *major* climate groups are identified by capital letters as follows:

A   *Tropical rainy climates*. The average monthly temperature always exceeds 18°C. There is no winter season. Annual rainfall is large, exceeding annual evaporation to give a water surplus.

B   *Dry climates*. On average, potential evaporation always exceeds precipitation throughout the year. Hence there is a net water deficit.

C   *Warm temperate climates*. The mean temperature of the coldest month lies between 18°C and −3°C. At least one month has a mean temperature above 10°C. Seasonal differences from winter to summer are clear.

D   *Cold, boreal forest climates*. The mean temperature of the coldest month is below −3°C, but the mean temperature of the warmest month is above 10°C. This isotherm was used as it appears to coincide approximately with the poleward limit of forest growth.

E   *Polar climates*. Mean temperature of the warmest month is below 10°C, so even summers are cool.

Four of the groups, A, C, D and E, are defined in terms of temperature; the fifth, B, is defined by the deficiency of precipitation in relation to potential evaporation.

Sub-groups of the major types are identified by a second letter:

-f   Moist climates have adequate precipitation throughout the year for vegetative growth. It is used in types A, C and D.

-w   Winter is the main dry season.

-s   Summer is the main dry season.

-m   Used only with A types, it indicates rain forest climate with a brief dry season in a monsoon type of precipitation cycle.

Two capital letters, S and W, refer only to B climates.

BS   Steppe or semi-arid climate with mean annual precipitation lying between 380 mm (15 in) and 760 mm (30 in) depending upon mean annual temperature.

BW   Desert or arid climate. Mean annual precipitation is less than 250 mm (10 in).

E climates are separated into:

ET   Tundra climate where the mean temperature of the warmest month is between 0°C and 10°C.

EF   Ice climates where mean monthly temperatures are all below 0°C. Unless the areas are arid, these areas are normally ice-covered.

To denote further details of climatic characteristics, Köppen added a third letter. It is used to identify particular temperature variations.

-a   Hot summers in which the warmest month has

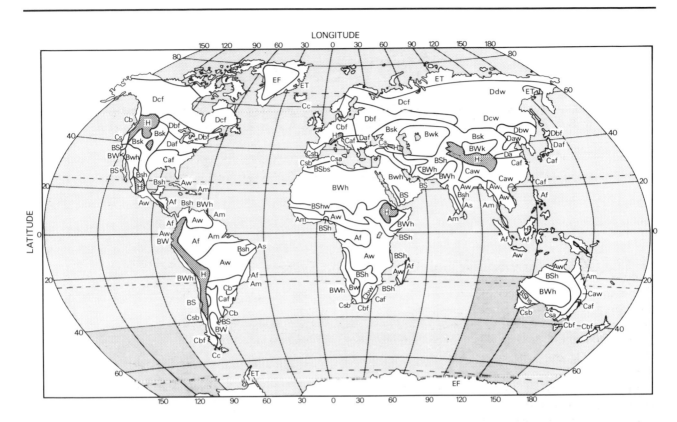

Figure A1.1 Climates of the continents according to the Köppen Classification

a mean temperature above 22°C. Used in C and D climates.

-b Warm summers in which the warmest month has a mean temperature below 22°C. Used in C and D climates.

-c Cool, short summers with fewer than four months having a mean temperature above 10°C (50°F). Used in C and D climates.

-d Very cold winters in which the mean temperature of the coldest month is below −38°C. Used in D climate only.

-h Dry and hot where the mean annual temperature is above 18°C. Used in B climate only.

-k Dry and cool where the mean annual temperature is below 18°C. Used in B climates only.

H Highlands

The world map of climates using this system is shown in Figure A1.1.

Obtain average climatic data for your own area, and confirm into which climatic type your area falls. If annual information is available as well take a sample of thirty years and see if it fell within the same type each year or whether it changed.

Appendix II
Soil classification systems

1 The US Comprehensive Soil Classification System

Soil orders

Name	Formative element	Meaning	Description
Entisols	*ent*	meaningless	Soils lacking horizons.
Inceptisols	*inceptum* (L)	beginning	Soils with weakly developed horizons containing weatherable materials.
Histosols	*histos* (Gr)	tissue	Soils with thick surface organic horizon.
Oxisols	*oxide* (Fr)	oxide	Very old, intensely weathered soils, with an oxic horizon and low cation exchange capacity.
Alfisol	*alf*	meaningless	Soils of humid and subhumid climates with high base status and a clay-enriched B horizon.
Ultisols	*ultimus*	last	Soils of warm climates with a low base status and a clay-enriched horizon.
Vertisols	*verto* (L)	turn	Soils of tropical and subtropical zones, with high clay content, cracking when dry.
Spodosols	*spodos* (Gr)	wood ash	Soils with a bleached, illuviated horizon with low cation exchange capacity and lacking carbonates.
Mollisols	*mollis* (L)	soft	Mid-latitude soils, with a thick, dark coloured surface horizon of high organic matter content and base status.
Aridisols	*aridus* (L)	dry	Soils of dry climates with accumulations of carbonates or soluble salts.

Soil suborders

Name	Formative element	Meaning	Description
Aquents	*aqua* (L)	water	ENTISOLS of wet places; permanently saturated.
Arents	*arare* (L)	to plough	ENTISOLS mixed by ploughing.
Fluvents	*fluvius* (L)	river	ENTISOLS formed from recent alluvium.

Orthents	*orthos* (Gr)	true	ENTISOLS on recent erosional surfaces; shallow; stony, loamy or clayey.
Psamments	*psammos* (Gr)	sand	ENTISOLS with sandy texture throughout; usually on dune or beach sand.
Andepts	*andosol*	soils of volcanic ash	INCEPTISOLS with a dark surface horizon, chiefly derived from recent volcanic ash.
Aquepts	*aqua*	as above	INCEPTISOLS of wet places; seasonally saturated; dark organic topsoil.
Ochrepts	*ochros* (Gr)	pale	INCEPTISOLS with a pale brown topsoil of low organic matter content.
Plaggepts	*plaggen* (G)	turf	INCEPTISOLS with man-made topsoil.
Tropepts	—	—	INCEPTISOLS of low latitudes with brownish or reddish topsoil.
Umbrepts	*umbra* (L)	shade	INCEPTISOLS with dark, acid topsoils, in moist mid-latitude areas.
Fibrists	*fibra* (L)	fibre	HISTOSOLS composed mainly of *Sphagnum* moss; permanently saturated or artificially drained.
Folists	*folia* (L)	leaf	HISTOSOLS composed of forest litter on bedrock, freely drained.
Hemists	*hemi* (Gr)	half	HISTOSOLS which are almost permanently saturated and contain partially decomposed plant debris.
Saprists	*sapros* (Gr)	rotten	HISTOSOLS saturated with water most of the year or artificially drained; containing decomposed plant debris.
Aquox	*aqua*	as above	OXISOLS of wet places; seasonally saturated; mottled.
Humox	*humus* (Gr)	earth	OXISOLS of cool moist regions with high organic carbon contents.
Orthox	*orthos*	as above	OXISOLS of warm, humid regions with short dry seasons; moist most of the year.
Torrox	*torridus* (L)	hot, dry	OXISOLS of arid areas, dry for more than six months, lacking organic matter.
Ustox	*ustus* (L)	burnt	OXISOLS with marked seasonal water regime.
Aqualfs	*aqua*	as above	ALFISOLS of wet places; grey; seasonally saturated.
Boralfs	*boreas* (Gr)	northern	ALFISOLS of boreal forests and high mountains; grey topsoils.
Udalfs	*udus* (L)	humid	ALFISOLS in areas with short dry season; formed under deciduous forest.
Ustalfs	*ustus*	as above	ALFISOLS in areas with marked seasonal moisture regime; brownish or reddish throughout.
Xeralfs	*xeros* (Gr)	dry	ALFISOLS of Mediterranean areas; brownish-reddish throughout.

Aquults	*aq*ua	as above	ULTISOLS of wet places; seasonally saturated; dominantly grey throughout.
Humults	*hum*us	as above	ULTISOLS with organic-rich topsoils; in mid and low latitudes.
Udults	*ud*us	as above	ULTISOLS with marked, short dry season; reddish or yellowish B horizons.
Ustults	*ust*us	as above	ULTISOLS with marked seasonal moisture regime; brownish to reddish throughout.
Xerults	*xer*os	as above	ULTISOLS of Mediterranean areas; brownish to reddish throughout.
Torrerts	*torr*idus	as above	VERTISOLS of dry areas; cracks remain open through year.
Uderts	*ud*us	as above	VERTISOLS with short dry seasons; cracks are open only briefly.
Usterts	*ust*us	as above	VERTISOLS with marked seasonal water regime; cracks remain open for long periods.
Xererts	*xer*os	as above	VERTISOLS of Mediterranean areas; cracks close in winter.
Aquods	*aq*ua	as above	SPODOSOLS of wet places; seasonally saturated; topsoil lacks free iron; B horizon often cemented.
Ferrods	*ferr*um (L)	iron	Freely drained SPODOSOLS with iron-enriched B horizons.
Humods	*hum*us	as above	SPODOSOLS in which the B horizon is enriched with organic matter and aluminium but not iron.
Orthods	*orth*os	as above	Freely drained SPODOSOLS with B horizons enriched with organic matter, iron and aluminium.
Albolls	*alb*us (L)	white	MOLLISOLS with bleached horizon over slowly permeable clay; seasonally saturated.
Aquolls	*aq*ua	as above	MOLLISOLS of wet places; black organic-rich topsoil on mottled B horizon; seasonally saturated.
Borolls	*bor*eas	as above	MOLLISOLS of cold-winter arid plains (steppe) or high mountains.
Rendolls	*rend*zina	from rendzina	MOLLISOLS on calcareous parent material; high carbonate contents in subsoil; formed in moist climates under forest.
Udolls	*ud*us	as above	MOLLISOLS with short dry season; no carbonate accumulation; brownish throughout.
Ustolls	*ust*us	as above	MOLLISOLS with marked seasonal moisture regime; marked carbonate accumulation in subsoil.
Xerolls	*xer*os	as above	MOLLISOLS of Mediterranean areas.
Argids	*arg*illa	white clay	ARIDISOLS with clay enriched horizon.
Orthids	*orth*os	as above	ARIDISOLS with no clay enrichment but evidence of horizonation to at least 25 cm.

2 Soil classification in England and Wales

Major soil group	Soil group	Description
1 Terrestrial raw soils	*a* Raw sands	Poorly developed soils with a non-alluvial sandy topsoil; normally formed on dunes.
	b Raw alluvial soils	Poorly developed soils with alluvial topsoils; normally as sand or gravel banks along stream channels or beaches.
	c Raw skeletal soils	Poorly developed soils over bedrock (e.g. screes).
	d Raw earths	Poorly developed soils in non-alluvial loamy or clayey material; often on Pleistocene sediments.
	e Man-made raw soils	Poorly developed soils on rock waste (e.g. mine or quarry spoil).
2 Raw gley soils	*a* Raw sandy gley soils	Soils with poorly developed sandy topsoils, normally on intertidal sand flats.
	b Unripened gley soils	Gleyed soils with an unweathered (unripened) mineral soil within 20 cm of the surface.
3 Lithomorphic soils	*a* Rankers	Non-calcareous soils with a shallow organic topsoil over bedrock.
	b Sand rankers	Non-calcareous soils with a shallow organic topsoil over sandy non-alluvial deposits.
	c Ranker-like alluvial soils	Non-calcareous soils with a shallow organic topsoil over recent alluvium.
	d Rendzina	Shallow, organic rich topsoils over calcareous bedrock.
	e Pararendzinas	Shallow, calcareous, organic rich topsoils over non-carbonate bedrocks (e.g. calcareous mudstone or sandstone).
	f Sand-pararendzinas	Shallow, organic rich topsoils over calcareous sand (e.g. dune sand).
	g Rendzina-like alluvial soils	Shallow, organic rich topsoils over calcareous alluvium, lake marl or tufa.
4 Pelosols	*a* Calcareous pelosols	Slowly permeable clayey soils with calcareous subsoil.
	b Non-calcareous pelosols	Slowly permeable clayey soils with a non-calcareous subsoil; normally developed in shales and mudstones.
	c Argillic pelosols	Slowly permeable clayey soils with a clayey subsoil, often slightly calcareous; normally developed over calcareous mudstone.
5 Brown soils	*a* Brown calcareous earths	Brownish-reddish soils with a weathered subsoil, formed above calcareous bedrock.

	b Brown calcareous sands	Brownish-reddish soils with a weathered subsoil formed in sandy calcareous deposits other than recent alluvium.
	c Brown calcareous alluvial soils	Brownish-reddish soils with a weathered subsoil over calcareous recent alluvium.
	d Brown earths	Brownish-reddish soils with a weathered subsoil over non-calcareous bedrock; normally loamy.
	e Brown sands	Brownish-reddish soils with at least half the upper 80 cm sandy; non-calcareous; mainly formed on Pleistocene deposits.
	f Brown alluvial soils	Brownish-reddish soils with a non-calcareous weathered subsoil over loamy or clayey recent alluvium.
	g Argillic brown earths	Brownish-reddish soils with a clay-enriched subsoil; usually loamy throughout; often formed in Pleistocene till, head or loess.
	h Palaeo-argillic brown earths	Brownish-reddish soils with a clay enriched subsoil which was formed in the Ipswichian Interglacial.
6 Podzolic soils	*a* Brown podzolic soils	Soils with an eluviated topsoil horizon over a sesquioxide enriched (Bs) horizon.
	b Humic cryptopodzols	Soils with an organic rich or peaty topsoil and a subsoil horizon enriched with illuviated iron or sesquioxides, but with high organic matter contents throughout.
	c Podzols	Freely drained soils with a bleached topsoil (E) horizon, a marked humus-enriched (Bh) horizon and often an organic-rich surface layer.
	d Gley-podzols	Podzols (as above) with a gleyed subsoil horizon due to impeded drainage; normally with an organic-rich surface layer.
	e Stagnopodzols	Gleyed podzols (as above) with either a thin ironpan or with a gleyed bleached (E) horizon; common in uplands.
7 Surface-water gley soils	*a* Stagnogley soils	Poorly drained soils with impermeable subsoil horizons, and a distinct topsoil containing relatively little organic matter (i.e. non-humose); subsoils normally clayey.
	b Stagnohumic gley soils	Stagnogley soils as above with an organic-rich (humose or peaty) topsoil.
8 Ground-water gley soils	*a* Alluvial gley soils	Permeable soils developed above impermeable alluvium and with gleyed subsoil horizons.
	b Sandy gley soils	Permeable, sandy soils developed from aeolian or glacio-fluvial sands overlying impermeable strata; with gleyed, poorly structured subsoil horizons.

	c Cambic gley soils	Permeable soils with a distinct topsoil and gleyed subsoil (B) horizons, formed in Pleistocene or older deposits; loamy or sandy.
	d Argillic gley soils	Permeable soils with a clay-enriched subsoil horizon overlying permeable parent materials; gleying in subsoil horizons is due to a seasonally high regional water table; formed in drier lowlands.
	e Humic-alluvial gley soils	Alluvial gley soils (cf. 8a) with a peaty or organic rich (humose) topsoil, developed in clayey alluvium.
	f Humic-sandy gley soils	Sandy gley soils (cf. 8b) with an organic-rich (humose or peaty) topsoil.
	g Humic gley soils	Permeable soils with a humose or peaty topsoil formed in low-lying depressions on Pleistocene or older deposits, giving high regional water tables and gleyed subsoil horizons.
9 Man-made soils	*a* Man-made humus soils	Soils with a thick man-made topsoil, often due to spreading of manure and wastes.
	b Disturbed soils	Mineral soils consisting of artificially displaced materials; e.g. on restored opencast mining or quarried land.
10 Peat soils	*a* Raw peat soils	Organic soils lacking an earthy (decomposed and mineral-enriched) topsoil; formed in undrained bogs and active fens.
	b Earthy peat soils	Organic soils with an earthy topsoil; formed in drained wetlands and in fens with well-aerated surface layers.

Further reading

Chapter 1 Physical geography: a starting point

Chorley, R. J. and Kennedy, B. A. (1971), *Physical geography: a systems approach*, Prentice-Hall. An advanced and in some ways difficult text, but useful both for its methodical treatment of the theory of systems, and for the examples of systems in physical geography.

Chorley, R. J. and Haggett, P. (1969), *Physical and information models in geography*, Methuen University Paperbacks. Another advanced text which is not always easy going for the uninitiated reader, but it provides good examples of the use of models to understand systems in physical geography.

Chapter 2 The global energy system

Neiburger, M., Edinger, J. G. and Bonner, W. D. (1982), *Understanding our atmospheric environment*, San Francisco: W. H. Freeman (chs. 3 and 4). A good introductory textbook on meteorology which provides a clear account of the principles of radiation and energy.

Lockwood, J. G. (1979), *Causes of climate*, Edward Arnold (ch. 1). Chapter 1 gives a useful survey of the climatic system and the role of energy. The remainder of the book surveys the controls of climate and how they may control changes in climate. Intended for readers with only a limited knowledge of physics.

Chapter 3 The global atmospheric system

Chorley, R. J. and Kennedy, B. A. (1971), *Physical Geography: a systems approach*, Prentice-Hall. Few books deal explicitly with the atmospheric system. Although this is an advanced text it does provide useful examples of different scales of operation within the atmospheric system.

Chapter 4 Energy balance of the atmosphere

Barry, R. G. and Chorley, R. J. (1982), *Atmosphere, weather and climate*, Methuen (ch. 1). A popular book, now in its fourth edition, which covers the whole field of climatology in considerable detail. Chapter 1 is not always easy but provides good coverage on atmospheric composition and energy.

Lamb, H. H. (1972), *Climate: present, past and future Volume 1: Fundamentals and Climate now*, Methuen (ch. 2). An encyclopedic work but written in a clear style. Chapter 2 has a strong emphasis on radiation input aspects; the units are calories per square centimetre.

Lockwood, J. G. (1979), *Causes of Climate*, Arnold (ch. 2). Brief and a little advanced on certain aspects of radiation but useful.

Chapter 5 Heat and moisture in the atmosphere

Crowe, P. R. (1971), *Concepts in climatology*, Longman (chs. 2 and 3). The author has taken an unusual approach to some aspects of climatology to produce a distinctive book. Parts are now a little dated but ch. 2 is an interesting model for relating radiation and temperature.

Mason, B. J. (1975), *Clouds, rain and rain-making*, Cambridge University Press. A combination of elementary and advanced ideas about rain formation. Only for those who want to go into more detail especially about the microphysics of clouds and precipitation.

Schaefer, V. J. and Day, J. A. (1981), *A field guide to the atmosphere*, Boston: Houghton Mifflin. A very interesting book which stresses the visual approach to the atmosphere, hence the emphasis on clouds. Gives a nice balance between description and explanation.

Chapter 6   The atmosphere in motion

Bowen, D. Q., Atkinson, B. W., Davies, B. E. and Simmons, I. G. (1972), *A concise physical geography*, Hulton Educational Publications (chs. 3 and 5). An introductory text which gives a good framework for the explanation of motion in the atmosphere.

Neiburger, M., Edinger, J. G. and Bonner, W. D. (1982), *Understanding out atmospheric environment*, San Francisco: W.H. Freeman (chs. 7, 8 and 9). Clear introduction to the causes of atmospheric motion at all scales.

Lutgens, F. K. and Tarbuck, E. J. (1982), *The atmosphere*, Englewood Cliffs: Prentice-Hall (chs. 6 and 7). A modern, up-to-date presentation of the basic principles of air motion. Elementary level requiring no mathematics.

Chapter 7   Weather-forming systems

Chang, J.-H. (1972), *Atmospheric circulation systems and climates*, Honolulu: Oriental Publishing Co. A comprehensive book concentrating on the relationships between climates and the atmospheric circulation systems which produce them. Largely descriptive but requires some understanding of climatology.

Barry, R. G. and Chorley, R. J. (1982), *Atmosphere, weather and climate*, Methuen (ch. 4). The main emphasis of this chapter is on the weather forming systems. Basically introductory level but goes into detail on some features.

Riehl, H. (1978), *Introduction to the atmosphere*, Tokyo: McGraw-Hill Kogakusha (chs. 7 and 8). Clear account at an elementary level of the weather systems in tropical and temperate latitudes.

Chapter 8   Climates of the world

Boucher, K. (1975), *Global Climate*, English Universities Press. After a brief survey of atmospheric processes this book attempts to explain the global distribution of climates. Relatively elementary level. An unusual feature is that climatic values are given for a specific year as well as their means.

Critchfield, H. J. (1983), *General Climatology*, Englewood Cliffs: Prentice-Hall (chs. 7, 8 and 9). A rather dated approach but still provides a useful survey of global climates.

Lockwood, J. G. (1974), *World climatology. An environmental approach*, Edward Arnold (chs. 3 to 11). Despite the title, a traditional text on climatology with a stress on the interaction between circulation systems and climate. Intermediate level.

Chapter 9   Micro and local climates

Oke, T. R. (1978), *Boundary-layer climates*, Methuen. An intermediate to advanced level book demonstrating the significance of the ground surface in determining microclimates. Very clearly presented but still needs careful reading.

Smith, K. (1975), *Principles of applied climatology*, McGraw-Hill (UK) Ltd (chs. 2 and 3). Introductory and descriptive but it provides a useful reference list of examples of microscale climates, both natural and man-made.

Hanwell, J. D. and Newson, M. D. (1973), *Techniques in Physical Geography*, Macmillan (chs. 2 and 3). An elementary and practical book demonstrating techniques in meteorology and local-scale climatology.

Chapter 10   Climatic change

Lamb, H. H. (1977), *Climate: present, past and future; Volume 2, Climatic history and the future*, Methuen. Dating a little now but nevertheless a very useful work for any references concerning climatic history, both evidence and examples. Intermediate to advanced level.

Goudie, A. (1983), *Environmental change*, Clarendon Press. Elementary to intermediate text which includes sections on the causes of climatic change as well as a variety of examples through time and across the globe. Especially useful for tropical areas.

Imbrie, J. and Imbrie, K. P. (1979), *Ice ages: solving the mystery*, Macmillan. A history of the ideas about the origins of recent ice ages. Provides a clear account of the Milankovitch effect though gives the impression that this must be the dominant mechanism and other possibilities are subordinate.

Chapter 11 The global water balance

Chorley, R. J. (1969), *Water, earth and man*, Methuen (chs. 1.1 and 1.2). Few recent texts on this subject where the general principles are understood but the precise values of the components of the water balance are still open to debate. This book provides an adequate description of the water balance but the figures should be treated with caution.

Sellers, W. D. (1965), *Physical climatology*, Chicago: University of Chicago Press (ch. 7). More advanced approach to the subject and, with even older values than Chorley, even more caution is required over the figures.

Chapter 12 Precipitation

Chorley, R. J. (1969), *Water, earth and man*, Methuen (ch. 3.1). The examination of precipitation from a climatological and hydrological point of view. Rather brief but a useful survey.

Jackson, I. J. (1977), *Climate, water and agriculture in the tropics*, Longman (chs. 1 to 4). Primarily concerned with the tropics, rainfall is examined in terms of its origins, seasonality, variability and intensity. Intermediate level.

Chandler, T. J. and Gregory, S. (1976), *The climate of the British Isles*, Longman. The main aspects of the climate of Britain are given a chapter each. Precipitation is examined at all time scales from instantaneous rates to annual totals. Intermediate level but useful for examples on British precipitation.

Chapter 13 Evapotranspiration

Chorley, R. J. (1969), *Water, earth and man*, Methuen (ch. 4.1). A brief survey of the principles behind evapotranspiration together with methods of measurement and calculation. Introductory level.

Shaw, E. M. (1983), *Hydrology in practice*, Van Nostrand Reinhold (UK) Co. Ltd (chs. 4 and 11). A recent book giving a practical approach to the problems of measuring and calculating evapotranspiration. Intermediate level and requires some mathematical expertise.

Ward, R. C. (1975), *Principles of Hydrology*, McGraw-Hill (chs. 4 and 5). Intermediate level but non-mathematical. An extensive survey of measurement and calculation of evapotranspiration. Clearly written.

Chapter 14 Runoff and storage

Chorley, R. J. (1969), *Water, earth and man*, Methuen. Despite its date, still a useful book on the role of water in the landscape.

Knapp, B. J. (1979), *Elements of geographical hydrology*, George Allen & Unwin. One of very few texts to consider hydrology at an introductory level, so a useful starter; but watch out for the errors!

Ward, R. C. (1975), *Principles of hydrology*, McGraw-Hill. Fairly advanced but probably the best available book on hydrological processes.

Chapter 15 Oceans and their circulation

Harvey, J. G. (1976), *Atmosphere and oceans: our fluid environments*, Artemis Press (chs. 9 and 14). Elementary level text which demonstrates the interactions between atmosphere and ocean as well as examining the processes operating in both environments.

Pirie, R. G. (1973), *Oceanography: contemporary readings in ocean sciences*, New York: Oxford University Press (Part 3: Ocean currents, chemistry and weather). Chapters on specific subjects written by experts in the field. Intermediate to advanced level.

Turekian, K. K. (1968), *Oceans*, Englewood Cliffs, NJ: Prentice-Hall. Introductory text covering briefly the wide field of oceanography.

Chapter 16 Landscape form and process

Chorley, R. J. and Kennedy, B. A. (1971), *Physical geography: a systems approach*, Prentice Hall (chs. 2, 3 and 4). An advanced and in some ways difficult text, but good for the examples of geomorphological systems.

Bennison, G. M. and Wright, A. E. (1970), *The

geological history of the British Isles, Arnold. Rather dated in that it was published before the implications of plate tectonics were appreciated, but a useful fairly simple review of the stratigraphic history of Britain.

Chapter 17 The formation of rocks

Holmes, A. (1965), *Principles of physical geology*, Nelson. Very dated but still a useful reference, especially on aspects such as rock types and their formation.

Sparks, B. W. (1973), *Rocks and relief*, Longman. Again a classic text which has not really been replaced by more modern books; provides a good summary of the formation of rocks and the relationship between rock types and landscape in Britain.

Chapter 18 Earth-building

Decker, R. and Decker, B. (1982), *Volcanoes and the earth's interior*, Readings from *Scientific American*, San Francisco: Freeman. A well-illustrated and readily comprehensible review of volcanoes and their relationships with plate tectonics and earth structure.

Oxburgh, E. R. (1974), 'The plain man's guide to plate tectonics', *Proceedings of the Geologists Association* **85** no. 3, 299–357. For a brief and informative summary of plate tectonics it can't be bettered.

Wilson, J. T. (1976), *Continents adrift and continents aground*, Readings from *Scientific American*, San Francisco: Freeman. Beautifully illustrated compilation of many key references, including several of the articles referred to in this chapter.

Wyllie, P. J. (1976), *The way the earth works: an introduction to the new global geology and its revolutionary development*, New York: Wiley. A compellingly written account of the development of and evidence for theories of plate tectonics, complete with cartoons.

Chapter 19 Weathering

Embleton, C. E. and Thornes, J. B. (1979), *Process in geomorphology*, Arnold (ch. 4). Rather advanced, but a good summary of weathering processes.

Derbyshire, E. (1979), *Geomorphology and climate*, London: Wiley (chs. 2, 3, 13 and 14). Good chapters (2 and 3) on weathering processes, and (13 and 14) on effects on landscape in limestone and granite regions.

Ollier, C. (1984), *Weathering*, Longman. Considerably improved new edition of a traditional text, which nevertheless tends to over-emphasize the range of weathering processes rather than the common factors between them.

Trudgill, S. T. (1983), *Weathering and erosion*, Butterworth. An introductory book, outlining lots of techniques; ideal for project or fieldwork.

Clark, M. J. and Small, R. J. (1982), *Slopes and weathering*, Topics in Geography, Cambridge University Press. An introductory book, but ideal for background material.

Chapter 20 Erosion and deposition

Allen, J. R. L. (1970), *Physical processes of sedimentation*, George Allen & Unwin. An advanced text, explaining the principles of sediment transport and the characteristics of sediments in different depositional environments.

Briggs, D. J. (1976), *Sediments*, Butterworths. An introductory book, mainly considering techniques of sediment analysis, but also including useful discussion of the characteristics of sediments and sediment transport in different environments.

Statham, I. (1977), *Earth surface sediment transport*, Oxford University Press. An intermediate level text which concentrates mainly on slope processes but gives a useful general explanation of properties of materials and mechanisms of transport.

Trudgill, S. T. (1983), *Weathering and erosion*, Butterworth. Useful introduction to field techniques.

Whalley, W. B. (1976), *Properties of materials and geomorphological explanation*, Oxford University Press. A brief explanation of the properties of materials and the way in which they affect geomorphological processes; intermediate level except for those with a knowledge of physics.

Chapter 21 Hillslopes

Embleton, C. E. and Thornes, J. B. (1979), *Process in geomorphology*, Arnold (ch. 5). Detailed and informative summary of mass movement processes; intermediate.

Finlayson, B. and Statham, I. (1980), *Hillslope analysis*, Butterworth. A valuable introductory text, concentrating on techniques of hillslope analysis but also providing useful background material.

Statham, I. (1977), *Earth surface sediment transport*, Oxford University Press. Probably the best available intermediate-level text on hillslope processes.

Clark, M. J. and Small, R. J. (1982), *Slopes and weathering*, Topics in Geography, Cambridge University Press. An introductory text, but ideal for background reading.

Selby, M. J. (1982), *Hillslope materials and processes*, Oxford University Press.

Chapter 22 Streams

Gregory, K. J. and Walling, D. E. (1973), *Drainage basin form and process*, Arnold. As the title implies, concentrates especially on drainage basins, rather less emphasis than other texts on channel processes; but a good, intermediate level book.

Knighton, A. D. (1984), *Fluvial forms and processes*, Macmillan. Fairly advanced, but it provides a good, modern and largely European alternative to the classic Leopold, Wolman and Miller (see below).

Leopold, L. B., Wolman, M. G. and Miller, J. P. (1964), *Fluvial processes in geomorphology*, San Francisco: Freeman. Dated, and American in content, but still a definitive text by several of the fathers of modern fluvial geomorphology; advanced.

Lewin, J. (ed.) (1981), *British rivers*, George Allen & Unwin. A rather unbalanced collection of papers, but with lots of information, maps and diagrams describing characteristics of rivers in Britain.

Petts, G. E. (1983), *Rivers*, Butterworth. Useful introductory text, concentrating mainly on techniques of analysis but with much background material.

Richards, K. (1982), *Rivers. Form and process in alluvial channels*, Methuen. Advanced text, concerned mainly with channel processes; another alternative to Leopold, Wolman and Miller.

Chapter 23 Glacial and periglacial systems

Derbyshire, E., Gregory, K. J. and Hails, J. R. (1979), *Geomorphological processes*, Butterworth (ch. 5). A good summary of glacial and periglacial processes at an intermediate level.

Embleton, C. E. and King, C. A. M. (1975), *Glacial geomorphology*, Arnold.

Embleton, C. E. and King, C. A. M. (1975), *Periglacial geomorphology*, Arnold. Two advanced but useful texts providing a good overview of glacial and periglacial processes and landforms.

Sugden, D. E. and John, B. S. (1977), *Glaciers and landscape*, Arnold. By some distance the most useful intermediate text on glacial processes and landforms but little on periglacial environments.

Chapter 24 Aeolian systems

Bagnold, R. A. (1954), *The physics of blown sand and desert dunes*, Methuen. Very dated but still the 'bible' on this topic.

Cooke, R. U. and Warren, A. (1973), *Geomorphology in deserts*, Batsford. Detailed, advanced text.

Derbyshire, E., Gregory, K. J. and Hails, J. R. (1979), *Geomorphological processes*, Butterworth (ch. 4). A good, intermediate-level summary of aeolian processes and bedforms.

Embleton, C. E. and Thornes, J. B. (1979), *Process in geomorphology*, Arnold (ch. 10). Brief but useful chapter on aeolian processes, well-illustrated.

Chapter 25 Lacustrine, coastal and marine systems

Derbyshire, E., Gregory, K. J. and Hails, J. R. (1979), *Geomorphological processes*, Butterworth (ch. 3). Excellent summary of coastal processes.

Embleton, C. E. and Thornes, J. B. (1979), *Process in geomorphology*, Arnold (ch. 11). Good, detailed discussion of breaker type and relationships with beach characteristics.

Hutchinson, G. E. (1967), *A treatise on limnology* (vol. 1), New York: Wiley. Detailed, definitive text on lakes and lacustrine processes.

Pethick, J. (1984), *An introduction to coastal geomorphology*, Arnold. Excellent and extensive text on coastal processes and landforms; intermediate level.

Bird, E. C. F. (1984), *Coasts: an introduction to coastal geomorphology*, Blackwell. Third edition of a standard reference on coastal geomorphology; rather traditional in its approach but a good introductory text.

Chapter 26 Ecosystems

Gilbertson, D. D., Kent, M. and Pyatt, F. B. (1985), *Practical Ecology for Geography and Biology: Survey, mapping and data analysis*, Hutchinson. Useful introduction to techniques of ecological and bio-geographical studies, with lots of data and examples.

Smith, R. L. (1966), *Ecology and field biology*, New York: Harper & Row. A beautifully illustrated and readable book; strongly American in scope but with a clear ecological approach.

Whittaker, R. H. (1975), *Communities and ecosystems*, New York: Macmillan. Fairly advanced but very useful text, stressing the dynamics of ecosystems.

Chapter 27 Soil formation

Briggs, D. J. (1977), *Soils*, Butterworth. Basic text introducing selected techniques of soil analysis; ideal for projects and fieldwork.

Courtney, F. M. and Trudgill, S. T. (1984), *The soil. An introduction to soil study*, Arnold. Second edition of a basic text, useful for those who have not studied soils before.

Paton, T. R. (1978), *The formation of soil material*, George Allen & Unwin. An intermediate text looking at processes of soil formation and their relationship with soil forming factors.

Pitty, A. F. (1979), *Geography and soil properties*, Methuen. An intermediate text giving a good, systematic coverage of soil properties and the processes affecting them.

White, R. E. (1979), *Introduction to the principles and practice of soil science*, Blackwell. Useful intermediate text which stresses the applied aspects of soils.

Chapter 28 Development of the vegetation layer

Miles, J. (1978), *Vegetation dynamics*, Outline Studies in Ecology, Chapman & Hall. Excellent, brief introduction to processes of vegetation change; highly recommended.

Whittaker, R. H. (1975), *Communities and ecosystems*, New York: Macmillan (chs. 3 and 4). Detailed discussion of principles of ecosystem interactions and theories and processes of ecosystem change.

Chapter 29 Biogeochemical weathering

Trudgill, S. T. (1977), *Soil and vegetation systems*, Oxford University Press. A text which is explicitly aimed at demonstrating the links between the soil and vegetation through the examination of nutrient flows; advanced but useful.

Whittaker, R. H. (1975), *Communities and ecosystems*, New York: Macmillan (ch. 6). Good summary of the processes of nutrient cycling within ecosystems.

Chapter 30 World ecosystems

Collinson, A. S. (1977), *An introduction to world vegetation*, George Allen & Unwin. Good introductory-intermediate level summary of major world biomes, plus background on environmental relationships, nutrient cycling etc.

Bridges, W. M. (1970), *World soils*, Cambridge University Press. Basic, and in some ways over-simplistic, but a useful summary of the character and distribution of world soils; good photos.

Walter, H. (1973), *Vegetation of the earth in relation to climate and eco-physiological conditions*, English Universities Press. Despite the title a quite useful intermediate-advanced text on world vegetation.

Riley, D. and Young, A. (1983), *World vegetation*, Cambridge University Press. Simple, pictorial introduction to world vegetation types.

Chapter 31 Man and the ecosystem

Detwyler, T. R. (1971), *Man's impact on environment*, New York: McGraw Hill. Rather dated now but contains many of the classic examples of human impact on the environment; readable and comprehensive.

Goudie, A. S. (1981), *The human impact. Man's role in environmental change*, Blackwell. Good introductory text summarizing impacts on vegetation, animals, the soil, waters and landscape.

Simmons, I. (1979), *The ecology of natural resources*, Arnold.

Simmons, I. (1979), *Biogeography: natural and cultural*, Arnold. Two interesting and useful books which look at the ecological effects of man.

Bibliography

Arkley, R. J. (1963), 'Calculation of carbonate and water movement in soil from climatic data', *Soil Science*, **96**, 239–48

Atkinson, B. W. and Smithson, P. A. (1972), 'An investigation into meso-scale precipitation distributions in a warm sector depression', *Quart. Jl. Roy. Meteor. Soc.*, **98**, 353–68

Atkinson, B. W. and Smithson, P. A. (1976), 'Precipitation', in T. J. Chandler and S. Gregory (eds.), *The Climate of the British Isles*, Longman, 129–82

Atkinson, T. C. (1971), 'Hydrology and erosion in a limestone terrain' (Unpublished PhD thesis, University of Bristol)

Austin, P. M. and Houze, R. A. (1972), 'Analysis of the structure of precipitation patterns in New England', *J. Appl. Meteor.*, **11**, 926–35

Bagnold, R. A. (1954), *The physics of blown sand and desert dunes*, Methuen

Balme, E. O. (1953), 'Edaphic and vegetational zoning on the Carboniferous Limestone of the Derbyshire Dales', *J. Ecology*, **41**, 331–44

Barry, R. G. (1969), 'Evaporation and transpiration', in R. J. Chorley (ed.), *Water, Earth and Man*, Methuen, 169–84

Barry, R. G. (1979), 'Recent advances in climate theory based on simple climate models', *Progr. in Phys. Geography*, **3**, 119–31

Barry, R. G. and Chorley, R. J. (1982), *Atmosphere, Weather and Climate* (4th edn), Methuen

Beeton, A. M. (1971), 'Eutrophication of the St. Lawrence Great Lakes', in T. R. Detwyler (ed.), *Man's impact on the Environment*, New York: McGraw-Hill, 233–45

Bennett, H. H. (1939), *Soil Conservation*, New York: McGraw-Hill

Bennett, R. J. and Chorley, R. J. (1978), *Environmental Systems*, Methuen

Bonython, C. W. and Mason, B. (1953), 'The filling and drying of Lake Eyre', *Geogr. J.*, **119**, 321–30

Bormann, F. H. and Likens, G. E. (1979), *Pattern and process in a forested ecosystem*, New York: Springer-Verlag

Boucher, K. (1975), *Global Climate*, English Universities Press

Boulton, G. S., Jones, A. S., Clayton, K. M. and Kenning, M. J. (1977), 'A British ice-sheet model and pattern of glacial erosion and deposition in Britain', in F. W. Shotton (ed.), *British Quaternary Studies: recent advances*, Clarendon Press, 231–46

Brady, N. C. (1973), *Nature and properties of soils*, Macmillan

Bridgman, H. A. (1969), 'The radiation balance of the southern hemisphere', *Arch. Meteor., Geophys. u. Bioklim.*, Ser. B, **17**, 325–44

Briggs, D. J. (1977), *Soils*, Butterworth

Brown, L. (1976), *British birds of prey*, Collins

Bruce, J. P. and Clark, R. H. (1966), *Introduction to Hydrometeorology*, Pergamon

Budyko, M. I., Yefimova, N. A., Aubenok, L. I. and Strokhina, L. A. (1962), 'The heat balance of the surface of the earth', *Soviet Geogr.*, **3**, 3–16

Bullen, K. E. (1955), 'The interior of the earth', *Scientific Amer.*, report no. 804, San Francisco: Freeman

Burnham, C. P. (1970), 'The regional pattern of soil formation in Great Britain', *Scott. Geogr. Mag.*, **86**, 25–34

Carson, M. A. and Petley, D. J. (1970), 'The existence of threshold slopes in the denudation of the landscape', *Institut. Brit. Geogr., Trans.*, **49**, 71–95

Chandler, T. J. (1965), *The climate of London*, Hutchinson

Changnon, S. A. (1971), 'A note on hailstone size distributions', *J. Appl. Meteorol.*, **10**, 168–70

Changnon, S. A. (1976), 'Inadvertent weather modification', *Water Res. Bull.*, **12**, 695–718

Chapman, S. B. (1967), 'Nutrient budgets for a dry heath ecosystem in the south of England', *J. Ecol.*, **55**, 677–89

CLIMAP Project Members (1976), 'The surface of the ice age earth', *Science*, **191**, 1131–7

Collinson, A. S. (1977), *Introduction to world vegetation*, George Allen and Unwin

Cooke, G. W. (1983), *Fertilizing for maximum yield*, Granada

Coope, G. R. (1975), 'Climatic fluctuations in northwest Europe since the Last Interglacial, indicated by fossil assemblages of coleoptera', in A. E. Wright and F. Moseley (eds.), *Ice Ages: Ancient and Modern*, *Geological Journal*, spec. issue no. 6. Liverpool, 153–68

Critchfield, H. J. (1983), *General Climatology* (4th edn), Englewood Cliffs, NJ: Prentice-Hall Inc.

Crowe, P. R. (1951), 'Wind and weather in the equatorial zone', *Instit. Brit. Geogr., Trans.*, **17**, 23–76

Crowe, P. R. (1971), *Concepts in climatology*, Longman

Davenport, A. G. (1965), 'Relationship of wind structure to wind loading', *Proc. Conf. Wind Effects on Structures*, Symp. 16, vol. 1, HMSO, 53–102

Davies, E. B., Hogg, D. E. and Hopewell, H. G. (1962), 'Extent of return of nutrient elements by dairy cattle: possible leaching losses', Jt Meeting of International Soil Science Soc., Commission IV & V, New Zealand, 715–20

Decker, R. and Decker, B. (1982), 'Volcanoes and the earth's interior', *Readings from Scientific American*, San Francisco: Freeman

Department of Navigation and Ocean Development (1977), *Assessment and atlas of shoreline erosion along the Californian coast*, State of California Dept of Navigation and Ocean Development

Department of Environment (Water Data Unit) (1983), *Surface water: United Kingdom, 1977–1980*, HMSO

Derbyshire, E., Gregory, K. J. and Hails, J. R. (1979), *Geomorphological processes*, Butterworth

Dietz, R. S. and Holden, J. C. (1970), 'The break-up of Pangaea', *Scientific American*, **223**, 30–41

Duckham, A. N. and Masefield, G. B. (1970), *Farming systems of the world*, Chatto & Windus

Duvigneaud, P. and Denaeyer-De Smet, S. (1970), 'Biological cycling of minerals in temperate deciduous forests', in D. E. Reichle (ed.), *Analysis of temperate forest ecosystems*, Berlin: Springer-Verlag, 199–225

Eastwood, T. (1964), *Stanford's geological atlas of Great Britain*, Edward Stanford Ltd

Edwards, A. M. C. and Thornes, J. B. (1973), 'Annual cycle of river water quality: a time series approach', *Water Res. Research*, **9**, 1286–95

Edwards, C. A. (1969), 'Soil pollutants and soil animals', *Scientific American*, **220**, 88–99

Eichenlaub, V. L. (1970), 'Lake effect snowfall to the lee of the Great Lakes: its role in Michigan', *Bull. Amer. Meteorol. Soc.*, **51**, 403–12

Embleton, C. E. and Thornes, J. B. (1979), *Process in geomorphology*, Arnold

Ericson, D. B. and Wollin, G. (1968), 'Pleistocene climates and chronology in deep-sea sediments', *Science*, **162**, 1227–34

Evans, G. H. (1970), 'Pollen and diaton analysis of late-glacial Quaternary deposits in the Blelham Basin, North Lancashire', *New Phytologist*, **69**, 821–74

Eyre, S. R. (1968), *Vegetation and soils: a world picture*, Arnold

Fairbridge, R. W. (1961), *Eustatic changes in sea-level. Phys. & Chem of the Earth*, IV: New York: Pergamon, 99–185

Finlayson, B. and Statham, I. (1980), *Hillslopes*, Butterworth

Fleagle, R. G. and Businger, J. A. (1963), *An introduction to atmospheric physics*, Int. Geophysics Ser., vol. 5, New York: Academic Press

Flint, R. F. (1971), *Glacial and Quaternary geology*, New York: Wiley

Frissel, M. J. (ed.) (1978), 'Cycling of mineral nutrients in agricultural ecosystems', Proc. First Environmental Symp., Royal Netherlands Land Development Society, Amsterdam. 31 May–4 June 1976. Amsterdam: Elsevier. Reprinted from *Agroecosystems*, **4**, 1–354

Fritts, H. C., Smith, D. G. and Stokes, M. A. (1965), 'The biological model for paleoclimatic interpretation of Mesa Verde tree-ring series', *Amer. Antiquity*, **31**, pt 2, 101–21

Fuggle, R. G. and Oke, T. R. (1970), 'Infra-red flux divergence and the urban heat island', in *Urban Climates*, WMO technical note no. 108, Geneva: World Meteorological Organization, 70–8

Fuh, Baw-puh (1962), 'The influence of slope orientation on micro-climate', *Acta Meteorologica Sinica*, **32**, 71–86

Gabites, J. F. (1943), 'Flying conditions in the tropical South Pacific', New Zealand Meteorological Office, series A, no. 3

Gardiner, V. (1983), 'The relevance of geomorphology to studies of Quaternary morphogenesis', in D. J. Briggs and R. S. Waters (eds.), *Studies in Quaternary geomorphology*, Geo Books, 1–18

Gay, L. W., Knoerr, K. N. and Braaten, M. O. (1971), 'Solar radiation variability on the floor of a pine plantation', *Agricultural Meteorol.*, **8**, 39–50

Gerard, R. D. (1966), 'Salinity in the ocean', in R. W. Fairbridge (ed.), *Encyclopedia of Oceanography*, New York: Van Nostrand-Reinhold, 758–63

Gimingham, C. H. (1972), *Ecology of heathlands*, Chapman & Hall

Godwin, H. (1975), *The history of the British flora* (2nd edn), Cambridge University Press

Goldthwait, R. P. (1976), 'Frost sorted patterned ground: a review', *Quaternary Res.*, **6**, 27–35

Golley, F. (1978), 'Decomposition and biogeochemical cycles', in *Tropical forest ecosystems*, Paris: UNESCO, 270–85

Goltsberg, I. A. (1969), *Microclimate of the USSR*, Jerusalem: Israel Program for Scientific Translation

Goudie, A. S. (1983), *Environmental Change* (2nd edn), Oxford University Press

Goudie, A. S., Cooke, R. U. and Evans, I. S. (1970), 'Experimental investigation of rock weathering by salts', *Area*, **4**, 42–8

Gray, W. M. (1978), 'Hurricanes: their formation, structure and likely role in the tropical circulation', in D. B. Shaw (ed.), *Meteorology over the tropical oceans*, Royal Meteorological Society, 155–218

Gregory, K. J. and Walling, D. E. (1973), *Drainage basin form and process*, Arnold

Gregory, S. (1968), *Statistical methods and the geographer* (2nd edn), Longman

Griggs, D. T. (1936), 'The factor of fatigue in rock weathering', *J. Geology*, **44**, 781–96

Grindley, J. (1972), 'Estimation and mapping of evaporation', in *Symposium on World Water Balance*, publ. no. 92, IASH–UNESCO, 200–13

Harvey, J. (1976), *Atmosphere and ocean: our fluid environments*, Artemis

Hastenrath, S. and Lamb, P. J. (1977), *Climatic atlas of the Tropical Atlantic and Eastern Pacific Ocean*, Madison, Wisconsin: University of Wisconsin Press

Hastenrath, S. and Lamb, P. J. (1978), *Heat budget atlas of the Tropical Atlantic and Eastern Pacific Ocean*, Madison, Wisconsin: University of Wisconsin Press

Hebert, P. J. (1980), 'A "normal" year for hurricanes', *Weatherwise*, **33**, 26–30

Hebert, P. J. and Taylor, G. (1979), 'Hurricanes', *Weatherwise*, **32**, 60–7, 100–7

Hilder, E. J. (1964), 'The distribution of plant nutrients by sheep at pasture', *Proc. Australian Soc. of Animal Production*, **5**, 241–8

Hjulstrom, F. (1935), 'Studies of the morphological activity of rivers as illustrated by the River Fyris', *Bull. Geol. Instit., Uppsala*, **25**, 227–527

Holmes, A. (1965), *Principles of physical geology*, Nelson

Huff, F. A. and Shipp, W. L. (1969), 'Spatial correlations of storm, monthly and seasonal precipitation', *J. Appl. Meteorol.*, **8**, 542–50

Hurley, P. M. (1968), 'The confirmation of continental drift', *Scientific American*, **218**, 52–64

Huss, E. and Stranz, D. (1970), 'Die windverhältnisse am Bodensee', *Pure and Appl. Geophys.*, **81**, 323–56

Imbrie, J. and Imbrie, K. P. (1979), *Ice Ages: solving the mystery*, Macmillan

Ingmanson, D. E. and Wallace, W. J. (1973), *Oceanology: an introduction*, Belmont, California: Wadsworth

Israelsen, O. W. and Hansen, V. E. (1962), *Irrigation principles and practices* (3rd edn), New York: Wiley

Jackson, C. I. (1963), 'Some climatological grumbles', *Weather*, **18**, 278–82

Jackson, I. J. (1977), *Climate, water and agriculture in the tropics*, Longman

James, D. E. (1973), 'The evolution of the Andes', *Scientific American*, **229**, 60–9

Jarvis, R. A. (1968), *Soils of the Reading district. Memoir of the Soil Survey of Great Britain*, Harpenden: Rothampsted Experimental Station

Johnston, D. W. and Odum, E. P. (1956), 'Breeding bird populations in relation to plant succession on the piedmont of Georgia', *Ecology*, **37**, 50–62

Jones, P. D. and Kelly, P. M. (1983), 'The spatial and temporal characteristics of northern hemisphere surface air temperature variations', *J. Climatol.*, **3**, 243–52

Kellogg, W. W. and Schneider, S. H. (1974), 'Climate stabilization: for better or for worse', *Science*, **186**, 1163–72

King, C. A. M. (1962), *Oceanography for Geographers*, Arnold

Kirkby, A. (1969), 'Primitive irrigation', in R. J. Chorley (ed.), *Water, Earth and Man*, Methuen, 209–12

Kittredge, J. (1973), *Forest influences*, New York: Dover

Knighton, A. D. (1981), 'Longitudinal changes in the size and shape of stream bed material: evidence of variable transport conditions', *Catena*, **9**, 25–34

Knighton, A. D. (1984), *Fluvial forms and processes*, Arnold

Kononova, M. M. (1966), *Soil organic matter*, Pergamon

Koteswaram, P. and George, C. A. (1958), 'On the formation of monsoon depressions in the Bay of Bengal', *Indian Jl Meteorology and Geophysics*, **9**, 9–22

Kozarski, S. and Rotnicki, K. (1983), 'Changes of river channel patterns and the mechanism of valley floor construction in the north Polish plain during the Late Weichsel and Holocene', in D. J. Briggs and R. S. Waters (eds.), *Studies in Quaternary geomorphology*, Geo Books, 31–48

Lagrula, J. (1966), 'Hypsographic curve', in R. W. Fairbridge (ed.), *Encyclopedia of Oceanography*, New York: Van Nostrand Reinhold, 364–6

Lamb, H. H. (1977), *Climate: present, past and future*, vol. 2, *Climatic history and the future*, Methuen

Lamb, H. H. (1982), *Climate, history and the modern world*, Methuen

Lauenroth, W. K. (1979), 'Grassland primary production: North American grasslands in perspective', in N. French (ed.), *Perspectives in grassland ecology*, New York: Springer-Verlag, **32**, 3–24

Leopold, L. B. (1973), 'River channel change with time: an example', *Bull., Geol. Soc. America.*, **84**, 1845–60

Leopold, L. B. and Langbein, W. B. (1979), 'River meanders', in *The physics of everyday phenomena*, Readings from *Scientific American*, 28–35, New York: Freeman

Leopold, L. B., Wolman, M. G. and Miller, J. P. (1964), *Fluvial processes in geomorphology*, San Francisco: Freeman

Lieth, H. (1964–5), 'Versuch einer kartographischen Darstellung der Productivat der Pflanzendecke auf die Erde', *Geographisches Taschenbuch*, 72–80

Ludlam, F. H. (1961), 'The hailstorm', *Weather*, **16**, 152–62

Lundberg, J. (1977), 'Karren of the littoral zone, Burren District, Co. Clare, Ireland', in *Proceedings 7th Int. Speleological Congress*, Sheffield 1977, **2** Karst morphology, British Cave Research Association, 291–3

Lvovitch, M. I. (1972), 'World Water Balance (general report)', in *Symposium on World Water Balance*, publ. no. 92, IASH–UNESCO, 401–15

McGowan, A. and Derbyshire, E. (1977), 'Genetic influences on the properties of tills', *Quarterly Jl Engineering Geol.*, **10**, 391

Malkus, J. S. (1958), 'Tropical weather disturbances – why do so few become hurricanes?', *Weather*, **13**, 75–89

Marshall, J. S. and Palmer, W. McK. (1948), 'The distribution of raindrops with size', *J. Meteorol.*, **5**, 165–6

Mason, B. J. (1970), 'Future developments in meteorology: an outlook to the year 2000', *Quart. Jl Roy. Meteorol. Soc.*, **96**, 349–68

Mason, B. J. (1976), 'Towards the understanding and prediction of climatic variations', *Quart. Jl Roy. Meteorol. Soc.*, **102**, 473–98

Mather, J. R. (1974), *Climatology: fundamentals and applications*, New York: McGraw-Hill

Mather, K. B. and Miller G. S. (1967), Notes on topographic factors affecting the surface wind in Antarctica, with special reference to katabatic winds, Univ. Alaska, Tech. Rep. UAG–K–189

Mattingly, G. E. G., Chater, M. and Johnston, A. E. (1975), 'Experiments made on Stackyard Field, Woburn, 1867–1974. III Effects of NPK fertilizers and farmyard manure on soil carbon, nitrogen and organic phosphorus', Report Rothamsted Experimental Station for 1974, pt 2, 61–77

Meyer, L. D. and Wischmeier, W. H. (1969),

'Mathematical simulation of the process of soil erosion by water', *Trans. Amer. Soc. Agricult. Engnr.*, **12**, 754–8; 762

Miller, A. (1966), *Meteorology*, Columbus, Ohio: Merrill Physical Science Series

Miller, D. H. (1965), 'The heat and water budget of the earth's surface', *Advances in geophysics*, **11**, 175–302

Miller, D. H. (1977), *Water at the surface of the earth*, New York: Academic Press

Mitchell, J. M. (1968), Concluding remarks, in 'Causes of Climatic Change', Amer. Meteorological Soc., *Meteorol. Monog.*, **8** no. 30, 155–9

Moffitt, B. J. and Ratcliffe, R. A. S. (1972), 'Northern hemisphere monthly mean 500mb and 1000–500mb thickness charts and some derived statistics', *Geophysical Memoir*, 117

Monkhouse, F. J. (1975), *Principles of physical geography* (8th edn), Hodder & Stoughton.

More, R. J. (1967), 'Hydrological models and geography', in R. J. Chorley and P. Haggett (eds.), *Physical and information models in geography*, Methuen, 145–85

Munn, R. E. (1966), 'Descriptive micrometeorology', *Advances in Geophysics*, supplement 1, New York: Academic Press

National Academy of Sciences (1975), *Understanding climatic change: a program for action*, Washington: US National Academy of Sciences

National Environment Research Council (1975), *Flood Studies Report*, vol. 2, *Meteorological Studies*, NERC

Neiburger, M., Edinger, J. G. and Bonner, W. D. (1982), *Understanding our atmospheric environment* (2nd edn), San Francisco: Freeman

Newell, R. E., Vincent, D. G., Dopplick, T. G., Ferruza, D. and Kidson, J. W. (1969), 'The energy balance of the global atmosphere', in G. A. Corby (ed.), *The global circulation of the atmosphere*, Royal Meteorological Society, 42–90

Newell, R. E., Kidson, J. W., Vincent, D. G. and Boer, G. J. (1974), *The general circulation of the tropical atmosphere*, vol. 2, Cambridge, Mass.: MIT Press

Nicholson, S. E. (1983), 'Sub-Saharan rainfall in the years 1976–1980', *Monthly Weather Review*, **111**, 1646–54

Nieuwolt, S. (1968), 'Uniformity and variation in an equatorial climate', *J. Tropical Geogr.*, **27**, 23–39

Nieuwolt, S. (1977), *Tropical climatology: an introduction to the climates of low latitudes*, Wiley

Nkemdirim, L. C. (1976), 'Dynamics of an urban temperature field – a case study', *Jl Appl. Meteorol.*, **15**, 818–28

Nye, D. H. and Greenland, D. J. (1960), *The soil under shifting cultivation*, Commonwealth Bureau of Soils, Technical Communication 51, Harpenden: Commonwealth Agricultural Bureaux

Oke, T. R. (1978), *Boundary-layer climates*, Methuen

Oke, T. R. and Maxwell, G. B. (1975), 'Urban heat island dynamics in Montreal and Vancouver', *Atmospheric Environment*, **9**, 191–200

Ovington, J. D. (1962), 'Quantitative ecology and the woodland ecosystems concept', in J. B. Craggs (ed.), *Advances in ecological research*, New York: Academic Press, 103–91

Oxburgh, E. R. (1974), 'The plain man's guide to plate tectonics', *Proc. Geol. Assoc.*, **85**, 299–357

Palmén, E. and Newton, C. W. (1969), 'Atmospheric circulation systems', Int. Geophys. Ser. **15**, New York: Academic Press

Palmer, R. C. (1976), 'Soils in Herefordshire IV', Sheet SO 74 (Malvern), Soil Survey Record no. 36., Rothamsted Experimental Station

Paterson, J. H. (1975), *North America*, New York: Oxford University Press

Peltier, L. C. (1950), 'The geographic cycle in periglacial regions as it is related to climatic geomorphology', *Annals Ass. Amer. Geographers*, **40**, 214–36

Penman, H. L. (1948), 'Natural evaporation from open water, bare soil and grass', *Proc. Roy. Soc.* **193A**, 120–48

Petterssen, S. (1950), 'Some aspects of the general circulation of the atmosphere', *Centenary Proc. Roy. Meteorol. Soc.*, Royal Meteorological Society, 120–55

Phillpot, H. R. and Zillman, J. W. (1969), *The surface temperature over the Antarctic continent*, Commonwealth Bureau of Meteorology, Melbourne

Pirie, R. G. (1973), 'Oceanography', *Contemporary Readings in Ocean Science*, New York: Oxford University Press

Pittock, A. B., Frakes, L. A., Jenssen, D., Peterson, J. A. and Zellman, J. W. (1978), *Climatic change and variability: a southern perspective*, Cambridge University Press

Plumley, W. (1948), 'Black Hills terrace gravels: a study in sediment transport', *Jl Geology*, **56**, 526–77

Richey, J. E. (1964), *Scotland: the Tertiary volcanic districts* (3rd edn), Geological Survey of GB

Richter, D. A. and Dahl, R. A. (1958), 'Relationship of heavy precipitation to the jet maximum in the eastern U.S.', *Monthly Weather Review*, **86**, 368–76

Riehl, H. (1979), *Climate and weather in the tropics*, New York: Academic Press

Rodda, J. C. (1969), 'An assessment of precipitation', in R. J. Chorley (ed.), *Water, Earth and Man*, Methuen, 130–4

Rodin, L. E. and Bazilevich, N. I. (1965), *Production and mineral cycling in terrestrial vegetation*, Oliver & Boyd

Rona, P. A. (1973), 'Plate tectonics and mineral resources', *Scientific American*, **229**, 86–95

Rorison, I. (1969), 'Ecological inferences from laboratory experiments of mineral nutrition', in I. Rorison (ed.), *Ecological aspects of the mineral nutrition of plants*, Symposia of British Ecological Society, **9**, 155–75, Blackwell

Rubin, J. and Bielorai, H. (1957), 'A study of the irrigation requirements and consumptive water use by sugar beet in the northern Negev', in *Final Report of the Ford Foundation Israel Project*, A5, spec. bull. Volcani Inst. Agric. Res. no. 6, 20–4

Rusin, N. P. (1964), *Meteorological and radiational regime of Antarctica*, Jerusalem: Israel Program for Scientific Translations

Russell, E. W. (1974), *Soil conditions and plant growth*, Longman

Saugier, B. (1977), 'Micrometeorology on crops and grassland', in J. J. Landsberg and C. V. Cutting (eds.), *Environmental effects on crop physiology*, Academic Press

Savage, J. C. and Paterson, W. S. B. (1963), 'Borehole measurements in the Athabasca glacier', *Jl Geophysical Res.*, **68**, 4521–36

Sawyer, J. S. (1956), 'Rainfall of depressions which pass eastward over or near the British Isles', *Professional Notes*, **7** (118), HMSO

Schmidt, W. (1930), 'Der tiefsten minimum temperatur in Mitteleuropa', *Naturwissenschaft*, **18**, 367–9

Schumm, S. A. (1977), *The fluvial system*, New York: Wiley

Schytt, V. (1981), 'The net mass balance of Storglaciären, Kebnekaise, Sweden, related to the height of the equilibrium line and the height of the 500mb surface', *Geografiska Annaler*, **63A**, 219–23

Sellers, W. D. (1965), *Physical climatology*, Chicago: University of Chicago Press

Semb, A. (1978), 'Sulphur emissions in Europe', *Atmospheric Environment*, **12**, 455–60

Shellard H. C. (1976), 'Wind', in T. J. Chandler and S. Gregory (eds.), *The Climate of the British Isles*, Longman, 39–73

Smith, D. I. (1981), 'Actual and potential flood damage: a case study for urban Lismore, NSW, Australia', *Applied Geography*, **1**, 31–9

Smith, K. (1975), *Principles of applied climatology*, McGraw-Hill

Smith, R. L. (1966), *Ecology and field biology*, New York: Harper & Row

Sparks, B. W. (1971), *Rocks and relief*, Longman

Stearns, C. R. (1969), 'Surface heat budget of the Pampa de La Joya, Peru', *Monthly Weather Review*, **97**, 860–6

Steiner, J. T. (1980), 'The climate of the South-West Pacific region', New Zealand Meteorological Service, misc. publ., 166

Stewart, J. H. and Lamarche, V. C. (1967), 'Erosion and deposition produced by the floods of December 1964 on Coffee Creek, Trinity County, California', *US Geol. Survey Professional Paper*, 422–K

Strahler, A. N. and Strahler, A. H. (1979), *Elements of Physical Geography* (2nd edn), New York: Wiley

Strakhov, N. M. (1967), *Principles of lithogenesis*, vol. 1. Oliver & Boyd

Stride, A. H. (1982), 'Sand transport', in A. H. Stride (ed.), *Offshore tidal sands – processes and deposits*, Chapman Hall, 58–94

Sugden, D. E. and John, B. S. (1977), *Glaciers and landscape*, Arnold

Sverdrup, H. U., Johnson, M. W. and Fleming, R. H. (1970), *The oceans: their physics, chemistry and general biology* (2nd edn), Englewood Cliffs, NJ: Prentice-Hall Inc.

Thompson, B. W. (1965), *The climate of Africa*, Nairobi: Oxford University Press

Thornbury, W. D. (1954), *Principles of geomorphology*, New York: Wiley

Toksoz, M. N. (1975), 'The subduction of the lithosphere', *Scientific American*, **233**, 88–98

Trafford, B. D. (1970), 'Field drainage', *Jl Roy. Agricult. Soc. England*, **131**, 129–52

Tsung-lien Chou (1976), 'The Yellow River: its unique features and serious problems', Rivers '76, American Society of Civil Engineers, 507–26

Turner, C. (1970), 'The Middle Pleistocene deposits of Marks Tey, Essex', *Phil. Trans. Roy. Soc.*, **257B**, 373–437

Van Bavel, C. H. M., Fritschen, L. J. and Reeves, W. E. (1963), 'Transpiration by sudangrass as an externally controlled process', *Science*, **141**, 269–70

Visher, S. S. (1945), 'Climatic maps of geologic interest', *Bull. Geol. Soc. America*, **56**, 713–76

Walling, D. E. (1974), 'Suspended sediment and solute yields from a small catchment prior to urbanisation', in K. J. Gregory and D. E. Walling (eds.), *Fluvial processes in instrumented watersheds*, Institute of British Geographers, spec. publication no. 6, 169–92

Walling, D. E. (1978), 'Suspended sediment and solute response characteristics of the River Exe, Devon, England', in R. Davidson-Arnott and W. Nickling (eds.), *Research in fluvial systems*, Geobooks, 169–97

Ward, R. C. (1975), *Floods: a geographical perspective*, Macmillan

Weyl, P. K. (1970), *Oceanography: an introduction*, New York: Wiley

Whittaker, R. H. (1975), *Communities and ecosystems*, New York: Macmillan

Whittaker, R. H. and Likens, G. E. (1975), 'The biosphere and man', in H. Lieth and R. H. Whittaker (eds.), *Primary productivity of the biosphere*, *Ecological Studies*, **14**, 305–28, New York: Springer-Verlag

Wiman, S. (1963), 'A preliminary study of experimental frost weathering', *Geografiska Annaler*, **45A**, 113–21

Wise, A. F. E. (1971), 'Effects due to groups of buildings', *Phil. Trans. Roy. Soc.*, **269A**, 469–85

Wolman, M. G. (1955), 'The natural channel of Brandywine Creek, Pennsylvania'. US Geol. Survey, professional paper, no. 271

Wolman, M. G. (1967), 'A cycle of sedimentation and erosion in urban river channels', *Geografiska Annaler*, **49A**, 385–95

Woodwell, G. M. (1963), 'The ecological effects of radiation', *Scientific American*, **208**, 40–9

Yoshino, M. M. (1975), *Climate in a small area*, Tokyo: University of Tokyo Press

Young, C. P. and Gray, E. M. (1978), *Nitrate in groundwater*, Technical Report TR69, Medmenham: Water Research Centre

Index

Personal acknowledgements

There are many people to whom we are indebted for help, advice, encouragement and forbearance in producing this book. Not least, we are grateful to our colleagues in the Department of Geography at Sheffield University for providing information, reference material, valuable comments and photographs: in particular Rick Cryer, Stan Gregory, Jim Hansom, David Knighton, Steve Trudgill and David Thomas. In addition, many others responded to our request for illustrations, not all of which, unfortunately, could be used: thanks are thus due to Trevor Elkington, Gerry Garland, Dave Gilbertson, Dr A. Giordano, R. J. MacRae, Nigel Pears, Brian Pyatt, Allan Straw, Jim Rose, C. J. Richards, Peter Crabb, Brian Wheeler, Liz Rollin, Gwyn Rowley, John Dixon, Tony Waltham and Ron U. Cooke.

Our thanks similarly go to David Maddison and John Owen for producing the half-tone prints from the diverse range of photographs with which we presented them. The artwork included in a book such as this is also a major undertaking, and we are especially grateful to Paul Coles for his immensely patient work on the diagrams. Additionally, we thank the many typists who struggled with the various and constantly changing manuscripts, notably Pauline Chedgzoy, Anita Fletcher, Carol Ellis and Penny Shamma.

Finally, our greatest debt is to our families for their tolerance and support throughout the long history of this book: we promise not to start writing another one (at least for a month or two).

Acknowledgements

The authors and publishers would like to thank the copyright holders below for their kind permission to reproduce the following material. Every effort has been made to obtain permission from the relevant copyright holders, but where no reply was received we have assumed that there was no objection to our using the material.

Academic Press Inc for Figures, 2.6, 6.12, 7.16, 8.18, 9.7, and Tables 4.3, 12.2, 13.1; Allan Cash Photolibrary for Plates 7.6, 22.5, 23.9; American Association for the Advancement of Science for Figures 2.11 © 8, 10.6 © 1976, after CLIMAP, 13.15 © 1963; American Meteorological Society for Figures 3.2(c), 6.24, 7.10, 10.10, 10.16, 12.4, 12.20, 12.21, 13.5; American Society of Civil Engineering for Figure 22.29; American Water Resources Association for Table 9.4; Annals of the Association of American Geographers for Figures 19.12, 19.13; Archive for Meteorologie, Geophysics u Bioklim for Figure 4.9; Artemis for Figures 15.10, 15.12, 15.15; Atmospheric Environment for Figure 9.13; Barnabys Picture Library for Plate 7.5; Berkhauser Verlag for Figure 9.24; Blackwell Scientific Publications Ltd for Figure 30.6; British Cave Research Association for Figure 19.14; Bulletin of the Geological Society of America for Figure 22.31; Butterworths for Figures 14.13, 14.15, 21.11, 21.22, 23.26, 25.10; Cambridge University Press for Figures 10.5, 28.2; Catena for Figure 20.8(c); Chapman and Hall for Figures 25.20, 28.16; Charles E. Merrill Publishing Co. for Figure 2.14; Clarendon Press for Figures 23.26, 23.27; Collins for Figure 31.8; G. R. Coope for Figure 10.7; CSIRO, Canberra for Plate 11.3; Deutscher Wetterdienst for Figure 6.20; Edward Arnold for Figures 19.17, 24.2, 30.1, and Tables 3.1, 20.1; Frank Lane Picture Agency Ltd for Plate 8.1; W. H. Freeman and Co. for Figures 2.3, 2.4, 2.10, 22.7, 22.14; Geo Books for Figures 17.6, 22.5, 22.18; Geographical Annals for Figures 22.30 (originally in Geological Society of America), 23.4; Geological Association for Figure 18.6; George Allen and Unwin for Figures 28.1, 30.10; © Dr Georg Gerster/John Hilleson Agency for Plate 18.6; George Philip & Son Ltd for Figure 18.15; Harper and Row for Figures 26.4, 26.7; Harvard University Press for Figure 19.11; The Controller of Her Majesty's Stationery Office for Figures 6.22, 6.23, 7.8, 14.7; Hodder & Stoughton Ltd for Figure 3.10; Hutchinson for Figure 9.11; IASH for Figure 13.13; IASH–Unesco for Table 11.1; Government of India for Figure 8.16; Institute of British Geographers for Figures 6.15, 19.4, 22.6; Israeli Program for Scientific Translation, Keter Publishing Ltd, for Figures 8.31, 8.32, and Table 9.1; Journal of Applied Meteorology for Figure 9.14; Journal of Geology for Figure 20.8(a); Journal of Meteorology for Figure 12.1; C. A. M. King for Figure 15.9; Leonard Hill for Figure 13.8; Longman for Figures 6.16, 6.19, 7.1, 12.18, 12.23, 18.13, and Table 12.5; McGraw Hill for Figures 8.2 © J. R. Mather, 9.10, 13.2, and Table 22.2; Macmillan, London and Basingstoke for Figures 10.11, 10.13, 19.6; Macmillan Publishing Co. Inc. (NY) for Figures 14.11, 28.9, 29.5, 29.10; Methuen for Figures 3.12, 7.14, 7.15, 8.1, 9.3, 9.8, 10.4, 10.8, 11.4, 12.5, 12.22, 13.6, 13.7, 13.9, 13.11, 25.4; MIT Press for Table 3.2; National Physical Laboratory for Figure 9.15; Nelson for Figures 18.12, 25.22, and Table 17.1; Director of the New Zealand Meteorological Service for Figure 8.4; Open University for Figure 31.2; Oxford University Press for Figures 8.9, 8.10, and Table 10.1; Oxford University Press Inc. (NY) for Figure 14.12; Pergamon Press for Figure 27.14, and Table 12.3; Prentice–Hall Inc. for Figures 5.1, 6.1, 11.2 (partly after Critchfield), and Table 15.2; Rothampstead for Figures 17.5, 31.6; Royal Geographical Society for Figure 25.5; Royal Meteorological Society for Figures 4.6, 6.21, 7.13, 8.6, 8.21, 10.4, 10.9, 10.12, 12.8, and Table 7.2; Royal Society for Figures 9.16, 10.1 and Table 13.3; Scottish Geographical Magazine for Figure 27.18; Scientific American Inc. for Figures 18.2 © 1970, 18.3 © 1968, 18.5 © 1982, 18.7 © 1970, 18.8 © 1973, 18.9 © 1975, 18.10 © 1975, 18.18 © 1973, 18.19 ©

1982, 22.15 © 1979, 30.15 © 1963; Singapore Journal of Tropical Geography for Figure 12.10; Soil Survey Directory of USDA for Figure 30.4; Society for American Archaeology from *American Antiquity* for Figure 10.2; Space Department, Royal Aircraft Establishment, Farnborough for the cover illustration; Springer Verlag for Figures 26.8, 31.17, 31.18, and Table 26.1; University of Chicago Press for Figures 2.15, 3.6, 3.7, 4.3, 11.3, 11.7, 13.16; University of Dundee for Plates 5.8, 6.1, 7.2, 7.3; University of Sheffield, Dept of Botany for Plate 10.1; University of Tokyo Press for Table 9.3; University of Wisconsin Press for Figures 13.4, 15.3; USGS Professional Papers for Figures 22.2, 22.10; US National Academy of Science for Figure 3.11; US Naval Institute, Maryland for Figure 15.11; Van Nostrand Reinhold for Figures 15.2, 15.6, 15.7; Wadsworth, Calif. for Figures 15.5., 15.16; Water Research Centre, Bucks for Figure 14.22; Weatherwise for Tables 8.2, 8.3; John Wiley and Sons Ltd for Figures 8.8, 13.10, 15.8, 15.14, 22.12, 25.3, and Table 7.1; the Williams & Wilkins Co., Baltimore for Figure 27.6; V. H. Winston & Sons (Soviet Geography) for Figures 4.7, 4.10, 4.12; World Meteorological Organization for Figure 9.12.